Textbooks in Telecommunication Engineering

Era of the Internet and Network Science/Engineering

Series Editor

Tarek S. El-Bawab, Boston, MA, USA

Telecommunication and networks have evolved to embrace all aspects of our everyday life, including smart cities and infrastructures, healthcare, banking and businesses, manufacturing, space and aviation, meteorology and climate change, oceans and marine life, Internet of Things, defense, homeland security, education, research, social media, entertainment, and many others. Network applications and services continue to expand, virtually without limits. Therefore, specialized telecommunication and network engineering programs are recognized as a necessity to accelerate the pace of advancement in this field, and to prepare a new generation of engineers for imminent needs in our modern life. These programs need curricula, courses, labs, and textbooks of their own.

The IEEE Communications Society's Telecommunication Engineering Education (TEE) movement, led by Tarek S. El-Bawab- the editor of this Series, resulted in recognition of this field of engineering by the Accreditation Board for Engineering and Technology (ABET) on November 1, 2014. This Springer Series was launched to capitalizes on this milestone. The Series goal is to produce high-quality textbooks to fulfill the education needs of telecommunication and network engineering, and to support the development of specialized undergraduate and graduate curricula in this regard. The Series also supports research in this field and helps prepare its scholars for global challenges that lay ahead. The Series have published innovative textbooks in areas of network science and engineering where textbooks have been rare. It is producing high-quality volumes featuring innovative presentation media, interactive content, and online resources for students and professors.

Book proposals are solicited in all topics of telecommunication and network engineering including, but not limited to: network architecture and protocols; traffic engineering; network design, dimensioning, modeling, measurements, and analytics; network management and softwarization; cybersecurity; synchronization and control; applications of artificial intelligence in telecommunications and networks; applications of data science in telecommunications and networks; network availability, reliability, protection, recovery and restoration; wireless communication systems; cellular technologies and networks (through 5G, 6G, and beyond); satellite and space communications and networks; optical communications and networks; heterogeneous networks; broadband access and free-space optical communications; Multiple system operators (MSO)/cable networks; storage networks; optical interconnects; and data centers; social networks; transmission media and systems; switching and routing (from legacy to today's paradigms); network applications and services; telecom economics and business; telecom regulation and policies; standards and standardization; and laboratories.

Proposals of interest shall be for textbooks that can be used to develop university courses, either in full or in part. They should include recent advances in the field while capturing whatever fundamentals that are necessary for students to understand the foundations of the topic and appreciate its evolution trends. Books in this series will provide high-quality illustrations, examples, end-of-chapters' problems/exercises and case studies.

This series is indexed in Scopus. For further information and to submit proposals, please contact the Series Editor, Dr. Tarek S. El-Bawab, telbawab@ieee.org; or Mary James, Executive Editor at Springer, mary.james@springer.com.

Vivek Ashok Bohara • Swaminathan R

Optical and Wireless Access Networks

Springer

Vivek Ashok Bohara
IIIT-Delhi
New Delhi, Delhi, India

Swaminathan R
IIT Indore
Indore, Madhya Pradesh, India

ISSN 2524-4345 ISSN 2524-4353 (electronic)
Textbooks in Telecommunication Engineering
ISBN 978-3-031-89695-8 ISBN 978-3-031-89696-5 (eBook)
https://doi.org/10.1007/978-3-031-89696-5

This Springer imprint is published by the registered company Springer Nature Switzerland AG
The registered company address is: Gewerbestrasse 11, 6330 Cham, Switzerland

Note from the Series Editor

This Series has its roots in the Telecommunication Engineering Education (TEE) movement in the USA, which I had the honor of leading from 2008 to 2015. The movement was a result of remarkable impacts telecom networks had already started to have on society, and many more that were looming on the horizon at that time. Major changes in the theoretical foundations of this field of engineering, its design strategies, and its practical applications were unfolding at the same time. The industries we had known as the Telecom, Cable, and Data-Communication industries through the twentieth century were going through technical innovations and were reshuffling, repositioning themselves in the market, and converging, in preparation for the global economy of the twenty-first century.

The TEE movement was prompted by a personal experience I went through as I moved from industry to academia. *Telecom engineering, or Network Engineering*, since the network is centric to the evolution telecom has been through for three to four decades so far, became a distinct multidisciplinary field of study. This field blends ingredients from electrical engineering, computer engineering, computer science, and other disciplines, but may not be taught merely as a branch of any one of them, per se. It became evident that society would be better served by adding specialized undergraduate/graduate courses and programs in telecom/network engineering to our higher-education system, especially in the USA where this was not the norm.

The story of the TEE movement has been told and also discussed within broader contexts in the literature (e.g., *IEEE Communications Magazine, Vol. 53, No. 11, pp. 35–9, November 2015*). The movement had accomplished its main goal of adding telecom/network engineering to ABET's portfolio of recognized engineering disciplines. ABET is the Accreditation Board of Engineering and Technology, Inc., which accredits engineering programs in the USA and more than 40 other countries. A lot of work had to be done to capitalize on that milestone, however. The technical literature of this discipline was rich in scholarly research papers, but not as rich in textbooks designed to support the development of new university courses and curricula. These textbooks are needed to facilitate rigorous academic training for a new generation of network engineers. The idea of this Series came up subsequently to help the telecom/ networking community overcome this shortcoming. Since its inception, Springer's TTE (*Textbooks in Telecommunication Engineering*) has been doing this job. The proximity of the acronyms herein—between TEE and TTE—is coincidental, but worth noting!

I am proud of the inroads TTE has made in publishing quality textbooks and in fulfilling the goals and objectives the Series has had since its beginning. By the end of 2024, we approach the milestone of having 35 books, including 20 published and about 15 in the making (whether in the writing, reviewing, or production phases). More than 50 authors have contributed to this Series so far and they come from about 20 different countries and cultures all over the world.

The topics covered by this Series so far scan the entire landscape of telecom and network engineering, including wireless, optical, satellite; access and core; devices, systems, and networks; space-aerial-terrestrial integration; queuing theory and traffic engineering; network management; Internet and Internet of Things (IoT); signal processing; applied electromagnetics; and economy/business aspects of telecom. The Series includes a cellular-generations primer and has a Wi-Fi primer project under progress. It covers the most cutting-edge topics with textbooks that are carefully designed to strike the right balance between the need to create new courses thereto today and the fact that progress is imminent in each of these topics, and in the short term. These include books that discuss the roles of edge and cloud computing; of blockchains, and of artificial intelligence and machine learning in network engineering. Special topics such as under-water communications, railways telecom, networked navigation systems, semantic communications, video analytics, and broadband to combat the digital divide are also covered by TTE. I have tried my best to make this Series a top de facto global home for quality telecom/network engineering textbooks, and am hoping that many in our community can start to consider it as such now. The fruits of the hard work put by many people into fulfilling this goal are now handy to pick by our community.

I want to thank my wife Shahira and my daughter Nadine for their support of my work in this Series over nearly ten years. I have put a lot of time into this task and they have understood how this work is important to me. I owe a big deal to all the authors of this Series. Words cannot describe my appreciation of their contributions and talent. I would like to also express my thanks to Springer in general, and to Mary James in particular. Mary has been my main partner throughout this journey, and I couldn't have thought of a better partner than her. Finally, I thank all my colleagues in the field of telecom/network engineering and our readership for their continued support of the mission and the endeavor of this great project.

Tarek El-Bawab, PhD
December 2024

Preface

A comprehensive, flexible, and efficient communication infrastructure that meets the diverse and critical needs of the modern society—optical and wireless access networks. The purpose of this book is to disseminate knowledge and expertise gained through the years of teaching, consulting, and working on optical and wireless access networks to the targeted beneficiaries. This book covers the basics of optical access communications, wireless multiple access communications, and optical and wireless convergence toward 6G networks along with the discussions on various key enabling technologies such as Free space optical (FSO), Visible light communication (VLC) (exemplified by Light fidelity (LiFi)), power-over-fiber, and integration of optical and wireless access networks. The readers can benefit with sufficient in-depth knowledge in designing both optical and wireless access networks for applications such as multiuser access, wireless backhaul, intelligent transportation systems, internet-of-things (IoT), etc. The intended readers/targeted beneficiaries of this book include UG and PG telecommunication engineering students, faculty who are offering courses in telecommunication, application engineers, researchers, space enthusiasts, scientists and anyone who are interested in learning more about sustainable, reliable, and spectral-efficient integrated optical and wireless access networks.

The integration of optical fiber and wireless technologies provides a comprehensive, reliable, and efficient communication infrastructure that meets the diverse and evolving needs of the modern society. The key reasons optical and wireless access networks are important and relevant to the series is:

1. High-Speed Connectivity: Optical fibers offer high data transmission rates for transporting large volumes of data across long distances. When integrated with wireless technologies, this high-speed backbone is extended to users' devices, enabling seamless and faster connectivity.
2. Flexible Deployment: Integrated networks provide flexibility in network deployment. Optical fibers can be laid in areas where it's feasible, while wireless technologies cater to areas with limited infrastructure or challenging terrain.
3. Urban and Rural Connectivity: Integrated networks bridge the digital divide by offering high-speed broadband access to both urban and rural areas.
4. 5G and Beyond: Integrated networks offer high capacity fronthaul and backhaul connections required for ultra-high speeds and low latency.
5. Cost-Effective Solutions: Deploying optical fibers throughout an entire area can be costly. Integrating wireless technologies allows for a more cost-effective approach, targeting connectivity where it's needed the most.

With the above advantages of optical and wireless access networks, this book introduces the basics of optical fibers and the evolution of wireless communications in a simple language for ease of understanding. This book will touch upon the technologies that bridge the digital divide by offering high-speed broadband access to both urban and rural areas. This book will also discuss some technological solutions that offer high capacity fronthaul and backhaul connections required for ultra-fast speeds and low latency. Strong emphasis is provided on various key enabling technologies for integrated optical and wireless networks along with their potential applications in beyond 5G (B5G) communications. It also provides detailed description on various wireless/optical/hybrid network models, which are expected to be deployed in B5G communications.

Further, this book presents a blend of theoretical and problem-solving approaches to enhance understanding.

New Delhi, India
Indore, India
July, 2025

Vivek Ashok Bohara
Swaminathan R

Acknowledgments

We would like to extend our sincere gratitude to Prof. Tarek El-Bawab, IEEE fellow, for his consistent perseverance and invaluable assistance in reviewing multiple drafts of our proposal regarding the content of this book. His thoughtful suggestions and unwavering support have been instrumental, and we are deeply appreciative of his contributions. This book is also grateful to thousands of researchers, scientists, professionals, and students who are working tirelessly for the advancement of integrated optical and wireless technology in order to make the world a better place.

We would like to acknowledge the following people from IIIT-Delhi and IIT-Indore who have contributed immensely to helping us in preparing some of the contents of the book as well as painstakingly taking a lot of effort in reviewing the final contents as well:

1. IIIT-Delhi
   a. Dr. Akshita Gupta
   b. Dr. Saumya Chaturvedi
   c. Saswati Paramita
   d. Priyanka Singh
2. IIT-Indore
   a. Prashant Sharma
   b. Dr. Deepshikha Singh
   c. Manojkumar Kokare
   d. Kishan Tripathi

This book has also benefited immensely from the critical feedback from students who have taken our wireless communication and networks courses in IIIT-Delhi and IIT-Indore. Further we are also indebted to our research scholars in Wirocomm Research group and CoE-LiFi in IIIT-Delhi, and Future Generation Communication Systems Research group in IIT-Indore for their research and development work, which has shaped some of the contents of the book.

In addition to the above, we would also like to thank all members of our family who have been very supportive of the time we have spent writing this book. Without their help, support, and understanding, this book wouldn't have been possible.

Competing Interests The authors have no competing interests to declare that are relevant to the content of this manuscript.

Contents

Abbreviations

1G	First generation
2G	Second generation
3D	Three-dimensional
3G	Third generation
3GPP	Third Generation Partnership Project
3R	Re-amplification, re-shaping, and re-timing
4G	Fourth generation
5G	Fifth generation
6G	Sixth generation
10G-EPON	10-Gigabit ethernet PON
ACK	Acknowledgment
ACO-OFDMA	Asymmetrically clipped optical OFDMA
ADDCP	Advanced data communications control procedure
AES	Advanced encryption standard
AF	Amplify-and-forward
ANSI	American national standard interim
AON	Active optical network
AOA	Angle of arrival
AOS	Active optical splitter
AP	Access point
APON	ATM-based PON
ARP	Address resolution protocol
ATM	Asynchronous transfer mode
AWGN	Additive white Gaussian noise
B5G	Beyond 5G
BER	Bit error rate
BGP	Border gateway protocol
BMFCB	Baseband modulated fiber coax bus
BPON	Broadband PON
BW	Bandwidth
C-V2X	Cellular-V2X
CA	Coexistence assurance
CAPEX	Capital expenditures
CATV	Cable television
CAV	Connected autonomous vehicle
CAVE	Cellular authentication and voice encryption
CB	Channel busy
CBP	Contention-based protocol
CC	Central controller
CCDF	Complementary cumulative distribution function
CD-NOMA	Code-domain NOMA
CDF	Cumulative distribution function

CDMA	Code division multiple access
CF	Characteristic function
CM	Cable modem
CMMW	Chinese millimeter wave
CMP	Cloud management platform
CO	Central office
CP	Cyclic prefix
CS	Central station
CSI	Channel state information
CSMA	Carrier-sense multiple access
CSMA/CA	Carrier sense multiple access with collision aggregation
CSMA/CD	Carrier sense multiple access with collision detection
CTS	Clear to send
D-AMPS	Digital advanced mobile phone system
D2D	Device-to-device
DBA	Dynamic wavelength allocation
DCF	Distributed coordination function
DCO	Digital coherent optics
DCO-OFDMA	Direct current-biased optical OFDMA
DD	Direct detection
DF	Decode-and-forward
DFB	Distributed-feedback
DFT	Discrete Fourier transform
DHCP	Dynamic host configuration protocol
DIFS	Distributed inter-frame space
DL	Deep learning
DMT	Discrete multi-tone
DOR	Delay outage rate
DRA	Distributed Raman amplifiers
DSD	Dynamic spot-diffusing
DSL	Digital subscriber line
DSLAM	DSL access multiplexer
DSP	Digital signal processors
DSRC	Dedicated short-range communication
DWDM	Dense WDM
E2E	End-to-end
ECSA	Extended channel switch announcement
EE	Energy efficiency
EDFA	Erbium-doped fiber amplifier
EDGE	Enhanced data rates for GSM evolution
ELH	Extended long-haul
EMI	Electromagnetic interference
EPON	Ethernet PON
ETSI	European telecommunications standards institute
EVDO	Evolution-data optimized
F5G	Fifth-generation fixed network
FDM	Frequency division multiplexing
FDMA	Frequency division multiple access
FEC	Forward error correction
FeMBB	Further enhanced mobile broadband
FiWi	Fiber-wireless
FSOC	Free-space optical communication
FOV	Field of view

FSAN	Full-service access network
FSO	Free space optical
FTP	File transfer protocol
FTTH	Fiber-to-the-home
FTTB	Fiber-to-the-building
FTTC	Fiber-to-the-curb
FTTE	Fiber-to-the-edge
FTTCab	Fiber-to-the-cabinet
FTTP	Fiber-to-the-premises
FTTN	Fiber-to-the node
FTTR	Fiber-to-the-room
FTTX	Fiber-to-the-everything
FW	Fixed-wing
GEO	Geostationary earth orbit
GEM	GPON encapsulation method
GEPON	Gigabit ethernet over PON
GFP	Generic framing procedure
GG	Gamma-gamma
GMPLS	Generalized multiprotocol label switching
GPON	Gigabit PON
GPRS	General packet radio service
GSM	Global system for mobiles
GTC	G-PON transmission convergence
HAPS	High-altitude aerial platforms stations
HD	Heterodyne detection
HDLC	High-level data link control
HDTV	High-definition television
HetNet	Heterogeneous network
HFC	Hybrid fiber-coaxial
HiperLAN	High-performance local area network
HLWN	Hybrid LiFi/WiFi network
HLWNets	Hybrid LiFi-WiFi networks
(HSUPA/HSDPA)	High-speed uplink/downlink packet access
HTPPS	Hypertext transfer protocol secure
HWMP	Hybrid wireless mesh protocol
IC	Integrated circuit
ICI	Inter-carrier interference
ICMP	Internet control message protocol
IDF	Intermediate distribution frame
IDGN	Input-dependent Gaussian noise
IDMA	Interleave division multiple access
IEEE	Institute of Electrical and Electronics Engineers
IGGG	Inverted gamma-gamma
IGMA	Interleave-grid multiple access
IM	Intensity modulated
IMA	Intelligent mirror array
IMR	Intelligent metasurface reflector
IOR	Information outage rate
IoT	Internet of Things
IOWA	Integrated optical wireless access
IP	Internet protocol
IPACT	Interleaved polling with adaptive cycle time
IPS	Indoor positioning system

IR	Infrared
IRC	Infra-red communications
IS	Interim standard
ISM	Industrial, scientific, and medical
ISO	International Organization for Standardization
ISRO	Indian Space Research Organisation
IrDA	Infrared Data Association
IRS	Intelligent reflecting surface
IT	Information technology
ITS	Intelligent transportation system
ITS-G5	Intelligent transport systems-5th generation
ITU	International telecommunication union
ITU-T	ITU-Telecommunication Standardization Sector
KPIs	Key performance indicators
LA	Link aggregation
LAN	Local area network
LCRS	Low code rate spreading
LCs	Liquid crystals
LD	Laser diode
LDR	Light detector
LED	Light-emitting diode
LEO	Lower earth orbit
LiFi	Light fidelity
LTE	Long term evolution
LTE-A	Long-term evolution-advanced
LT	Laplace transform
LLC	Logical Link Control
LMSC	LAN/MAN Standards Committee
LSTM	Long short-term memory
LoS	Line-of-sight
LO	Local oscillators
M2M	Machine-to-machine
MA	Multiple access
MAN	Metropolitan area network
MAC	Media access control
MBBLL	Mobile broadband and low-latency
MCD	Mixed centralized/distributed
MCM	Multi-carrier modulation
MED	Minimum Euclidean distance
MEMS	Micro-electro-mechanical systems
MET	Maximum entropy throughput
MGF	Moment generating functional
MHCP	Matern hard-core point process
MIMO	Multiple input multiple output
ML	Machine learning
mLLMT	Massive low-latency machine-type communication
MMF	Multi mode fiber
mmWave	Millimeter wave
MPCP	Multipoint control protocol
MPSK	M-ary phase-shift keying
MRC	Maximal-ratio combining
MSO	Multiple system operators
MTT	Minimum transmission time

MUX	Multiplexer
NA	Numerical aperture
NAT	Network address translation
NDP	Neighbor discovery protocol
NG-PON	Next generation-PON2
NG-PON2	Next generation PON2
NGA	Next-generation access
NICT	National Institute of Information and Communications Technology
NIU	Network interface unit
NLoS	Non-line-of-sight
NMT	Nordic mobile telephone
NOMA	Non-orthogonal multiple access
NR	New Radio
NTNs	Non-terrestrial networks
OADM	Optical add-drop multiplexer
OAM	Operation and management
OC	Optical carrier
OCC	Optical camera communication
ODN	Optical distribution network
OEO	Optical-electrical-optical
OFDM	Orthogonal frequency division multiplexing
OFDMA	Orthogonal frequency division multiple access
OLT	Optical line terminal
OMA	Orthogonal multiple access
ONT	Optical network terminal
ONU	Optical network unit
ONU-RSU	Optical network unit-roadside unit
OOK	On-off keying
OSI	Open systems interconnection
OSM	Optical spatial modulation
OSPF	Open shortest path first
OSSK	Optical space shift keying
OTL	Optical trunk line
OW	Optical wireless
OWC	Optical wireless communication
P2MP	Point-to-multipoint
P2P	Point-to-point
PAM	Pulse amplitude modulation
PAPR	Peak to-average power ratio
PC	Personal computer
PD	Photodetector
PDF	Probability distribution function
PD-NOMA	Power-domain NOMA
PDC	Personal digital cellular
PEP	Pair-wise error probability
PGFL	Probability generating functional Laplace
PIN	Positive-intrinsic-negative
PLOAM	Physical layer operations, administration and maintenance
POE	Power over ethernet
POF	Plastic optical fiber
PON	Passive optical network
PPM	Pulse position modulation
PSK	Phase shift keying

POP	Point of presence
PSD	Power spectral density
PS	Private stream
PSTN	Public switched telephone network
PWM	Pulse width modulation
POP	Points-of-presence
POF	Polymer optical fibers
PUP	PARC universal packet
QAM	Quadratic amplitude modulation
QPSK	Quadratic phase shift keying
QoS	Quality-of-service
RW	Rotary-wing
RL	Reinforcement learning
RSU	Road side unit
RED	Rotated Euclidean distance
R-MED	Rotated minimum Euclidean distance
RS	Reed–Solomon
RN	Resource node
RMS	Root-mean-square
RR-TAG	Radio Regulatory Technical Advisory Group
RTMI	Radio telefono mobile integrato
RSMA	Rate splitting multiple access
RoF	Radio-over-fiber
RSS	Received signal strength
RTS	Request to send
RV	Random variable
RWP	Random waypoint
RDMA	Repetition division multiple access
RWP	Random way point
Rx	Receiver
RS	Rate-splitting
REs	Resource elements
RIS	Reconfigurable intelligent surface
RB	Resource block
RWA	Routing and wavelength assignment
ROADM	Reconfigurable OADM
RF	Radio frequency
STA	Station
SSL	Solid-state lighting
SC	Selection combining
SSC	Switch-and-stay combining
SEC	Switch-and-examine combining
SECps	Switch-and-examine combining with post-selection
SCMA	Sparse code multiple access
SC-FDMA	Single-carrier frequency division multiple access
SIFS	Short inter-frame space
SDN	Software-defined networking
SNR	Signal-to-noise ratio
SONET	Synchronous optical networking
SDH	Synchronous digital hierarchy
SNR	Signal-to-noise ratio
SER	Symbol error rate
SDMA	Space division multiple access

SAGIN	Space-air-ground integrated network
SIC	Successive interference cancellation
SD	Standard deviation
SIM	Sub-carrier intensity modulation
SSMF	Standard single-mode fiber
SWIPT	Simultaneous wireless information and power transfer
SDV	Switched digital video
SONET	Synchronous optical network
SDH	Synchronous digital hierarchy
THz	Terahertz
TACS	Total access communications system
Tx	Transmitter
T-CONT	Transmission container
TCP/IP	Transmission control protocol/Internet protocol
TIR	Total internal reflection
TPC	Transmit power control
TAG	Technical Advisory Group
TDM	Time division multiplexing
TDMA	Time division multiple access
TPON	Telephony PON
TC	Transmission convergence
TWDM	Time and wavelength multiplexing
TMA	Triangular moving average
UV	Ultraviolet
UE	User equipment
UDP	User datagram protocol
UWB	Ultra-wide band
UART	Universal asynchronous receiver transmitter
URLLC	Ultra-reliable low latency communication
UAV	Unmanned aerial vehicles
USB	Universal serial bus
UMTS	Universal mobile telecommunications system
UHF	Ultra-high frequency
ULH	Ultra-long-haul
VHF	Very-high frequency
VL	Visible light
VR	Virtual reality
VLC	Visible light communication
V2P	Vehicle-to-pedestrian
V2I	Vehicle-to-infrastructure
V2V	Vehicle-to-vehicle
V2N	Vehicle-to-network
VN	Variable node
V-VLC	Vehicular VLC
V-RF	Vehicular RF
VECN	Vehicular edge computing networks
VR	Virtual reality
V2X	Vehicle-to-everything
WiMAX	Worldwide interoperability for microwave access
WDM	Wavelength division multiplexing
WCDMA	Wideband code division multiple access
WADM	Wavelength add-drop multiplexer
WEP	Wired equivalent privacy

WPA	Wi-Fi protected access
WiFi	Wireless fidelity
WLAN	Wireless local area network
WDM	Wavelength division multiplexing
WS-ODN	Wavelength switching Optical distribution node (ODN)
WR-ODN	Wavelength routing ODN
XGS-PON	10-Gigabit symmetrical PON
XG-PON	Gigabit capable PON
XGTC	XG-PON transmission convergence

1 Introduction to Access Networks

1.1 Optical Access Network

The "last mile" connectivity of the information flow, which links subscribers and service providers, is another name for the access networks. Service providers need to enhance their networks in order to meet the growing quality of service (QoS) demands of the users, which is usually measured in terms of high reliability, high throughput, and low error rates. The conventional approach of the service providers has been on boosting the backhaul network's bandwidth in order to satisfy the subscribers' demand of high QoS. However, in the case of fibered networks, the last mile is a potential bottleneck [1]. Utilizing optical access networks can be one of the promising solutions to the access technology problems. Thus, fiber to the curb (FTTC), fiber to the building (FTTB), or fiber to the home (FTTH) will be part of the next wave of access network deployment, according to [2], enabling Gbps speeds at prices that are equivalent to those of other technologies like hybrid fiber coax (HFC) and digital subscriber line (DSL).

For the next-generation access networks, three optical access technology candidates show promise. The following is a list of these candidates [3]:

1. Point-to-point topologies: Dedicated fiber cables that op erate point to point can link every customer to the telecom central office (CO). This architecture's vast fiber rollout makes it simple but is also expensive. Using an active star topology, which multiplexes and demultiplexes signal between the subscribers and the CO by placing a curb switch close to the subscribers, is an alternate strategy. In such strategies, the curb switch is an active component that needs backup power at the curb-unit site in addition to electrical power, which is a drawback of this method.
2. Passive optical networks: The curb switch is substituted by a passive optical component, such as an optical splitter, in passive optical networks (PONs). The topology described is one of several viable configurations for PONs, which also include tree-and-branch, ring, and bus topologies. PON reduces the footprint of fiber deployment, the overall number of optical transceivers in the system, and the electrical power consumption. At present, there is ongoing research on two PON technologies: Asynchronous Transfer Mode (ATM)-based PONs (APON) and Ethernet PONs (EPON). The ATM layer-2 protocol employed by APON enables the provision of quality-of-service capabilities. EPON transports exclusively the data contained within Ethernet frames and offers a comparatively cost-effective alternative to APON. EPON is increasingly popular and is being standardized as a solution for accessing networks within the IEEE 802.3ah community. The design concerns and the proposal of a new protocol interleaved polling with adaptive cycle time (IPACT) for EPON are addressed in [4].
3. Optical wireless technology (Free-space optics): Low-power infrared lasers are capable of high-speed data transmission through point-to-point (up to 10 Gbps) or meshed (up to 622 Mbps) topologies. An optical data link can be built, transmitting data via air using lasers positioned on rooftops and directed toward a receiver. Optimal meteorological conditions enable this technology to achieve a transmission range of up to 4 kilometers. Optical wireless technology must overcome several obstacles, such as adverse weather conditions, building mobility, airborne items, and safety concerns.

1.2 Wireless Networks

Typically, the term "wireless connectivity" is used in reference to a capability of a mobile phone (nowadays exemplified by smartphones). However, there plethora of wireless technology exists that extends beyond the smartphone. Moreover, a mobile phone typically connects to the Internet via two means: the cellular network and the WiFi network, which is a component of the local area network (LAN). Wireless fidelity

V. A. Bohara, Swaminathan R, *Optical and Wireless Access Networks*, Textbooks in Telecommunication Engineering, https://doi.org/10.1007/978-3-031-89696-5_1

(WiFi) which works on the industrial, scientific, and medical (ISM) bands is a wireless communication technology primarily meant for small distance connectivity or indoor connectivity. It has been widely used in offices, colleges, airports etc. However, WiFi connectivity in a bus or train is contingent upon its connection to the Internet via a cellular or satellite backhual network. Therefore, cellular is the more prevalent wireless network that supports wide range of devices both indoors and outdoors as well as in the surrounding areas.

In this context, it is important to look into the proprietary standards (sometimes referred as cellular standards) developed to enable services long time back to network users. In the first generation (1G) of the cellular standards, there were 12 distinct standards in various European countries (nordic mobile telephone (NMT), total access communication system (TACS), C-Netz's C-450, triangular moving average (TMA), radio telefono mobile integrato (RTMI), etc.), whereas Japan and North America had their own proprietary standards for cellular devices [5]. Each of these standards was incompatible with one another. The evolution of cellular network standards till date is also summarized in Fig. 1.1. In the 1980s, the first mobile systems were launched. These systems were entirely analog and provided voice services having a data rate of approximately 2.4 kilobits per second (kbps) [3]. Early multiple access communication systems employed FDMA. Typically, the first-generation (1G) phones weighed 5 kg emitted 5 W and were primarily used for voice communication [3]. The battery, which could only sustain a 30-minute phone call, required nearly 10 hours to charge. On the other hand, the current fifth-generation (5G) smartphone typically weighs less than 1 kg and produces a power output of less than 1 W while carrying out other functions beyond voice call and internet connectivity. The transformation over a span of thirty-five years is truly remarkable.

In the early 1990s, the second generation (2G) of cellular digital standards emerged in North America (digital advanced mobile phone system (D-AMPS) or interim standard (IS)-136), in Japan personal digital cellular (PDC), and in majority of countries in Europe, Global System for Mobile Communications (GSM) [6]. The 2G system, marking the transition to digital cellular technology, offered a stable data rate of 64 kbps while ensuring a reliable QoS. It introduced features such as short message service (SMS) and email. Prominent 2G standards included the interim standard 95 (IS-95), GSM, and IS-136 [7]. The GSM standard rapidly emerged as a popular standard. This decision made by the majority of European countries had a substantial worldwide influence. Voice and low-speed data became crucial applications of cellular technology. Time division multiple access (TDMA) and frequency division multiple access (FDMA) were the access technologies that supported all the digital systems. Given that digital technology introduced a markedly different period from previous 1G analog technology, it was officially named 2G or the second-generation cellular standard.

In addition to the prevalent GSM, another technology called code division multiple access (CDMA) became popular as 2G standard in North America. These two technologies weren't compatible with each other.

Due to demand for higher data rates, in late 1990s, a wider bandwith CDMA systems were explored. This systems were later known as third generation (3G) systems. 3G systems provided transmission rates of up to 2 megabits per second (Mbps), significantly enhancing QoS. These systems combined high-speed mobile access with internet protocol (IP) services, offering superior voice quality and global roaming, marking a notable progression in mobile communication technology. Key technologies in 3G included CDMA 2000, universal mobile telecommunications system (UMTS), and wideband code division multiple access (WCDMA) [8]. Additionally, evolution-data optimized (EVDO) and high-speed uplink/downlink packet access (HSUPA/HSDPA) technolo-

Features	1G	2G	3G	4G	5G
Standards/Technology	AMPS, NMT, TACS	GSM, IS-136, PDC, TACS	WCDMA, UMTS	LTE, LTE-Advanced, WIMAX	New Radio (Technologies such as mu-MIMO, mmWave)
Data Bandwidth	~2kbps	~64 kbps	~42 Mbps	~1 Gbps	~10-20 Gbps
Multiplexing	FDMA	TDMA/ FDMA, CDMA	CDMA	CDMA, OFDMA, SC-FDMA	OFDMA, Massive MIMO, Beamforming
Latency	~1 sec	~500-1000msec	~200 ms	~50-100 ms	~1ms
Capabilities	Analog Voice	Digital voice, SMS, Limited roaming	Mobile broadband, Multimedia streaming, Global roaming, VoIP, GPS	High speed mobile internet, HD multimedia streaming	Ultra-Reliable Low-Latency Communication (URLLC), Massive IoT, Enhanced Mobile Broadband (eMBB), Private Networks

Fig. 1.1 Evolution of wireless standards

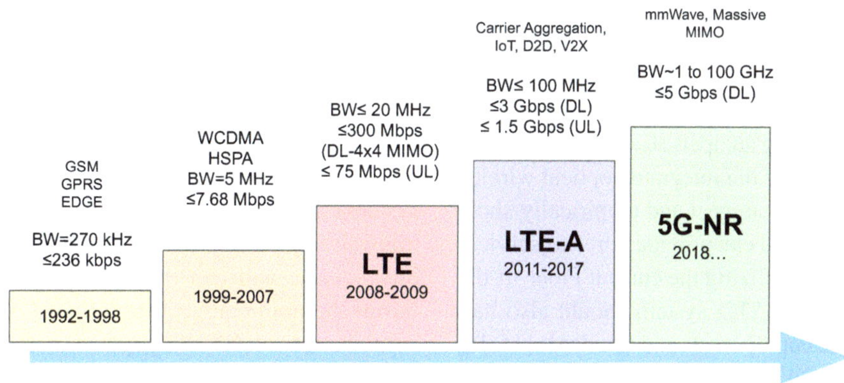

Fig. 1.2 Evolution of wireless standards

gies enabled the transition from 3G to 4G, offering enhanced data rates ranging from 5 to 30 Mbps. Despite these advancements, 3G systems consumed more power compared to many 2G systems, resulting in higher operational costs [9]. The advent of 3G standard facilitated the formation of the Third-Generation Partnership Project (3GPP) as a significant standards organization that assumed responsibility for subsequent generations of the cellular architecture. Consequently, due to formation of 3GG, cellular was perceived as as a universally accepted worldwide communication system with well-defined connections to all other communication technologies such as satellite, Internet protocol (IP) network, and public switched telephone network (PSTN). Arguably, the most significant accomplishment was in ensuring the security of this technology, therefore instilling trust among its user community, as earlier iterations of cellular connectivity were susceptible to interception [10].

The establishment and adoption of a universally accepted fourth-generation (4G) standard called long-term evolution (LTE) by 3GPP in the early 2000 has made it a fundamental feature of all mobile phones worldwide. With its foundation in orthogonal frequency division multiple access (OFDMA) technology in downlink, and DFT precoded OFDM in uplink, it represents the fourth generation or 4G of cellular technology. Indeed, LTE and LTE-Advance (LTE-A) had become the universal standard by which every cell phone functioned worldwide till 2020. With the advent of 5G, which is backward compatible with 4G LTE, its popularity has further surged. It emphasizes the need of considering the cell phone as the key tool for accessing the Internet in all applications. Every website, news outlet, and social media platform must acknowledge the fact that while everyone in the twenty-first century has cell phones, not everyone necessarily has access to a wired LAN or a large-screen personal computer (PC). The 4G systems incorporate sophisticated technologies such as multiple-input multiple-output (MIMO), OFDMA, and link adaptation, which facilitate data rates of up to 100 Mbps for high mobility conditions and up to 1 gigabit per second (Gbps) for low mobility scenarios [11]. Despite these improvements, 4G networks are nearing their theoretical limits and are not fully equipped to handle future traffic demands. The long-term evolution advanced (LTE-A) standard has addressed some of these issues by implementing multi-hop in-band and out-of-band relays to extend coverage. Nevertheless, 4G systems encounter significant challenges, including energy inefficiencies due to cell-specific reference signals and limitations in spectrum availability. Consequently, the emergence of the 5G standard has become essential to meet the growing demands of mobile communication [12]. Presently, the 3GPP standards group has a clear vision for the implementation of fifth generation (5G), which is being carried out in stages [13], with the first operations commencing in 2019. The 5G standard also known as New Radio (NR) has three primary verticals, enhanced mobile broadband (eMBB), massive machine type communication (mMTC), and ultra reliable low latency communication (URLLC). 5G networks are characterized by high device densities, advanced evolved NodeBs with numerous antennas, and ultra-high carrier frequencies. The primary objective of 5G is to provide comprehensive connectivity and support seamless, continuous communication between individuals and devices, irrespective of time, location, or service type [14]. Although the first phase of standardization of 5G typically focused on eMMBB, remaining phases are expected to cater to mMYc and URLLC as well. Release 18 catering to the 5G standard by 3GPP is expected to be finalized in mid-2024 with Release 19 following in late 2025 (Fig. 1.2).

1.3 Integrated Optical and Wireless Access Networks

Mobility and high bandwidth are essential prerequisites for the next-generation access networks to enable the provision of novel applications and to support multitude of services.

Furthermore, from an economic standpoint, it is advantageous to aim for smooth integration in order to establish a feasible business model for the implementation of new fiber infrastructure. This would involve mobile carriers and other service providers sharing an open-access infrastructure. Therefore, the primary goal of the integrated optical wireless access (IOWA) solution is to suggest and empirically showcase a transmission system that can provide wireless services, both indoors and outdoors, utilizing the current Fiber-to-the-home (FTTH) infrastructure. This system should also have compatibility with future deployments and standards [15, 16].

The development of IOWA is motivated by the concept of providing a convergent access network capable of delivering both current FTTH services and new wireless services, such as WiFi/worldwide interoperability for Microwave Access (WiMAX) spectrum regulations. The utilization of a shared infrastructure allows for the sharing of operational expenses (OPEX) and the exploitation of any existing FTTH infrastructure. The provision of wired and wireless services on the same access infrastructure at Greenfield locations will also save capital expenditures (CAPEX). In order to effectively control the available bandwidth in a combined fiber-wireless medium, IOWA develops both integrations of fiber with radio systems in radio over fiber (RoF) approach and advanced media access control (MAC) solutions.

Presently, Radio frequency (RF)-based technologies are extensively employed for most of the applications of wireless communication. The RF band is located within the frequency range of 30 kHz to 300 GHz in the electromagnetic spectrum [10], and its usage is subject to rigorous regulations [6]. Most RF communications use frequencies below 10 GHz due to its advantageous communication characteristics such as favorable propagation environment, manageable antenna heights. However, this range of the spectrum is nearly depleted. Recent projections indicate that the RF wireless technology alone will not be sufficient to satisfy the future requirements for wireless data traffic [13]. Furthermore, RF systems experience significant interference from neighboring RF Access points (APs) operating on the same or adjacent bands. The utilization of the optical segment of the spectrum for wireless connectivity has recently garnered attention from researchers as a possible strategy to meet the increasing need for high-data-rate wireless services. Infrared (IR), Visible light (VL), and Ultraviolet (UV) spectra-based Optical wireless communication (OWC) is regarded as the most effective alternative to the existing RF-based communications, which are inadequate to satisfy the increasing throughput need. Offloading significant data traffic from RF networks is theoretically possible with OWC. Data rate of 100 Gbps has been previously shown using OWC [13]. It is capable of delivering effective functionality for both indoor and outdoor communication services. Some notable benefits of OWC include (i) extensive unregulated bandwidth, (ii) a very high degree of security, (iii) excellent power efficiency, (iv) absence of licensing fees, and (v) absence of interference with the RF spectrum [4, 13]. Emerging OWC technologies include photonic Light fidelity (LiFi), Visible light communication (VLC), Optical camera communication (OCC), and Free-space optical communication (FSOC). The primary physical challenges faced by OWC systems are (i) susceptibility to obstruction, (ii) restricted transmission power, (iii) atmospheric radiation loss, intercell interference, and (iv) interference from ambient light [8]. Given the numerous constraints of OWC technologies, the combination of RF and optical wireless systems appears to be a promising strategy for attaining optimal services in Fifth generation (5G) communications and beyond. The successful functioning of the RF and optical wireless systems as a unified system necessitates successful coordination. This integration can be achieved by implementing hybrid or coexisting networks. Within a hybrid system, a wireless connection is provided by an existing alternative network, and both RF and optical wireless networks collaborate to offer wireless connections in the coexisting system.

1.3.1 Advantages of Integrated Optical and Wireless Network

The integrated optical and wireless access networks have the following advantages:

1. High-speed connectivity: Optical fibers provide excellent data transmission rates for efficiently moving substantial amounts of data across extended distances. By incorporating wireless technologies, this high-speed backbone is expanded to the devices of the consumers, therefore facilitating uninterrupted and rapid connectivity.
2. Flexible deployment: Integrated networks have the advantage of greater flexibility in network implementation. Wireless and optical technologies are designed to operate in regions with inadequate infrastructure or difficult terrain, where installation of optical fiber may not be possible. Hence, wireless can extend the coverage and flexibility of conventional optical fiber installations.
3. Urban and rural connectivity: It refers to the provision of high-speed internet to both urban and rural areas, hence bridging the digital divide.
4. Integrated networks for 5G and beyond: It provides the necessary high capacity fronthaul and backhaul links to achieve ultrafast speeds and minimal latency.
5. Cost-effective solutions: The deployment of optical fibers over a whole region can incur significant expenses. The integration of wireless technologies enables a more economically efficient strategy, focusing on connectivity in areas with the highest need.

References

1. El-Bawab, T.S.: Fttx: The rise of broadband optical access. In: 2007 IEEE Globecom Workshops, pp. 1–2 (2007)
2. Hoven, G.: Fttx-how close to the end user should fiber come? In: Conference Proceedings - Lasers and Electro-Optics Society Annual Meeting-LEOS pp. 621–621 (2009)
3. Senior, J.M.: Optical Fiber Communications: Principles and Practice, 3rd edn. Pearson Education, London (2009)
4. Kramer, G., Mukherjee, B., Pesavento, G.: IPACT a dynamic protocol for an ethernet pon (EPON). IEEE Commun. Mag. **40**(2), 74–80 (2002)
5. Raghunandan, K.: Introduction, pp. 1–14. Springer International Publishing, Cham (2022)
6. Network interworking between gsm-map and ansi-41 map. https://www.3gpp2.org/Public_html/Specs/N.S0028-0_v1.0.pdf
7. Schwarz, R..: IS-136 technology. https://www.rohde-schwarz.com/us/technologies/cellular/is-136/is-136-technology/is-136-technology_55931.html
8. Kukushkin, A.: Introduction to Mobile Network Engineering: GSM, 3G-WCDMA, LTE and the Road to 5G. Wiley Telecom (2018)
9. Perrucci, G., Fitzek, F., Sasso, G., Kellerer, W., Widmer, J.: On the impact of 2G and 3G network usage for mobile phones' battery life, pp. 255–259 (2009). https://doi.org/10.1109/EW.2009.5357972
10. Eavesdropping on gsm: State of affairs, by fabian van den broek, radboud university, nijmegen, institute for computing and information sciences (icis). http://www.cs.ru.nl/~fabianbr/pub/WISSec2010_GSM_Eavesdropping.pdf
11. Service requirements for enhanced V2X scenarios (3GPP TS 22.186 version 15.3.0 release 15) (2018). https://www.etsi.org/deliver/etsi_ts/122100_122199/122186/15.03.00_60/ts_122186v150300p.pdf
12. IMT vision-framework and overall objectives of the future development of IMT for 2020 and beyond (2015)
13. Tentative timeframe for 5g. www.3gpp.org/news-enews/1674-timeline
14. del Peral-Rosado, J.A., Raulefs, R., López-Salcedo, J.A., Seco-Granados, G.: Survey of cellular mobile radio localization methods: From 1g to 5g. IEEE Commun. Surv. Tutor. **20**(2), 1124–1148 (2018). https://doi.org/10.1109/COMST.2017.2785181
15. van den Hoven, G.: Fttx-how close to the end user should fiber come? In: 2009 IEEE LEOS Annual Meeting Conference Proceedings, pp. 621–621 (2009)
16. Kiycioglu, D., Unverdi, N.O.: Applications of fttx technology in optical communication systems. In: 2018 26th Signal Processing and Communications Applications Conference (SIU), pp. 1–4 (2018)

Optical Networks

2

2.1 Introduction

Light communication, also known as visible optical wave communication, has been an essential part of the communication system for a long time. Historically, optical communication was done through fire signals, reflecting mirrors, or signaling lamps. In the early stage, a photophone was developed by Alexander Graham Bell in 1880. This invention was four years after the invention of the telephone. This system used modulated sunlight with a diaphragm, giving speech transmission over a distance of 200 m [1]. Later, in the twentieth century, the low-capacity optical communication links were developed [2]. However, due to a lack of suitable light sources and the effect of atmospheric disturbances, these researches were limited to low-capacity and line-of-sight transmission. Further, the researchers find out that various external factors also affects light transmission, such as rain, fog, atmospheric turbulence, and dust. The focus, thus, shifted toward low-frequency electromagnetic waves. Depending on their wavelengths, these electromagnetic carriers can be transmitted over considerable distances but are limited in the amount of information they can convey through their frequencies. This is because the information-carrying capacity is directly related to the bandwidth or frequency extent of the modulated carrier, which is generally limited to a fixed fraction of the carrier frequency. The low-frequency bands were unable to meet the ever-increasing demand for higher communication data rates. Thus, higher-frequency radio communication bands, such as Very-high frequency (VHF) and Ultra-high frequency (UHF) bands, were explored. This also led to the further development of Millimeter wave (mmWave), Terahertz (THz), and sub-THz transmissions. The relative frequencies and wavelengths of these types of electromagnetic waves can be observed from the electromagnetic spectrum [29] shown in Fig. 2.1. Toward this end, the researchers realize the potential optical frequencies can offer. They can increase the usable bandwidth by a factor of around 10^4 as compared to conventional microwave transmission [3]. An optical network has vast network capacities in addition to a shared infrastructure that may be used to deliver a range of services. Additionally, these networks are getting better at providing flexible bandwidth delivery where and when it's needed.

In the context of optical networks, two main generations are commonly discussed. In the first generation, optics were mostly employed for gearbox and only served for capacity enhancement. Compared to traditional copper lines, optical fiber offered greater capacity and reduced bit error rates. In this generation, the switching and other intelligent network activities are handled electronically. First-generation optical networks include the Synchronous optical network (SONET) and the essentially similar Synchronous digital hierarchy (SDH) networks, which are the backbone of the networks supporting enterprise applications like fiber channels and the telecom infrastructure in North America, Europe, and Asia, respectively [28]. The second generation of fiber-optic communication was developed for commercial use in the early 1980s, operated at 1.3 μm and used InGaAsP semiconductor lasers. In the second generation, optical networks have routing, switching, and intelligence in the optical layer. It is evident from literature that optics is the preferred method of transmission, and Wavelength division multiplexing (WDM) transmission is extensively employed in networks. Optical networks are capable of performing a variety of functions beyond point-to-point transmission. The optical portion of the network can benefit significantly from the integration of certain switching and routing functions. For instance, the processing of data becomes increasingly challenging as data rates increase. Assume that the electronics are required to process data in segments of 35 bytes each, such as a small Ethernet packet. The processing time for a block in a 50 Mb/s data stream is 2.8 μs, whereas at 5 Gb/s, the same block must be processed within 28 ns [3]. In first-generation networks, the electronics at a node are required to manage all data that is destined for that node, as well as all data that is

V. A. Bohara, Swaminathan R, *Optical and Wireless Access Networks*, Textbooks in Telecommunication Engineering, https://doi.org/10.1007/978-3-031-89696-5_2

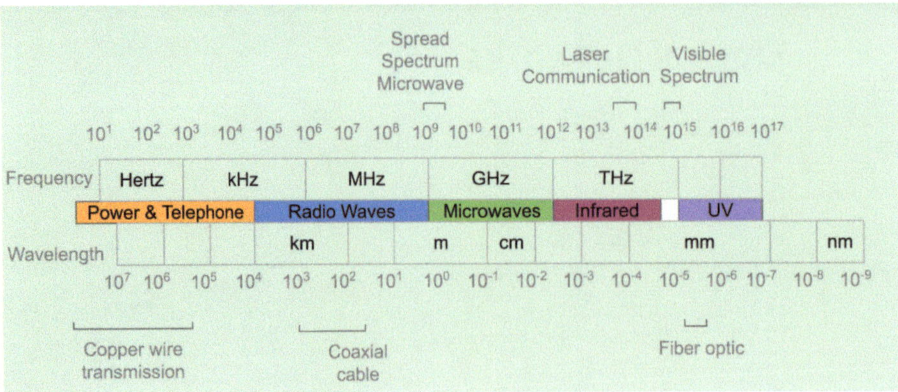

Fig. 2.1 The electromagnetic spectrum showing the region used for optical fiber communications

being passed through that node to other nodes in the network. The burden on the underlying electronics at the node would be considerably reduced if the latter data could be routed through in the optical domain. This is one of the primary factors that is driving the development of second-generation optical networks.

2.2 Optical Fibers and Cables

The choice between copper and fiber is a trade-off between the technology's cost and the energy it can save, as well as the volume of information it can transport. Experimental demonstrations have achieved bandwidth-distance products exceeding 1000 Tbit/s × km in single-mode optical fiber [4]. In contrast, the bandwidth-distance product for twisted pair copper and coaxial cables falls within the range of 100 Mbit/s × km [4]. Historically, this calculation has indicated that fiber optics were employed for long-haul transmission networks (subsea and terrestrial) and, more recently, for metropolitan area networks. However, the trend is toward the use of fiber optics in a growing number of local applications, such as interconnecting data centers, enterprises and homes, as a result of innovation and refinements in optical technology. However, even today, the most common physical media for transmitting network information are copper cabling, which was first used for the telegraph in the 1820s; radio spectrum, which is the backbone of the cellular and WiFi networks, is the fastest growing medium today; fiber-optic cables, which have been in use since 1977 [5].

Fiber optics has the greatest potential to address the majority of the networking requirements among the three media, while the Radio frequency (RF) offers mobility and flexibility to the network. Optical transport networks are founded on using glass strands of optical fiber. These strands are capable of transmitting light pulses and, therefore information, with virtually no restrictions on capacity or distance. In the following subsections, we will examine fiber-optic networks, their operation, and technological advancements over the years.

Due to its potentially infinite capabilities, massive bandwidth (nearly 50 terabits per second [Tbps] for single-mode fiber), low signal attenuation (as low as 0.2 dB/km), low signal distortion, low power requirement, low material usage, small space requirement, and low-cost, fiber-optic technology can meet the aforementioned demands of the network. Taking advantage of the significant optic-electronic bandwidth mismatch is essential, as the maximum bandwidth of a single-mode fiber generally exceed electronic data rates by over three orders of magnitude [6]. Concurrency between various user transmissions should be implemented to take advantage of the fiber's enormous bandwidth, as the maximum rate at which an end user—which can be a workstation or a gateway that interacts with lower-speed subnetworks—can access the network is restricted by the speed of the conventional electronics in the equipment. WDM is a popular multiplexing technology in optical communication networks due to its ability to allow several transmissions in the wavelength domain concurrently at a lower cost. Multiple simultaneous transmissions are made possible by multiplexing several communication channels—each carried by a separate wavelength—into a single fiber strand at one end and demultiplexing them at the other. Every communication channel (wavelength), such as Optical carrier (OC)-192 or OC-768, can function at any electronic processing speed [7]. An aggregate capacity of 6.4 Tbps can be produced, for instance, by a fiber strand supporting 160 communication channels or 160 wavelengths working at 40 Gbps apiece. These subnetworks are each responsible for executing a distinct set of functions; consequently, each presents a distinct set of technological requirements, research problems, and challenges. For instance, carriers prioritize capacity, protection, and restoration in the long-haul net-work, ensuring robust network performance and resilience against failures. Simultaneously, they prioritize service provisioning/monitoring, flexibility, and other factors in the metro or access network. The following section offers a concise overview of the critical components of optical WDM networks. This is succeeded by a comprehensive examination of access, metropolitan, and long-haul networks in isolation. The long-haul network has received particular attention due to the significant requirement for the development of new intelligent algorithms and methodologies to facilitate the efficient design and operation of these wide-area optical-mesh networks, which are constructed on emergent technologies.

2.3 Optical Fiber Network

The initial proposal and investigation of the transmission of light through a dielectric waveguide structure occurred at the outset of the twentieth century. For instance, [1] conducted a theoretical study in 1910, and [2] reported experimental work in 1920. Nevertheless, a transparent dielectric rod, which is typically composed of silica glass with a refractive index of approximately 1.5 and is surrounded by air, was not a practical waveguide due to its unsupported structure (particularly when very thin waveguides were considered in order to restrict the number of optical modes propagated) and the excessive losses at any discontinuities of the glass-air interface. However, in the mid-1950s, proposals for a clad dielectric rod were made in order to address these issues as a result of the interest in the application of dielectric optical waveguides in fields such as optical imaging and medical diagnosis (e.g., endoscopes) [8]. In Fig. 2.2, this structure is depicted as a transparent core with a refractive index of n_1 that is encircled by a transparent cladding with a slightly reduced refractive index of n_2. The waveguide structure is supported by the cladding, which also significantly reduces the radiation loss into the ambient air when it is sufficiently thick. Basically, the light propagation in optical fibers involves the light energy traveling through both the core and the cladding, thereby enabling the associated fields to decay to a negligible value at the cladding-air interface due to lower refractive index of the cladding due to the core. The buffer is the outer layer, which functions as a "shock absorber" to safeguard the core and cladding from injury. In order to safeguard the fiber from the physical environment, the coating typically consists of one or more shells of a plastic material. Occasionally, metallic sheaths are incorporated into the coating to provide additional physical protection.

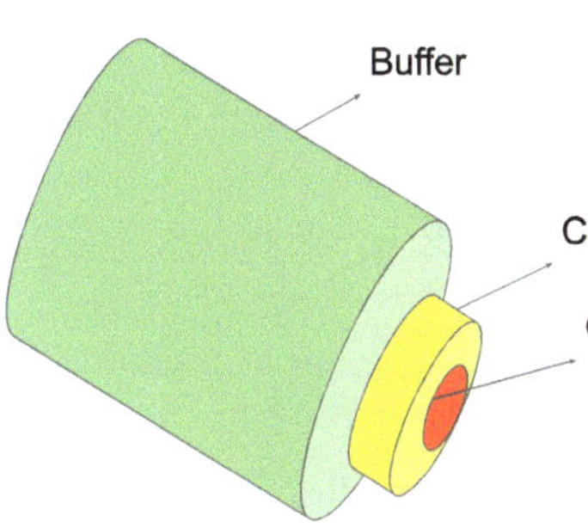

Fig. 2.2 Optical fiber waveguide showing the core of refractive index n_1, surrounded by the cladding of slightly lower refractive index n_2

2.3.1 Total Internal Reflection

In order to represent the propagation of light within an optical fiber using the ray theory model, it is essential to account for the refractive index of the dielectric medium. The refractive index of a medium is the ratio of the velocity of light in a vacuum to the velocity of light in the medium. A ray of light travels at a slower pace in an optically dense medium than in one that is less dense, and the refractive index is a metric that quantifies this phenomenon. Refraction occurs when a ray is incident on the interface between two dielectrics with varying refractive indices (e.g., glass-air), as demonstrated in Fig. 2.3a. The ray that is approaching the interface is propagating in a dielectric with a refractive index of n_1 and is at an angle ϕ_1 to the normal at the interface's surface. In the event that the dielectric on the opposite side of the interface has a refractive index n_2 that is less than n_1, the refraction will result in the ray path in this lower index medium being at an angle ϕ_2 to the normal, and ϕ_2 is greater than ϕ_1. The refractive indices of the dielectrics and the angles of incidence ϕ_1 and refraction ϕ_2 are related by Snell's law of refraction [9], which is as follows:

$$n_1 \sin(\phi_1) = n_2 \sin(\phi_2). \tag{2.1}$$

This is equivalent to the following:

$$\frac{\sin(\phi_1)}{\sin(\phi_2)} = \frac{n_2}{n_1}. \tag{2.2}$$

It is also possible to observe in Fig. 2.3a that a small quantity of light is reflected back into the originating dielectric medium (partial internal reflection). The angle of refraction is always greater than the angle of incidence, as n_1 is greater than n_2. Therefore, the angle of incidence must be less than 90° when the refracted ray emerges parallel to the interface

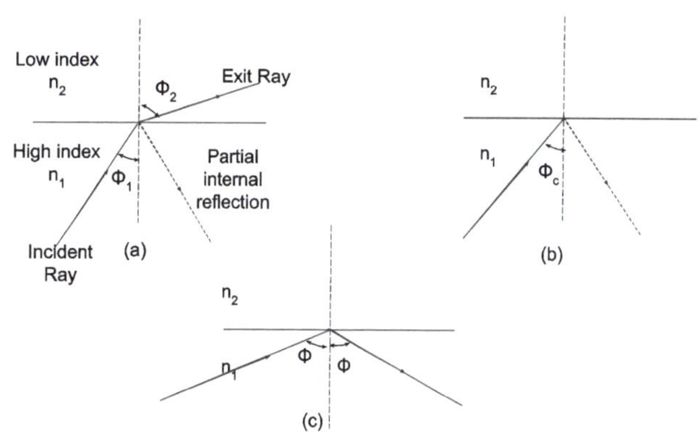

Fig. 2.3 Light ray incident on a high to low refractive index interface (e.g., glass-air): (**a**) refraction; (**b**) the limiting case of refraction showing the critical ray at an angle ϕ_c; (**c**) total internal reflection where $\phi > \phi_c$

between the dielectrics, and the angle of refraction is 90°. The angle of incidence is now referred to as the critical angle ϕ_c, and this is the limiting case of refraction, as illustrated in Fig. 2.3b. The critical angle's value is determined by (2.2):

$$\sin(\phi_c) = \frac{n_2}{n_1}, \tag{2.3}$$

Light is reflected back into the originating dielectric medium, known as Total internal reflection (TIR), with a high efficiency (approximately 99.9%) at angles of incidence that exceed the critical angle. Therefore, Fig. 2.3c illustrates that TIR occurs at the interface between the two dielectrics of varying refractive indices when light is incident on the dielectric of the lower index from the dielectric of the higher index and the angle of incidence of the ray exceeds the critical value. This is the mechanism by which light at an adequately shallow angle (less than 90°–ϕ_c) can be considered to propagate down an optical fiber with minimal loss. The transmission of a light ray in an optical fiber is depicted in Fig. 2.4, which is the result of a series of TIRs at the interface of the silica core and the slightly lower refractive index silica cladding. The angle of incidence ϕ of the ray at the interface is greater than the critical angle, and it is reflected at the same angle as the normal. The light ray depicted in Fig. 2.4 is referred to as a meridional ray due to its passage through the fiber core's axis. This form of the ray is the most straightforward to describe and is typically employed to demonstrate the fundamental transmission properties of optical fibers. It is also important to point out that the light transmission depicted in Fig. 2.3 is based on a perfect fiber. Any imperfections or discontinuities at the core-cladding interface would likely lead to refraction rather than total internal reflection, resulting in the subsequent loss of the light ray into the cladding.

2.3.2 Acceptance Angle

After considering the propagation of light in an optical fiber through total internal reflection at the core-cladding interface, it is beneficial to elaborate on the geometric optics approach in relation to light rays entering the fiber. Geometric optics helps visualize this by treating light as rays. It is evident that not all rays entering the fiber core will continue to be propagated down its length, as only rays with a sufficiently shallow grazing angle (i.e., an angle to the normal greater than ϕ_c) at the core-cladding interface are transmitted by total internal reflection.

Figure 2.5 shows the geometry of a meridional ray "A" at the critical angle ϕ_c within the fiber at the core-cladding interface, which is responsible for launching a light ray into an optical fiber. It is possible to observe that this ray enters the fiber core at an angle θ_a to the fiber axis and is refracted at the air-core interface before being transmitted to the core-cladding interface at the critical angle. As a result, any beam incident into the fiber core at an angle greater than θ_a will be transmitted to the core-cladding interface at an angle less than ϕ_c and will not be completely internally reflected. This situation is also illustrated in Fig. 2.5, where the incident ray "B" is refracted into the cladding and subsequently lost by radiation at an angle greater than θ_a. Consequently, in order for rays to be transmitted by total internal reflection within the fiber core, they must be incident on the fiber core within an acceptability cone defined by the conical half angle θ_a. Therefore, θ_a is the maximum angle to the axis at which light can penetrate the fiber in order to propagate, and it is frequently referred to as the acceptance angle for the fiber.

In the event that the fiber has a regular cross section, meaning that the core-cladding interfaces are parallel and there are no discontinuities, an incident meridional ray at a greater angle than the critical angle will continue to be reflected and pass through the fiber. It is possible to observe that the output angle to the axis will be equivalent to the input angle for the ray, provided that the ray emerges into a medium with the same refractive index as the medium from which it was sent, based on symmetry considerations.

2.3.3 Numerical Aperture

In the preceding section, the acceptance angle of an optical fiber was established. Nevertheless, it is feasible to continue the ray theory analysis in order to establish a correlation be-

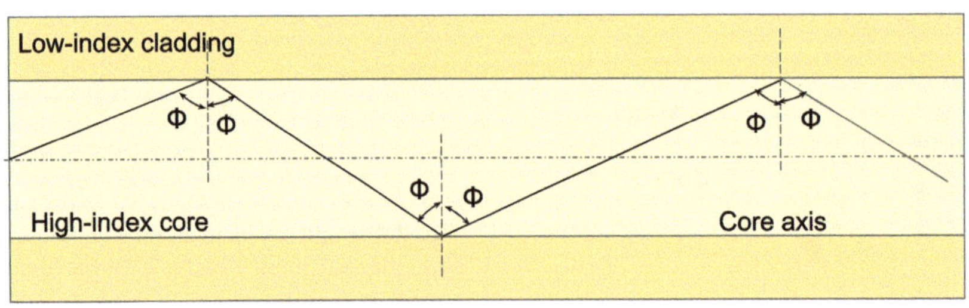

Fig. 2.4 The transmission of a light ray in a perfect optical fiber

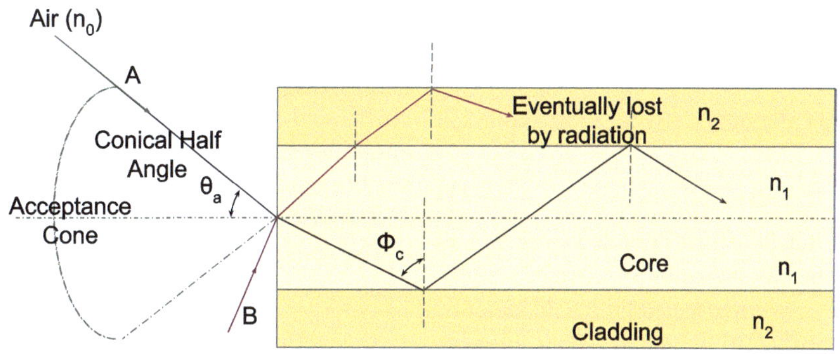

Fig. 2.5 The acceptance angle θ_a when launching light into an optical fiber

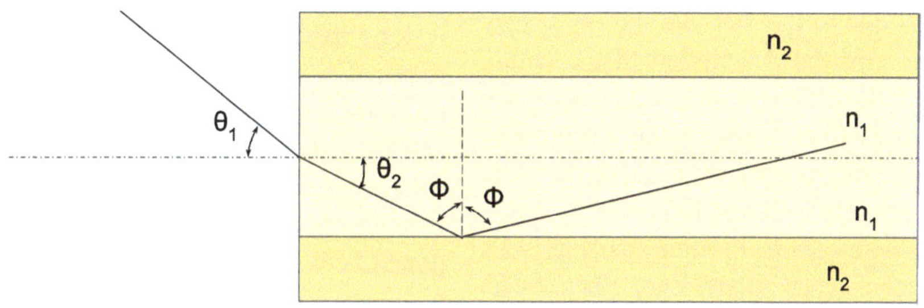

Fig. 2.6 The ray path for a meridional ray launched into an optical fiber in air at an input angle less than the acceptance angle for the fiber

tween the refractive indices of the three media in question—namely, the core, cladding, and air—and the acceptance angle. This results in the definition of a more widely used term, the numerical aperture of the fiber. In this analysis, as in the previous discussion of acceptance angle, the focus is on meridional beams within the fiber.

A light ray incident on the fiber core at an angle θ_1 to the fiber axis is depicted in Fig. 2.6. This angle is less than the acceptability angle for the fiber, which is θ_a. The ray enters the fiber from a medium (air) with a refractive index of n_0. The fiber core has a refractive index of n_1, which is marginally higher than the cladding refractive index of n_2. Assuming that the entrance face at the fiber core is perpendicular to the axis, the refraction at the air-core interface is taken into account, and Snell's law is applied in accordance with (2.2):

$$n_0 \sin(\theta_1) = n_1 \sin(\theta_2). \quad (2.4)$$

Considering the right-angled triangle ABC indicated in Fig. 2.6, then

$$\phi = \frac{\pi}{2} - \theta_2, \quad (2.5)$$

where ϕ is greater than the critical angle at the core-cladding interface. Hence, (2.4) can be written as

$$n_0 \sin(\theta_1) = n_1 \cos(\phi). \quad (2.6)$$

Using the trigonometrical relationship $\sin^2(\phi) + \cos^2(\phi) = 1$, (2.6) may be written in the form:

$$n_0 \sin(\theta_1) = n_1(1 - \sin^2(\phi))^{\frac{1}{2}}. \quad (2.7)$$

As ϕ is equivalent to the critical angle for the core-cladding interface when the limiting case for total internal reflection is taken into account, as indicated by (2.3). In this limiting case, the approval angle for the fiber θ_a is also θ_1. Combining these limiting cases into (2.7) results in

$$n_0 \sin(\theta_a) = (n_1^2 - n_2^2)^{\frac{1}{2}}. \quad (2.8)$$

Equation (2.8) is the foundation for the specification of the Numerical aperture (NA), an essential optical fiber parameter, in addition to relating the acceptance angle to the refractive indices. Therefore, the NA is defined as

$$NA = n_0 \sin(\theta_a) = (n_1^2 - n_2^2)^{\frac{1}{2}}. \quad (2.9)$$

The NA is equivalent to $\sin(\theta_a)$ because it is frequently employed in conjunction with the fiber in air, where n_0 is unity. It is also worth noting that incident meridional radiation within the range of $0 \leq \theta_1 \leq \theta_a$ will propagate within the fiber. The NA may also be given in terms of the relative refractive index difference Δ between the core and the cladding which is defined as

$$\Delta = \frac{n_1^2 - n_2^2}{2n_1^2} = \frac{n_1 - n_2}{n_1}, \quad \text{for } \Delta << 1. \qquad (2.10)$$

Therefore, by combining (2.9) and (2.10), it can be expressed as

$$NA = n_1(2\Delta)^{\frac{1}{2}}. \qquad (2.11)$$

The numerical aperture relationships provided in (2.9) and (2.11) are a highly valuable indicator of a fiber's light-collecting capacity. They are not affected by the diameter of the fiber core and are capable of maintaining their integrity for diameters as small as 8 μm. However, the geometric optics approach is invalid for lower diameters, resulting in their breakdown. This is due to the fact that the ray theory paradigm only provides a partial description of the nature of light. It delineates the direction in which a plane wave component travels within the fiber, while ray theory correctly predicts the direction of light propagation, it fails to account for interference effects between different light rays. It is discovered that only beams with specific discrete characteristics propagate in the fiber core when interference phenomena are taken into account. Consequently, the fiber will only accommodate a limited number of guided modes. This is particularly important in fibers with a small core diameter, as they can only accommodate one or a few modes.

2.4 Core Optical Network

Optical networks are dependent on the use of advanced fiber transmission systems, developments in optical component enabling technologies, and evolution in the network topological strategies together with the associated protocols to provide efficient, high-capacity operation. Significant progress in all these aspects has been made in recent years, and therefore, current optical networks operate at higher speed with increased functionality and reliability. The major deployed optical fiber networks are in the area of the public telecommunications network together with on-site or on-premises local area networks (LANs). The modern telecom network can be divided into three subnetworks: access, which is roughly 1 to 10 km in length, metropolitan, which is roughly 10 to 100 km in length, and long haul, which is around 100s or 1000s of km in length. Fiber is being widely implemented in all three subnetworks. A star, bus, or ring can typically be used as the network topology for access, a ring for metro, and a mesh for long-haul networks [10]. These subnetworks are each responsible for executing a distinct set of functions; consequently, each presents a distinct set of technological requirements, research problems, and challenges. Optical network deployments, particularly in the public telecommunications network and its crossover with LANs, are shaped by distinct priorities in different network layers. For instance, Long-haul networks, carriers prioritize capacity, protection, and restoration, due to their role in transporting large data volumes over long distances. Conversely, metro and access networks focus on service provisioning, monitoring, and flexibility to meet diverse user demands and enable rapid service deployment. The below section further elaborates on the present and future deployment strategies within these areas, considering the unique requirements of each.

2.4.1 Long-Haul Network

A long-haul network is a network that connects multiple regional or national networks, as its name suggests. These networks are also known as core or backbone networks, and they also communicate to other long-haul networks to enhance global interconnectivity between national domains. Erbium-doped fiber amplifiers (EDFAs) are typically positioned between the end terminals of a current long-haul optical network, which consists of point-to-point Dense WDM (DWDM) connections with optical regenerators at the endpoints. Moreover, an optical Re-amplification, re-shaping, and re-timing (3R) regenerator is frequently employed at 600 km intervals to mitigate the overall signal degradation on the link. To enhance connectivity, these point-to-point DWDM links can be interconnected in a mesh topology.

The continuous evolution of long-haul optical networks is primarily driven by three key factors: evolving traffic patterns, advancements in optical device technologies, and economic pressures from the service providers. These factors necessitate increased network capacity, speed, and efficiency to meet growing data demands. Long-haul optical fiber networks are now categorized according to their utmost achievable distance without optical signal regeneration: long-haul, Extended long-haul (ELH), and Ultra-long-haul (ULH). The transmission distances for these designations are characterized by the following ranges [11, 12]:

1. Long-haul optical fiber networks from 600 to 1000 km
2. Extended long-haul (ELH) from 1000 to 2000 km
3. Ultra long-haul (ULH) from 2000 to 4000 km

2.4.2 Metropolitan Area Networks

Metropolitan area networks (MANs), also known as metro networks, serve as the regional interface that connects the access network end users (i.e., commercial or residential customers) with the long-haul networks. Metro networks are subject to unique traffic demands and requirements that are distinct from those of long-haul or access networks. For instance, their traffic patterns are subject to rapid change, necessitating that the networks be dynamic, scalable, modular, and reconfigurable. Furthermore, the MAN must be highly cost-effective in terms of both operation and maintenance. Ring topologies are the typical structure of optical MANs in order to accomplish these features [13].

2.4.3 Access Networks

A public telecommunications access network connects users or smaller networks (like LANs or MANs) to the larger, core telecommunications network. This network acts as the "last mile" or the final link connecting individual devices or local networks to the broader network infrastructure or other long-distance communication systems. The access network can be provided from a single point that can be located at the local exchange, a regional hub (metro), or at Points-of-presence (POP) for a long-haul network connection to service end users. In subsequent sections, the specifics of the access network will be discussed.

2.5 Introduction to Optical Access Network

This section emphasizes several emerging and innovative technologies that have the potential to transform the design and efficacy of WDM networks, despite the ongoing development of new technologies and approaches. The optical components that are utilized in the construction of a typical point-to-point optical WDM transmission system are illustrated in Fig. 2.7. A wavelength multiplexer is employed to combine multiple optical signals transmitted by transmitters (lasers) into a fiber. EDFAs are employed to compensate for signal attenuation by amplifying signals when necessary. Optical add-drop multiplexers (OADMs) can be employed to add new signals and remove these signals at intermediate nodes. The individual wavelengths that arrive on the fiber are segregated at the receiving end by a (wavelength) demultiplexer, which is subsequently transmitted into the receivers (filters). The optical network components are defined in detail below:

1. Transmitter: The transmitter of the optical fiber network consists of the following units:
   a. Message: Typically, the origin of the message is a transducer that transforms a non-electrical message into an electrical signal. Microphones are frequently employed to convert sound impulses into currents, while video cameras are employed to convert videos/images into current. For data which is already in a digital form, transducers are not required.
   b. Modulator: The function of modular in an optical communication system is converting the electrical message signal to a proper format (symbol mapping) and modulating this signal onto the wave generated by the carrier source. Depending on the signals, the modulator can be an analog or digital modulator.
   c. Carrier source: The wave on which the information is transmitted is generated by the carrier source. The carrier is the term used to describe this wave. A Laser diode (LD) or a Light-emitting diode (LED) is employed in a fiber-optic system. They can be referred to as optic oscillators, and they generate stable, single-frequency waves with an adequate amount of power for long-distance transmission. The differences between LD and LED are summarized in Table 2.1.
   d. Channel coupler: The power required by the information channel is provided by the coupler. In an atmospheric optic system, the channel coupler is a lens that is employed to collimate the light emitted by the source and direct it toward the receiver. The modulated light beam must be transferred from the source to the optic fiber in an effective manner by the coupler. The channel coupler design is a critical component of the fiber system due to the potential for significant losses.
2. Multiplexer: A fiber-optic multiplexer is a hardware device that integrates numerous analog or digital input signals into a transmission line. It is typically operated through software and rotator blocks and employs fiber-optic technology.
3. Erbium-doped fiber amplifier (EDFA): The gain bandwidth of the EDFAs can be expanded by interconnecting them, which typically fall within the 1530 to 1560 nm range and have a gain spectrum of 30 to 40 nm each [3]. This ultrawide-band EDFA is capable of fully utilizing the expanded low-attenuation region of the new fiber network [3].
4. Optical add-drop multiplexers (OADMs): OADMs, which are also known as Wavelength add-drop multiplexers (WADMs), are used to add and remove individual wavelengths from an optical fiber in the optical domain

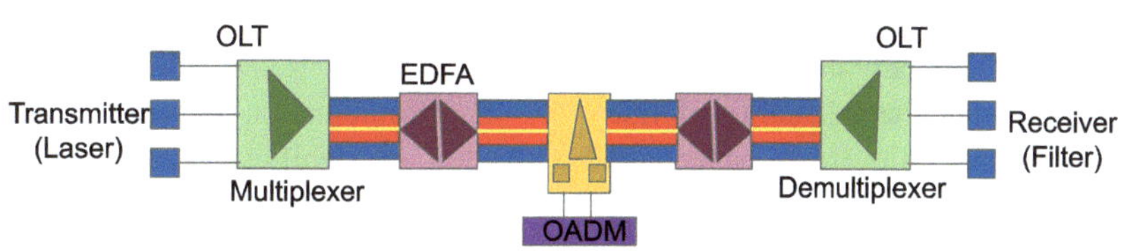

Fig. 2.7 Optical fiber network architecture

Table 2.1 Comparison of LED and laser diode [10]

S. No.	Parameter	LED	LD
1	Principle of operation	Spontaneous emission	Stimulated emission
2	Output beam	Noncoherent	Coherent
3	Spectral width	Board spectrum (20 nm–100 nm)	Much narrower (1–5 nm).
4	Data rate	Low	Very high
5	Transmission distance	Smaller	Greater
6	Temperature sensitivity	Less sensitive	More temperature sensitive
7	Coupling efficiency	Very low	High
8	Compatible fibers	Multimode step index multimode GRIN	Single-mode SI multimode GRIN
9	Circuit complexity	Simple	Complex
10	Life time	105 hours	104 hours
11	Cost	Low	High
12	Output power	Linearly proportional to drive current	Proportional to current above the threshold
13	Current required	Drive current 50 to 100 mA peak	Threshold current 5 to 40 ma
14	Applications	Moderate distance low data rate	Long distance high data rates

without converting the optical signal to the electronic domain. OADMs are classified as follows [3]:

a. Fixed-tuned OADMS: They are also known as static OADM. They are capable of adding and removing a specific predefined channel (or a group of channels).
b. Reconfigurable OADMs (ROADMs): The add-drop wavelengths in ROADMs are referred to as tunable (or dynamic) if they can be regulated by an external stimulus or software. ROADMs are more potent due to their ability to adjust to the changing traffic demand; however, the technology for their construction is still in its infancy.

5. Demultiplexer: A fiber-optic demultiplexer is the inverse of a multiplexer. The data stream is separated by the demultiplexer upon the signal's arrival at its destination.

2.5.1 Data Transmission Over Optical Fiber

Optical fibers are strands of glass filaments that have been specifically prepared to enable the transmission of light over extended distances. Light pulses are transmitted through the material using lasers. However, a multitude of difficult obstacles need to be catered to in order to turn it into a reality. The initial obstacle was signal attenuation, which was caused by noise or imperfections in the glass. Corning Glass resolved this issue in 1975 by being the first to manufacture fiber-optic cables that were ultrapure and had low loss [14]. The first optical networks were commercially used for telephone traffic in the late 1970s, marking the beginning of fiber-optic communications [3]. Further, to overcome attenuation and dispersion of the light signal, regardless of the minimal noise generated by the glass fiber, optical amplification was the subsequent step. A light signal can only travel a certain distance before it becomes unreadable without amplification. The distance is contingent upon a variety of factors, but in the majority of network spans, amplification is necessary at some point, typically after 80 to 100 km [3].

In the initial phase, the sole method of enhancing the signal was to convert the light signal back to an electrical signal, retransmit it electrically, and then convert it back to light. These Optical-electrical-optical (OEO) repeaters were expensive, power-hungry, and occupied a significant amount of space. Additionally, they were exceedingly inefficient, particularly when simultaneously amplifying numerous channels. Optical amplifiers, which do not require a conversion to electrical power, are more energy efficient. Additionally, a single amplifier can amplify multiple channels simultaneously, which is particularly crucial when multiplexing multiple channels on a single fiber. Consequently, WDM is a technology that has been extensively implemented since 1995 [10]. It enables the transmission of multiple light signals in parallel down a single fiber. Initially, lasers transmitted their pulses using a single color of light. However, in the 1980s, researchers at Nokia Bell Labs and other institutions discovered a method to transmit multiple colors or wavelengths of light through a single glass fiber [15]. As a consequence, a single fiber could transmit multiple channels in both directions on multiple distinct colors rather than a single channel on a single color. The information capacity of a single fiber increased by 40 or 80 times by DWDM, a later development that can transmit 40 or 80 duplex channels or colors [16].

The combination of optical amplification and WDM resulted in significant order-of-magnitude efficiencies in information-carrying capacity. This made fiber optics more cost-effective than copper-based electrical transmission over virtually all distances. Since the 1980s, Passive optical networks (PONs) have been increasingly employed to provide fiber-to-the-home broadband access to the residential and enterprise markets. Enterprise LANs began to utilize fiber optics in the 1990s to connect IP routers and Ethernet switches. Additionally, the proliferation of internet traffic has resulted in the dominance of fiber optics in long-haul and metropolitan transport networks since the 1990s, particularly in the context of DWDM.

2.5.2 Coherent Transmission Enabled Massive Scaling of Optical Networks

The data capacity of the transmitting lasers at one end and the receivers at the other end of the glass strand must be further scaled by a variety of technologies in addition to improved glass fiber. Researchers in institutions and industry have been conducting research and development on these for the past few decades. Coherent transmission techniques have enabled significant increase in data transmission velocities over the past few decades. Since its introduction a decade ago, coherent technology has enabled a 100-fold increase in wavelength speed by integrating the technology of Digital signal processors (DSP) with the application of phase modulation techniques (also prevalent in cellular radio networks) to optical signals.

Prior to discussing coherent transmission, it is imperative to examine the manner in which all optical transport networks previously transmitted information. It was accomplished through On-off keying (OOK), which involved the activation of a light signal to transmit a "one" or its deactivation to communicate a "zero." This may appear straightforward, and it was implemented for an extended period of time; however, there is a restriction on the rate at which a laser can be activated and deactivated. The limit for OOK transmission was reached in the late 1990s at a data rate of 10 Gb/s per wavelength (OC-192), and workarounds were devised, but they only ever provided incremental speed increases. This was still substantial and resulted in the extensive network enhancements that occurred from the late 1990s to approximately 2010.

However, the revolutionary transition to the de facto optical transmission technology for 400G, 600G, 800G, and shortly terabit solutions was facilitated by the emergence of coherent transmission.

Coherent transmission enables the generation of a binary signal by employing phase modulation, as opposed to turning the laser on or off. Light waves are similar to other waves in the radio spectrum in that they can be modulated to increase or decrease in size or to pause and resume at various times. For example, it is feasible to represent a 0 with a small wave and a 1 with a large wave by altering their phase and amplitude. By employing phase modulation, it is possible to advance or recede the wave, thereby incorporating additional information that is derived from various locations within the waveform, rather than solely the crest and trough. Additionally, the intensity of the light can be adjusted by adjusting its vertical and horizontal polarization.

All of these techniques are combined in Quadratic amplitude modulation (QAM). In order to transmit a significantly greater amount of information than OOK, coherent transmission employs a lower variant of QAM known as Quadratic phase shift keying (QPSK). Today's fiber-optic networks are capable of transmitting nearly 80 terabits per second of data over hundreds of kilometers of fiber without the use of electrical repeaters when repeated across 128 channels.

2.5.3 Coherent Transmission to the Network Edge

The optical components that transmit and detect light, such as lasers and detectors, have been miniaturized to the extent that they can be incorporated into small, low-power, and pluggable packages as optical transmission technologies have advanced. This enables the integration of all necessary optoelectronic devices into a single module that can be plugged into an optical interface port of a router. The advantage of using pluggable transceivers is that in the event of a failure, they can be replaced individually, which is a simpler and less costly process than replacing an entire line card. They also enable the interchangeable use of a variety of optics with varying characteristics, eliminating the need to create specialized line cards for each. In either scenario, fiber optics have become sufficiently affordable to be deployed for a diverse array of applications beyond metro and long-haul transmission. They are supplanting copper-based cabling in a variety of settings, including the data center Local area network (LAN), last-mile access, and short-and-medium-haul applications. Even at the micro level, numerous industry initiatives are attempting to supplant copper wire on circuit boards with optical fiber circuits that connect silicon Integrated circuits (ICs) to one another. This is known as "on-board optics" or "co-packaged optics."

These new pluggable coherent transceivers are currently employed in carrier networks for router interfaces at speeds of up to 400 Gbit/s and will soon be able to scale to even higher rates. Nokia's Coherent Routing solution, for example, allows for direct router-to-router data center interconnection or other metro-level applications by integrating pluggable coherent transceivers with its high-speed routers [15]. This conserves power by omitting the use of dedicated optical transport systems in fiber-optic transmission networks to perform the transmission and reception.

2.6 PON Technologies

PONs utilize a passive component, such as an optical star coupler or a static wavelength router, as the remote node. The reliability, simplicity of maintenance, and absence of the need to power the field-deployed network are the primary advantages of passive architectures in this scenario. Furthermore, the fiber infrastructure is not susceptible to modulation formats and bit rates, and it is possible to expand the network in the future without compromising the infrastructure. The various architectures are compared in Table 2.2.

The Telephony PON (TPON) architecture, which is originally referred to as PON for telephony, is the most prevalent PON architecture, as illustrated in Fig. 2.8. A passive star coupler is used to broadcast the downstream traffic to all Optical network units (ONUs) from a transmitter located at the Central office (CO). Switched services can be accommodated by allocating specific time slots to individual ONUs in accordance with their bandwidth requirements despite the broadcast architecture. The ONUs share a channel for the upstream channel, which is combined using a coupler. This is achieved through fixed Time division multiplexing (TDM) or another multiple access protocol. The ONUs must be synchronized to a shared schedule in the TDM approach. This is accomplished through a process known as ranging, in which each ONU measures its delay from the CO and

Table 2.2 PON technologies

Technology	Upstream bandwidth (Gbps)	Downstream bandwidth (Gbps)
Gigabit PON (GPON) [17]	1.25	2.5
10 Gigabit PON (XG-PON) [18]	2.5	10
10 Gigabit symmetrical PON (XGS-PON) [19]	10	10
Next-generation PON2 (NGPON2) [20]	40	40

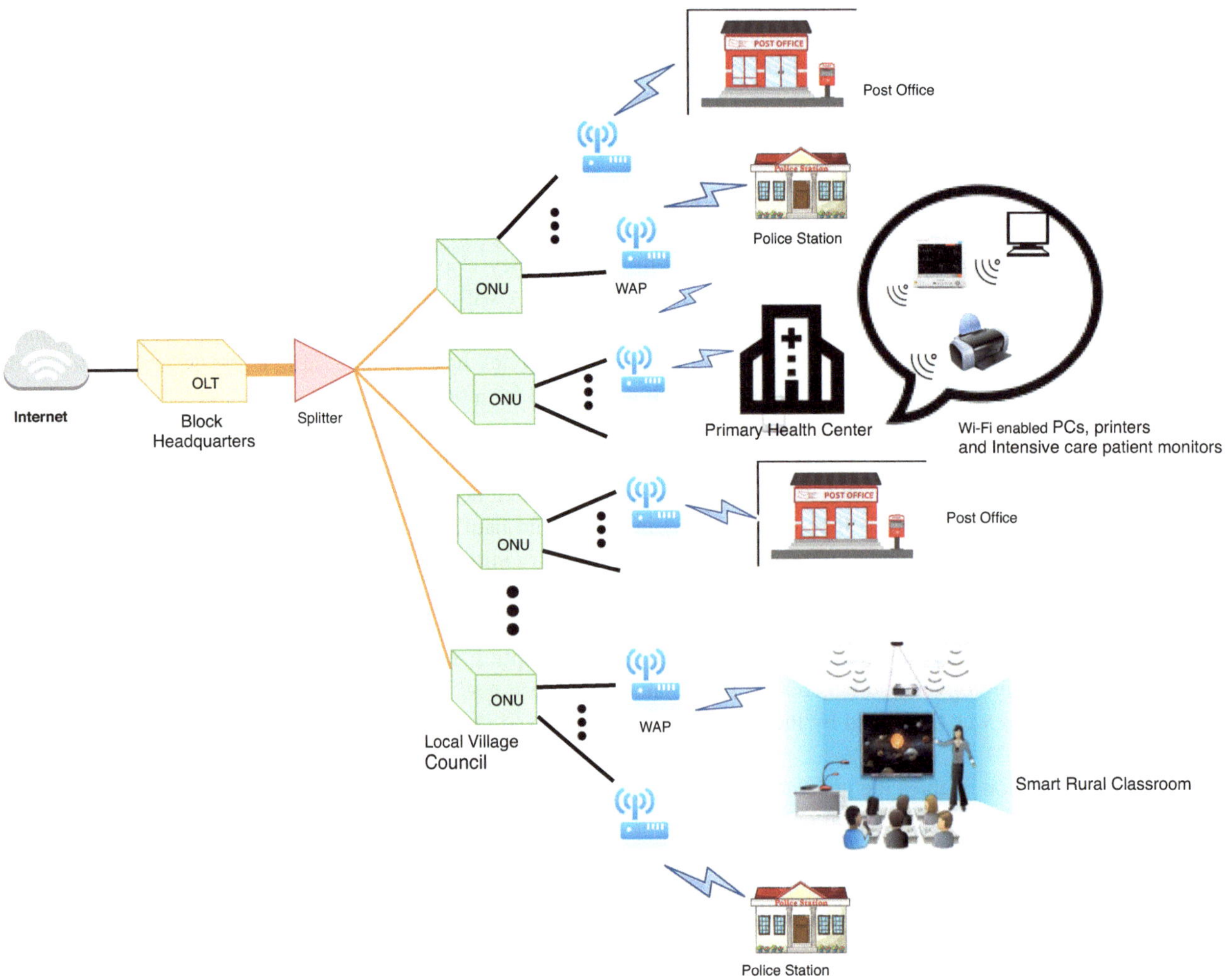

Fig. 2.8 Passive optical network architecture

modifies its clock to ensure that all ONUs are synchronized relative to the CO. The CO then allocates time periods to each ONU as required.

This architecture enables the relatively costly CO apparatus to be shared among all ONUs and utilizes relatively mature low-cost optical components. An LED or a Fabry-Perot laser may serve as the CO transmitter, and the ONUs may incorporate inexpensive, uncooled pinFET receivers and LEDs/Fabry-Perot lasers [10]. The dividing loss in the star coupler restricts the number of ONUs that can be supported. Each ONU must operate at the aggregate bit rate of all ONUs. There is a trade-off between the total distance traversed and the transmit power, receiver sensitivity, bit rate, and number of ONUs (which determines the splitting loss).

TPONs may be more cost-effective than SONET/ SDH circuits when providing lower-speed services. In a TPON, the failure of one subscriber's equipment does not impact other subscribers, whereas the failure of a SONET/SDH ring node affects all the nodes on the ring. Nevertheless, SONET/SDH is equipped with protection mechanisms that are designed to quickly restore services in the event of both apparatus failures and fiber cuts, as well as to reroute traffic. On the other hand, the TPON architecture presents a challenge in addressing fiber cuts, as it necessitates the fiber infrastructure to be duplicated. In the same vein, the TPON architecture enables the addition of additional subscribers without disrupting any of the existing subscribers. This is a more intricate process in SONET/SDH circuits.

2.6.1 TPON Standards

This section briefly describes the various TPON standards.

The Full-service access network (FSAN) working group, which was composed of numerous service providers and equipment manufacturers, was responsible for the establishment of the ATM-based PON (APON). Upstream bandwidths of up to 155 Mb/s and downstream bandwidths of up to 622 Mb/s are specified by this standard [10]. The desired distance is 20 km, with a total fiber attenuation within the 10–30 dB range. This approach enables a 16-to-32-way division in practical link budgets that utilize lasers at the CO and ONU. Using a 32-way splitter, a TPON that operates at 622 Mb/s can offer each subscriber approximately 20 Mb/s of bandwidth [3]. The International telecommunication union (ITU) devised Broadband PON (BPON), which is essentially an APON with minor enhancements. These enhancements encompass the dynamic allocation of upstream bandwidth and the support of survivability.

TPON standards at gigabit rates include the IEEE's Ethernet PON (EPON), which is Gigabit Ethernet over PONs (GEPON), and the ITU's Gigabit PON (GPON), which is an upgrade of APON. The upstream bandwidth is shared by Time division multiple access (TDMA), while the downstream bandwidth is point to multipoint. The upstream bandwidth for GPON is limited to 155 Mb/s, 622 Mb/s, 1.2 Gb/s, or 2.5 Gb/s, and it cannot exceed the downstream bandwidth [17]. The downstream bandwidth for GPON is either 1.2 Gb/s or 2.5 Gb/s. The downstream and upstream bandwidths of EPON are 1.2 Gb/s. The G-PON encapsulation mode (GEM), which is analogous to the Generic framing procedure (GFP) and efficiently transports user traffic, is supported by GPON, as well as ATM. A 10-Gigabit ethernet PON (10G-EPON) has an asymmetric architecture that relies on more mature technologies, with 10 Gb/s downstream and 1 Gb/s upstream, and a symmetric architecture that has 10 Gb/s in both the upstream and downstream [21].

The BPON, GPON, and EPON protocols all support Dynamic bandwidth allocation (DBA), which allows ONUs to communicate their upstream bandwidth requirements to the CO. For instance, in GPON, ONUs transmit the backlogs of their upstream packet queues. The CO establishes the time intervals during which each ONU is permitted to transmit upstream and transmits this information to the ONUs in grants. The bandwidth can be oversubscribed due to the fact that DBA enables the statistical multiplexing of upstream traffic.

2.7 PON Architecture

The passive optical network commences at the Optical line terminal (OLT). It is connected to a core switch via Ethernet pluggables. The OLT's primary function is to coordinate the Optical network terminal (ONT) multiplexing for shared upstream transmission and to convert, frame, and transmit signals for the PON network. The ONT is the powered device located at the opposite end of the passive optical network system. It is equipped with Ethernet connectors for in-home device or network connectivity.

The PON comprises of CO occasionally referred to as an exchange or headend, the OLT at the service provider's origination location is the starting point of a PON network. The CO houses a series of large switches that establish temporary connections for the duration of a requested connection time between subscriber lines which terminate at the switch. Subsequently, the fiber-optic feeder cable (or feeder fiber) is directed to a passive coupler, which may also include a backup fiber. The splitter is connected to a drop terminal by distribution fibers, which can be situated on a telegraph pole, on the side of a structure, in a ruggedized housing mounted in a pit, or in a street cabinet. The final one-to-one connection from a drop terminal port to an end user ONT/ONU is then provided by drop fibers. In certain instances, multiple splitters are employed in series. This architecture is known as a cascaded splitter.

The feeder fiber can divide the signals to provide service to up to 256 users, with an ONU or ONT converting the signals and providing users with internet access. The splitter or split ratio, which is the number of ways in which the downstream OLT signal is divided or split before reaching the end user (e.g., 1:32 or 1:64), is a term to describe this.

2.7.1 Time Division Multiplexing-Passive Optical Network (TDM-PON) vs. Wavelength Division Multiplexing-Passive Optical Network (WDM-PON)

The architecture of a time division multiplexing PON (TDM-PON) and a WDM PON is illustrated in Fig. 2.9 [22,23]. The fiber infrastructure from the OLT at a CO to the ONT at the customer premise is entirely passive in both scenarios. As the remote node, a passive power converter is employed in a TDM-PON. The power splitter transmits the identical signal from the OLT to various ONTs. In the time domain, signals from various ONTs are multiplexed. The address identifiers that are embedded in the signal enable ONTs to identify their own data. As the remote node, a passive WDM coupler is employed in a WDM-PON. The WDM coupler is a passive optical device that possesses the unique property of periodicity. This property is characterized by the cyclic nature of the device, which enables the routing of multiple spectral orders to the same output port from an input port [23]. This enables the spatial reuse of the wavelength channels. Consequently, the WDM coupler routes signals for various ONTs to the appropriate ONT despite the fact that the wavelengths are distinct. WDM-PON offers superior scalability and privacy as each ONT exclusively receives its own wavelength. Additional TDM techniques may be implemented in conjunction with WDM-PONs (refer to Fig. 2.9b). This results in hybrid WDM-TDMA PONs and enhances scalability by enabling division ratios of up to 1:1000 [24].

2.7.2 Types of PON Architectures

The PON network topology has undergone numerous iterations since its introduction in the 1990s, as PON technology has continued to evolve. The bandwidth and overall performance advantages of the more recent versions have progressively replaced the original passive optical network standards, BPON and APON.

2.7.2.1 G-PON

The ITU-T's Gigabit-capable PON, or G-PON, is a protocol that is IP based and is renowned for its exceptional adaptability to a variety of traffic types, including "triple-play" applications for voice, internet, and television. The generic G-PON encapsulation method is capable of encapsulating a wide range of data types, including IP, Ethernet, and Voice over Internet Protocol (VoIP).

G-PON is currently the de facto PON standard in use, with networks that span distances of 20 to 40 km, contingent upon the split ratio used, over single-mode fiber [25]. A downstream wavelength of 1490 nm is configured, while an upstream wavelength of 1310 nm is used. The downstream speed is 2.4 Gbps, while the upstream speed is 1.2 Gbps [17].

2.7.2.2 EPON

Ethernet PON, or E-PON, is an additional passive optical network standard from the IEEE that has been developed to ensure seamless compatibility with Ethernet devices. E-PON does not require any additional encapsulation or conversion protocols to establish a connection to Ethernet-based networks, as it is compliant with the IEEE 802.3 standard [26]. This is applicable to both the upstream and downstream directions of data transfer.

Symmetrical frequencies of up to 1.25 Gbps downstream and upstream are feasible with conventional E-PON. In the same way as G-PON, E-PON offers a range of 20 to 40 km, contingent upon the split ratio, and employs comparable 1310 nm upstream and 1490 nm downstream wavelengths [26]. Consequently, E-PON and G-PON cannot be installed on the same PON network. In 2020, a new amendment to the IEEE E-PON standard was implemented to facilitate the operation of 25 Gbps and 50 Gbps E-PON over a single strand of optical fiber.

2.7.2.3 10G-EPON

Additionally, the 10G-EPON standard, which is more sophisticated, increases speeds to a symmetrical 10 Gbps upstream and downstream. It operates at different wavelengths than E-PON, using 1577nm downstream and 1270 nm upstream [21]. This allows for the simultaneous use of the same PON for both E-PON and 10G-EPON, facilitating a seamless service upgrade and capacity increase on an existing PON network.

2.7.2.4 XGS-PON

The 10G variant of G-PON is referred to as XG-PON. The downstream and upstream rates of this new protocol are 10 Gbps and 2.5 Gbps, respectively. The wavelengths have been adjusted to 1577 nm for the downstream and 1270 nm for the upstream, similar to 10G-EPON, despite the fact that the physical fiber and data formatting conventions are identical to the original G-PON [19]. Once more, this modification allows for the concurrent use of the same PON network for both G-PON and XG-PON. 10-Gigabit symmetrical PON (XGS-PON) is the upgraded variant of XG-PON, which

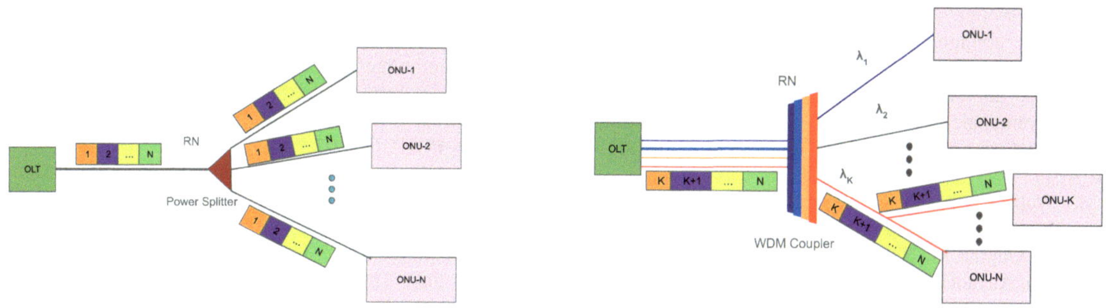

Fig. 2.9 Architecture of (**a**) TDM-PON and (**b**) (hybrid) WDM-(TDM) PON

operates on the same wavelengths as XG-PON and offers a symmetrical 10 Gbps downstream and upstream.

2.7.2.5 NG-PON2

In addition to XG(S), NG-PON2 implements WDM with multiple 10G wavelengths, both upstream and downstream, to provide a symmetrical 40 Gbps service [20]. NG-PON2 again employs distinct wavelengths from G-PON and XG/XGS-PON to facilitate the coexistence of all three services on the same PON network.

XG-PON, XGS-PON, and NG-PON2 will offer an upgrade path that is expected to be particularly advantageous in large multi-tenant or business client contexts and as a component of next-generation wireless networks, as speed demands continue to rise year over year.

2.7.2.6 50G-PON

The ITU-T has chosen a 50 Gbps line rate for the next generation of G-PON. The first standard for asymmetric 50G-PON (50 Gbps/12.5 Gbps, 50 Gbps/25 Gbps) was published in 2021, and an amendment defining symmetrical 50 Gbps service will be published in 2022. The new standard is designed to coexist with both G-PON and XG(S)-PON by operating at wavelengths of 1286 nm upstream and 1342 nm downstream [27].

2.8 Summary

This chapter discussed different types of optical network and its key features. The various concepts of optical fiber cables are discussed, such as total internal reflection, acceptance angle, and numerical aperture. These concepts are then utilized for signal transfer in three types of optical fiber networks, namely, long-haul networks, metropolitan networks, and access networks. Further, the various kinds of standardized PON architectures and their specifications are discussed in the later part of the chapter.

Homework Problems

1. Explain the evolution of optical networks.
2. Explain the concept of total internal reflection in optical fiber. The index of refraction of the core of a piece of fiber-optic cable is 1.42. If the index of the surrounding cladding is 1.28, what is the critical angle for total internal reflection for a light ray in the core incident on the core-cladding interface?
3. The index of refraction of the core of a piece of fiber-optic cable is 1.42. If the index of the surrounding cladding is 1.28, calculate the numerical aperture of the fiber.
4. Define the types of core optical networks.
5. Define the key components of the point-to-point WDM transmission system.
6. Define the architecture of a passive optical network. Explain the term "passive" in PON.
7. Define the types of multiplexing used in PON.
8. Define the types of PON architectures in detail.

References

1. Bell, A.G.: Alexander graham bell invents the photophone, the first wireless communication system (1880). https://www.historyofinformation.com/detail.php?id=2539#::text=%22The%20photophone%20used%20crystalline%20selenium,it%20is%20exposed%20to%20light
2. Huxford, W.S., Platt, J.R.: Survey of near infra-red communication systems. J. Opt. Soc. Am. **38**(3), 253–268 (1948). https://opg.optica.org/abstract.cfm?URI=josa-38-3-253
3. Senior, J.M.: Optical Fiber Communications: Principles and Practice, 3rd edn. Pearson Education, London (2009)
4. Hoven, G.: Fttx-how close to the end user should fiber come? In: Conference Proceedings - Lasers and Electro-Optics Society Annual Meeting-LEOS, pp. 621–621 (2009)
5. Labs, N.: Optical networks explained (2019). https://www.nokia.com/about-us/newsroom/articles/fiber-optic-networks-explained/#::text=There%20are%20primarily%20three%20physical,optic%20cables%2C%20used%20since%201977

6. Wikipedia: Fiber-optic communication. https://en.wikipedia.org/wiki/Fiber-optic_communication
7. Cisco: Introduction to DWDM technology. https://www.cisco.com/c/dam/global/de_at/assets/docs/dwdm.pdf
8. Visible light communications illuminates the future (2021). http://apc.aast.edu/ojs/index.php/ACE/article/download/ace.2021.01.2.033/pdf_4
9. Born, M., Wolf, E.: Principles of Optics, 7th edn. Cambridge University Press, Cambridge (1999)
10. Patibandla, A., Guptha, M., Nagma, M.: Optical Fiber Communication: Overview, Advantages, and Signal Distortion, Study Guides, Projects, Research of Communication
11. Saleh, Simmons: Evolution toward the next-generation core optical network. J. Lightwave Technol. **24**(9), 3303–3321 (2006)
12. Chatterjee, S., Yang, S.: Managing the optical networking solution: is increased bandwidth enough to reduce delays? IT Professional **7**(1), 46–50 (2005)
13. Li, M.J., Soulliere, M., Tebben, D., Nederlof, L., Vaughn, M., Wagner, R.: Transparent optical protection ring architectures and applications. J. Lightwave Technol. **23**(10), 3388–3403 (2005)
14. Corning: 50 years of fiber. https://www.corning.com/worldwide/en/markets/Optical-Communications-Market/streamlined-connectivity/timeline-1970.html
15. Nokia: Coherent routing: Build application-optimized ip-optical networks using coherent optics. https://www.nokia.com/networks/ip-networks/coherent-routing/
16. Jeon, S.W., Kim, Y., Park, C.S.: Long-reach transmission experiment of a wavelength division multiplexed-passive optical networks transmitter based on reflective semiconductor optical amplifiers. Optical Eng. **51**(1), 015008–1–015008–5 (2012). https://ui.adsabs.harvard.edu/abs/2012OptEn..51a5008J/abstract
17. Ieee standard for information technology– local and metropolitan area networks– specific requirements– part 3: CSMA/CD access method and physical layer specifications amendment 1: Physical layer specifications and management parameters for 10 Gb/s passive optical networks. IEEE Std 802.3av-2009 (Amendment to IEEE Std 802.3-2008), pp. 1–227 (2009)
18. G.987.3: 10-gigabit-capable passive optical networks (XG-PON): Transmission convergence (TC) layer specification (2020). https://www.itu.int/rec/T-REC-G.987.3
19. Series G: Transmission systems and media, digital systems and networks, 10-Gigabit-capable symmetric passive optical network (XGS-PON), G.9807.1, ITU-T (2016)
20. G.989 NG-PON2 passive optical networks. ITU-T Study Group 15 (2020). https://www.itu.int/en/ITU-T/studygroups/2017-2020/15/Documents/flyers/Flyer_ITU-T_G.989.pdf
21. 10g-epon (2010). https://www.zeitgeistlab.ca/doc/10G-EPON.html#::text=The%2010G%2DEPON%20standard%20provides,1%20Gbps%20upstream%20data%20rates
22. Frigo, N.J.: A survey of fiber optics in local access architectures. Opt. Fiber Telecommun., 461–522 (1997)
23. Banerjee, A., Park, Y., Clarke, F., Song, H., Yang, S., Kramer, G., Kim, K., Mukherjee, B.: Wavelength-division-multiplexed passive optical network (wdm-pon) technologies for broadband access: a review. J. Optical Networking **4**(11), 737–758 (2005)
24. Grobe, K., Elbers, J.P.: Pon in adolescence: from tdma to wdm-pon. IEEE Commun. Mag. **46**(1), 26–34 (2008)
25. Altabas, J.: Passive optical networks, introduction. https://upcommons.upc.edu/bitstream/2117/121280/5/Consolidated_Version_Passive%20Optical%20Networks%20Introduction_JAL_SSH_final.pdf
26. Ieee draft standard for service interoperability in ethernet passive optical networks. IEEE P1904.1/D3.2, January 2013, pp. 1–826 (2013)
27. IEEE/ISO/IEC international standard-telecommunications and exchange between information technology systems — requirements for local and metropolitan area networks — part 3: Standard for ethernet amendment 9: Physical layer specifications and management parameters for 25 gb/s and 50 gb/s passive optical networks. ISO/IEC/IEEE 8802-3:2021/Amd.9:2021(E), pp. 1–270 (2021)
28. Ramaswami, R., Sivarajan, K.N.: Optical Networks: A Practical Perspective. Morgan Kaufmann, San Francisco (2018)
29. Hossain, M., Anam, M.K.: Analysis of data rate, SNR & BER improvement of free space optical communication system under different weather conditions & wave length variation (2015). https://doi.org/10.13140/RG.2.2.36027.23841

3 Fiber-to-the-X Networks

3.1 Introduction

The optical fiber access networks, as discussed in the previous chapter, possess the ability to transmit vast amounts of bandwidth over extended distances. The performance metric used to evaluate this capability is the bandwidth-distance product. Given the rapidly growing demand for bandwidth—where the average bandwidth consumption per end user typically doubles every two years—optical fiber represents the only feasible long-term solution for delivering bandwidth directly to homes [1]. This chapter explores advancements in optical fiber technology that have driven down component and system costs while meeting the growing demand for high-bandwidth services. Additionally, it examines optical access solutions designed to deliver triple-play services—voice, data, and video—to customers. These developments in optical fiber technology are motivating service providers, Telcos, and Cable Multiple system operatorss (MSOs) alike to invest in fiber-to-the-everything (FTTX). FTTX signifies a number of optical access technologies, such as fiber-to-the node (FTTN), fiber-to-the-curb (FTTC), fiber-to-the-building (FTTB), fiber-to-the-home (FTTH), and fiber-to-the-premises (FTTP).

The evolution of broadband infrastructure toward improved bandwidth and Quality-of-service (QoS) is characterized by the central component of Next-generation access (NGA) networks, which is Fiber-to-the-everything (FTTX) network. The boundaries of this technology are already being pushed by high-definition television (HDTV), Virtual reality (VR), and other bandwidth-intensive applications. Blockchain technology, smart cities, fifth generation (5G), Sixth generation (6G), and the Internet of Things (IOT) are all experiencing accelerated growth as the number of high-speed, low-latency applications for FTTX continues to increase.

From a topological point of view, FTTX solutions can be categorized into two types: Point-to-point (P2P) systems and Point-to-multipoint (P2MP) systems [2]. In a P2P topology, a distinct fiber is routed from an optical line terminal (OLT) in the central office (CO) to an optical network unit (ONU) at a specific endpoint (node, curb, residence, or other). This is occasionally referred to as "home-run fiber." A remote Ethernet-based aggregation node can be positioned between the OLT and the ONU to distribute the bandwidth of the feeder fiber cable among the multiple end consumers. In this scenario, each client is still served by a P2P link and has an ONU. The Ethernet node is electrically powered by active electronics. Consequently, this solution is sometimes referred to as an active Ethernet star and is said to be found on an active optical network (AON). Subscribers may be situated at a distance of several tens of kilometers from the remote node in an active FTTX architecture of this nature. An AON can accommodate numerous hundreds of consumers [3]. The feeder fiber extends from the OLT of the CO to a dividing point in the field in a P2MP topology. From this point, shorter fibers extend to multiple ONUs at end points, including the node, curb, house, or any other location. Passive optical splitters are employed at the splitting point to distribute the intensity of the input signal among multiple outputs. The network is referred to as a passive optical network (PON) because it does not employ any active electronics between the OLT and optical network terminal (ONT) units.

3.1.1 Network Designing and Planning

Well-coordinated design and planning are the initial steps in the implementation of this technology. The quantity and location of users, fiber distribution, and access points, as well as architectural elements such as Passive optical network (PON) technologies, are all preliminary planning considerations that

V. A. Bohara, Swaminathan R, *Optical and Wireless Access Networks*, Textbooks in Telecommunication Engineering, https://doi.org/10.1007/978-3-031-89696-5_3

must be addressed before the detailed FTTX network design is established. Splice locations, precise distribution patterns, and loss budget calculations are among the detailed design elements at the "micro" level. Other factors that are incorporated into a comprehensive FTTX network design include the avoidance of existing utilities and the establishment of equipment locations.

A successful FTTX deployment necessitates meticulous network planning, attention to detail, precise labeling, efficient routing, and prudent test practices which are essential to prevent delays during cable splicing and splitter installation, as the timelines can be quite aggressive.

3.1.2 Customer Collection and Installations

The increasing demand for high-bandwidth services like streaming video, online gaming, and virtual reality, coupled with the introduction of new service offerings like multichannel HD television and enhanced internet speeds, has made traditional copper-based networks insufficient. This, combined with advancements in FTTX presents a strong argument for its implementation. Service providers worldwide have made progress in the direction of the widespread deployment of FTTX. The United States is experiencing an increase in subscriber numbers and accept rates, while China has begun to experience a decline in digital subscriber line (DSL) subscriber numbers due to optical competition. Additionally, European incumbents and municipalities are making progress in the deployment of FTTX [3]. The development of a new generation of broadband optical access networks has a short-term impact, as it enables unprecedented services and applications. In the long term, these advancements have the potential to facilitate the resumption of optical networking in core and metro networks. Ultimately, the type of service demands and the magnitude of bandwidth facilitated by optical broadband access will present challenges in other segments of carriers' networks. Reconfigurable optical networking is the sole solution to these obstacles. Consequently, FTTX may serve as a critical component of the optical resurgence [3].

Service providers and contractors are generally under considerable pressure to ensure that fiber is deployed in a timely and cost-effective manner while also guaranteeing the quality and long term reliability of the installations. It is often perceived as a viable approach to minimize deployment costs and time by conducting no or limited tests [4]. However, it has been established that a lack of testing results in activation delays, excessive troubleshooting, and revenue losses. The sole method of certifying an FTTX link and establishing a record of conformance to ensure successful first-time turn-up and minimize unnecessary truck rolls or repeats is through proper testing during the construction and installation phases [3].

3.1.3 Wavelength Allocation

Voice and data transmission from the OLT are generated at a downstream wavelength of 1490 nm for the majority of FTTX applications. Wavelength division multiplexing (WDM) allows for bidirectional transmission over the same fiber by establishing an upstream connection wavelength at 1310 nm. At the Optical network terminal (ONT) (user) end, laser optic technology can be implemented at a reasonable cost. Next generation PON2 (NG-PON2) and 10-Gigabit symmetrical PON (XGS-PON), which are more recent and faster iterations of PON technology, operate at slightly different downstream and upstream wavelengths. This enables simultaneous or interchangeable use of multiple PON architecture varieties, contingent upon the service level requirements.

3.2 Fiber-to-the-Curb (FTTC)

In contrast to Hybrid fiber coax (HFC), Fiber-to-the-curb (FTTC) refers to the digital transmission of data from the hub, or central office, to fiber-terminating terminals such as Optical network units (ONUs). The anticipated result of this architecture is that the fiber would be significantly closer to the subscriber. Various terms are used to characterize this architecture, contingent upon the proximity of the fiber to an individual subscriber. In the most optimistic scenario, fiber would be connected to each residence, resulting in an architecture known as Fiber-to-the-home (FTTH). The ONUs would assume the role of the Network interface units (NIUs). This can be referred to as FTTC or Fiber-to-the-building (FTTB) in the event that ONUs serve a limited number of dwellings or buildings, such as 8–64. The fiber is typically located within approximately 100 meters of the end consumer in FTTC. In this instance, an additional distribution network is present from the ONUs to the NIUs. The fiber is terminated in a cabinet in the neighborhood and is within approximately 1 km of the end user with the fiber-to-the-cabinet (FTTCab) approach.

The network from the Central office (CO) to the ONU is typically a PON in order to render the FTTC architecture viable. The remote node is a passive device, such as an optical star coupler, and it may occasionally be situated in the CO rather than in the field. The term "FTTC" is generally used to refer to a version of the technology in which the signals are broadcast from the central office to the ONUs, and the ONUs share a common total bandwidth in a time division multiplexed fashion. The feeder network is the section of the network that connects the central office to the remote node in the context of FTTC, while the distribution network is the portion of the network that connects the remote node to the

ONUs. A diverse array of architectures can be achieved by combining various categories of sources at the central office with various types of remote nodes.

The FTTC architecture is occasionally referred to as the baseband modulated fiber coax bus (BMFCB) or switched digital video (SDV). The subsequent section will focus on various options for implementing the optical element of the access network. The network must be simple to operate and service, and optical access network architectures must be reliable. This implies that passive architectures are more advantageous than active ones, as the network itself does not require any toggling and does not require control. Additionally, passive networks necessitate no power, with the exception of the endpoints, which result in substantial cost reductions for operators. The ONU must be kept as simple as possible to enhance reliability and reduce costs. This precludes the utilization of advanced lasers and other optical components within the ONU. It is preferable that the components utilized in the ONU be capable of functioning without the need for temperature regulation. This is because temperature control adds complexity, cost, and potential points of failure to the device. The cost of the CO equipment can be amortized over the numerous subscribers it serves, as it is located in a controlled environment. Consequently, the equipment may be somewhat more sophisticated.

3.3 Fiber-to-the-Edge (FTTE)

Fiber-to-the-edge (FTTE), or fiber to the edge, is an architecture for Local area networks (LANs) that employs optical fiber to transport data to the network's periphery. At the edge, the network interfaces with Internet of Things (IoT) devices, such as cell phones, laptops, security cameras, machine-to-machine controls, building management systems, automated guided vehicles, and the applications that support them through ports or wirelessly through cellular or wireless fidelity (WiFi). An FTTE-based LAN is composed of active optical components (e.g., routers, switches) and passive optical components (e.g., cables, connectors) in the same way as any enterprise information technology (IT) network.

In a conventional enterprise LAN IT architecture, data is transmitted by copper or through a combination of a fiber backbone and copper to the edge. This necessitates equipment closets throughout the network. The entire data transmission is carried by optical fiber in an FTTE architecture. This eliminates the necessity for Intermediate distribution frames (IDFs) by enabling the network to operate from a central equipment room. The FTTE architecture's exceptionally high bandwidth enables the wireless operation at edge devices, thereby reducing the necessity for individual wiring and porting.

Separate copper electrical wiring or composite cabling that incorporates both copper and fiber can be employed to supply power to an FTTE-based LAN. This method delivers both data and power to the periphery, such as wireless antennas, cameras, and point-of-sale systems, at distances exceeding 100 meters (328 feet). Traditional, decentralized hardware options or software can be employed to manage an FTTE-based LAN. Some of the benefits of fiber-to-the-edge network includes:

1. Extremely high bandwidth and low latency
2. Increased security
3. Reduced complexity
4. Lower installation and lifetime cost
5. Space efficiency
6. Reduced environmental impact
7. Power over distances beyond 100 meters (328 feet)
8. Scalability for a variety of building and campus sizes
9. Flexibility for the future

3.4 Fiber-to-the-Premises (FTTP)

In this section, a variety of optical techniques that are currently being implemented for shared-fiber multiple access are discussed. These techniques are based on time slot multiplexing, frequency slot multiplexing, code division multiplexing, and wavelength multiplexing. The application of these techniques in FTTH/Fiber-to-the-premises (FTTP) networks for fast data transfer (Ethernet based) and for broadband service distribution is also discussed. Research laboratories are currently investigating techniques that are designed to facilitate next-generation optical access. These techniques include wavelength routing, which enables the centralized processing of radio signals and facilitates the flexible allocation of capacity and the adaptability of services and service providers. Additionally, radio-over-fiber (RoF) techniques are being developed to foster a powerful symbiosis between the fiber and wireless worlds.

Residential FTTP is a term that denotes the equipment utilized in fiber access implementations. This term signifies that optical fiber extends all the way to the individual building or dwelling, like a house or apartment, providing high-speed internet access. This equipment is specifically designed and optimized for residential applications, with fibers that extend all the way to the end-user premises. This segment does not encompass equipment that is specifically designed and optimized for fiber-to-the-business applications; the optical transport segment encompasses such equipment.

FTTP is further subdivided into P2P and P2MP FTTP. The deployment of P2P FTTP technologies is more costly due to the increased number of fibers. However, they have the advantage of providing a dedicated fiber to each end-user location, which enables the sharing of P2P infrastructure between various carriers. Optical Ethernet switches are frequently employed at both the user premises and the central office in a P2P FTTP deployment.

3.4.1 Architecture and Standards

In the pre-fiber era, telecommunications (telco) Central Offices (COs) were primarily connected using coaxial cables and microwave links. Cable television headends, on the other hand, relied on satellite or microwave links. Hybrid Fiber-Coaxial (HFC), which is still widely used in the cable industry, and various types of DSL, which are typically supplied to residences over twisted pair from a centrally located DSL access multiplexer (DSLAM), a remote DSLAM, or an ONU, were the initial steps in "fiberizing" the entire system.

In the deployment of broadband in general and fiber to the household in particular, other countries have advanced significantly beyond the United States [5]. In terms of per capita broadband deployment, the United States is ranked 11th, trailing Korea, Canada, Sweden, and other countries. Although cable is more prevalent than DSL in North America, the opposite is true in Korea, Japan, and the European countries. This is partially due to the fact that in North America, over 50% of the subscribers are located more than 15 km from the central office, whereas in more compact countries, the figure is closer to 1 km [6].

3.4.2 Fiber-to-the-Room

One of the most significant emerging fifth-generation fixed network (F5G) scenarios is Fiber-to-the-room (FTTR), which involves the extension of the optical network to each room of a household that is equipped with multiple ONTs [7]. FTTR is founded on optical technology for the optical home networking segment and on IEEE 802.11ax (WiFi6) for the wireless access segment [8]. The optical technology has a shared capacity of up to 10 Gb/s downstream and up to 2.5 Gb/s upstream. New levels of precise coordination between multiple APs are made possible by such an optical backhaul network.

FTTR is a P2MP topology that comprises a master ONT, one or more peripheral ONTs, an Optical distribution node (ODN), and a home Cloud management platform (CMP). The Master ONT functions as the intermediary between the residential networks and the carrier. The edge ONTs establishes "cellular" system in which each room serves as a cell and offers wireless access to home user devices via WiFi 6. Optical splitters and fiber-optic circuits comprise the ODN. Running in a remote cloud, the Home CMP is an application that automates the management of the FTTR network. In order to monitor user performance and status, FTTR establishes a new channel between the optical network and the WiFi6 network. Operators can expand their business reach into home networks by detecting and managing failures in the home network through a dedicated management channel. The coexistence of multiple ONTs on an FTTR network presents a variety of opportunities and challenges in the areas of interference management, resource allocation, handovers, and RF design. A few of the opportunities and challenges are listed below:

1. The problem of user-ONT association is to ensure that WiFi resources are allocated fairly and equitably among all the users.
2. Power control and a novel antenna pattern design are employed to mitigate co-channel and adjacent channel interference in order to accommodate dense ONT deployments.
3. In FTTR networks, roaming events are more frequent than in typical residential networks, i.e., the possibility of mobile users moving across different rooms with mobile devices, this necessitates fast and seamless roaming.
4. The potential for Radio frequency (RF) designs that are more compact and cost-effective. In comparison to conventional WiFi, the target performance can be achieved with lower transmit power and receiver sensitivity, as each ONT essentially encompasses only its own room. This facilitates the development of RF designs that are highly integrated on a single semiconductor technology.
5. Centralized processing of remote communications. The opportunity to supplant the edge ONTs with simpler, centrally processed remote antenna units is provided by the high-speed optical network.

Fourteen typical commercial application scenarios for gigabit networks are presented by F5G [9]. These cases demonstrate a distinct trend of escalating data rates and more stringent delay requirements. For instance, highly interactive VR services may necessitate data rates of up to 1 Gb/s and a latency of less than 10 ms. In the interim, energy consumption has emerged as a significant concern, primarily due to the escalating cost of energy and the growing concern regarding climate change. In the context of home networking, FTTR has the capacity to simultaneously resolve all of these issues.

The FTTR architecture enables each ONT to operate at a comparatively low power level, as per [7]. Generally, the transmit power is typically between 0 dBm and 20 dBm, although the precise level may vary slightly due to regulations in different countries. Nevertheless, the transmit power is optimized by a suitable power control algorithm that is contingent upon the configuration of the specific residence. The purpose of power control is to maintain an acceptable level of interference between devices operating in different locations while ensuring that the entire house is covered.

3.5 Fiber-to-the-Node (FTTN)

One of the numerous alternatives for delivering broadband and other data services to numerous destinations is FTTN. In FTTN, fiber optic cables are run to a central distribution point, often called a node, and then the connection to individual premises is completed using existing copper infrastructure.

The capacity to transmit data over fiber-optic lines that are more efficient, as opposed to other lines that are subject to greater bandwidth restrictions, is one of the primary advantages of FTTN as compared to FTTH and similar systems. Copper or other wire types which is employed to cover the remaining distance from the node to an individual destination, is frequently referred to as "last mile" service. In order to facilitate the distribution of services to numerous customers, FTTN systems frequently employ coaxial or twisted-pair cables.

3.6 Distribution Networks

The ODN is the fiber-based infrastructure that interconnects the OLT and ONUs, and it consists of a feeder fiber, passive optical power splitter(s), and fiber drops to each ONU. Generally, a PON consists of multiple ONUs connected to an OLT via an ODN. The PON ODN is commonly deployed in a treelike topology with one or several stages of optical power splitters and/or wavelength filters. In the FTTH architecture, the ONUs are located at the customer premises, while the OLT is typically hosted at the CO. The ODN splitters or filters are located at distribution points in the CO service area. In PONs, the ODN is a shared transmission medium, and depending on the type of PON, different mechanisms are employed for sharing of the medium (i.e., time domain, wavelength domain, etc.). The ODN is a long-term investment that should support future technology generations. Considering the existing base of Time division multiplexing (TDM)-PON deployments (i.e., EPON and GPON), the most common ODN infrastructure today is a splitter-based PON ODN. Besides splitter-based ODNs, other possible ODN deployments are P2P fiber and filter-based PON ODNs. For splitter-based ODNs, careful planning of the splitter stages and splitter locations is needed to facilitate upgrades with increasing traffic and data rates. The ODN is owned by the network operator, so in order to enable interoperability between different PON component vendors, the ODN needs to comply with a standard as well. The main parameters of the ODN are optical power loss (dB) due to splitter and fiber attenuation and fiber distance (km). Maximum and minimum ODN losses are defined for different classes of optical power budgets, which correspond to attenuation ranges that need to be supported, as defined in the available IEEE and International telecommunication union (ITU)-T PON standards. The earlier ITU-T standards allow the implementation of bidirectional transmission using WDM on a single fiber or unidirectional transmission over two fibers. The majority of available PON standards set the upstream and downstream wavelengths with a large separation, which enables low-loss and low-cost 45 angle filters to multiplex and demultiplex the downstream and upstream wavelengths onto a single fiber. Therefore, the loss budget must be divided between fiber, splitter, and WDMs (in the single-fiber case). The reach parameter of the ODN determines the maximum fiber length. The ODN must use Standard single-mode fiber (SSMF) as described in ITU-T G.652. Three original power budget classes are defined for B-PON/G-PON, namely, class A, class B, and class C. The power budget includes fiber loss, splitter loss, and WDM losses to combine and separate up and downstream wavelengths. Class B+ and class C+ were later added to the standard after realizing that three additional decibels were needed to accommodate WDM losses [10]. In general, the IEEE PON standards have more relaxed parameters than the ITU-T PON standards.

3.7 Use Case: QoS-Aware Reliable Broadband Access Networks

This section discusses a wireless path routing technique for the interrupted traffic to OLT via backup ONU-Access point (AP) without deploying redundant fibers. Furthermore, the access network's performance also depends upon the position of the backup ONU-AP. A stochastic optimization algorithm is utilized to find the optimal position of the backup ONU-AP [11]. Moreover, a performance comparison is presented by incorporating backup ONU-AP with three different antenna configurations, namely, omnidirectional antenna, sector antenna, and parabolic grid antenna, for establishing a P2P link between backup ONU-AP and active ONU-AP.

3.7.1 System Model

The Fiber wireless (FiWi) system architecture integrates Gigabit capable PON (XG-PON) with an Institute of Electrical and Electronics Engineers (IEEE) 802.11ac-based wireless access network. Conventionally, in XG-PON, Optical line terminal (OLT) is located at the central office and serves N number of ONU-APs, which are connected to OLT through a 1:N optical splitter. Here, N can be 2, 4, 8, 16, 32, etc. The survivable FiWi architecture considered in this section consists of a splitter of 1:16 split ratio [12]. Further, 16 ONU-APs are divided into four clusters of four ONU-APs. In each cluster, three ONU-APs act as active ONU-APs, and one ONU-AP will act as a backup ONU-AP, which provides connectivity to the users of the other three active ONU-APs in the case of a failure. The users are distributed according to the Poisson point process with a density of λ users/km^2 in the coverage region of active ONU-APs. The coverage region of active ONU-AP is assumed to be a circular cell. Here, circular cell means that the radiation pattern of a circularly polarized omnidirectional antenna is circular in shape and provides 360° full coverage [13] up to the radius of 1000 m [14]. Whenever a failure occurs, backup ONU-AP provides the required services to the users through the wireless path. Since it is an XG-PON-based back-end network, there is no bandwidth limitation from the network up to the ONU level, and each backup ONU-AP has sufficient bandwidth to guarantee the QoS to users in case of failures.

In case of failures, disconnected users must remain connected to the backhaul network via backup ONU-AP to prevent significant traffic loss. The performance of the survivable FiWi network is critically determined by the placement of backup ONU-AP. Hence, it is essential to optimize the position of backup ONU-AP in such a way so that it will provide coverage to the users of active ONU-APs in case of failure. In order to optimize the position of backup ONU-AP such that the average distance between backup ONU-AP and the users in the cluster is minimized and stochastic gradient method is employed. Specifically, the stochastic optimization problem is defined as a minimization of the average distance between backup ONU-AP and the users in the cluster by optimizing the position of the backup ONU-AP and radius of the backup ONU-AP in (3.1):

$$(P1): \min_{R_{B_{ONU}}, P_{B_{ONU}}} \quad \mathbb{E}_{P_u}[D_{B_{ONU},u}], \tag{3.1}$$

$$\text{s.t.} \quad R_{B_{ONU}} \in R^+, \tag{3.2}$$

$$P_{B_{ONU}} \in \mathcal{A}, \tag{3.3}$$

where $\mathcal{A}$ is the area of the cluster of active ONU-APs, the position of backup ONU-AP is $P_{B_{ONU}}$, radius of the circular cell of backup ONU-AP is $R_{B_{ONU}}$, and $D_{B_{ONU},u}$ is average distance between backup ONU-AP and uth user. In practice, due to the mobility of the users, P_u varies from time to time within radius of the circular cell of active ONU-APs.

3.7.2 Methodology

To solve the Problem $(P1)$, one way is to calculate $\mathbb{E}_{P_u}[D_{B_{ONU},u}]$ and turn it into a deterministic optimization problem. However, there is one more efficient way to solve this problem using a stochastic optimization techniques [15].

This stochastic optimization problem can be solved through the stochastic gradient method, in which decision variables iteratively update along a stochastic gradient. It is a more efficient and less complex approach to solve this type of stochastic optimization problem [16, 17]. Using the stochastic gradient method, the position of a backup ONU-AP is optimized as shown in Algorithm 3.1.

Let the total number of users in the network be U. The set of the users $u = \{1, 2, ..., U\}$, $P_u = (X_u, Y_u)$ denotes the position of the uth user, where X_u and Y_u are the X and Y coordinates of the position of uth user. $P_{B_{ONU}} = (X_{B_{ONU}}, Y_{B_{ONU}})$ denotes the position of backup ONU-AP, where $X_{B_{ONU}}$ and $Y_{B_{ONU}}$ are the X and Y coordinates of backup ONU-AP. In order to find the optimal position of backup ONU-AP that solves problem P1, the position, and radius of backup ONU-AP need to be updated iteratively as

$$[P^{(n+1)}_{\mathrm{B_{ONU}}}, R^{(n+1)}_{B_{ONU}}] = ([P^n_{\mathrm{B_{ONU}}}, R^n_{B_{ONU}}] - \times \mu^n \Delta^n_{U_o}) \tag{3.4}$$

where n $= 0, 1, \ldots$ is iteration index, $\mu^n > 0$ is step size, $P^n_{\mathrm{B_{ONU}}}$, and $R^n_{B_{ONU}}$ are initial position and radius of backup ONU-AP, respectively.

For updating the position of backup ONU-AP, the stochastic gradient of the numbers of users outside the radius of backup ONU-AP, U_o is calculated with respect to position $P_{B_{ONU}}$ and radius $R_{B_{ONU}}$ of backup ONU-AP, respectively. However, the sigmoid function is used to make U_o differentiable [18]. U_o can be calculated as

$$U_o = \sum_{u=1}^{U} S(D_{\mathrm{B_{ONU},u}} - R_{B_{ONU}}), \tag{3.5}$$

where $S(.)$ is the sigmoid function defined as

$$S(D_{\mathrm{B_{ONU},u}} - R_{B_{ONU}}) = \frac{1}{1 + e^{-\beta(D_{\mathrm{B_{ONU},u}} - R_{B_{ONU}})}}, \tag{3.6}$$

where $\beta > 0$ is the steepness factor that controls the accuracy of the sigmoid function. The average distance between the backup ONU-AP and uth user $(D_{B_{ONU},u})$ is given by

$$D_{B_{ONU},u} = ||P_{B_{ONU}} - P_u||. \tag{3.7}$$

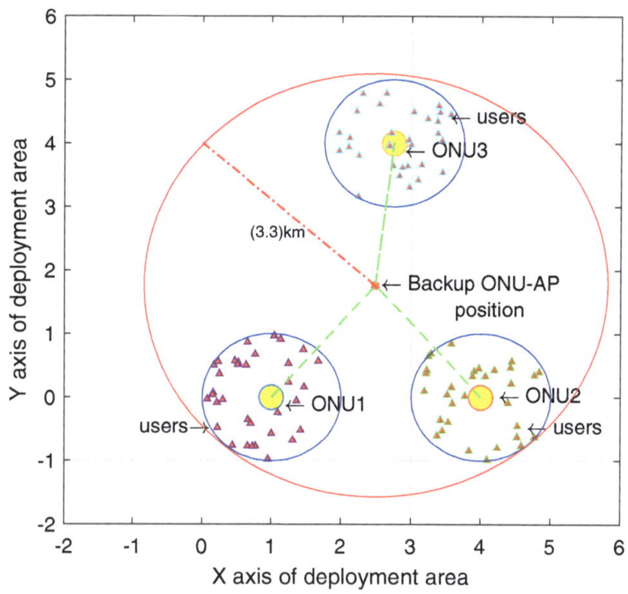

Fig. 3.1 Optimal position for backup ONU-AP in case of uniform distribution of users

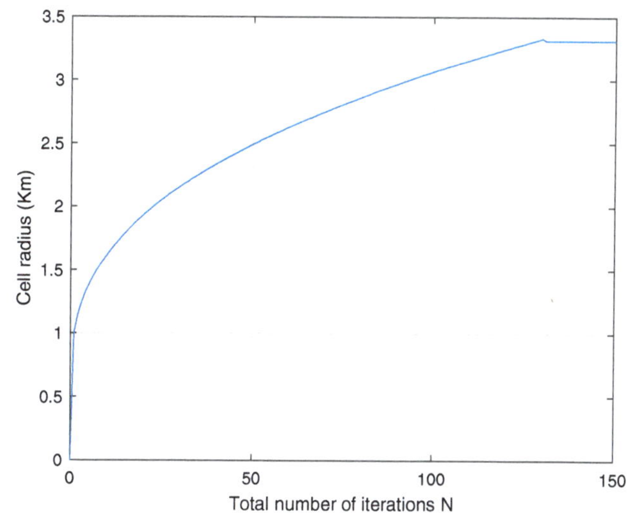

Fig. 3.3 Cell radius of backup ONU-AP versus number of iterations, N

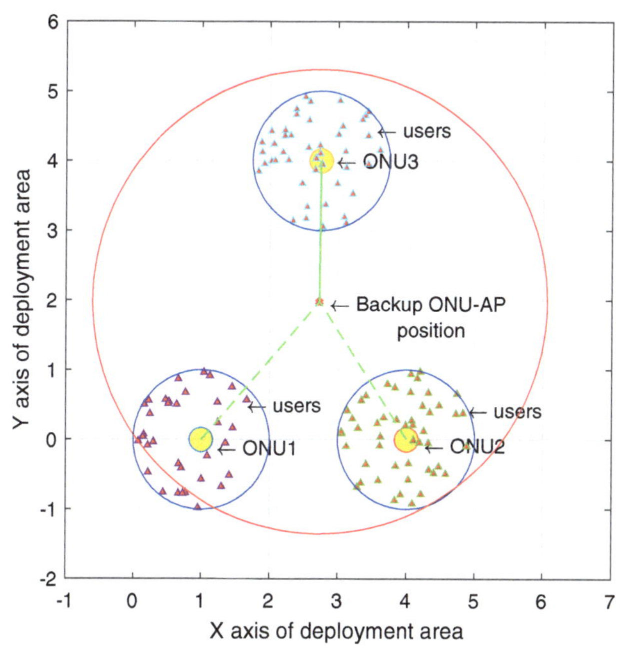

Fig. 3.2 Optimal position for backup ONU-AP in case of nonuniform distribution of users

Now, to enhance the convergence rate of optimal positioning of backup ONU-AP algorithm (OPA), the learning rate μ^n of the algorithm must be set [16, 17]. The learning rate is normalized by the norm of the stochastic gradient for the number of users outside the radius of backup ONU-AP:

$$\mu^n = \frac{L}{||\Delta^n_{Uo}||}, \tag{3.8}$$

where $L = \frac{a}{n^b}$ and set positive constants, $a = 0.15$ and $b = 0.6$ for simulations. To analyze the performance of the placement algorithm, the uniform distribution of users is considered wherein an equal number of users are distributed inside active ONU-AP. In Fig. 3.1, using the algorithm, the position of the ONU-AP backup is optimized so that the average distance between the ONU-AP backup and the users is minimized. Moreover, it can be observed from Fig. 3.1 that the circular cell of the backup ONU-AP touches the circumference of the circular cell of active ONU-APs, when the considered backup ONU-AP is equipped with an omni-directional antenna with antenna height of 30 m. In contrast, Fig. 3.2 shows the optimal placement of the backup ONU-AP for the case where users are not uniformly distributed in the cluster, which primarily means that the number of users inside each active ONU-AP is not equal. It can be seen that the number of users inside active ONU-AP 3 is more than that of active ONU-AP1 and ONU-AP 2. Therefore, in such a case, the placement algorithm provides the location for backup ONU-AP nearer to the position of active ONU-AP with higher user density. It may be noted that even with variations in the number of users in OPA, there is only a slight difference in backup ONU-AP's positions. Figure 3.3 illustrates the convergence performance of the algorithm 3.1. It can be seen that the cell radius of the backup ONU-AP increases with increasing number of iterations until the placement algorithm reaches convergence at N $\geq$ 140. It is worth mentioning that in the stochastic optimization algorithm, at each iteration the cost function is updated recursively by moving a distance step size in the direction opposite to the gradient [19]. The size of the step on each iteration to reach the local minimum is determined by the learning rate. For the algorithm, a variable step size $\mu^n = \frac{a}{n^b||\Delta^n_{Uo}||}$ is

Algorithm 3.1 Algorithm for finding optimal position for backup ONU-AP (OPA)

Input:
Given set of users in circular cell of active ONU-APs in a cluster, initial location of backup ONU-AP is the centroid of the three active ONU-APs and radius of circular cell of backup ONU-AP is 1000 m [14]

for n=1, ..., maximum iteration N **do**
  $\Delta_{Uo}^{n} = (\sum_{u=1}^{U} S'(D_{\mathrm{BONU,u}} - R_{\mathrm{BONU}})\frac{\partial D_{\mathrm{BONU,u}}}{\partial P_{\mathrm{BONU}}}, - \sum_{u=1}^{U} S'(D_{\mathrm{BONU,u}} - R_{\mathrm{BONU}}))$
  Calculate learning rate, μ^n using (8)
  Update the position and radius of backup ONU-AP

$$P_{\mathrm{BONU}}^{(n+1)}, R_{BONU}^{(n+1)} = (P_{\mathrm{BONU}}^{n}, R_{BONU}^{n} - \mu^n \times \Delta_{Uo}^{n})$$

  if $D_{BONU,u} \leq R_{BONU}^{n+1}$ **then**
    $P_{\mathrm{BONU}}^{final} = P_{\mathrm{BONU}}^{(n+1)}$
  end if
end for
return $P_{\mathrm{BONU}}^{final}$

utilized. Hence, the overall time complexity of the algorithm is $O(1/n)$, where n is the number of iterations.

3.7.3 Performance Analysis

There can be different kinds of antenna configurations used at backup ONU-AP. This section considers three typical antenna configurations: omnidirectional antenna, sector antenna, and parabolic grid antenna (P2P link). It can be observed in Fig. 3.4 that the backup ONU-AP position can be optimized depending on the above antenna configurations. For instance, in Fig. 3.4a, the position of backup ONU-AP is optimized when it is equipped with an omnidirectional antenna to provide services to the disconnected users via P2MP links in the case of failures. Similarly, Fig. 3.4b shows that optimal position of the ONU-AP backup in which the ONU-AP backup is equipped with a sector antenna. Users get connectivity from the backup ONU-AP via P2MP wireless links in case of failure. Figure 3.4c presents the optimal position for backup ONU-AP, when it is equipped with P2P link. For P2P link, the position is optimized with respect to positions of three active ONU-APs as it is assumed that the users are not directly connected to backup ONU-AP; they get connectivity from active ONU-APs via P2MP wireless links. In case of P2P link, the co-channel interference is alleviated as only one P2P link from backup ONU-AP is activated which is directed toward disconnected users in case of failures unlike the case of omnidirectional antenna. The antenna configurations are discussed in detail in the following subsections.

3.7.3.1 Omnidirectional Antenna

In this case, the backup ONU-AP is equipped with an omnidirectional antenna. In any case of failures such as distribution fiber cut, ONU, and AP failure, disconnected users will remain connected to the backbone network through backup ONU-AP via P2MP links. To evaluate the coverage obtained through an omnidirectional antenna, the simplified two-ray ground reflection model [20] is utilized. By utilizing the above model, the radius of circular cell of backup ONU-AP in terms of critical distance is calculated as

$$d_c = \frac{4h_t h_r}{\lambda} \tag{3.9}$$

where h_r is the height of the antenna at the user equipment, h_t is the height of the omnidirectional antenna at backup ONU-AP, and λ is the wavelength. For analysis, $h_r = 1.5$ m, $\lambda = 0.0547$ m,[1] and h_t varies from 0 m to 100 m is considered.

In a survivable FiWi network, it is essential to achieve maximum user coverage even in the case of failures. In this context, user coverage is defined as the number of users present inside the cell radius of backup ONU-AP or active ONU-AP. Figure 3.5 shows the user coverage with respect to the height of the omnidirectional antenna at backup ONU-AP. It can be observed that there is steep increase in user coverage with the increase in antenna height. Further, it can be observed that no user is covered with 10 m of antenna height. However, user coverage gradually increases with an increase in antenna height. Further, for antenna height ≥ 30 m, 100% user coverage is achieved.

3.7.3.2 Sector Antenna

In the case of sector antenna, backup ONU-AP is equipped with the one sector antenna for each active ONU-AP. Sector-based APs have focused radiation patterns in a particular direction in the form of the beam with specific azimuth and elevation half-power beamwidth [22]. For the coverage analysis, the radiation pattern of each sector antenna at backup ONU-AP consists of the main lobe with a horizontal and vertical beamwidth of 120° and $\theta_{BW} = 6.5°$, respectively. According to [23], the tilt angle, θ_{tilt} can be calculated as $\theta_{tilt} = \tan^{-1}\left(\frac{\mathrm{h_t - h_r}}{D}\right)$, where D is distance between backup ONU-AP and the farthest user of active ONU-AP.

Figure 3.6 depicts the inner radius (C_{in}) and outer radius (C_{out}) of the ellipsoid region of backup ONU-AP equipped with sector antenna as shown in Fig. 3.4b. The C_{in} and C_{out} can be rewritten as

$$C_{in} = \frac{h_t - h_r}{\tan(\theta_{tilt} + 0.5.\theta_{BW})}, \tag{3.10}$$

[1]Corresponding to IEEE 802.11ac frequency of 5.48 GHz [21].

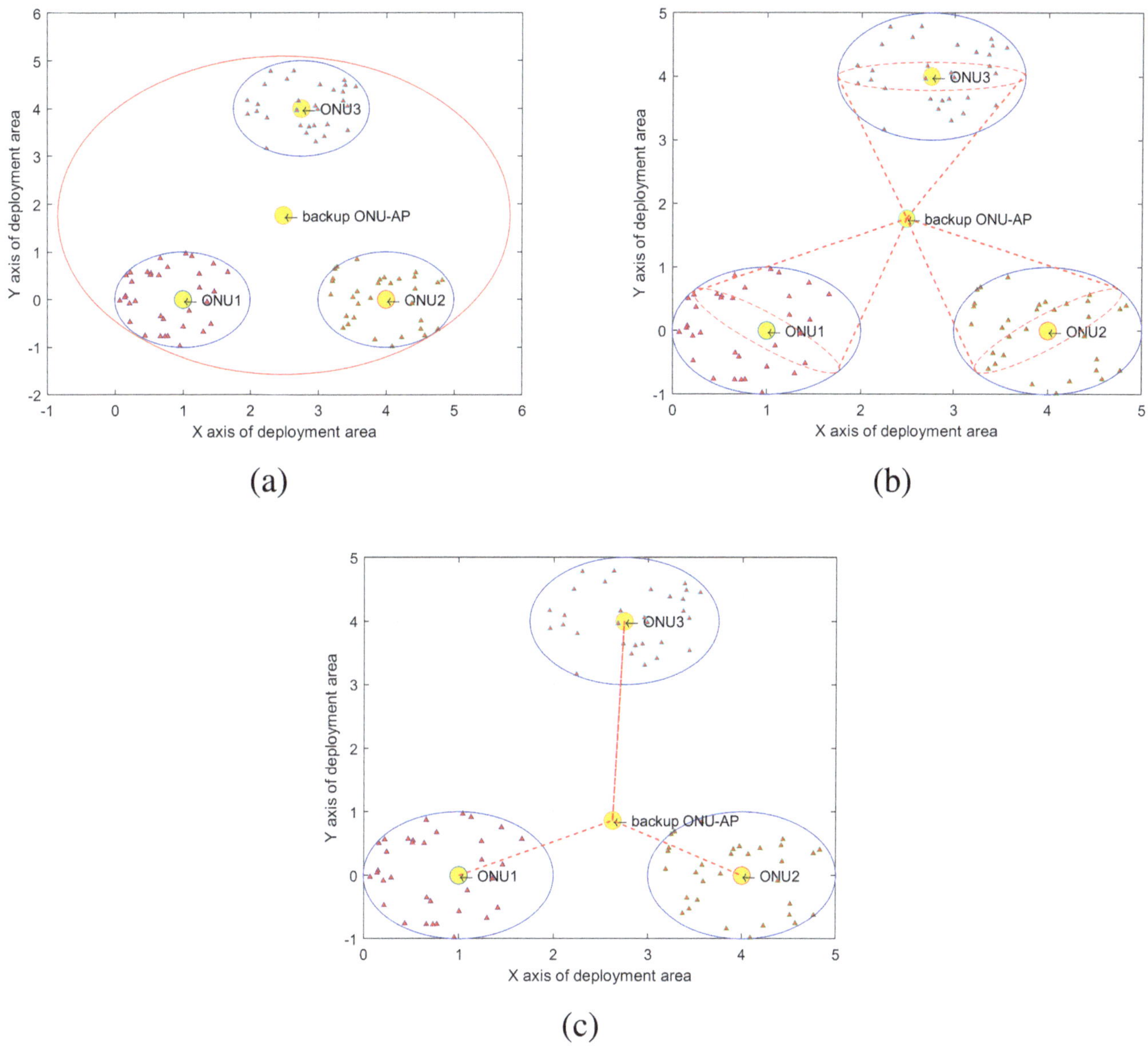

Fig. 3.4 Placement of backup ONU-AP for different antenna configurations (**a**) omnidirectional antenna, (**b**) sector antenna, and (**c**) P2P link

$$C_{out} = \frac{h_t - h_r}{\tan(\theta_{tilt} - 0.5.\theta_{BW})}. \tag{3.11}$$

For each sector, antenna radiates only in one direction; therefore, the path loss for one sector can be calculated as

$$P_r^{sec} = P_d^{sec} \times A_{eff}, \tag{3.12}$$

where P_r^{sec} is the received power, P_d^{sec} is transmitted power density for a particular sector, and A_{eff} is effective area of antenna. Using [24], (3.12) can be rewritten as

$$P_r^{sec} = \frac{P_t G_t}{\int_0^{\frac{2\pi}{3}} \int_0^{\pi} d^2 \sin\theta d\theta d\phi} \times \frac{G_r \lambda^2}{4\pi}, \tag{3.13}$$

where d is the distance between backup ONU-AP and the users of active ONU-AP, P_t is transmit power from backup ONU-AP, and G_t and G_r are gains of antenna at backup ONU-AP and users, respectively (3.13), can be rewritten as

$$P_r^{sec} = \frac{P_t G_t}{(4/3)\pi d^2} \times \frac{G_r \lambda^2}{4\pi}. \tag{3.14}$$

Using (2.14), the path loss, PL_{sec} for the sector antenna can be expressed as

$$PL_{sec} = 10\eta \log\left(\frac{4\pi d}{\lambda}\right) - 10\log(3), \tag{3.15}$$

where $\eta = 2$ denotes the path loss exponent as given in [21].

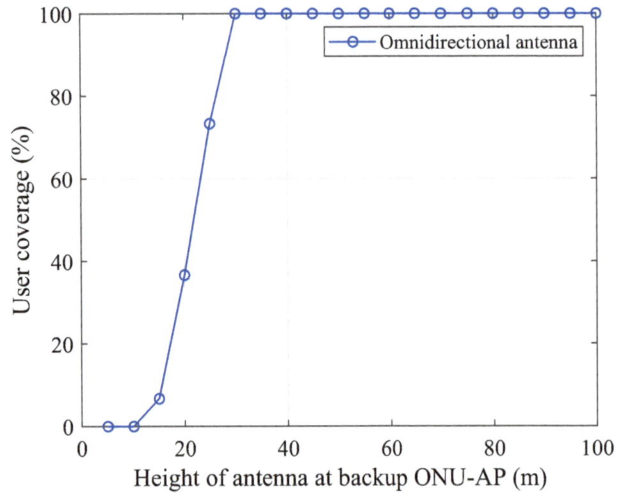

Fig. 3.5 User coverage (%) with respect to height of antenna at backup ONU-AP equipped with omnidirectional antenna

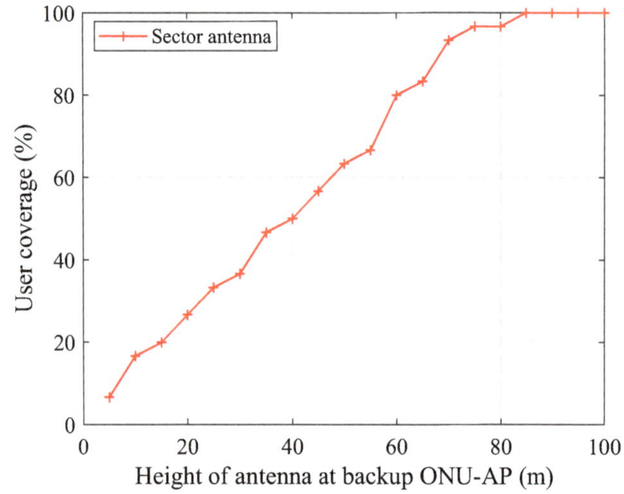

Fig. 3.7 User coverage (%) with respect to height of antenna at backup ONU-AP (in case of sector antenna)

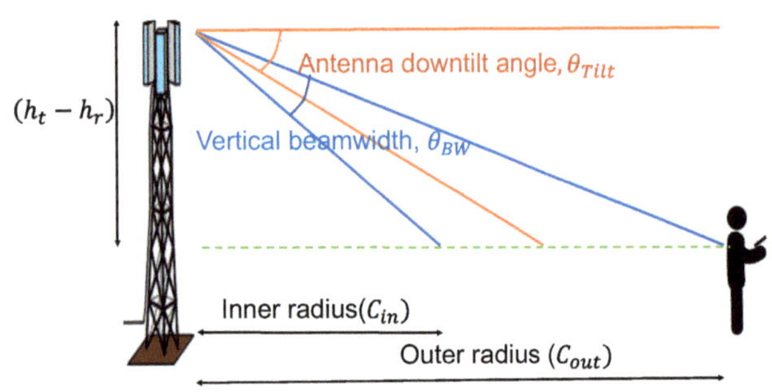

Fig. 3.6 Illustration of inner and outer radius an of ellipsoid region of backup ONU-AP, when it is equipped with sector antennas

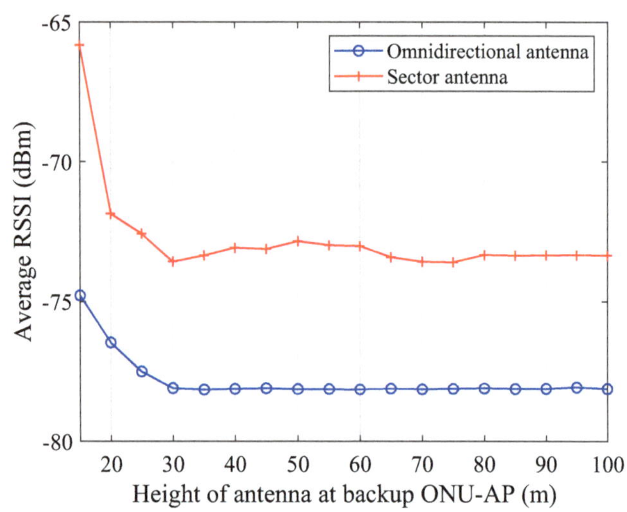

Fig. 3.8 Average RSSI (dBm) with respect to height of antenna at backup ONU-AP (m)

Figure 3.7 shows the user coverage with respect to sector antenna height at backup ONU-AP. It is observed that as antenna height increases, the user coverage also increases. It is seen that only 20% user coverage is achieved with 15 m of antenna height, when this height is further increased to 30 m, user coverage is also increased to 36.6%. If antenna's height is further increased up to 80 m, 100% user coverage is possible.

3.7.3.3 P2P Link

In the survivable FiWi network in [11], P2P link is considered between backup ONU-AP and active ONU-AP. It is also assumed that active ONU-AP consists of two APs, where one AP is equipped with parabolic antenna used for establishing a P2P link between backup ONU-AP and active ONU-AP and another AP is used for providing connectivity to the users via P2MP wireless links. In this type of scenario, parabolic grid antennas are used to make a P2P link between backup ONU-AP and active ONU-AP. Whenever failure occurs in FiWi network, the disconnected users are reconnected to the backbone network through the P2P link (Figs. 3.8 and 3.9).

3.7.3.4 Reliability

This subsection evaluates the reliability of the survivable FiWi network with respect to distribution fiber cut, ONU, and AP failures. In the FiWi network, the splitter is more reliable than other components as it is a passive component [25]. Reliability is already investigated in some previous literature [25–27]. In this section, three survivable schemes in which backup ONU-AP can be equipped with either omnidirectional antenna or sector antenna or parabolic grid antenna for reconnecting the disconnected users to the backbone network in case of failures is evaluated. Further, it is also assumed that backup ONU-AP is connected to OLT through a distribution fiber.

Let P_C denote the probability of distribution fiber cut between ONU-AP and the splitter, P_F is the failure probabil-

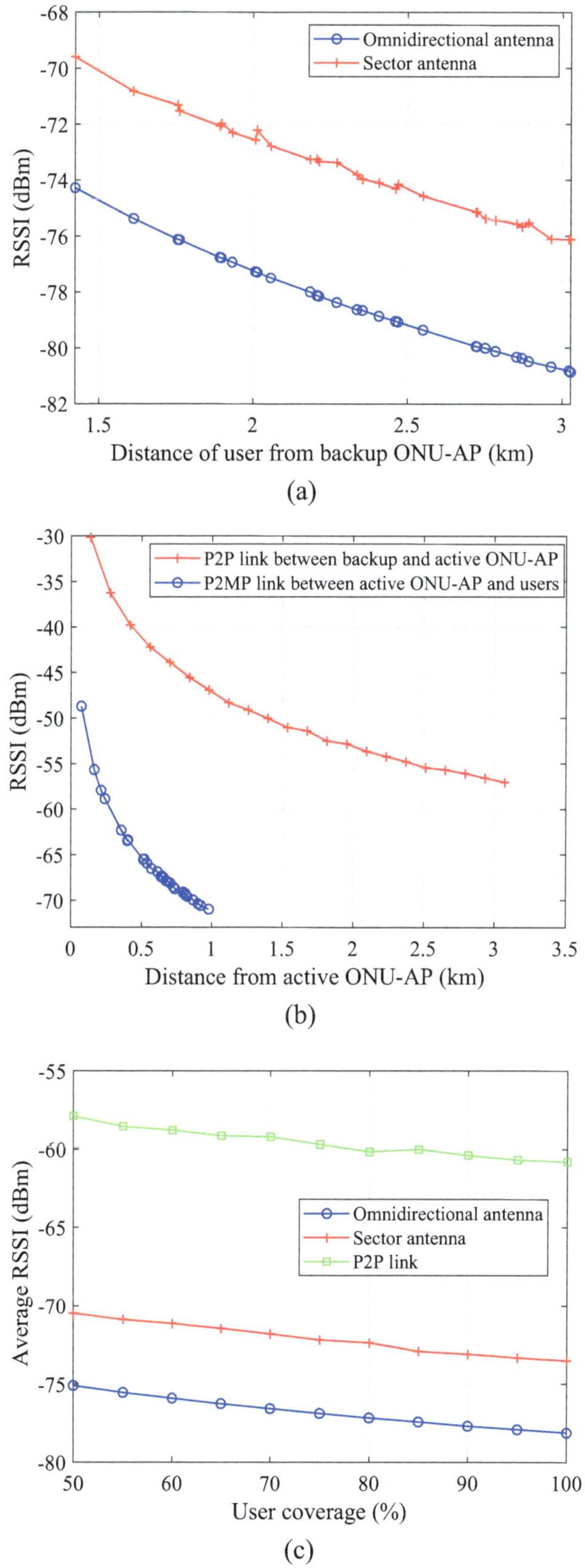

Fig. 3.9 RSSI of the users with respect to distance, when backup ONU-AP is connected to active ONU-AP via (**a**) omnidirectional antenna and sector antenna, (**b**) P2P link, and (**c**) average RSSI (dBm) with respect to user coverage (%)

ity of components (active ONU, backup ONU, and AP). To evaluate the FiWi network's reliability, the ONU and AP's failure probability is considered as 10^{-6} [26]. The first part of the reliability analysis considers no backup ONU-AP is deployed in the FiWi network. In this case, the probability of the users connected to the backbone network is given as

$$P^U_{con} = (1 - P_C) \times (1 - P^{AO}_F) \times (1 - P^{AOAP}_F), \quad (3.16)$$

where P^{AO}_F and P^{AOAP}_F denote the failure probability of active ONU and AP equipped with active ONU, respectively. Figure 3.10a depicts that the probability of user connectivity decreases with an increase in distribution fiber cut probability in the FiWi network without any backup ONU-AP.

When backup ONU-AP is deployed in the cluster of active ONU-APs for providing connectivity to the disconnected users in case of distribution fiber cut, ONU, and AP failures, the probability of user connectivity is shown in Fig. 3.10b and computed as follows:

$$P^U_{con} =(1 - P^{WP}_F) \times (1 - (P^{AOAP}_F \times P^{AO}_F \times P_C)), \quad (3.17)$$

where P^{WP}_F is the probability of failure of wireless protection scheme in which backup ONU-AP is deployed in the cluster to provide network connectivity to disconnected users via the wireless path in case of failures and is given by

$$P^{WP}_F = 1-[(1-P_C)\times(1-P^{B_{ONU}}_F)\times(1-P^{BAP}_F)], \quad (3.18)$$

where P_C denotes distribution fiber cut probability and $P^{B_{ONU}}_F$ and P^{BAP}_F denote failure probability of backup ONU and backup AP, respectively. With the deployment of backup ONU-AP, the probability of user connectivity at 0.1 distribution fiber cut probability is enhanced from 0.90 to 0.99, as can be seen from Fig. 3.10b.

In the survivable FiWi network, when a P2P link is employed between active ONU-AP and backup ONU-AP, then the probability of user connectivity to the backbone network is estimated as follows:

$$P^U_{con} =(1 - P^{WP}_F) \times (1 - (P^{AOAP}_F \times P^{AO}_F \times P_C)), \quad (3.19)$$

where P^{WP}_F is the probability of failure of wireless protection scheme given by

$$P^{WP}_F =1 - [(1 - P_C) \times (1 - P^{B_{ONU}}_F)(1 - P^{BAP}_F)^2]. \quad (3.20)$$

Figure 3.10c depicts that in a P2P link, with a distribution fiber cut probability of 0.1, the probability of user connectivity of 0.92 is achieved. Therefore, the user connectivity

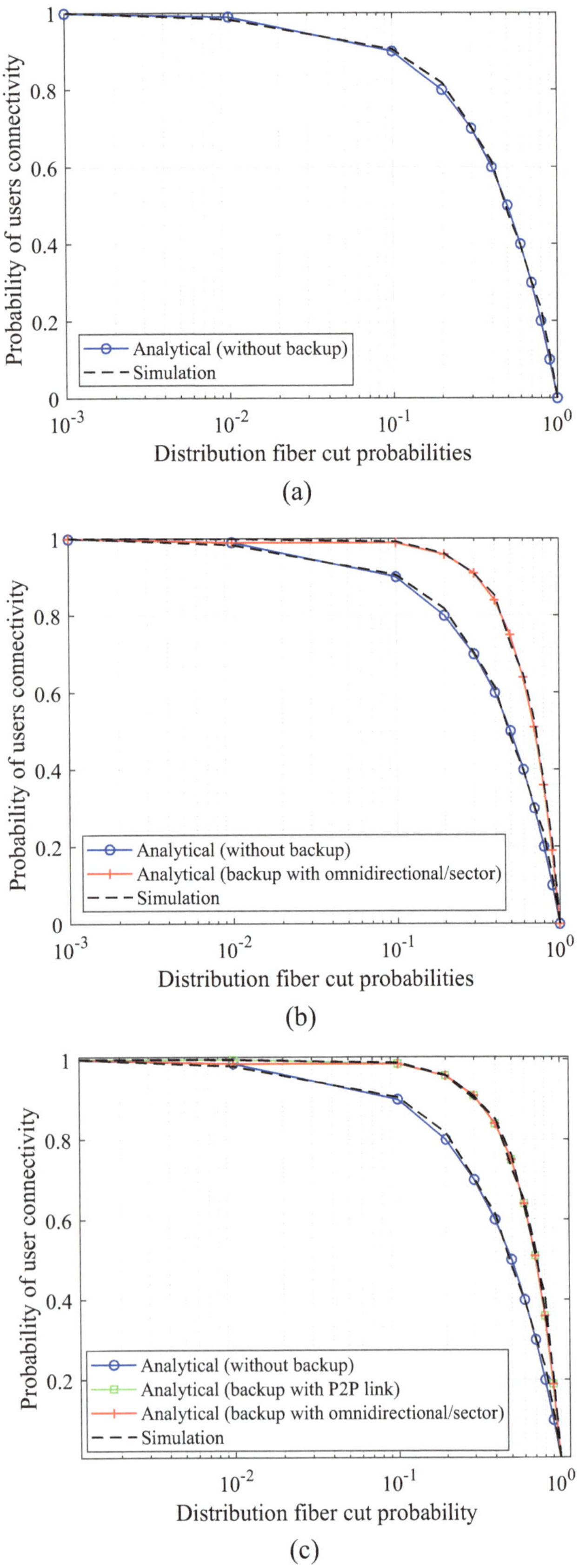

Fig. 3.10 Probability of user connectivity with respect to distribution fiber cut probabilities (**a**) without backup ONU, (**b**) with backup ONU-AP (equipped with omnidirectional/sector antenna), and (**c**) with backup ONU-AP (P2P link)

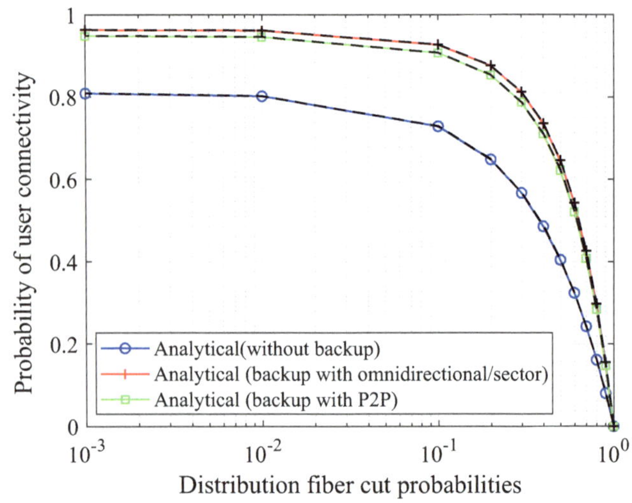

Fig. 3.11 Probability of user connectivity with respect to distribution fiber cut probabilities (where ONU and AP's failure probability is 10^{-1} is considered)

probability of the P2P link is improved with respect to the FiWi network having no backup ONU-AP. If the failure probability of ONU and APs is considered high as 10^{-1}, then from Fig. 3.11, it is clearly observed that P2P link is less reliable than backup ONU-AP equipped with either omnidirectional or sector antenna due to the addition of one AP to provide a P2P connectivity of active ONU-AP with the backup ONU-AP. As compared to omnidirectional and sector antenna, in the case of the P2P link, the achieved probability of user connectivity of 0.91 at a distribution fiber cut probability of 0.1.

3.8 Summary

This chapter discusses the progress in optical fiber technologies, which brought down component and system costs, coupled with demand for new bandwidth-hungry services. Moreover, data service delivering to customers (voice, data, and video) is changing the economics of optical access solutions. The various system architectures for FTTX are discussed in this chapter, such as fiber-to-the-node (FTTN), fiber-to-the-curb (FTTC), fiber-to-the-business—or-building-(FTTB), fiber-to-the-home (FTTH), and fiber-to-the-room (FTTR). Further, a performance evaluation of a use case of QoS-aware reliable broadband access networks is also presented in this chapter.

Homework Problems

1. Define the different optical access technologies in fiber-to-the-everything (FTTX) network.
2. Define fiber-to-the-curb (FTTC) networks.

3. Define the advantages of fiber-to-the-edge network.
4. Explain wavelength allocation in FTTX network.
5. Define the opportunities and challenges of fiber-to-the-room network.
6. Explain distribution networks in detail.
7. Calculate the inner radius (C_{in}) and outer radius (C_{out}) of the ellipsoid region of backup ONU-AP equipped with sector antenna with height of transmit antenna is 30 m and height of receiving antenna is 15 m. Consider the antenna down tilt angle $\theta_{tilt} = 30°$ and vertical beamwidth $\theta_{BW} = 15°$.
8. Find the relation between the received power and the path loss for one sector of sector antenna.
9. Find the path loss, PL_{sec} for the sector antenna. Consider the frequency of operation to be 20 MHz, the distance between transmitter and receiver to be 20 m, and the path loss exponent $\eta = 2$.
10. Consider in a FiWi network, the probability of active ONU and AP failure is 0.1 and 0.3, respectively. The probability of distribution fiber cut is 0.0001. What is the probability of the users being connected to the backbone network?

References

1. Hoven, G.: Fttx-how close to the end user should fiber come? In: Conference Proceedings - Lasers and Electro-Optics Society Annual Meeting-LEOS, pp. 621–621 (2009)
2. Koonen, T.: Fiber to the home/fiber to the premises: What, where, and when? Proc. IEEE **94**(5), 911–934 (2006)
3. El-Bawab, T.S.: Fttx: The rise of broadband optical access. In: 2007 IEEE Globecom Workshops, pp. 1–2 (2007)
4. Senior, J.M.: Optical Fiber Communications: Principles and Practice, 3rd edn. Pearson Education, London (2009)
5. The state of US broadband in 2022: Reassessing the whole picture (2022). https://itif.org/publications/2022/12/05/state-of-us-broadband-in-2022-reassessing-the-whole-picture/
6. Green, P.: Fiber to the home: the next big broadband thing. IEEE Commun. Mag. **42**(9), 100–106 (2004)
7. Chang, T., Si, X., Wang, X., Chen, Y., Gan, S., Caire, G.: Wireless access technology in the next generation home networks: an overview. IEEE Wireless Commun. **30**(5), 44–49 (2023)
8. Khorov, E., Kiryanov, A., Lyakhov, A.: Ieee 802.11ax: How to build high efficiency wlans. In: 2015 International Conference on Engineering and Telecommunication (EnT), pp. 14–19 (2015)
9. ETSI: Fifth generation fixed network (F5G) generation definition release #1 (2020)
10. Kani, J.i., van Veen, D.: Current TDM-PON Technologies, pp. 849–870. Springer International Publishing, Cham (2020)
11. Singh, P., Gupta, A., Bohara, V.A., Srivastava, A.: QoS-aware reliable architecture for broadband fiber-wireless access networks. IEEE Syst. J. **16**(4), 5753–5764 (2022). https://doi.org/10.1109/JSYST.2022.3149854
12. Kaur, R., Gupta, A., Srivastava, A., Chatterjee, B.C., Mitra, A., Ramamurthy, B., Bohara, V.A.: Resource allocation and QoS guarantees for real world IP traffic in integrated XG-PON and IEEE802.11e EDCA networks. IEEE Access **8**, 124883–124893 (2020)
13. Yu, Y., Shen, Z., He, S.: Compact omnidirectional antenna of circular polarization. IEEE Antennas Wireless Propagat. Lett. **11**, 1466–1469 (2012)
14. Yu, Y., Ranaweera, C., Lim, C., Guo, L., Liu, Y., Nirmalathas, A., Wong, E.: Hybrid fiber-wireless network: An optimization framework for survivable deployment. J. Optical Commun. Networking **9**(6), 466–478 (2017)
15. Zhang, Y., Dai, L.: On the optimal placement of base station antennas for distributed antenna systems. IEEE Commun. Lett. **24**(12), 2878–2882 (2020)
16. Goodfellow, I., Bengio, Y., Courville, A., Bengio, Y.: Deep Learning, vol. 1. MIT Press, Cambridge (2016)
17. Bach, F., Moulines, E.: Non-asymptotic analysis of stochastic approximation algorithms for machine learning. In: Neural Information Processing Systems (NIPS) (2011)
18. Yan, L., Dodier, R.H., Mozer, M., Wolniewicz, R.H.: Optimizing classifier performance via an approximation to the wilcoxon-mann-whitney statistic. In: Proceedings of the 20th International Conference on Machine Learning (ICML-03), pp. 848–855 (2003)
19. LeCun, Y., Bengio, Y., Hinton, G.: Deep learning. Nature **521**(7553), 436–444 (2015)
20. Goldsmith, A.: Wireless Communications. Cambridge University Press, Cambridge (2005)
21. Planning for BharatNet phase 2, IITB (2016). https://www.sc.iitb.ac.in/~vivekyogi/bharatnet.pdf
22. Unni, S., Raj, D., Sasidhar, K., Rao, S.: Performance measurement and analysis of long range Wi-Fi network for over-the-sea communication. In: 2015 13th International Symposium on Modeling and Optimization in Mobile, Ad Hoc, and Wireless Networks (WiOpt), pp. 36–41 (2015)
23. Yang, J., Ding, M., Mao, G., Lin, Z., Zhang, D.g., Luan, T.H.: Optimal base station antenna downtilt in downlink cellular networks. IEEE Trans. Wirel. Commun. **18**(3), 1779–1791 (2019)
24. Balanis, C.A.: Antenna Theory: Analysis and Design. Wiley, New York (2015)
25. Rimal, B.P., Van, D.P., Maier, M.: Mobile-edge computing versus centralized cloud computing over a converged FiWi access network. IEEE Trans. Network Service Manag. **14**(3), 498–513 (2017)
26. Beyranvand, H., Lim, W., Maier, M., Verikoukis, C., Salehi, J.A.: Backhaul-aware user association in FiWi enhanced LTE-A heterogeneous networks. IEEE Trans. Wirel. Commun. **14**(6), 2992–3003 (2015)
27. Ghazisaidi, N., Scheutzow, M., Maier, M.: Survivability analysis of next-generation passive optical networks and fiber-wireless access networks. IEEE Trans. Reliab. **60**(2), 479–492 (2011)

4 Free-Space Optical Communication

4.1 Introduction to Optical Wireless Technologies

Wireless devices and technologies have come a long way. About thirty years ago, few could imagine the advances we see today. Now, "wireless" often means RF communication, because of RF devices and systems that are so widespread. Current 5G technology offers many appealing services that includes huge system capacity, massive device connections, better security, ultralow latency, low power usage, and good quality of experience (QoE). Combined with ultradense heterogeneous networks, 5G will let us connect a hundred times more wireless devices and transmit data faster than today's networks. Because of this, 5G and beyond (5GB) networks need high-capacity backhaul connectivity. This is important for supporting very dense and high-speed access networks while keeping minimum end-to-end delay and power consumption. Further, handling the huge data volume in these networks is tough. Hence, we need solid technical solutions to make sure users get guaranteed Quality-of-service (QoS). As 5GB network demands keep growing, we expect that the current RF frequency band in the electromagnetic spectrum is not sufficient due to limited spectrum range with regulatory allocation constraints and high interference from nearby RF access points. Therefore, it is very crucial to look into other wireless communication options that can support data-heavy wireless devices. In recent years, there has been a lot of interest in developing fiberless optical wireless technologies in the visible light, infrared, and ultraviolet frequency bands [1].

OWC involves the transmission of signals through unguided propagation media using optical carriers. Historical examples of OWC include methods such as beacon fires, smoke signals, etc. Ever since ancient times, sunlight has been harnessed for long-distance communication. Fast forward to the modern era, OWC utilizes lasers or LEDs as transmitter. Over several decades, interest in OWC was only limited to military and space applications—like inter-satellite and deep-space communication links. In recent years, however, the scenario has changed. The increasing number of companies providing terrestrial OWC links along with the development of visible light communication (VLC) products signifies potential growth for the future market. OWC technologies boast features such as wide spectrum coverage, high data rates, low latency, and cost-effectiveness paired with reduced power consumption. These attributes seamlessly meet the extensive demands of 5G communications. Unlike RF-enabled networks, which often come with limitations, OWC technology has distinct advantages. It can transmit data at high rates over few kilometers and is suitable for both indoor and outdoor environments. Furthermore, it offers exceptional communication features including reliable security and less interference. In addition, utilizing a wide spectrum contributes to high system efficiency. OWC technology can achieve speeds of the order of Gbps while supporting energy-efficient communication. A significant advantage of OWC is its minimal infrastructure requirement compared to RF networks, leading to lower deployment costs. In essence, optical wireless communication continues to evolve and diversify its applications from historical roots to futuristic capabilities, all while promising robust performance and efficiency in modern contexts [2].

OWC offers better data security, since optical signals cannot pass through walls. OWC technologies such as free-space optical (FSO) communication, VLC, and light fidelity (LiFi) operate on different electromagnetic spectrum such as visible light, Infrared (IR), and ultraviolet. These technologies share some similarities but also have differences in terms of system architecture, channel modeling, and operating environment.

The near IR band is extensively used in FSO communication, whereas short IR is mainly used for long-distance FSO communications. Moreover, IR band is also commonly used in LiFi technology. The near IR band with the wavelength range of 750 nm to 1 mm makes it particularly suitable for line-of-sight (LoS) communication. However, it may pose

V. A. Bohara, Swaminathan R, *Optical and Wireless Access Networks*, Textbooks in Telecommunication Engineering, https://doi.org/10.1007/978-3-031-89696-5_4

safety concerns for humans and is less effective in non-LoS communication scenarios. On the other hand, the visible light band with the wavelength ranging from 360 nm to 760nm is often used in VLC and LiFi technologies. The benefits of the visible light spectrum include its safety for humans and its ability to serve both illumination and communication concurrently. Similar to near IR band, visible light band is optimal for LoS communication but experiences lower data rates for non-LoS scenario due to light reflection issues. Moreover, ultraviolet communication is gaining enhanced attention for supporting non-LoS and LoS communication scenarios with high speed. The ultraviolet spectrum ranging from 10 to 400 nm is also suited for LiFi technology [3].

4.2 Free-Space Optical Communication

FSO is an emerging technology, which is meeting the rising need for ultra-high data rates and extensive bandwidth. This technology transmits data through open space using optical carriers, operating in the unlicensed near IR spectrum at frequencies above 300 GHz. FSO offers a vast amount of optical bandwidth, facilitating data transmission rates of up to 10 Gbps. The primary benefits of FSO communication include minimal latency, enhanced security, and high dates. Further, the straightforward deployment without requiring optical fiber installation makes it even more appealing. FSO links are versatile and can be used in various applications such as backhaul connectivity in cellular networks, integrated space-air-ground networks, enterprise and campus area connectivity, extending metropolitan area networks, post-disaster/emergency communications, etc. [2]. Figure 4.1 illustrates the block diagram of FSO communication system.

Usually, information waveforms, which are generated by a source, are modulated using an optical carrier in FSO communications. This optical signal is transmitted through the atmosphere to a remote destination. Upon reaching the receiver, the optical signal is detected and converted into an electrical signal with the help of a photodetector. The receiver then processes this electrical signal to recover the originally transmitted information signal [4].

- **Transmitter:** As depicted in Fig. 4.2, the transmitter consists of several components: an optical source, a modulator, an optical amplifier, and optical beamforming. Initially, data bits from the information source are encoded before modulation. The modulated optical signal is then amplified by passing it through an optical amplifier, which increases its intensity. Optical beamforming technique steers and refocuses the light beam before transmission. Typically, FSO systems utilize a LD or LEDs with beam collimators as the optical source.
- **Receiver:** Direct detection and heterodyne detection are two types of detection techniques are usually employed for demodulating the FSO signals at the receiver [5]. Heterodyne detection, a type of coherent detection technique, requires a local oscillator at the receiver. The incoming optical signal is mixed with a coherent carrier signal generated by the local oscillator and then converted into an electrical signal, as illustrated in Fig. 4.3. This spatial mixing amplifies the weak incoming signal, thereby enhancing the quality of the signal reception. Direct detection, on the other hand, is a noncoherent

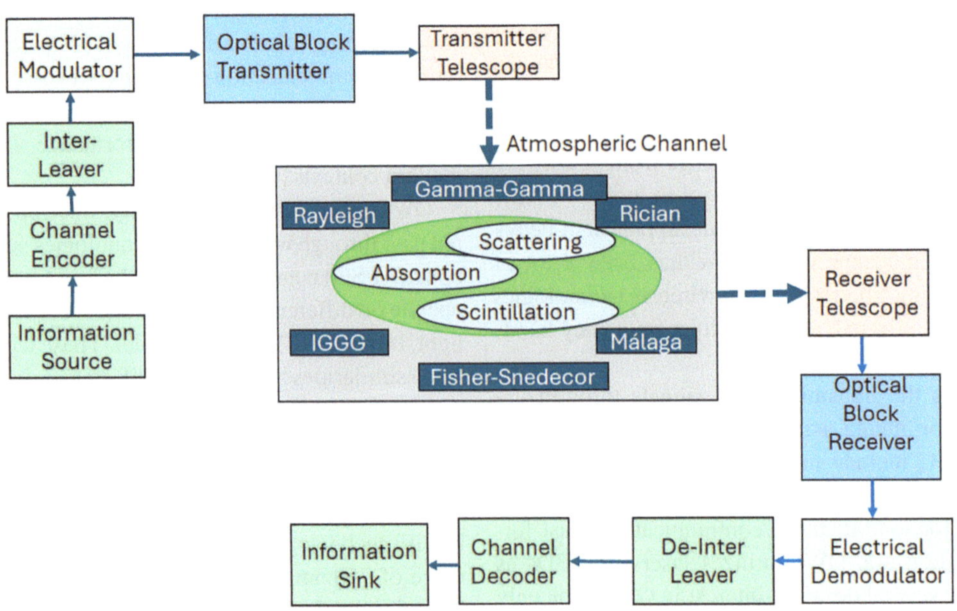

Fig. 4.1 Block diagram of an FSO communication system

technique where the received optical signals pass through a bandpass filter to limit background noise. The filtered optical signals are then directed onto a photodetector to produce an output electrical signal, as shown in Fig. 4.4. This electrical signal is proportional to the instantaneous intensity of the received optical signal. A low-pass filter is subsequently used to effectively retrieve the information signal. In Fig. 4.5, the block diagram of an FSO receiver has been illustrated.

- **Modulator:** Typically, the subcarrier intensity modulation (SIM) scheme is employed to modulate the FSO signals. In SIM, the RF subcarrier signal is pre-modulated with the information signal. The RF subcarrier can be modulated using various techniques, such as M-ary phase-shift keying (PSK) and M-ary quadrature amplitude modulation (QAM) techniques [2, 4]. After that, the intensity of the optical carrier is modulated by this pre-modulated signal. Another commonly used intensity modulation (IM) technique is on-off keying (OOK), which is a binary modulation scheme, where the data is encoded by the presence or absence of a light pulse during each symbol interval. However, OOK requires the detection threshold level to vary due to atmospheric turbulence and noise at the receiver. In addition, it has poor spectral and energy efficiency. To overcome some disadvantages, other IM schemes such as pulse position modulation (PPM), pulse width modulation (PWM), and pulse amplitude modulation (PAM), which encode the data by varying the position, width, and amplitude of the light pulse, respectively, within a predefined time slot, have been proposed. Each modulation technique has its unique advantages and is chosen based on the specific requirements of the communication system, such as power efficiency, data rate, and implementation complexity.

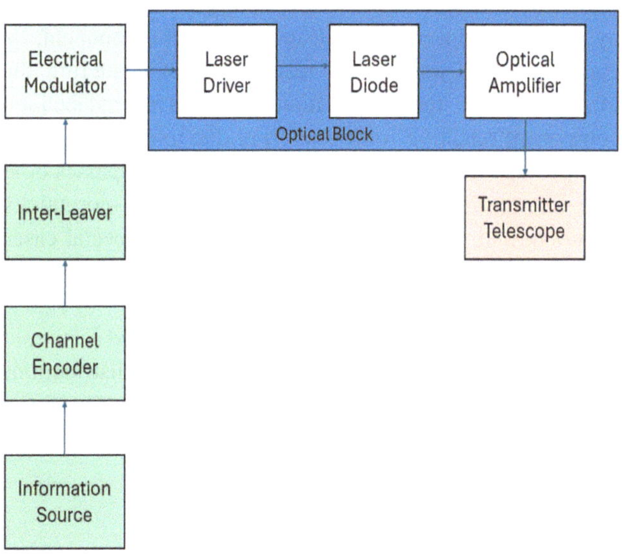

Fig. 4.2 Transmitter block FSO communication

4.2.1 FSO Channel Modeling

While FSO communication presents many benefits, it also encounters specific challenges that affect its effectiveness and dependability. During transmission, the FSO link is subject to several types of losses, including fading caused by atmospheric turbulence, misalignment or pointing errors due to displacement of laser beam, and attenuation due to adverse weather conditions like fog, snow, smog, etc. restricting its operational range to a few kilometers [2].

4.2.1.1 Atmospheric Turbulence

The primary cause of turbulence in FSO communication is the Earth's atmosphere. Variations in temperature, pressure, and wind movement change the air's refractive index, scattering light beams. This phenomenon, known as atmospheric turbulence or scintillation, causes random fluctuations in the intensity, phase, and angle of the light, leading to signal degradation, as shown in Fig. 4.6. As the laser beam travels through the atmosphere, it encounters varying refractive

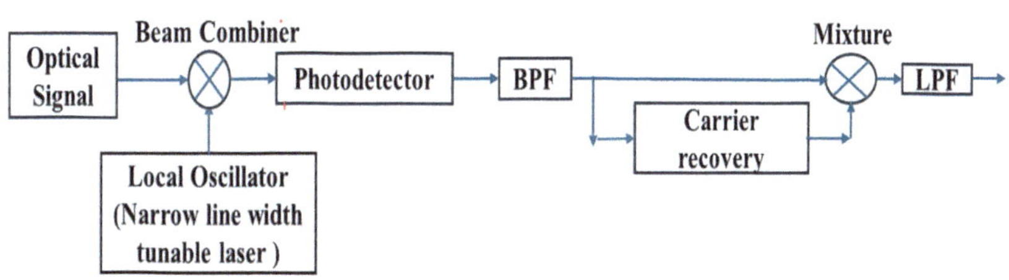

Fig. 4.3 Block diagram of the heterodyne detection technique

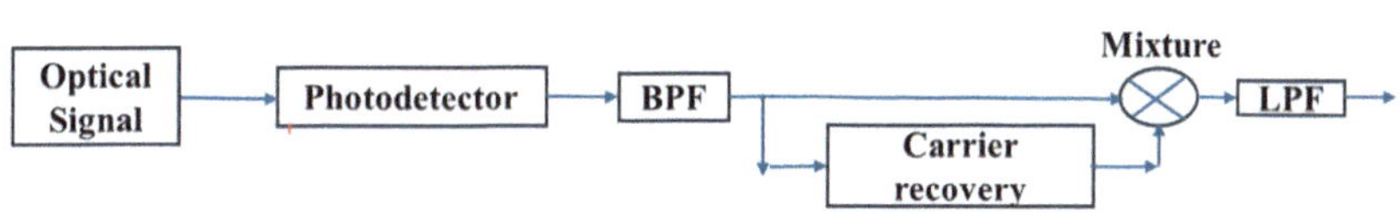

Fig. 4.4 Block diagram of the direct detection technique

indices, causing interference and scintillation effects. These effects result in signal fading, a degraded SNR, and errors in received data. Additionally, atmospheric turbulence can cause the transmitted laser beam to wander or deviate from its intended path, leading to signal loss or degradation in signal quality due to mialignment errors. The severity of atmospheric turbulence is measured by refractive index structure parameter C_n^2 [6]. The value of C_n^2 depends on both altitude and link distance, typically ranging from 10^{-17} to 10^{-13}. Additionally, the magnitude of intensity fluctuations caused by atmospheric turbulence can be quantified using the scintillation index, and it is expressed as $\sigma_I = \frac{E[I^2]}{E[I]^2} - 1$, where $\frac{E[I^2]}{E[I]^2}$ is the ratio of the standard deviation of intensity fluctuations to the average intensity. The scintillation index measures the severity of turbulence, with higher values indicating stronger turbulence.

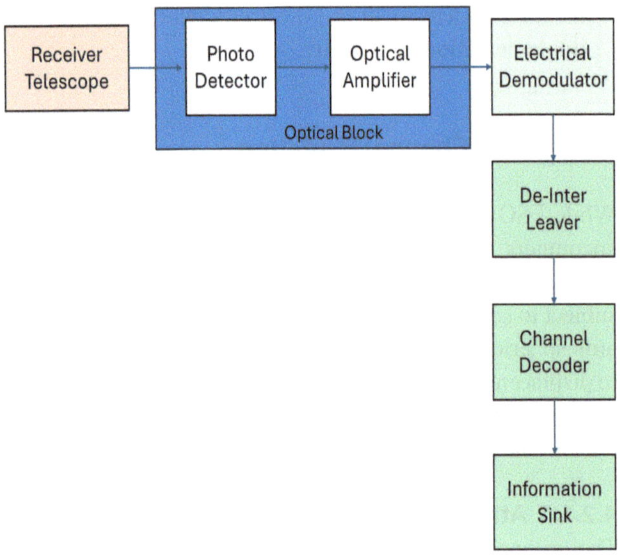

Fig. 4.5 Receiver block FSO communication

To comprehensively characterize the atmospheric turbulence phenomenon, different channel models are being proposed. The most widely accepted model for characterizing weak turbulence is the log-normal distribution [6]. However, this model is not suitable for characterizing moderate-to-strong turbulence in FSO channels. Instead, the gamma-gamma distribution [5, 7] is extensively used for moderate-to-strong turbulence due to its excellent fit with experimental data. The irradiance fluctuation due to turbulence is usually expressed as the product of large-scale and small-scale atmospheric effects. In case of gamma-gamma distribution, both the atmospheric effects are modeled using gamma distribution. Other distributions, such as the exponential distribution, K-distribution, Weibull, and I-K distribution, are also available for modeling turbulence. Recently, Malaga distribution [8] has been introduced to model fading in FSO links caused by atmospheric turbulence. It includes LoS component, the coupled-to-LoS component resulting from scattering due to eddies along the propagation axis, and an independent component resulting from scattering due to off-axis eddies. The generalized Malaga distribution also encompasses other well-known distributions like K-distribution, log-normal distribution, and gamma-gamma distribution as special cases. A new turbulence model, known as Fisher-Snedecor distribution [9], has been proposed that aligns well with experimental data across all turbulence conditions and is simpler compared to the gamma-gamma and Malaga distributions. In the proposed model, the small-scale irradiance fluctuation

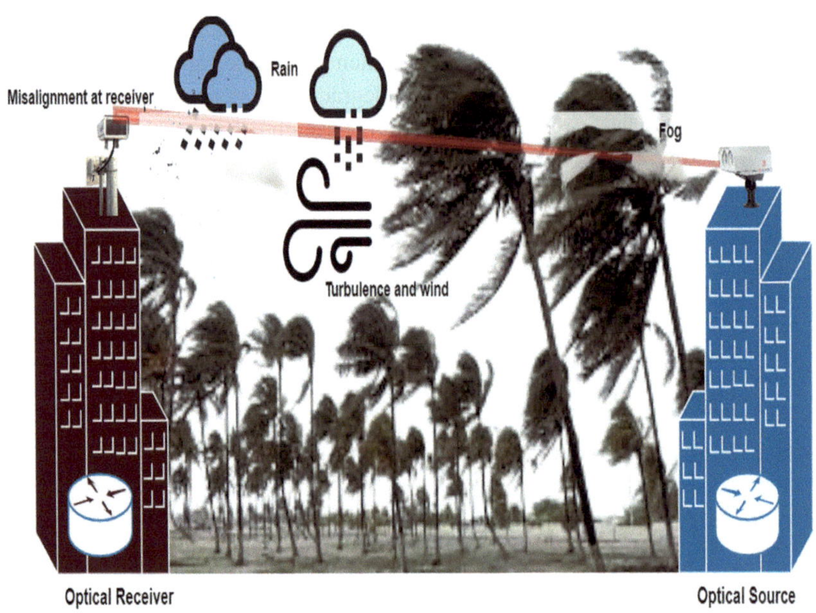

Fig. 4.6 Effect of atmospheric turbulence on FSO signal

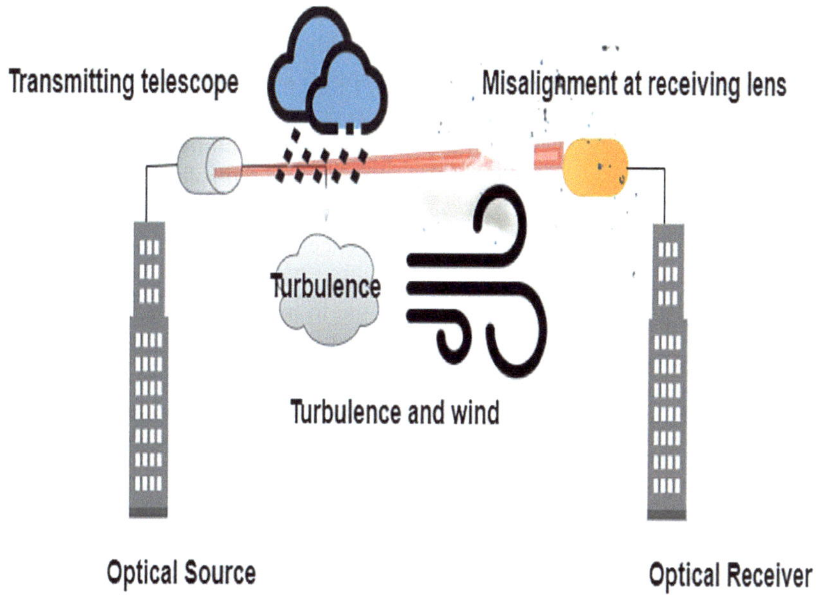

Fig. 4.7 Pointing error misalignment

is modelled using gamma distribution, and the large-scale fluctuation is modelled using inverse gamma distribution. The product of these two irradiance fluctuations corresponds to Fisher-Snedecor distribution. Recently, inverse Gaussian gamma (IGG) distribution [10] has been introduced as an alternative to the gamma-gamma distribution for effectively modeling moderate-to-strong turbulence scenario. Here, inverse Gaussian distribution is used to model large-scale atmospheric effects, and the small-scale irradiance fluctuation is modelled using generalized gamma distribution. IGG model aligns exceptionally well with the experimental data compared to gamma-gamma and log-normal models but is not mathematically simpler especially in computing error probability metrics. Though Fisher-Snedecor distribution is mathematically tractable, it aligns less with experimental data. Recently, a doubly stochastic process that uses an inverted gamma and gamma distribution to model large-scale turbulence and a single gamma distribution to model small-scale fading has been proposed. It is a three parameter model, termed as doubly inverted gamma-gamma (IGGG) distribution [11], which demonstrates an excellent fit to experimental data compared to Fisher-Snedecor, gamma-gamma, and log-normal distributions as well as mathematically tractable. Since all the existing turbulence models are two-parameter models, IGGG distribution, which is three-parameter model, provides more flexibility in modeling atmospheric turbulence scenarios compared to its counterparts. Finally, IGGG is also a generalized distribution comprising both gamma-gamma and Fisher-Snedecor distributions as a special case.

4.2.1.2 Pointing Errors

Pointing errors arise from misalignment between the transmitting and receiving terminals of an FSO system, significantly impacting the communication link's performance, as shown in Fig. 4.7. These errors can be caused by various factors, including mechanical vibrations due to environmental influences and building sway, which can induce slight movements in the transmitter or receiver aperture, leading to misalignment. Thermal effects are another cause of pointing errors in FSO communication; temperature changes can cause thermal expansion or contraction in system components, affecting the alignment between the transmitter and receiver apertures. Additionally, atmospheric conditions such as strong winds or turbulence can introduce pointing errors by altering the direction of the transmitted laser beam, causing misalignment at the receiving terminal [6].

Jitter and boresight are the two main components of pointing errors [5, 8]. Jitter is caused due to random variation of the beam's center on the detector plane, whereas boresight refers to the fixed offset between the beam's center and the detector's center. The statistical modeling of pointing errors plays a significant role in analyzing the system performance of FSO communication. By assuming that the statistical properties of pointing errors follow independent Gaussian distributions in both the horizontal and vertical directions, the radial displacement caused by pointing errors can be modelled using five different distributions depending upon the nature of jitter and boresight values as outlined in Table 4.1. The jitter categorization depends on whether it is single sided or double sided, where the double-sided

Table 4.1 Different pointing error models

Boresight and jitter values	Distribution
Zero boresight and identical jitters	Rayleigh distribution
Nonzero boresight and identical jitters	Rician distribution
Zero boresight and nonidentical jitters	Hoyt distribution
Zero boresight and single jitter	Central-chi distribution
Nonzero boresight and single jitter	Noncentral-chi distribution

jitter is further divided based on whether the variances of the jitter values in the horizontal and vertical directions are identical or not. The Beckman distribution exemplifies the general form of radial displacement with nonzero boresight and nonidentical variances for jitter, where it can be reduced to various distributions as shown in Table 4.1.

4.2.1.3 Atmospheric Attenuation

Atmospheric attenuation, or path loss, occurs in FSO communication due to the absorption and scattering of light by the atmosphere, leading to signal strength degradation, as depicted in Fig. 4.8. The atmosphere contains various aerosols, dust particles, and water vapor that can scatter and absorb light. Scattering happens when light interacts with small particles or molecules in the atmosphere, such as dust, smoke, fog, rain, and snow, causing the light beam to spread out and reducing the power density at the receiver. Absorption, on the other hand, results in a loss of energy in the signal. Both scattering and absorption contribute to signal attenuation and decrease the effective range of FSO communication. In the case of rain, the particle sizes are significantly larger than the wavelength of the optical signal, resulting in a relatively minimal impact on FSO transmission. When the diameter of the particles is comparable to the wavelength, the scattering coefficient increases significantly, making fog and haze the most detrimental environmental conditions for FSO transmission [2, 4].

4.3 Applications of FSO Communication

FSO systems initially garnered interest as an effective solution to the "last mile" problem, addressing the connectivity gap between end users and existing fiber-optic networks. Telecom carriers have heavily invested in expanding the capacity of their fiber backbones. To maximize this capacity, growth in backbone networks must be complemented by similar expansion at the network edge. Additionally, FSO systems are appealing for various applications [1, 2], several of which are discussed below:

- **Intercampus connectivity:** Buildings in various infrastructures, such as school campuses, offices, and corporations, are facing diverse traffic demands. Thus, FSO technology can establish connections between multiple buildings in these areas, providing extremely high data speeds at a lower cost than installing dedicated fiber links, as depicted in Fig. 4.9.
- **Video surveillance and streaming:** Surveillance cameras play a crucial role in public safety and law enforcement. Similarly, live streaming of various events such as sports, concerts, etc. requires signals to be transmitted to a central node. Traditional wireless technologies frequently struggle to meet the throughput demands for video streaming and surveillance, as illustrated in Fig. 4.10. Thus, FSO communication, because of its high throughput, can enable high-quality video transmission, supporting video surveillance and streaming-/broadcasting-based applications.
- **High-capacity backhaul network in 5G and B5G:** 5G communication offers extensive connectivity for smart devices, high throughput, very low latency, robust security, and high power efficiency. 5G networks can support downlink data rates of up to few Gbps, making them significantly faster than 4G long-term evolution (LTE). The number of connected devices in 5G networks is much higher compared to 4G LTE networks, leading to a massive increase in data volume. Additionally, beyond 5G (B5G) networks are projected to achieve better spectral efficiency than 4G and 5G networks as well as transfer large data volumes from macrocells to indoor small cells. A backhaul network connects the access network to the core network, facilitating the exchange of large data volumes in 5G and B5G communications. Traditionally, wireline connections link base stations to the mobile switching center in cellular systems. However, the rising demand for mobile services, which consumes huge bandwidth, now requires technologies like FSO that provide significantly higher throughput. The densification of cellular networks through the widespread deployment of small cells effectively improves the spectrum utilization and system throughput, particularly in densely populated areas with high traffic volumes. High-speed FSO connectivity is crucial for maintaining robust backhaul connections to support network densification. Without high-speed backhaul links, communication systems face bottlenecks, though the access and core networks support connections with Gbps speed. In a backhaul network scenario, FSO systems can manage high traffic volumes, consume less power, and offer high spectral efficiency. Although optical fiber network currently offers the highest throughput, both fiber-optic and FSO systems use similar transceiver architecture, suggesting potential for comparable performance in the future.
- **Disaster recovery:** In case of emergencies and natural disasters, temporary FSO links can be quickly installed, as shown in Fig. 4.11. Furthermore, FSO links can be easily

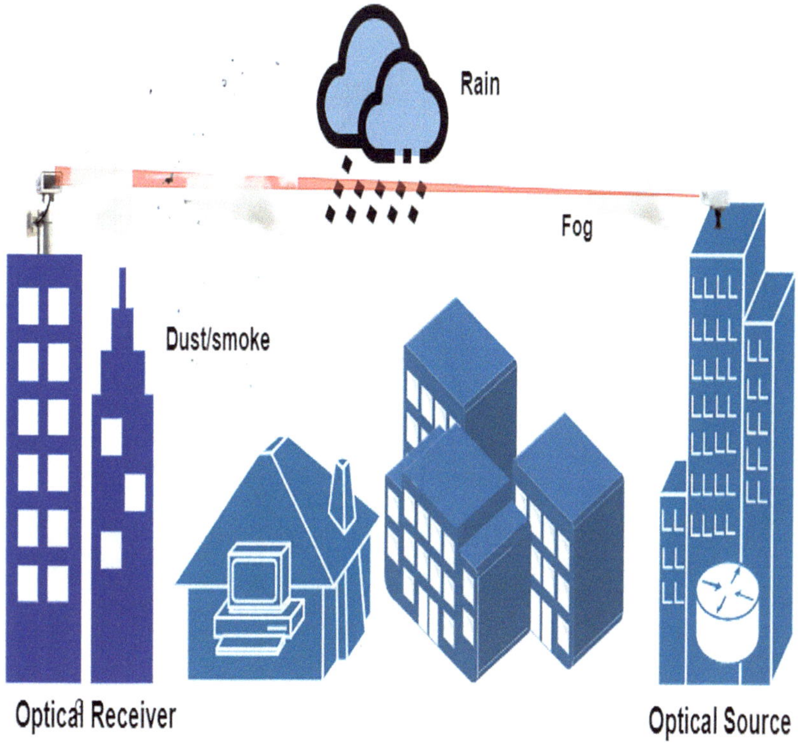

Fig. 4.8 Atmospheric attenuation of an FSO signal

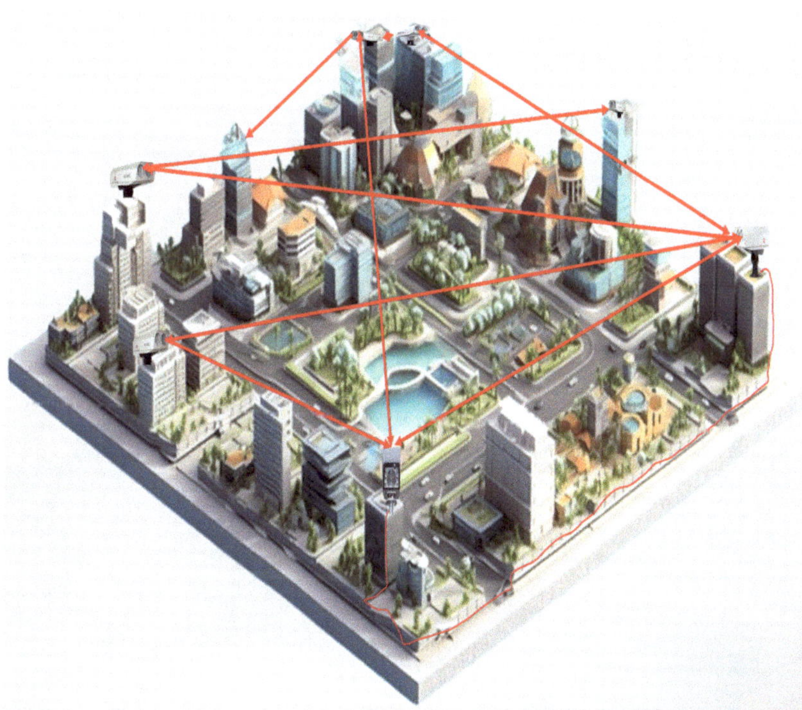

Fig. 4.9 Intercampus connectivity using FSO link

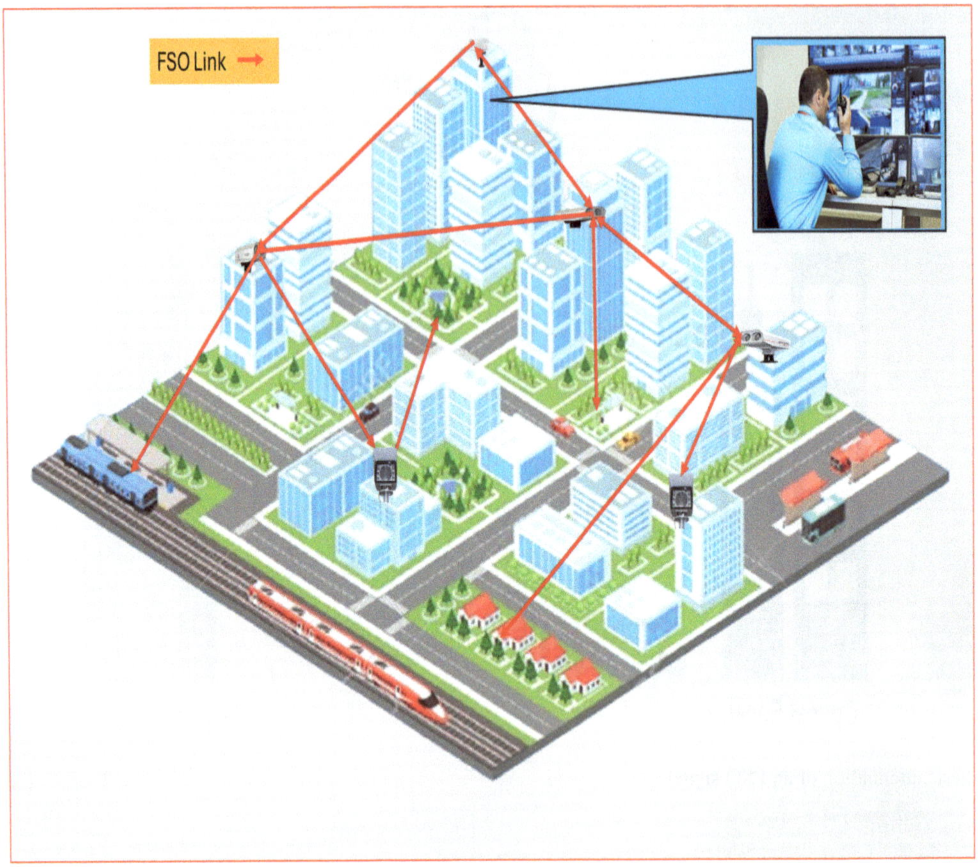

Fig. 4.10 Video surveillance using FSO communication

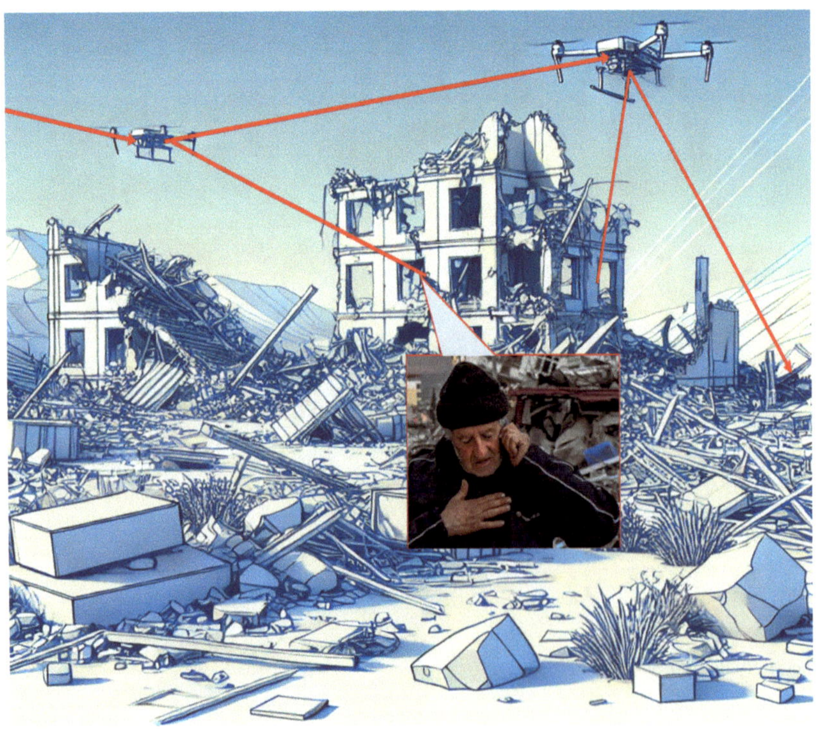

Fig. 4.11 Establishment of FSO communication in disaster-affected area

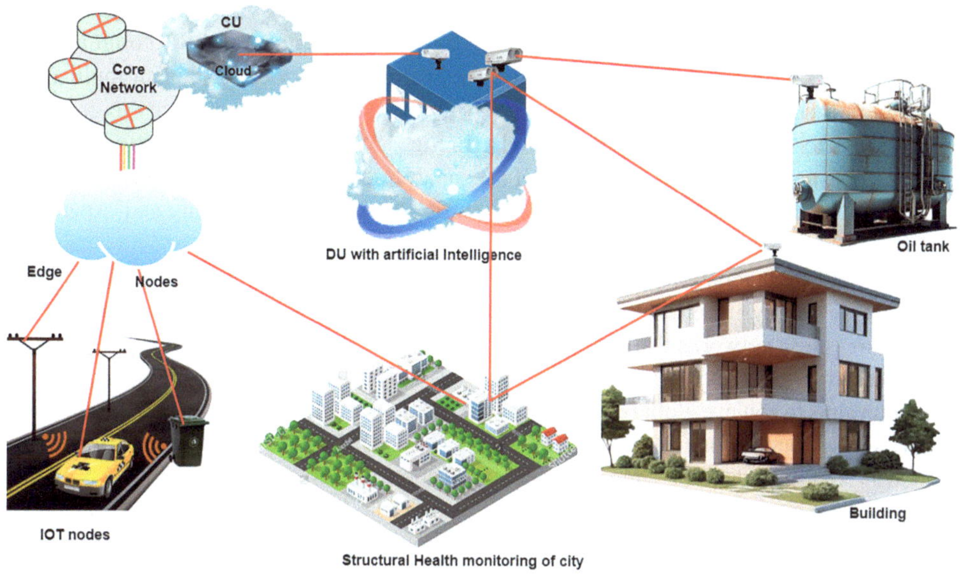

Fig. 4.12 Schematic diagram of FSO communication system for IoT application

installed in locations where wired infrastructure is not feasible.

- **Security:** The focus on cybersecurity has intensified with the advent of quantum computing, especially regarding electronic money transfers in a secured manner. Quantum cryptography promises absolute security, commonly associated with fiber-optic networks. FSO links present a versatile option when installing fiber optics is either too expensive or not feasible. Unauthorized access to information is prevented because FSO signals cannot pass through physical barriers.
- **Last mile access:** Despite most buildings being connected through optical fiber cables, many end users still lack access to fiber-optic infrastructure. FSO communication stands out as a vital solution for distant users, such as those in rural regions, providing connections in the Gbps speed over significant distances. FSO can swiftly enhance existing fiber networks during downtimes, maintaining continuous service for fiber-optic users. In addition, FSO links facilitate broadband internet connection to isolated areas where conventional access technologies encounter substantial difficulties.
- **Internet-of-things (IoT):** IoT extends internet connectivity beyond mobile devices, allowing for the exchange of information among various devices over communication networks. Energy efficiency is crucial in designing large-scale IoT infrastructures, and FSO technology aids in reducing energy consumption compared to traditional RF systems. FSO technology supports the extensive connection of smart devices in the IoT ecosystem using LED or laser, ensuring reliable connections with strong encryption capabilities. As illustrated in Fig. 4.12, the FSO link can serve as a robust backhaul connection for IoT devices, ensuring reliable data transmission over short to medium distances. The figure demonstrates how the line-of-sight nature of FSO links enables efficient data exchange between IoT nodes, significantly enhancing overall network performance.

4.4 FSO-Based Non-terrestrial Networks

The rapid growth of various emerging IoT applications has produced a massive volume of data traffic. These pose unprecedented challenges to the existing terrestrial network, limited by the network capacity and coverage, especially in remote areas. Regarding the coverage issue, the space-air-ground integrated network (SAGIN) composed of satellites, high-altitude aerial platforms stations (HAPSs)/ unmanned aerial vehicles (UAVs), and terrestrial systems, is a promising solution [12, 13]. As for the network capacity issue, FSO technology provides complementary solution to the current RF one. In recent years, space FSO communications have advanced remarkably. Esteemed organizations, e.g., NICT (Japan), UK Space Agency, German Aerospace Center, ISRO, and NASA, have been working for years on FSO satellite-related projects. Recent project updates highlighted the potential of space FSO networks in providing global coverage and high-speed internet from space. FSO leverages abundant unlicensed bandwidth and transmission capability at ultra-high data rates over long distances. As a result, the FSO-based SAGIN is a potential network paradigm for diverse applications, e.g., vertical backhaul solutions for terrestrial/aerial base stations (BSs), post-disaster/emergency communications, and remote/rural connectivity [12].

Non-terrestrial networks (NTNs) form the "Space and Air" layers of SAGIN [12]. NTNs, which include satellites, HAPSs, and UAVs, provide the space and air segments of the overall communication infrastructure. These layers are integral to SAGIN, which seeks to unify space, air, and ground networks into a single, cohesive communication system. In SAGIN, NTNs are integrated with terrestrial networks (fiber optic, cellular, Wi-Fi, etc.) to ensure global coverage, low-latency communications, and high reliability. While NTNs offer extended coverage, especially in remote and hard-to-reach areas, SAGIN ensures these networks work seamlessly alongside terrestrial counterparts. The deployment of NTNs with FSO communication aims to achieve global coverage for Sixth generation (6G) and beyond. Alongside space nodes, aerial NTN nodes such as HAPSs and rotary-wing UAVs can be deployed based on coverage needs and operational altitude requirements.

UAVs fly at lower altitudes, typically a few hundred meters, with a coverage radius of about 2 km. Thanks to their fully controllable Three-dimensional (3D) maneuverability, UAVs can be easily deployed by adjusting their height and horizontal position to fit various situations. However, UAVs have limited battery and onboard energy, resulting in a lifespan in communication networks ranging from a few minutes to a few hours. Over the past few years, UAVs have evolved from their initial military and aviation uses to being widely utilized in rapidly growing applications such as post-disaster rescue, agricultural practices, scientific research, and logistics. In wireless communication, UAVs can serve as aerial platforms by installing communication transceivers, enabling communication services to ground users in overloaded situations and during high traffic demands, known as UAV-assisted communications. UAV communications use LoS links, which support highly reliable FSO transmissions over long distances. Additionally, UAVs can adjust their hovering positions to maintain link quality. Unlike fixed ground infrastructures, UAVs can be dynamically deployed based on real-time demands, providing resilience against environmental changes for FSO communication. Furthermore, a swarm of UAVs can establish multi-UAV FSO networks, offering ubiquitous connectivity to ground stations as shown in Fig. 4.13. With their high flexibility and rapid deployment capabilities, multi-UAV FSO networks present a viable solution for quickly recovering and expanding communication networks efficiently.

Depending on their aerodynamic characteristics, UAVs can be classified into three main types as shown in Fig. 4.14: balloon, fixed wing, and rotary wing (quadcopter) [12].

- **Balloon UAVs:** These operate as quasi-stationary cellular towers, typically positioned in the stratosphere, usually above 20 km.
- **Fixed-wing (FW) UAVs:** These are deployed at altitudes between 1 to 10 km. They have strong turbine engines that can reach a wide range of altitudes at the fastest horizontal speeds. However, FW UAVs cruise at the lowest possible speed to minimize Doppler shift and the associated design challenges. They can carry high payloads and fly for several hours.
- **Rotary-wing (RW) UAVs:** These are deployed below 1 km. They can hover, have low speeds, and have a flight time of less than 1 hour. Due to their low cruise altitude, RW UAVs can be frequently and quickly recharged. However, compared to FW UAVs and balloon UAVs, RW UAVs have much lower overall energy efficiency.

HAPSs are aerial platforms located in the stratosphere, approximately 17 to 32 km above the Earth's surface. This atmospheric layer is essentially free from weather disturbances, making it ideal for deploying HAPS. Their relatively stationary nature prevents Doppler shift and allows for better capacity utilization. Due to low latency and dynamic deployment capabilities, HAPS can provide rapid network access and emergency communications to rural, urban, and remote

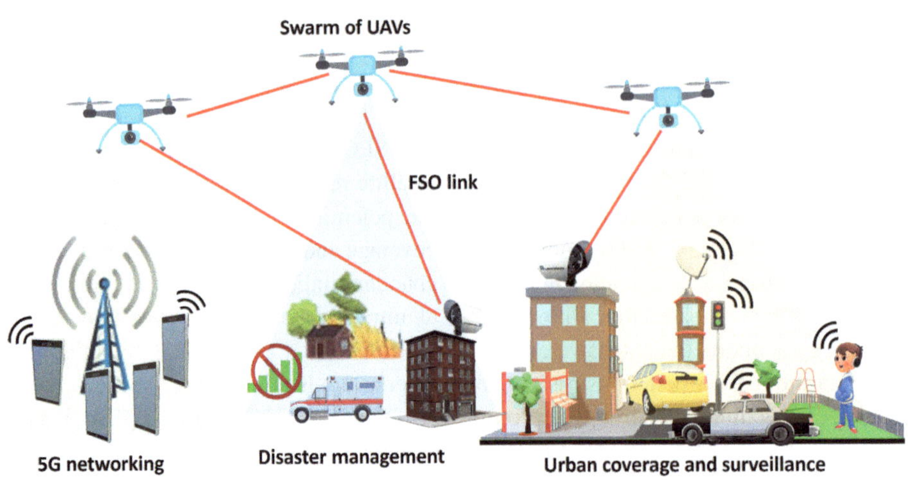

Fig. 4.13 Multi-UAV FSO network

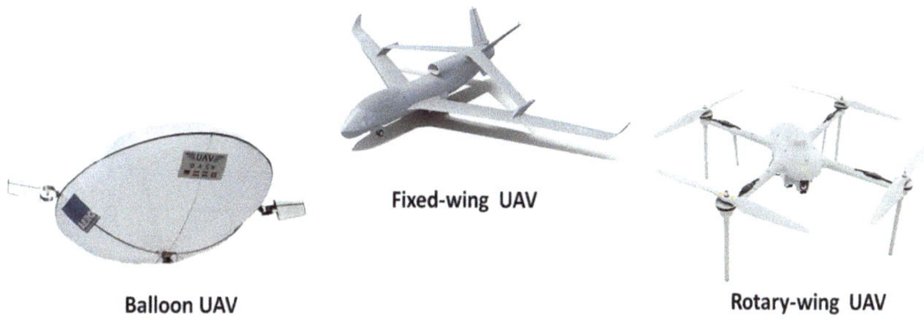

Fig. 4.14 Types of UAVs

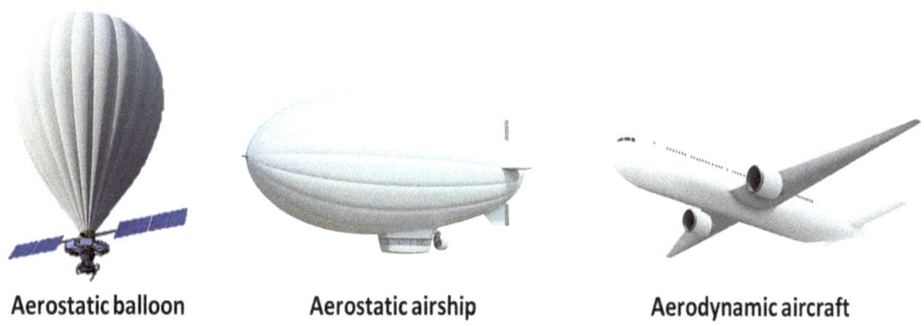

Fig. 4.15 Types of HAPS

areas. The primary source of energy for HAPS is solar energy for two main reasons. Firstly, HAPS can easily accommodate solar panels due to their large surface areas. Secondly, being deployed above the clouds, they have abundant access to natural solar energy. HAPS can support flight durations ranging from several hours to a few years.

HAPS can be categorized into three main types [12]: aerostatic balloons, aerostatic airships, and aerodynamic aircraft as shown in Fig. 4.15. The classification is determined by the principle that generates the lifting force.

- **Aerostatic Platforms (Lighter Than Air):** These rely on buoyancy to stay aloft and include balloons and airships. They use a lifting gas, typically helium or hydrogen, to provide buoyancy.
  - **Balloons:** Typically unpowered and can be tethered for flight control. However, stratospheric tethered balloons have been discontinued due to air safety concerns. These balloons are made from large polyethylene sheets equipped with antennas and solar panels, capable of remaining in the stratosphere for over 100 days. Tethered HAPS are now mostly limited to altitudes of up to 2 km. An example is the Google Loon project.
  - **Airships:** Powered by unmanned systems and can hover for several weeks or months. Their primary limitation is their size.
- **Aerodynamic Platforms (Heavier Than Air):** These platforms rely on dynamic forces generated through the air because of their movement to create lift. To maintain a nearly stationary position over the coverage area, continuous forward movement should be there, as these platforms cannot stay aloft without this forward motion.

This distinction between aerostatic and aerodynamic platforms highlights their different operational principles and capabilities in providing high-altitude platform services.

NTN nodes can support both backhaul and access FSO networks, where the end users can be connected with the onboard base station in NTN nodes through access link or the airborne base station can be connected to the core network through backhaul link [12]. NTNs can provide connectivity to end users in remote regions, where the terrestrial infrastructure is sparse or absent. In urban environments, NTNs can serve as an additional solution, supporting events like concerts, sports events, and other live gatherings. They are also capable of offering ubiquitous connectivity to users irrespective of their location. For instance, users on a cruise, flight, or high-speed train can connect to the network via NTN nodes even without terrestrial infrastructure. Additionally, NTNs are vital in disaster recovery scenarios, offering service to mobile users in affected zone, where terrestrial infrastructure is damaged or destroyed. FSO communication, known for its high data rates comparable to optical fiber communication, is ideally suited for backhauling in NTNs. The wide range of applications supported by FSO-based NTNs are shown in Fig. 4.16.

The backhaul system architecture is tailored to the specific use case. Single-layer backhaul systems are sufficient for rural coverage, while multilayer designs are often required to ensure connectivity in urban areas, where establishing LoS links are difficult. HAPSs provide extensive cover-

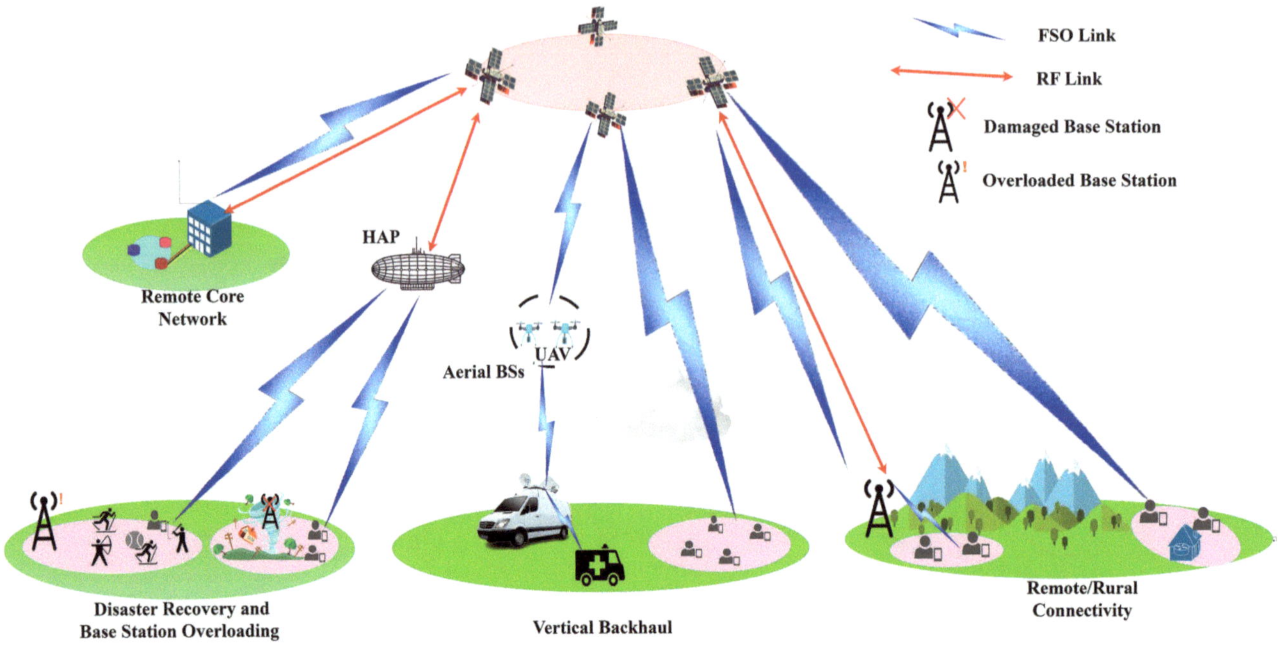

Fig. 4.16 Applications of SAGIN

age along predetermined paths, while UAVs that operate at low/medium altitudes complement the operation of HAPS in providing connectivity to the ground users.

Figure 4.17 illustrates a single-layer aerial backhaul FSO system [12]. In this configuration, HAPSs move along circular paths at a constant speed. Multiple circular paths are utilized to ensure continuous coverage. The entire cellular system coverage area is segmented into smaller cells, and within each cell, there is a base station that covers a designated area. The transmitter at HAPS is outfitted with multiple laser sources, each offering a dedicated steerable FSO link to the ground base station. Additionally, it includes a pointing acquisition-tracking (PAT) system that maintains the LoS link in FSO communication with the ground nodes.

Figure 4.18 illustrates a backhaul FSO network with multilayer designed for urban areas [12]. In addition to HAPSs, which operate at the stratosphere layer in the atmosphere, UAVs are utilized in the medium to lower altitude layers, which are limited to a few kilometers. In contrast to HAPSs, which travel in circular routes, rotary-wing UAVs are capable of hovering in relatively stable positions. The middle layer is comprised of UAVs that serve multiple ground stations. Additionally, low-altitude UAVs are used to manage particular links and overcome any obstructions.

It is to be noted that the critical concerns for practical FSO-based NTN are atmospheric turbulence, attenuation due to climatic conditions, and pointing misalignment/obstacle blockage due to narrow optical beams. These issues substantially degrade the network performance, necessitating extensive research efforts. On the other hand, unlike conventional single-tier networks, the inherent characteristics of heterogeneity and time variability in multitier network architectures pose various challenges for network design and optimization in FSO-based NTN. In general, the HAPSs in NTN functions as a relay station between satellite and other terrestrial/non-terrestrial nodes, employing relaying protocols. Moreover, FSO communication shall be used between HAPSs and satellites, as well as between HAPSs and UAVs or ground stations, due to the availability of LoS. While the reliability of FSO links between HAPS and satellites is not a significant concern, since HAPS operate above cloud-covered altitudes, the links between HAPS and UAVs or ground stations are more susceptible to turbulence and cloud interference [13, 14].

Atmospheric losses are notably greater for FSO uplink communications than for downlink due to the spreading and distortion of optical beams, since the optical beam size is smaller than the turbulent eddy size near the ground station [15]. Factors such as temperature variations and air pressure changes with altitude, which greatly impact the FSO signal propagation in the atmosphere. These losses include absorption and scattering, both dependent on wavelength and are measured in dB/km. Atmospheric loss due to scattering is more pronounced when the wavelength of FSO falls within the ultraviolet or visible spectrum. Thus, choosing the right wavelength is essential for FSO system design. Rayleigh scattering occurs due to smaller atmospheric particles compared to the optical wavelength, whereas Mie scattering happens with larger particles. Rayleigh scattering is typically caused by air molecules and haze particles, whereas Mie scattering is mainly due to aerosols, snow, and fog [16].

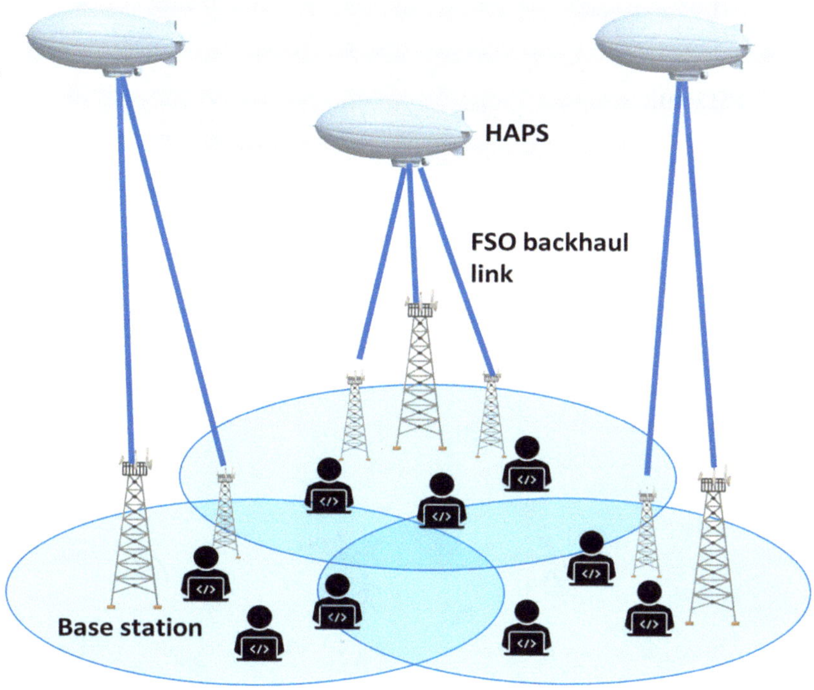

Fig. 4.17 FSO backhaul connectivity using HAPS

When dense fog reduces visibility to less than 50 meters, the signal attenuation can exceed 350 dB/km. In such scenarios, the reliability can be improved using high-power lasers. Heavy rain droplets also cause significant attenuation. For example, heavy rain (25 mm/hr) causes about ten times more attenuation than light rain (2.5 mm/hr), resulting in attenuation up to 10 dB/km at a wavelength of 1500 nm [16]. Hybrid RF/FSO communication with RF link serving as a backup for FSO link can effectively improve link availability under these conditions. Attenuation caused by snow falls between signal losses due to rain and fog. In space communication, the atmospheric turbulence effect will lead to beam wandering when the turbulence eddy size exceeds the beam radius, especially during the uplink scenario. The smaller turbulent eddies compared to the optical beam size cause irradiance fluctuations in the received signal strength at the receiver end.

Beam divergence is a crucial consideration in the FSO system design for uplink scenario in space communications. Due to light diffraction, beam divergence occurs causing a portion of the laser beam to be uncollected, resulting in geometrical loss. This loss substantially increases as the distance increases, irrespective of the size of receiver aperture. Therefore, using a narrow beam laser source is ideal, and employing diversity combining or spatial diversity techniques at the receiver can be beneficial [15].

Pointing errors can result from minor satellite vibrations or platform jitter. Even a small misalignment between transmitter and receiver can lead to performance degradation, making it essential to maintain LoS alignment. The beam wandering effect, where the optical beam deviates from its intended path, can also cause misalignment or pointing errors. Hence, the beam-wander-induced pointing errors will also be considered while modeling the atmospheric turbulence phenomenon in uplink scenario. These errors can significantly lead to FSO link failure, causing increase in the bit error rate (BER). Ensuring the proper installation of vibration-free FSO devices and implementing adequate bandwidth control to mitigate residual jitter are crucial. Pointing errors are more pronounced in lower earth orbit (LEO) satellites compared to geostationary earth orbit (GEO) satellites. Additionally, errors can also arise from satellite vibrations and noise associated with the tracking sensors. Therefore, a compensation method is necessary for FSO communications to mitigate misalignment and tracking errors in uplink scenario [13–16].

Inter-satellite space FSO links as shown in Fig. 4.19 are unaffected by atmospheric conditions, as the satellites orbit well above the atmospheric layer. However, inter-satellite links encounter other significant challenges, including acquisition and tracking, stability of satellite, Doppler effect, and background noise due to radiation. For FSO communication in inter-satellite links, coherent detection is preferable compared to direct (or noncoherent) detection, as they enable longer link distances. Both homodyne and heterodyne coherent techniques are power efficient and offer excellent receiver sensitivity while requiring only a minimal amount

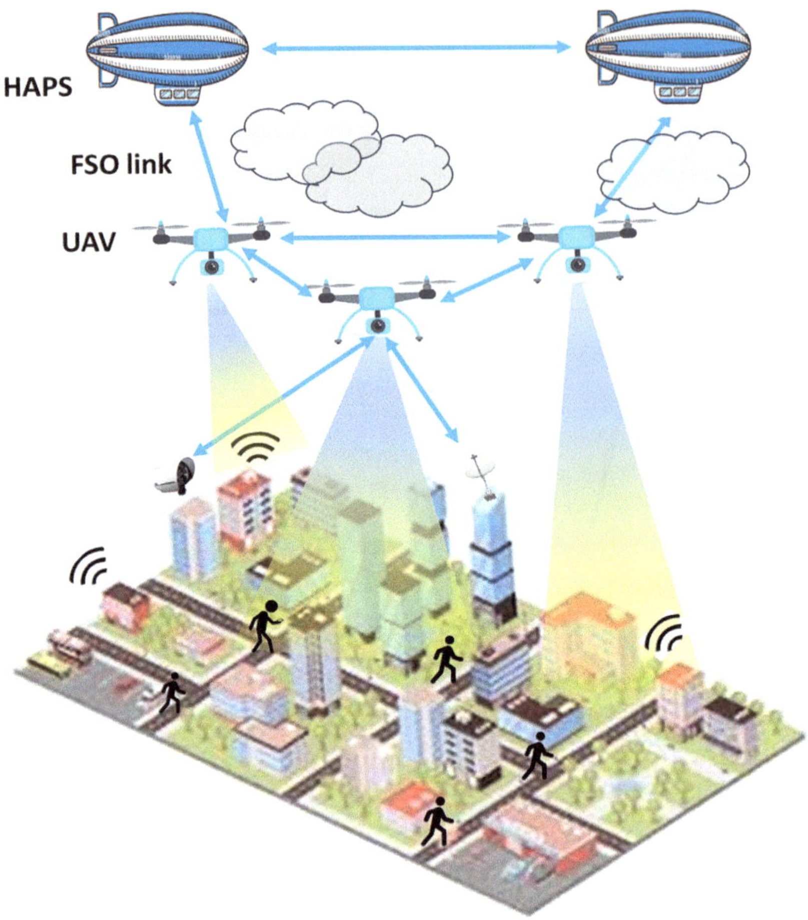

Fig. 4.18 FSO backhaul connectivity for urban areas

of additional power. In FSO space communications, detector arrays are used to collect light signals and mitigate the misalignment errors [12].

4.5 Techniques to Improve FSO Link Performance

Reliability is essential for any communication system. FSO systems achieve high SNR under indoor and outdoor channel conditions overcoming various challenges. However, atmospheric factors like turbulence and weather conditions limit the FSO link performance. Therefore, FSO systems must be designed to operate effectively under severe attenuation caused by weather conditions and signal fluctuations caused by atmospheric turbulence [1].

The FSO beam with minimal divergence and a limited field of view (FoV) at the receiver can result in temporary connectivity issues. As a result, this may cause pointing or misalignment errors, making PAT system crucial at the receiver. While estimating received optical power is straightforward in optical fiber communication, it becomes tedious in the FSO communication due to atmospheric effects, complicating the link budget.

Optical signal propagation through space is significantly impacted by atmospheric conditions, causing intensity fluctuations. This results in reduced signal-to-noise ratio (SNR) performance, decreased link availability, and increased BER. Fog and high humidity notably degrades FSO signal quality. Further, scintillation effects and molecular absorption also attenuate the optical signal. All these factors are also influenced by the operational wavelength. Thus, choosing the right wavelength is crucial for the efficient operation of FSO communication. The following sections discuss several viable approaches to improve FSO communication link performance.

4.5.1 MIMO Schemes

Multiple-input multiple-output (MIMO) schemes, which provide spatial diversity, are being proposed to address the limitations of FSO communication, as shown in Fig. 4.20. Spatial diversity techniques involve using multiple apertures

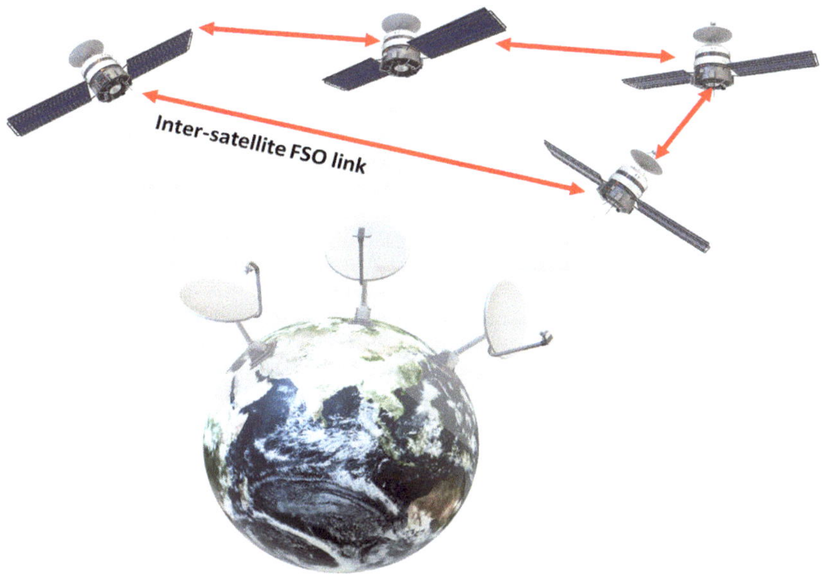

Fig. 4.19 Inter-satellite FSO communication

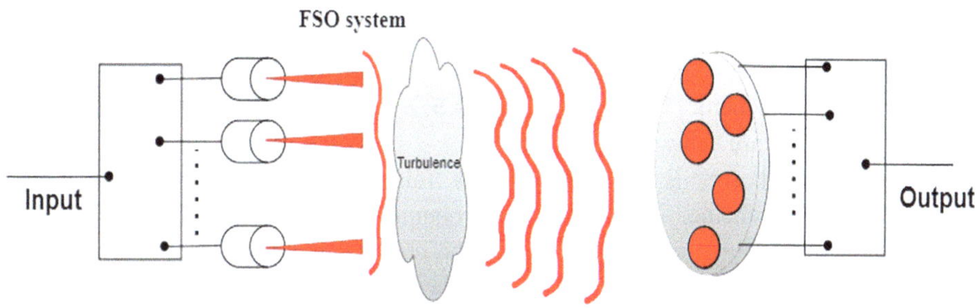

Fig. 4.20 MIMO scheme

at the transmitter, receiver, or both to achieve diversity benefits. For instance, a multiple-input single-output (MISO) FSO system can utilize a transmit laser selection scheme to achieve full diversity and enhance performance. The severity of turbulence-induced fading can be reduced by using a larger aperture at the receiver. This technique, known as aperture averaging, helps to average out the intensity fluctuations. Additionally, various diversity combining schemes can be utilized at the receiver for multiple beams to achieve maximum system diversity. These schemes include maximal-ratio combining (MRC), selection combining (SC), switch-and-stay combining (SSC), switch-and-examine combining (SEC), switch-and-examine combining with post-selection (SECps), etc. [17–19].

- **Selection Combining:** The receiver compares the signals from multiple apertures and selects the one with the highest instantaneous SNR. This reduces the impact of fading and improves overall system performance.
- **Maximal-Ratio Combining:** This scheme merges signals from multiple apertures to maximize the system's overall output SNR. Unlike SC, which selects only the strongest signal, MRC combines all received signals based on their instantaneous SNRs.
- **Switch-and-Stay Combining:** In this scheme, if the primary FSO link's instantaneous SNR falls below a specified threshold, message signal is transmitted through a backup FSO link, irrespective of its instantaneous SNR.
- **Switch-and-Examine Combining:** In the SEC scheme, if the instantaneous electrical SNR of the received FSO signal in the current branch is less than the threshold SNR, then the SNR of the next branch is checked. If it exceeds the threshold SNR, then the received FSO signal is processed and decoded. Otherwise, the process continues with subsequent branches until the second-to-last branch. If the SNR of the second-to-last branch is also below threshold SNR, then the received FSO signal is decoded from the last branch regardless of its SNR value.
- **Switch-and-Examine Combining with Post-Selection:** In the SECps scheme, the examination process continues

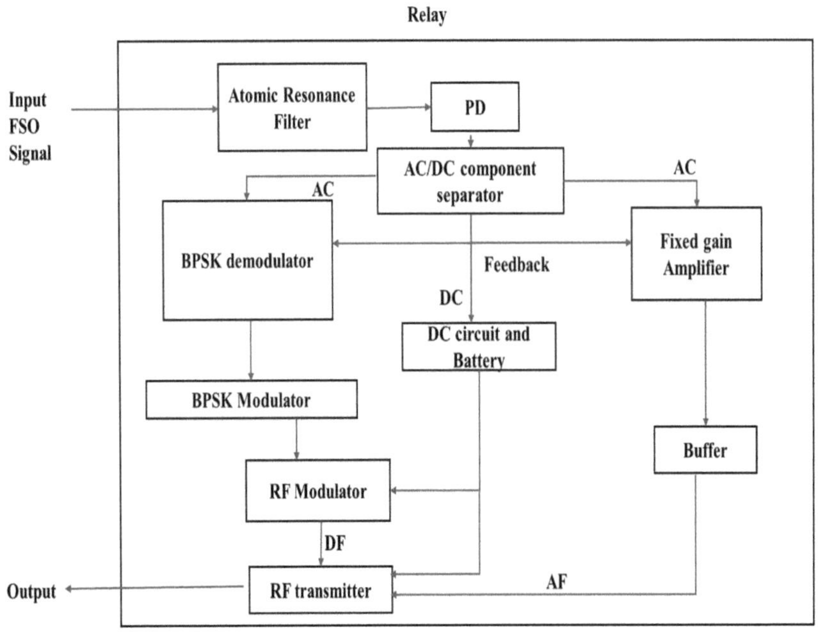

Fig. 4.21 Block diagram of DF and AF relaying technique

until the last branch unlike SEC scheme. If the instantaneous SNR values of all branches are below the threshold SNR, then the branch with the highest SNR is selected for decoding the FSO signal like SC scheme.

Unlike the conventional SC scheme, which requires the channel state information (CSI) of all branches at the receiver, switched diversity combining schemes such as SSC, SEC, and SECps need the CSI of only one branch at a time. The CSI of the next branch is required only if the instantaneous SNR of the current branch drops below a threshold value. Apart from enhancing the system reliability, the switched diversity combining techniques reduce the processing load in terms of the required number of CSI at the receiving end compared to the conventional diversity combining techniques such as SC and MRC.

A low-complexity modulation scheme known as optical spatial modulation (OSM) was introduced for FSO-based MIMO systems to enhance data rates, spectral efficiency, and energy efficiency. OSM employs index modulation, transmitting data through both aperture and signal dimensions. A variant, optical space shift keying (OSSK), activates only one transmitting aperture at a time, decoding data based on the active aperture's index. The OSM technique significantly boosts spectral and energy efficiency by implicitly conveying additional information bits through aperture index usage or activation patterns. This approach harnesses extra index bits alongside those carried by standard modulation schemes, without additional energy consumption.

4.5.2 Relaying Techniques

A relay-assisted FSO system and cooperative diversity schemes were introduced to extend coverage and address the challenges of FSO communication. In the cooperative relaying technique, during the initial transmission phase, a source node sends information symbols to a destination node, which are also intercepted by a cooperative relay node due to the broadcast nature of transmission. In the subsequent orthogonal transmission phase, the relay node utilizes its resources to forward the intercepted information to the destination node. As a result, the destination receives redundant copies of the same information symbols in two orthogonal time slots, thereby achieving diversity.

The relay nodes process received information symbols according to relaying protocols before forwarding them to the destination node. The error performance of a cooperative diversity system hinges on the combining schemes at the destination and the relaying protocols at the relay nodes. Two widely employed relaying protocols include decode-and-forward (DF) [7] and amplify-and-forward (AF) [20]. Figure 4.21 illustrates the block diagram of DF and AF relaying techniques.

- **Decode-and-Forward (DF):** In DF relaying, the received information signal from the source node is decoded at the relay node. Further, it is re-encoded and forwarded to the destination node. This approach increases computational complexity and results in error propagation, thereby compromising BER performance and diversity order.

- **Amplify-and-Forward (AF):** In AF relaying, the received information signal at the relay node is amplified by a certain factor and is forwarded to the destination node. However, a significant drawback of AF relaying is that it amplifies both the signal and the noise present at the relay node.

4.5.3 Hybrid FSO/RF Communication

To enhance reliability and reduce losses, an FSO link can be backed up with a more reliable RF link, which is less affected by turbulence, misalignment errors, and attenuation due to fog [1]. However, the RF link is susceptible to multipath fading and rain, which do not significantly impact the FSO link. Thus, it is noted that FSO and RF links enhance each other's performance under varying channel conditions, leading to improved overall system efficiency. Consequently, hybrid FSO/RF communication is seen as a promising solution for 5GB systems, especially for wireless backhaul connections. Moreover, this hybrid approach can be applied as backhaul links in various scenarios, including ship connectivity, cellular communications, and satellite communications.

In a standard hybrid FSO/RF system, data transmission can occur through the FSO link, the RF link, or a combination of both, depending on the switching scheme used. The RF link mainly acts as a backup to the FSO link to improve system performance. The following sections discuss two of the most commonly used switching schemes.

- **Hard-Switching Scheme:** When the FSO link's quality becomes unacceptable, the system transitions to the RF link, placing the FSO subsystem in standby mode which is illustrated in Fig. 4.22. This method requires frequent hardware switching, leading to suboptimal performance [3, 21–23].
- **Adaptive Combining Scheme:** As long as the FSO link maintains satisfactory quality, it remains active while the RF link is kept in standby mode. If the quality of the FSO link deteriorates, a feedback signal activates the RF link, and MRC is used to integrate both FSO and RF links at the receiver as shown in Fig. 4.23. This approach leverages the advantages of both switching and diversity combining, conserving RF link power when the RF subsystem is inactive. The adaptive combining scheme effectively overcomes the limitations of traditional diversity combining and hard-switching methods in hybrid FSO/RF systems, making it a more effective solution [24, 25].

Although switching schemes are efficient, they necessitate real-time CSI at the transmitter, complicating their implementation. To avoid relying on a feedback channel, diversity combining techniques such as SC and MRC can be used [26]. These methods combine signals received from each link directly in the electrical domain on a symbol-by-symbol basis. But the drawbacks of the diversity combining schemes for hybrid FSO/RF systems are as follows: (i) RF link power is wasted while the FSO link remains acceptable in quality, and (ii) the overall transmission data rate of the hybrid system is limited to the rate of the RF link because both the FSO and RF links transmit the message signal simultaneously.

4.5.4 Reconfigurable Intelligent Surfaces

A reconfigurable intelligent surface (RIS), also known as an intelligent reflecting surface (IRS), is a metasurface made up of an array of passive reflecting elements. These elements can reconfigure incident signals and intelligently manipulate them by reflecting, refracting, or focusing, thereby improving communication performance [27]. Each RIS element can be individually configured in real time to control various characteristics of the incident signal, such as amplitude, phase, and polarization. For instance, RISs can direct reflected waves to converge at a specific point instead of reflecting in a traditional manner. This is accomplished by modifying the

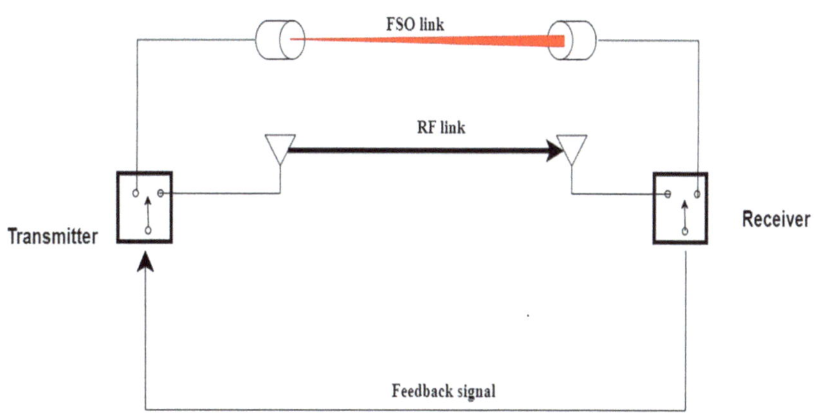

Fig. 4.22 Hard-switching scheme

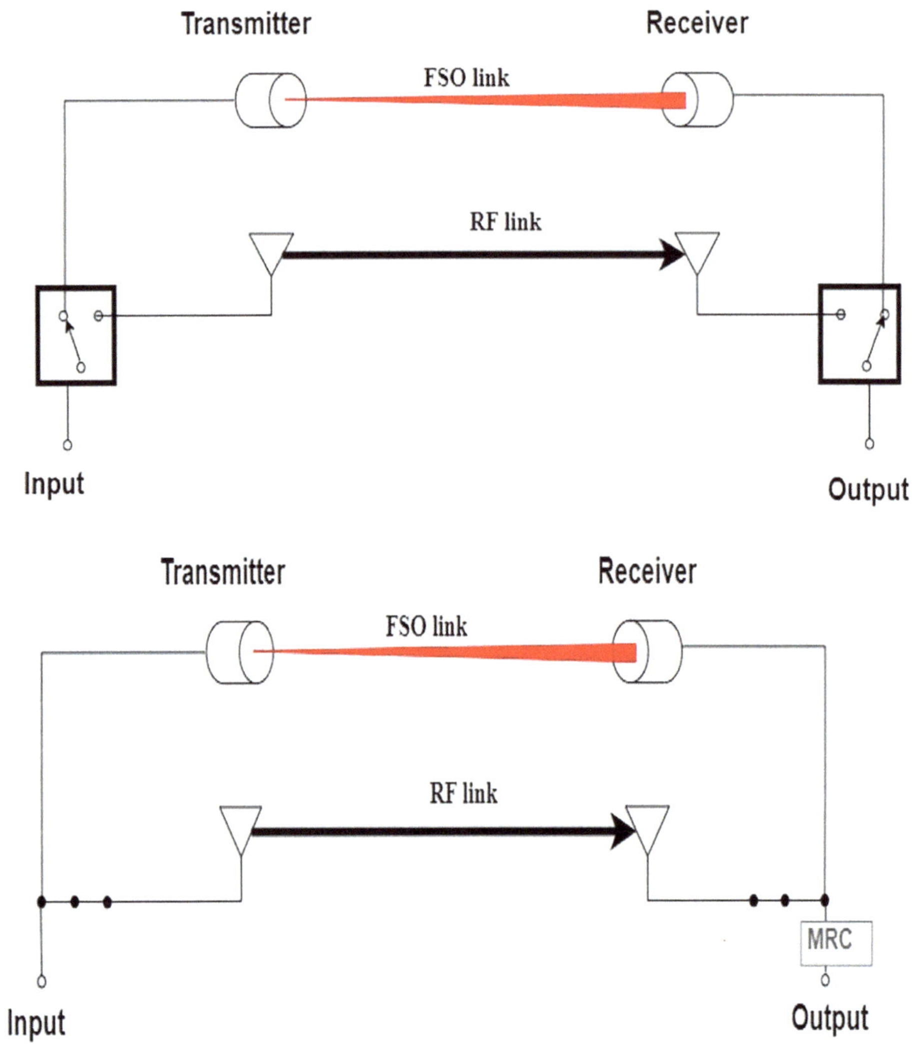

Fig. 4.23 Adaptive combining scheme

electromagnetic response of each reflecting element using electrical voltage stimulation, which induces a phase shift in the incoming signals and directs the main path of the reflected signals. Typically, RIS control mechanisms employ varactor diodes, positive-intrinsic-negative (PIN) diodes, or microelectromechanical systems (MEMS) switches, all of which communicate with a central controller. Unlike traditional metasurfaces that depend on human judgment and recognition, an RIS controller can sense the environment and utilize intelligent algorithms to autonomously identify and adapt to environmental changes. As a result, deploying RISs densely within a wireless communication network allows for comprehensive manipulation of transmitted and reflected waves, enabling intelligent control of the communication channel and signal propagation, thereby enhancing the end user's quality of experience [28].

RIS technology has recently garnered substantial research interest in wireless communications due to its numerous advantages, including (i) the ease of manufacturing metasurfaces for RISs using traditional methods like photolithography and electron-beam lithography, facilitated by advancements in the semiconductor industry; (ii) the straightforward deployment of RISs on existing infrastructures such as building exteriors and interiors, roadside billboards, and even clothing; (iii) low power consumption and reduced carbon footprint; (iv) enhancements in critical performance metrics such as outage probability, BER, and coverage probability, especially in scenarios lacking a LoS path between the transmitter and receiver; and (v) compatibility with current wireless standards [28]. Although extensive research has been conducted on its application in RF communications, RIS technology and its implementation in RF communication systems cannot be directly transferred to optical wireless systems due to differences in signal characteristics, typical functionalities, hardware, deployment locations, performance optimization, propagation models, technological maturity, and cost [29–31].

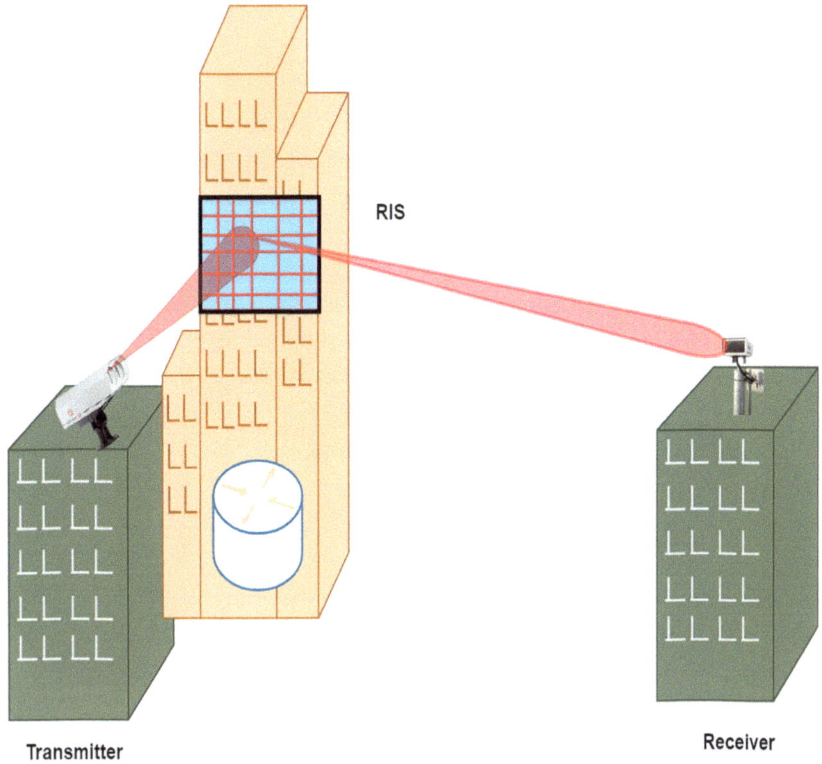

Fig. 4.24 Envisioned deployment scenarios of RIS at the transmitter

In FSO systems, there are two distinct configurations for RISs: the intelligent metasurface reflector (IMR) and the intelligent mirror array (IMA). For the IMR, common tuning mechanisms involve liquid crystals (LCs), adjustable chips, graphene, ferrite materials, and photoconductive semiconductors. In contrast, the design and production of the IMA predominantly employ MEMS for tuning, with actuators adjusting individual micro-mirrors via electrostatic actuation.

RIS can be deployed in FSO systems at the transmitter side, the receiver side, or anywhere in between [29, 31].

- **RIS at the transmitter:** In FSO systems, the most commonly used transmitters are LEDs and LDs. LEDs generally emit light over a broad surface area, providing a large field of view in nearly all directions and covering a wide spectrum of wavelengths. As a result, LEDs offer broader coverage and more consistent illumination, although the light intensity may vary across different regions. In contrast, LDs emit highly directional light beam from a small area at a single wavelength or a few closely spaced wavelengths. Considering the characteristics of LEDs and LDs, and the typical use of RISs for beam steering applications at the transmitter side, it is always important to pair RISs with transmitters that use LEDs. Figure 4.24 demonstrates envisioned deployment scenarios and functionalities of utilizing RISs at the transmitter side in FSO systems to improve performance aspects such as coverage and directionality.
- **RIS at the receiver:** In contrast, Fig. 4.25 shows an FSO system using an RIS positioned at the receiver. In this configuration, the RIS helps direct the incoming light and concentrate its beam onto the photodetector. It can function as a filter by allowing light signals of certain wavelengths to pass through while blocking unwanted light, thus minimizing interference. Furthermore, the RIS can either produce a single parallel beam or concentrate all light onto the photodetector surface, as depicted. The LC-enabled RIS is the most commonly used type at the receiver, providing several benefits, including filtering out unwanted light beam, enhancing the performance at the receiver, extending coverage, and crucially expanding the receiver's FoV.
- **RIS in between transmitter and receiver:** A frequent performance problem in FSO systems is the blockage of the LoS path. Hence, it is essential to investigate different methods for utilizing RISs to establish virtual LoS paths to ensure effective data transmission. Figure 4.26 depicts blockages caused by high-rise buildings in outdoor environments. For instance, a newly constructed structure situated in the LoS path of an FSO link can obstruct the light from reaching the photodetector on another building. In such cases, RIS can establish a virtual LoS path, enabling the optical signal to reach its destination.

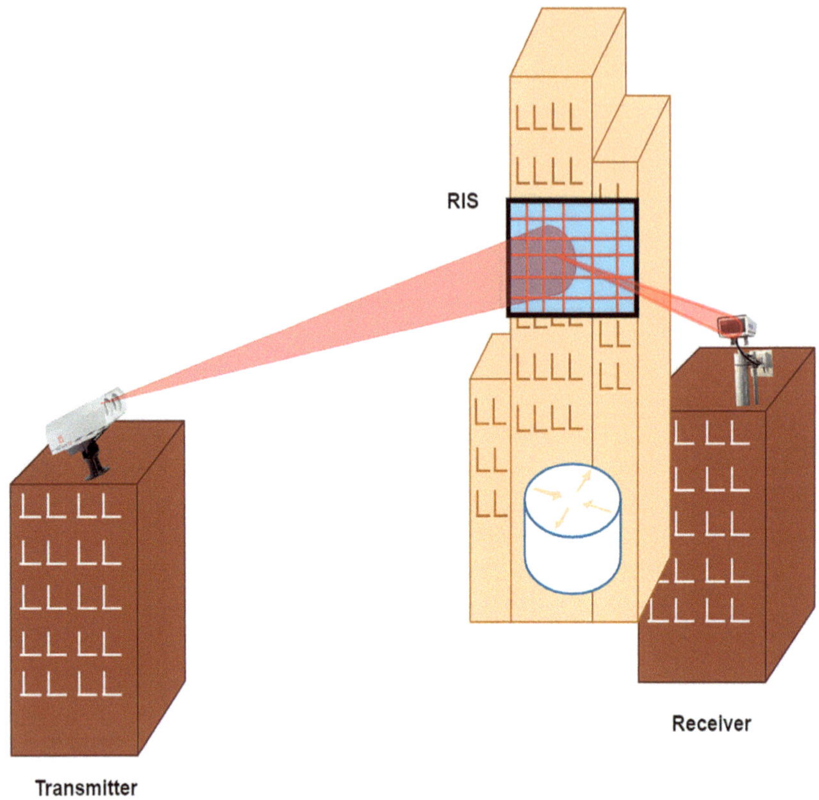

Fig. 4.25 Schematic of RIS directing an incoming beam to a photodetector

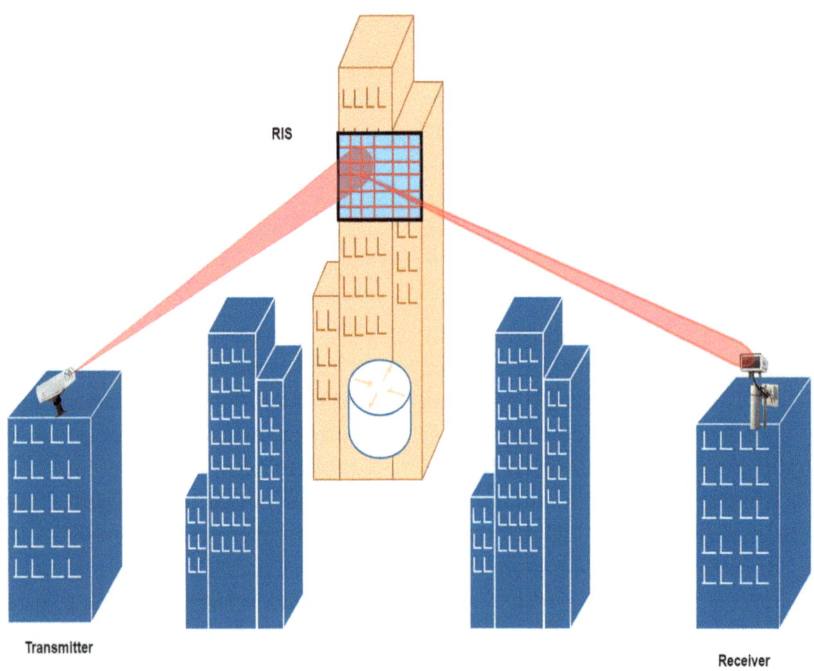

Fig. 4.26 RIS in between transmitter and receiver

Optical RIS and relay-assisted technologies can help lower the LoS requirements and improve the SNR of the OWC systems. Nevertheless, these two technologies vary considerably in several respects. Optical RISs generally comprise passive reflecting elements, which controls the incident light using reflection and refraction. In contrast, optical relays are active devices that typically amplify and retransmit or decode and retransmit the incoming signals, necessitating a complete transceiver setup and functioning through various relaying protocols. Although they differ in their approaches, both technologies strive to achieve similar functionalities using different methods [30, 31].

4.6 Summary

In a nutshell, FSO communication represents a transformative technology with the potential to revolutionize modern communication networks. By leveraging optical wireless technologies, FSO offers high-bandwidth, low-latency, and secure transmission solutions for various applications such as intercampus connectivity, video surveillance, 5G/6G backhaul, and IoT.

Throughout this chapter, we explored key aspects of FSO communication, including channel modeling, the challenges posed by atmospheric turbulence, pointing errors, and attenuation, as well as practical solutions like MIMO schemes, cooperative relaying, and hybrid FSO/RF communication. Additionally, FSO communication's integration into non-terrestrial networks, including UAVs, HAPS, and SAGIN, opens new avenues for remote connectivity, disaster recovery, and vertical backhaul networks.

As the demand for high-speed, secure, and efficient communication systems grows, FSO technology will play a critical role in meeting future needs, especially in 5G and beyond. By continually refining techniques such as diversity combining, reconfigurable intelligent surfaces, and hybrid systems, the potential of FSO communication will be fully realized, enabling reliable, high-capacity communication in a wide range of environments.

Homework Problems

1. What are the advantages of optical wireless communication (OWC) compared to RF communication
2. List the various optical wireless communication (OWC) technologies
3. What are the two types of detection techniques used to detect the received optical signal and explain them in brief?
4. What is meant by subcarrier intensity modulation (SIM) scheme ?
5. List various channel anomalies that affect the performance of FSO signal.
6. What is meant by misalignment or pointing errors in FSO communication?
7. List the various use cases and applications of FSO communication
8. What is the need for non-terrestrial networks (NTNs) in 6G communication?
9. What is the difference between UAVs and HAPSs?
10. What are the various techniques available to improve the performance of FSO communication?
11. What is the difference between selection and maximal-ratio combining schemes?
12. What are the drawbacks of AF and DF relaying techniques?
13. How does adaptive-combining-based hybrid FSO/RF communication overcome the drawback of MRC-based hybrid FSO/RF communication?
14. What is a reconfigurable intelligent surface (RIS), and how does it enhance communication performance in wireless networks?
15. What are the two distinct configurations of RIS, and where can RIS be deployed in FSO systems?

References

1. Jahid, A., Alsharif, M.H., Hall, T.J.: A contemporary survey on free space optical communication: potentials, technical challenges, recent advances and research direction. J. Netw. Comput. Appl. **200**(103311), 1–35 (2022)
2. Khalighi, M.A., Uysal, M.: Survey on free space optical communication: a communication theory perspective. IEEE Commun. Surveys Tuts. **16**(4), 2231–2258 (2014)
3. Vishwakarma, N., Swaminathan, R.: Performance analysis of hybrid FSO/RF communication over generalized fading models. Opt. Commun. **527**(128964), 1–14 (2022)
4. Kaushal, H., Jain, V.K. Kar, S.: Free Space Optical Communication. Springer, New Delhi (2017)
5. Jung, K.-J., Nam, S.S., Alouini, M.-S., Ko, Y.-C.: Unified finite series approximation of FSO performance over strong turbulence combined with various pointing error conditions. IEEE Trans. Commun. **68**(10), 6413–6425 (2020)
6. Farid, A.A., Hranilovic, S.: Outage capacity optimization for free-space optical links with pointing errors. J. Lightw. Technol. **25**(7), 1702–1710 (2007)
7. Bhatnagar, M.R.: Performance analysis of decode-and-forward relaying in gamma-gamma fading channels. IEEE Photon. Technol. Lett. **24**(7), 545–547 (2012)
8. Ansari, I.S., Yilmaz, F., Alouini, M.-S.: Performance analysis of free-space optical links over Málaga ($\mathcal{M}$) turbulence channels with pointing errors. IEEE Trans. Wireless Commun. **15**(1), 91–102 (2016)
9. Peppas, K.P., Alexandropoulos, G.C., Xenos, E.D. Maras, A.: The Fischer—Snedecor $\mathcal{F}$-distribution model for turbulence-induced fading in free-space optical systems. J. Lightw. Technol. **38**(6), 1286–1295 (2020)

10. Cheng, M., Guo, Y., Li, J., Zheng, X., Guo, L.: Inverse Gaussian gamma distribution model for turbulence-induced fading in free-space optical communication. Appl. Opt. **57**, 3031–3037 (2018)
11. Badarneh, O.S., Bouanani, F.E., Almehmadi, F.S., Silva, H.S.: FSO communications over doubly inverted Gamma-Gamma turbulence channels with nonzero-boresight pointing errors. IEEE Wireless Commun. Lett. **12**(10), 1761–1765 (2023)
12. Elamassie, M., Uysal, M.: Free space optical communication: an enabling backhaul technology for 6G non-terrestrial networks. Photonics. **10**(1210), 1–45 (2023)
13. Swaminathan, R., Sharma, S., Vishwakarma, N., Madhukumar, A.S.: HAPS-based relaying for integrated space—air—ground networks with hybrid FSO/RF communication: a performance analysis. IEEE Trans. Aerosp. Electron. Syst. **57**(3), 1581–1599 (2021)
14. Singh, D., Swaminathan, R.: Comprehensive performance analysis of hybrid FSO/RF space-air-ground integrated network. Opt. Commun. **527**(128964), 1–14 (2022)
15. Viswanath, A., Gopal, P., Jain, V.K., Kar, S.: Performance enhancement by aperture averaging in terrestrial and satellite free space optical links. IET Optoelectron. **10**(3), 111–117 (2016)
16. Viswanath, A., Jain, V.K.. Kar, S.: Analysis of earth-to-satellite free-space optical link performance in the presence of turbulence, beam-wander induced pointing error and weather conditions for different intensity modulation schemes. IET Commun. **9**(18), 2253–2258 (2015)
17. Bhatnagar, M.R., Ghassemlooy, Z.: Performance analysis of gamma gamma fading FSO MIMO links with pointing errors. J. Lightw. Technol. **34**(9), 2158–2169 (2016)
18. Kamboj, A., Mallik, R.K., Agrawal, M., Schober, R.: Diversity combining in FSO systems in presence of non-Gaussian noise. In: Proceedings of IEEE International Conference on Signal Processing and Communications, Bangalore, India, 2012, pp. 1–5
19. Das, A., Bag, B., Bose, C., Chandra, A.: Performance analysis of FSO links employing a transmit and receive diversity-based operating system under Málaga turbulence channels with pointing errors. Opt. Continuum. **1**(2), 366–387 (2022)
20. Trinh, P.V., Thang, T.C., Pham, A.T.: Two-way all-optical AF relaying FSO systems over Malaga ($\mathcal{M}$) channels with pointing errors. In: Proceedings of IEEE International Conference on Communications, Kuala Lumpur, Malaysia, 2016, pp. 1–7
21. Usman, M., Yang, H., Alouini, M.: Practical switching-based hybrid FSO/RF transmission and its performance analysis. IEEE Photon. J. **6**(5), 1–13 (2014)
22. Sharma, S., Madhukumar, A.S., Swaminathan, R.: Switching-based cooperative decode-and-forward relaying for hybrid FSO/RF networks. IEEE J. Opt. Commun. Netw. **11**(6), 267–281 (2019)
23. Sharma, S., Madhukumar, A.S., Swaminathan, R.: Effect of pointing errors on the performance of hybrid FSO/RF networks. IEEE Access **7**, 131418–131434 (2019)
24. Rakia, T., Yang, H.C., Alouini, M.S., Gebali, F.: Outage analysis of practical FSO/RF hybrid system with adaptive combining. IEEE Commun. Lett. **19**(8), 1366–1369 (2015)
25. Vishwakarma, N., Swaminathan, R.: On the maximal-ratio combining of FSO and RF links over generalized distributions and its applications in hybrid FSO/RF systems. Opt. Commun. **520**(128542), 1–17 (2022)
26. Chatzidiamantis, N.D., Karagiannidis, G.K., Kriezis, E.E., Matthaiou, M.: Diversity combining in hybrid RF/FSO systems with PSK modulation. In: Proceedings of IEEE International Conference on Communications, Kyoto, Japan, 2011, pp. 1–6
27. Basar, E., Di Renzo, M., De Rosny, J., Debbah, M., Alouini, M.S., Zhang, R.: Wireless communications through reconfigurable intelligent surfaces. IEEE Access **7**, 116753–116773 (2019).
28. Wu, Q., Zhang, S., Zheng, B., You, C., Zhang, R.: Intelligent reflecting surface-aided wireless communications: a tutorial. IEEE Trans. Commun. **69**(5), 3313–3351 (2021)
29. Aboagye, S., Ndjiongue, A.R., Ngatched, T.M., Dobre, O.A., Poor, H.V.: RIS-assisted visible light communication systems: a tutorial. IEEE Commun. Surveys Tuts. **25**(1), 251–288 (2023)
30. Najafi, M., Schmauss, B., Schober, R.: Intelligent reflecting surfaces for free space optical communication systems. IEEE Trans. Commun. **69**(9), 6134–6151 (2021)
31. Jamali, V., Ajam, H., Najafi, M., Schmauss, B., Schober, R., Poor, H.V.: Intelligent reflecting surface assisted free-space optical communications. IEEE Commun. Mag. **59**(10), 57–63 (2021)

5 Performance Analysis of Free-Space Optical Communication

5.1 Introduction

Atmospheric optical communication is gaining significant attention for high-data-rate wireless links. However, the reliability of optical wireless links can be greatly affected by turbulence-induced scintillation during beam propagation through the atmosphere [1].

FSO is utilized as a backhaul network by transmitting data through the atmosphere using laser beams, providing high-speed, high-bandwidth connections between points like cellular towers, data centers, and buildings. This technology is cost-effective and quick to deploy compared to laying fiber-optic cables, especially in urban areas or locations where physical infrastructure is challenging. Although FSO systems require a clear line of sight and are susceptible to weather conditions, they offer significant flexibility and scalability, making them ideal for rapidly setting up temporary links, enhancing existing networks, and providing redundancy in hybrid backhaul solutions [2]. Nevertheless, even under clear sky conditions, wireless optical links encounter fluctuations in the intensity and phase of the optical wave, as it propagates through the atmosphere [3]. These fluctuations are caused by time-varying inhomogeneities in the atmospheric refractive index. Assessing the reliability of an optical system in such an environment requires a statistical model to characterize the randomly fading irradiance signal (scintillation). Therefore, a key objective in studying optical wave propagation through turbulence is to identify a tractable statistical model that accurately describes the irradiance across all intensity fluctuation regimes.

Over the years, a variety of statistical models for irradiance have been developed, each with varying degrees of success. Among them, the most notable ones are the log-normal [1] and gamma-gamma (GG) [4] models. The log-normal model is particularly effective under weak irradiance fluctuations, while the gamma-gamma model is preferred for moderate-to-strong turbulence regimes due to simpler mathematical formulation. However, both the log-normal and GG models are special cases of the Málaga distribution model [5], which is useful for characterizing all turbulence regimes. Recently, a new model called Fisher-Snedecor $\mathcal{F}$ [6] was proposed to characterize atmospheric turbulence over FSO links. This $\mathcal{F}$ turbulence model offers a superior fit to experimental data across all turbulence conditions compared to the log-normal and GG models. The probability density function (PDF) of the $\mathcal{F}$ turbulence distribution includes basic elementary functions, making it mathematically simpler than other well-known distributions, such as the GG, log-normal, and Málaga distributions. More recently, a doubly inverted gamma-gamma (IGGG) turbulence model [7] was introduced in which small-scale eddies are modeled using the gamma distribution and large-scale eddies are modeled using inverse gamma-gamma distributions. The proposed IGGG model encompasses the Fisher-Snedecor $\mathcal{F}$ and GG models and provides the best fit to the experimental data compared to the $\mathcal{F}$, GG, and log-normal distributions.

The performance of FSO communication links can be significantly affected by various turbulence models and weather conditions, leading to path loss and misalignment due to pointing errors. Pointing error is particularly crucial in the analysis of FSO communication systems, as it directly impacts the reliability and efficiency of data transmission. Consequently, it is essential to accurately characterize FSO channels that take these pointing errors into account.

This chapter comprehensively considers all factors, including path loss, pointing error, and atmospheric turbulence, to provide a thorough analysis of FSO communication system performance under different environmental conditions. This detailed examination aims to enhance the understanding and mitigation of these effects, thereby improving the robustness and reliability of FSO communication links.

Building on the preceding discussion, this chapter examines the performance of an FSO communication system operating under different turbulence models with nonzero boresight-pointing errors. The primary contributions are as

V. A. Bohara, Swaminathan R, *Optical and Wireless Access Networks*, Textbooks in Telecommunication Engineering, https://doi.org/10.1007/978-3-031-89696-5_5

follows: (i) novel unified analytical expressions are derived for the envelope PDF and cumulative distribution function (CDF) of irradiance fluctuations considering pointing error impairments, and subsequently, these expressions have been used to obtain the PDF and CDF of the received instantaneous signal-to-noise ratio (SNR); (ii) unified analytical closed-form expressions are provided for outage probability, average bit error rate (BER), and ergodic capacity; (iii) simpler and more accurate asymptotic expressions for these performance metrics are also obtained, offering deeper insights into the effects of various FSO system and channel parameters.

5.2 System Model

An FSO backhauling system has been considered, where two base stations, denoted by optical source and optical receiver, will communicate over FSO links. It is assumed that the base stations have fixed nodes mounted on top of buildings, where line-of-sight (LoS) communication exists between the optical source and optical receiver, as shown in Fig. 5.1. The deviations of optical source and receiver due to winds and building sway are modeled using Gaussian random variables (RVs) [8]. The destination node is equipped with a single photodetector (PD) with a circular detection aperture of radius r_a as shown in Fig. 5.2.

Intensity modulation (IM) in conjunction with an M-ary phase-shift keying (MPSK) modulation technique at the transmitter has been employed while utilizing either the heterodyne detection (HD) or direct detection (DD) technique at the receiver.

For both IM/DD and HD detection scheme, the generalized expression for the received FSO signal is given by [9]:

$$Y_D = (\eta P_{tp} h_i)^{u/2} x_s + n_o \tag{5.1}$$

where $u = 1$ denotes the HD detection technique and $u = 2$ denotes the IM/DD detection technique. Here, P_{tp} represents the transmitted power, η is the efficiency of the photodetector at the receiver, x_s is the transmitted MPSK symbol, and n_o is the additive white Gaussian noise (AWGN) which is independent of the transmitted symbol.

The combined channel gain (h_i) of the FSO link is given by the product of three significant impairments, i.e.,

$$h_i = h_l h_{at,i} h_p, \tag{5.2}$$

Using the product of random variables technique, the PDF of $f_{h_i}(h_i)$ is written as

$$f_{h_i}(h_i) = \int f_{\frac{h_i}{h_{at,i}}}\left(\frac{h_i}{h_{at,i}}\right) f_{h_{at,i}} dh_{at,i} \tag{5.3}$$

The conditional PDF $f_{\frac{h_i}{h_{at,i}}}\left(\frac{h_i}{h_{at,i}}\right)$ in (5.3) can written using (5.2) as

$$f_{\frac{h_i}{h_{at,i}}}\left(\frac{h_i}{h_{at,i}}\right) = \frac{f_{h_p}\left(\frac{h_i}{h_{at,i}h_l}\right)}{\left(\frac{dh_i}{dh_p}\right)}, \tag{5.4}$$

where h_l represents atmospheric path loss, h_p denotes misalignment errors or pointing error, and $h_{at,i}$ indicates atmospheric turbulence, where $i = m$ for Málaga distribution, $i = f$ for $\mathcal{F}$ snedecor distribution, and $i = g$ for IGGG distribution.

5.3 Channel Model

In FSO communication, the channel model plays a critical role in determining system performance. The FSO channel is highly influenced by atmospheric turbulence, weather condi-

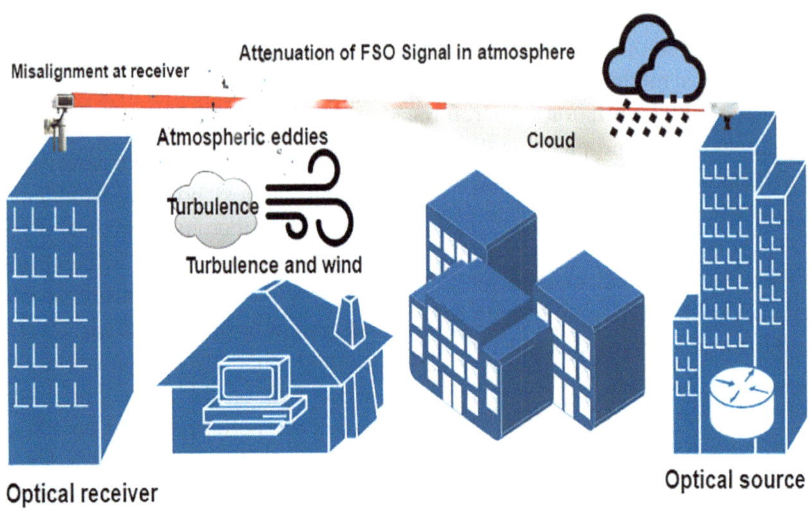

Fig. 5.1 System model of UAV-based FSO communication system

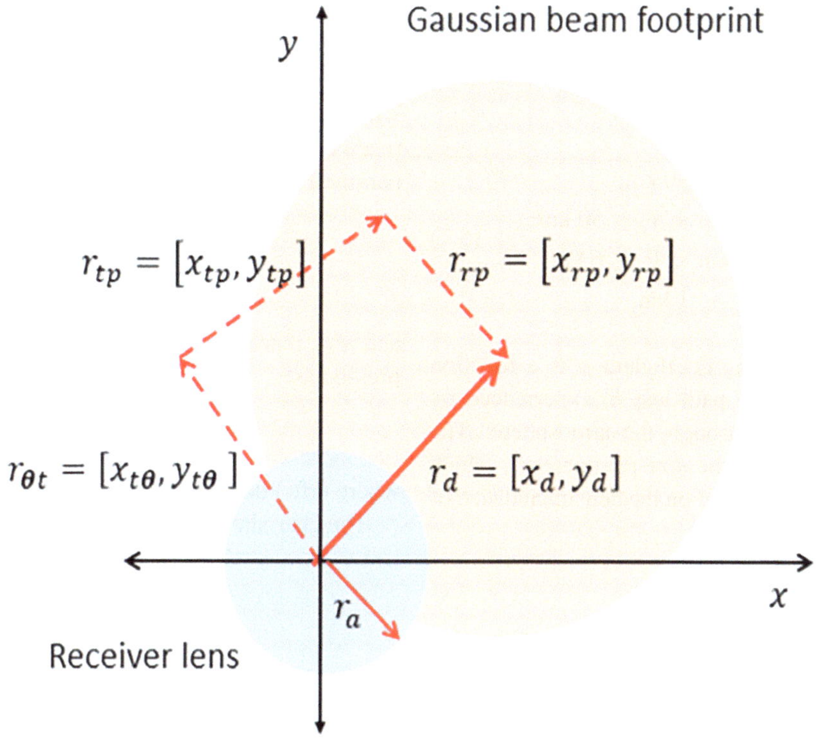

Fig. 5.2 Mismatch of the receiver lens and Gaussian beam footprint

tions, and misalignment or pointing errors, which can cause significant attenuation and variability in the received signal. Atmospheric turbulence, which is caused due to temperature and pressure variations in the atmosphere, leads to random fluctuations in the refractive index, causing scintillation and beam wander. Path loss due to weather conditions such as fog and cloud leads to signal attenuation at the receiver. Misalignment or pointing errors, arising due misalignment between the transmitter and receiver, further degrade the FSO signal quality. To accurately evaluate the performance of FSO systems, it is essential to model these effects using appropriate statistical distributions. Therefore, in this chapter, various statistical distributions are introduced to model the FSO channel impairments. These models help in obtaining key performance metrics like outage probability, average SER, and ergodic capacity, which facilitate the design and optimization of robust FSO communication systems.

5.3.1 Path Loss

In optical communication, signal degradation within the atmosphere is a significant concern, particularly due to the effects of geometric scattering. This phenomenon occurs when the optical signal encounters particles or droplets in the atmosphere, such as those found in fog or clouds, which are comparable in size to the signal's wavelength. This scattering disrupts the signal's path, leading to attenuation and dispersion. Fog composed of tiny water droplets suspended in the air scatters the light signal in various directions, causing substantial signal loss, especially in dense fog conditions. This scattering effect not only reduces the received signal power but also increases the likelihood of bit errors, thereby degrading the overall communication quality. Similarly, clouds consisting of water droplets or ice crystals scatter and absorb the optical signal. The impact of cloud-induced fading varies with the type and thickness of the cloud (i.e., thick or dense cloud) causing severe attenuation and significantly reducing the received signal strength. This scattering effect is particularly detrimental for long-distance FSO links, where the optical signal must traverse substantial atmospheric paths.

The combined effect of fog and cloud-induced fading can significantly impair the performance of FSO communication systems. According to Beer's-Lambert law [10], the atmospheric attenuation, or path loss, is modeled as

$$h_l = \exp(-\rho L) \tag{5.5}$$

where ρ is the scattering coefficient and L is the link distance from the source to the destination. The scattering coefficient, ρ, is given by

$$\rho = \frac{3.91}{V_r}\left(\frac{\lambda_{fso}}{550}\right)^{-r}. \tag{5.6}$$

Here, V_r denotes the visibility range in kilometers, λ_{fso} is the wavelength of the FSO signal, and r represents the size of the scattering particles. The value of q is defined according to Kim's model [10] as follows:

$$q = \begin{cases} 1.6 & \text{if } V_r > 50 \text{ km}, \\ 1.3 & \text{if } 6 \text{ km} < V_r \leq 50 \text{ km}, \\ 0.16R + 0.34 & \text{if } 1 \text{ km} < V_r \leq 6 \text{ km}, \\ V_r - 0.5 & \text{if } V_r \leq 1 \text{ km}. \end{cases} \quad (5.7)$$

In this model, the scattering coefficient ρ is a function of visibility, which affects the path loss h_l experienced by the signal as it propagates through the atmosphere. The visibility range V_r determines the size parameter q, which adjusts the scattering impact based on the density and size of atmospheric particles.

5.3.2 Pointing Errors

In the Cartesian coordinate system, we assume that the mean positions of the transmitter and receiver are located at [0, 0, 0] and [0, 0, Z], respectively. The position deviations in the transmitter are denoted by (x_{tp}, y_{tp}) and in the receiver are denoted by (x_{rp}, y_{rp}). According to the central limit theorem, the displacement vectors x_{tp}, y_{tp}, x_{rp}, and y_{rp} follow a zero-mean Gaussian distribution, specifically, $x_{tp} \sim \mathcal{N}(0, \sigma_{xtp}^2)$, $y_{tp} \sim \mathcal{N}(0, \sigma_{ytp}^2)$, $x_{rp} \sim \mathcal{N}(0, \sigma_{xrp}^2)$, and $y_{rp} \sim \mathcal{N}(0, \sigma_{yrp}^2)$. Here, σ_{xtp} and σ_{ytp} represent the standard deviations (SD) of the transmitter position in the x-z and y-z planes, respectively, while σ_{xrp} and σ_{yrp} denote the SD of the receiver position in the same planes. This displacement leads to a misalignment in FSO system, which is termed as pointing error misalignment.

Pointing errors in FSO communication links arise mainly from the displacement of the laser beam from the center of the receiving aperture, which can be categorized into two primary components: boresight error and jitter.

Boresight is the fixed displacement between the center of the laser beam and the center of the receiver aperture. This type of error is typically caused by factors such as the thermal expansion of the laser beam, which results in a consistent, non-varying offset. Jitter refers to the random fluctuations or offsets of the beam center at the detection plane. These random variations are often caused by dynamic environmental factors, including building sway and wind-induced vibrations. Unlike boresight error, jitter is unpredictable and varies over time.

These two factors result in a reduction in power at the receiver's aperture and can be computed by considering the Gaussian beam's footprint at the receiver, as depicted in Fig. 5.2. This power reduction is given by [11]

$$h_p = A_o \exp\left(-\frac{2r_d^2}{\omega_z^2}\right), \quad (5.8)$$

where A_o represents the amount of power gathered at the receiver in the absence of pointing error (i.e., $r_d = 0$). The parameter A_o is defined as $A_o = \text{erf}(t)^2$, where t is given by

$$t = \sqrt{\frac{\pi}{2}}\frac{r_a}{\omega_z}, \quad (5.9)$$

and ω_{zeq} represents the equivalent beamwidth, which is expressed as

$$\omega_{zeq} = \frac{\omega_z^2\sqrt{2}\text{erf}(t)}{2t\exp(-t^2)}, \quad (5.10)$$

where erf(·) denotes the error function.

The Gaussian beam waist, ω_z, can be approximated by

$$\omega_z \approx \omega_o\sqrt{1 + \Phi\left(\frac{\lambda_{fso}Z}{\pi\omega_o^2}\right)^2}, \quad (5.11)$$

where ω_o is the beamwidth at $Z = 0$, λ_{fso} is the optical wavelength, and Φ is defined as

$$\Phi = 1 + \frac{2\omega_o^2}{\rho^2(Z)}, \quad (5.12)$$

with the coherence length $\rho(Z)$ given by

$$\rho(Z) = (0.55C_n^2k^2fZ)^{-3/5}. \quad (5.13)$$

Here, C_n^2 is the refractive index structure parameter, k is the optical wave number, and f is the focal length.

In this context, r_d denotes the radial displacement vector from the center of the beam, r_a signifies the radius of the detecting aperture, and ω_z is the Gaussian beam waist. As illustrated in Fig. 5.2, the radial displacement vector $r_d = [x_d, y_d]$ results from the combination of three other vectors, i.e., the displacement vector due to transmitter orientation deviation $r_{t\theta} = [x_{t\theta}, y_{t\theta}]$, the displacement vector resulting from the transmitter position deviation $r_{tp} = [x_{tp}, y_{tp}]$, and the displacement vector arising from the receiver position deviation $r_{rp} = [x_{rp}, y_{rp}]$. The combined displacement along the x and y coordinates is given by

$$\begin{aligned} x_d &= x_{tp} + x_{rp} + x_{t\theta}, \\ y_d &= y_{tp} + y_{rp} + y_{t\theta}, \end{aligned} \quad (5.14)$$

with the beam position deviation along the x and y coordinates defined as $x_{t\theta} = Z\tan(\theta_{xt})$ and $y_{t\theta} = Z\tan(\theta_{yt})$. Given that θ_{xt} and θ_{yt} are small, using small angle approximation, we can write (5.14) as

$$\begin{aligned} x_d &= x_{tp} + x_{rp} + Z\theta_{xt}, \\ y_d &= y_{tp} + y_{rp} + Z\theta_{yt}. \end{aligned} \tag{5.15}$$

The orientation deviation or bias angle at the transmitter and receiver is modeled by Gaussian RVs, θ_{xt}, θ_{yt}, θ_{xr}, and θ_{yr}, with distributions $\theta_{xt} \sim \mathcal{N}(\theta'_{xt}, \sigma^2_{xto})$, $\theta_{yt} \sim \mathcal{N}(\theta'_{yt}, \sigma^2_{yto})$, $\theta_{xr} \sim \mathcal{N}(\theta'_{xr}, \sigma^2_{xro})$, and $\theta_{yr} \sim \mathcal{N}(\theta'_{yr}, \sigma^2_{yro})$. Here, σ_{xto} and σ_{yto} represent the orientation standard deviations at the optical source in the x-z and y-z planes, respectively, while σ_{xro} and σ_{yro} represent those at the optical receiver. The stability of FSO communication is highly susceptible to the hovering of the optical source and optical receiver caused by building sway and fluctuations due to various factors. Building sway is primarily influenced by high winds, which exert lateral forces that cause buildings to move. Seismic activity, such as earthquakes, induces oscillations that can misalign FSO systems. Temperature changes lead to thermal expansion and contraction of building materials, resulting in slight shifts. Additionally, the inherent structural flexibility of tall buildings allows for movement, preventing damage during extreme conditions, while vibrations from mechanical systems or nearby traffic further contribute to building movement. These factors can misalign FSO equipment, necessitating compensation techniques to maintain proper alignment. The misalignment caused by the hovering of the optical source and optical receiver has been accounted for in the cases outlined in Table 5.1.

The orientation deviations of the transmitter are denoted by θ_{xt} and θ_{yt}. Similarly, the orientation deviations of the receiver are represented by θ_{xr} and θ_{yr}. In case of zero boresight, $\theta'_{xt} = \theta'_{yt} = \theta'_{xr} = \theta'_{yr} = 0$. However, in case of nonzero boresight, $\theta'_{xt} \neq \theta'_{yt} \neq \theta'_{xr} \neq \theta'_{yr} \neq 0$.

For **Case 1**, x_d and y_d are modeled as zero-mean Gaussian RVs with variances $\sigma^2_{xtp} + \sigma^2_{xrp}$ and $\sigma^2_{ytp} + \sigma^2_{yrp}$, respectively, i.e., $x_d \sim \mathcal{N}(0, \sigma^2_{xtp} + \sigma^2_{xrp})$ and $y_d \sim \mathcal{N}(0, \sigma^2_{ytp} + \sigma^2_{yrp})$. For **Case 2** and **Case 3**, x_d and y_d are Gaussian RVs with means $Z\theta'_{xt}$ and $Z\theta'_{yt}$ and variances σ^2_{xd} and σ^2_{yd}, respectively, where $\sigma^2_{xd} = Z^2\sigma^2_{xto} + \sigma^2_{xtp} + \sigma^2_{xrp}$ and $\sigma^2_{yd} = Z^2\sigma^2_{yto} + \sigma^2_{ytp} + \sigma^2_{yrp}$.

Assuming x_d and y_d are independent RVs, the total radial displacement r_d is given by $r_d = \sqrt{x_d^2 + y_d^2}$. The displacement r_d is modeled using the Beckmann distribution, which is approximated by a modified Rayleigh distribution to obtain a closed-form expression. The PDF of r_d is given by

Table 5.1 Different cases arising from hovering of optical source and optical receiver

Cases	Conditions
Case 1	Optical source and optical receiver both are fixed
Case 2	Either optical source or optical receiver is hovering
Case 3	Both optical source and optical receiver are hovering

$$f_{r_d}(r_d) = \frac{r_d^2}{\sigma_t} \exp\left(-\frac{r_d^2}{2\sigma_t^2}\right), \quad r_d \geq 0, \tag{5.16}$$

where the total displacement σ_t is given by

$$\sigma_t^2 = \begin{cases} \left(\frac{3Z^2\theta^2_{xt}\sigma^4_{xd} + 3Z^2\theta^2_{yt}\sigma^4_{yd} + \sigma^6_{xd} + \sigma^6_{yd}}{2}\right)^{\frac{1}{3}}, & \textbf{Case2 \& Case3}, \\ \sigma^2_{xtp} + \sigma^2_{xrp} + \sigma^2_{ytp} + \sigma^2_{yrp}, & \textbf{Case1}. \end{cases} \tag{5.17}$$

Using (5.8) and (5.16), we obtain the PDF of h_p as

$$f_{h_p}(h_p) = \frac{\zeta^2}{A_o^{\zeta^2}} h_p^{\zeta^2 - 1}, \quad 0 \leq h_p \leq A_o, \tag{5.18}$$

where $\zeta = \frac{\omega_z}{2\sigma_t}$ denotes the severity parameter of the pointing error. Now, by substituting (5.18) in (5.4), the conditional PDF can be rewritten as

$$f_{\frac{h_i}{h_{at,i}}}\left(\frac{h_i}{h_{at,i}}\right) = \frac{\zeta^2}{A_o^{\zeta^2} h_{at,i} h_l}\left(\frac{h_i}{h_{at,i} h_l}\right)^{\zeta^2 - 1}, \quad 0 \leq h_p \leq A_o \tag{5.19}$$

By substituting (5.19) in (5.3), the combined FSO channel PDF can be written as

$$f_{h_i}(h_i) = \int_{\frac{h_i}{h_l A_o}}^{\infty} f_{\frac{h_i}{h_{at,i}} h_l}\left(\frac{h_i}{h_{at,i} h_l}\right) f_{h_{at,i}}(h_{at,i}) dh_{at,i} \tag{5.20}$$

5.3.3 Atmospheric Turbulence

Atmospheric turbulence in FSO communication links is caused by the uneven distribution of temperature, variations in wind velocity, and changes in atmospheric pressure due to irregular wind flow. These conditions lead to random fluctuations in both the amplitude and phase of the received FSO signal, resulting in the formation of turbulent cells, also known as eddies. The effect of these turbulent eddies on the FSO signal depends on their size relative to the laser beam. When the turbulent eddy size is larger than the beam size, beam wander becomes the dominant effect, causing a random beam deflection and variations in its pointing direction. Conversely, when the eddy size is smaller than the beam size, beam spreading occurs, which distorts the wavefront and broadens the beam, potentially degrading signal quality.

Here, the three most commonly used turbulence models in the literature, i.e., the Málaga distribution, Fisher-Snedecor $\mathcal{F}$ distribution, and the recently proposed IGGG distributions, are discussed in detail.

5.3.3.1 Málaga Distribution

The Málaga distribution is a contemporary model that has demonstrated excellent agreement with experimental data and is applicable across weak to strong turbulence regimes. To define the Málaga distribution, we consider an electromagnetic wave propagating through a turbulent atmosphere characterized by a random refractive index. As the wave moves through this medium, a portion of the energy is scattered, and the form of the irradiance probability distribution is determined by the type of scattering involved. The foundation of the Málaga distribution lies in conceptualizing an alternative physical model for generating small-scale fluctuations. The observed field at the receiver comprises three components, i.e., the LOS component U^L, the quasi-forward scattered component by the eddies along the propagation axis U_S^C, which is coupled to the LOS contribution, and the third component U_S^G, which is due to energy scattered to the receiver by off-axis eddies and is statistically independent of the first two terms.

Accordingly, we can express the total observed field as follows:

$$U_f = \left(U^L + U_S^C + U_S^G\right)\exp(\chi + jS) \tag{5.21}$$

where $U^L=\sqrt{G_a}\sqrt{\Omega_o}\exp(j\theta_A)$, $U_S^C=\sqrt{G_a}\sqrt{\rho^2 b_0}\exp(j\theta_B)$, and $U_{G_a}^S = \sqrt{(1-\rho)}U'^S$, with U_S^C and U_S^G being statistically independent stationary random processes. Additionally, U^L and U_S^G are also independent random processes. In this expression, G_a is a real variable following a gamma distribution with $E[G_a] = 1$, representing the slow fluctuation of the LOS component. According to [12], the parameter $\Omega_o = E[|U^L|^2]$ represents the average power of the LOS component, whereas the average power of the total scattered components is denoted by $2b_0 = E[|U_S^C|^2+|U_S^G|^2]$. The terms θ_A and θ_B are the deterministic phases of the LOS and the coupled-to-LOS scatter terms, respectively. The factor $0 \le \rho \le 1$ expresses the amount of scattering power coupled to the LOS component. Finally, U'^S is a circular complex Gaussian RVs, and χ and S are real random variables representing the log amplitude and phase perturbation of the field induced by atmospheric turbulence, respectively.

From (5.21), the observed irradiance of the proposed propagation model can be expressed as

$$h_{at,m} = \left|U^L + U_S^C + U_S^G\right|^2\exp(2\chi) = X_1X_2, \tag{5.22}$$

where

$$X_1 \triangleq \left|U^L + U_S^C + U_S^G\right|^2 \quad \text{(small-scale fluctuations)} \tag{5.23}$$

and

$$X_2 \triangleq \exp(2\chi) \quad \text{(large-scale fluctuations)}. \tag{5.24}$$

Small-scale fluctuations refer to the scintillation contributions from turbulent cells smaller than either the first Fresnel zone or the transverse spatial coherence radius, whichever is smaller. Conversely, large-scale fluctuations in irradiance are caused by turbulent cells larger than either the Fresnel zone or the "scattering disk," whichever is larger.

From (5.21), we can express the low-pass equivalent complex envelope as

$$\begin{aligned} R(t) &= U^L + U_S^C + U_S^G \\ &= \sqrt{G_a}\sqrt{\Omega_o}\exp(j\theta_A)+\sqrt{\rho}\sqrt{2b_0}\exp(j\theta_B)+\sqrt{1-\rho}U'^S, \end{aligned} \tag{5.25}$$

so that it will result in the shadowed Rice model, which is composed of the sum of a Rayleigh random phasor (the independent scatter component U'^S) and a Nakagami distribution $\sqrt{G_a}$. After that the procedure described in [12] is applied, which involves calculating the expectation of the Rayleigh component with respect to the Nakagami distribution. Hence, the PDF of X_1 is given by

$$f_{X_1}(y)=\frac{1}{g}\left(\frac{g\beta}{g\beta+\Omega'_o}\right)^{\beta}\exp\left(-\frac{y}{g}\right){}_1F_1\left(\beta;1;\frac{\Omega'_o}{g(g\beta+\Omega'_o)}y\right), \tag{5.26}$$

where $\beta \triangleq \frac{(E[G_a])^2}{\text{Var}[G_a]}$ is the amount of fading parameter, with Var[·] indicating the variance operator, and $|\cdot|$ denoting the absolute value. It is to be noted that $g = 2b_0(1-\rho)$ and $\Omega'_o = \Omega_o + \rho^2 b_0 + 2\sqrt{2b_0\Omega_o\rho}\cos(\theta_A - \theta_B)$. Finally, ${}_1F_1(a; c; x)$ is the Kummer confluent hypergeometric function of the first kind.

Large-scale fluctuations, denoted by $X_2 \triangleq \exp(2\chi)$, are often approximated by a gamma distribution for analytical convenience. Thus, the PDF of X_2 is given by

$$f_{X_2}(x) = \frac{\alpha^\alpha}{\Gamma(\alpha)}x^{\alpha-1}\exp(-\alpha x), \tag{5.27}$$

where α is a positive parameter related to the effective number of large-scale cells in the scattering process.

If X_2 and X_1 are random variables distributed according to (5.26) and (5.27), respectively, then $h_{at,m}$ follows Málaga distribution with α and β denoting the large- and small-scale scattering parameters, respectively [13]. The PDF of $h_{at,m}$ is given by

$$f_{h_{at,m}}(h_{at,m}) = A\sum_{p=1}^{\beta} a_p h_{at,m}^{\frac{\alpha+p}{2}-1} K_{\alpha-p}\sqrt{2\left(\frac{\alpha\beta h_{at,m}}{g\beta+\Omega'_o}\right)}, \tag{5.28}$$

where

$$A \triangleq \frac{2\alpha^{\frac{\alpha}{2}} g^{1+\frac{\alpha}{2}} \Gamma(\alpha)}{(g\beta)^{\frac{\beta+\alpha}{2}} (g\beta + \Omega'_o)^{\frac{\beta+\alpha}{2}}},$$
$$a_p \triangleq \binom{\beta-1}{p-1} (g\beta + \Omega'_o)^{1-\frac{p}{2}} \frac{(\Omega'_o/g)^{p-1}}{(p-1)!} \left(\frac{\alpha}{\beta}\right)^{\frac{p}{2}}. \tag{5.29}$$

In (5.28), $K_\nu(\cdot)$ is the modified Bessel function of the second kind and order ν.

The parameters α and β are essential in describing the atmospheric refractive-index structure parameter, the propagation path length, and the inner and outer scales of turbulence. These parameters can be defined as

$$\alpha = \frac{1}{\exp(\sigma^2_{\ln L}) - 1} \tag{5.30}$$

and

$$\beta = \frac{1}{\exp(\sigma^2_{\ln S}) - 1} \tag{5.31}$$

where $\sigma^2_{\ln S}$ and $\sigma^2_{\ln L}$ represent the log-irradiance variances for small-scale and large-scale turbulence, respectively. Considering the effects of the turbulent inner and outer scales, the large-scale log-irradiance variance of spherical waves can be expressed as the difference shown below:

$$\delta^2_{\ln x} = \delta^2_{\ln x}(l_0) - \delta^2_{\ln x}(L_0), \tag{5.32}$$

where the first term, $\delta^2_{\ln x}(l_0)$, and the latter term, $\delta^2_{\ln x}(L_0)$, are the large-scale log-irradiance variances due to the inner and outer scale turbulence effects, respectively. These are given by

$$\delta^2_{\ln x}(l_0) = 0.04\beta_0^2 \left(\frac{\eta_{X^d} Q_l}{\eta_{X^d} + Q_l}\right)^{7/6} \left[1 + 1.75\left(\frac{\eta_{X^d}}{\eta_{X^d} + Q_l}\right)^{1/2} - 0.25\left(\frac{\eta_{X^d}}{\eta_{X^d} + Q_l}\right)^{7/12}\right],$$

$$\delta^2_{\ln x}(L_0) = 0.04\beta_0^2 \left(\frac{\eta_{X^{d_0}}}{\eta_{X^{d_0}} + Q_l}\right)^{7/6} \left[1 + 1.75\left(\frac{\eta_{X^{d_0}}}{\eta_{X^{d_0}} + Q_l}\right)^{1/2} - 0.25\left(\frac{\eta_{X^{d_0}}}{\eta_{X^{d_0}} + Q_l}\right)^{7/12}\right],$$

$$\eta_{X^d} = \frac{8.56}{\left(1 + 0.18d^2 + 0.20\beta_0^2 Q_l^{1/6}\right)}, \quad \eta_{X^{d_0}} = \frac{Q_0 \eta_{X^d}}{Q_0 + \eta_{X^d}}, \quad Q_l = \frac{10.89L}{kl_0^2}, \quad Q_0 = \frac{64\pi^2 L}{kL_0^2},$$

where β_0^2 is the Rytov variance for a plane wave, representing the optical turbulence strength. Here, $k = \frac{2\pi}{\lambda}$ is the optical wave number, λ is the wavelength, L is the source-destination link distance, and C_n^2 is the refractive index structure parameter, which significantly determines the turbulence strength. The parameter C_n^2 depends on altitude and local conditions such as terrain type, geographic location, cloud cover, and daytime. Typically, C_n^2 ranges from 10^{-13} to 10^{-17} $\mathrm{m}^{-2/3}$ and $d = \sqrt{\frac{kD^2}{4L}}$ is the equivalent aperture diameter, and D is the receiving aperture diameter.

The small-scale log-irradiance variance of spherical waves can be written as

$$\delta^2_{\ln y} = \frac{0.51\delta^2_{SP}\left(1 + 0.69\delta_{SP}^{12/5}\right)^{-5/6}}{1 + 0.90d^2(\beta_0/\delta_{SP})^{12/5} + 0.62d^2\beta_0^{12/5}}, \tag{5.33}$$

where δ^2_{SP} is the spherical wave scintillation index in a weak turbulent atmosphere and is given by

$$\delta^2_{SP} = 9.65\beta_0^2 \left\{0.4\left(1 + \frac{9}{Q_l^2}\right)^{11/12} \left[\sin\left(\frac{11}{6}\tan^{-1}\left(\frac{Q_l}{3}\right)\right) + \frac{2.61}{(9 + Q_l^2)^{1/4}} \sin\left(\frac{4}{3}\tan^{-1}\left(\frac{Q_l}{3}\right)\right) - \frac{0.52}{(9 + Q_l^2)^{7/24}} \sin\left(\frac{5}{4}\tan^{-1}\left(\frac{Q_l}{3}\right)\right)\right] - \frac{3.5}{Q_l^{5/6}}\right\}.$$

According to [14], the normalized irradiance fluctuation σ_I^2, is written as

$$\sigma_I^2 = e^{\sigma^2_{\ln S} + \sigma^2_{\ln L}} - 1. \tag{5.34}$$

Alternatively, assuming a negligible inner scale, α and β can be directly associated with physical parameters using the following expression:

$$\beta_0^2 = 1.23k^{7/6}C_n^2 L^{11/6}. \tag{5.35}$$

It is important to note that α and β cannot be arbitrarily selected in FSO communications because these parameters are related to the Rytov variance. Additionally, the value of the Rytov variance can characterize different turbulence levels: weak turbulence for $\beta_0^2 \leq 0.3$, moderate turbulence for $0.3 < \beta_0^2 \leq 5$, and strong turbulence for $\beta_0^2 > 5$.

The expressions for α and β are given as [15]

$$\alpha = \left[\exp\left(-0.49\left(\beta_0^2\right)\left(1+0.56\left(\beta_0^2\right)^{12/5}\right)^{-7/6}\right)-1\right]^{-1}, \tag{5.36}$$

$$\beta = \left[\exp\left(-0.51\left(\beta_0^2\right)\left(1+0.69\left(\beta_0^2\right)^{12/5}\right)^{-5/6}\right)-1\right]^{-1}. \tag{5.37}$$

In the context of weak turbulence scintillation, it is commonly accepted that small-scale effects predominantly contribute to the observed fluctuations. However, there is no direct one-to-one correspondence between statistical models and the scale of these effects. Specifically, large-scale amplitude fluctuations are associated with large-scale refractive eddies, whereas small-scale amplitude fluctuations are linked to small-scale diffractive eddies.

Málaga distribution asymptotically reduces to a log-normal distribution when only considering the term X_1 and is related to small-scale fluctuations. Conversely, if we only consider large-scale effects by treating X_2 as a statistically distributed variable and X_1 as a constant, the Málaga distribution simplifies to a gamma distribution. This is due to the fact that a gamma distribution is a close approximation to a log-normal distribution.

Table 5.2 gives a comprehensive list of various existing turbulence distribution models and illustrate how they can be derived from the Málaga distribution model. Given that these models have been validated using different experimental and simulation data, the validity of the Málaga distribution is thereby thoroughly demonstrated.

Now, by using (5.20) and substituting $i = m$, the combined channel PDF of an FSO link $f_{h_m}(h_m)$ is obtained as the product of path loss h_l, pointing error h_P, and atmospheric turbulence considering Málaga distribution $h_{at,m}$. The expression for $f_{h_m}(h_m)$ is given by

$$f_{h_m}(h_m) = \int_{\frac{h_m}{h_l A_o}}^{\infty} \frac{1}{h_l h_{at,m}} f_{h_p}\left(\frac{h_m}{h_l h_{at,m}}\right) f_{h_{at,m}}(h_{at,m}) dh_{at,m} \tag{5.38}$$

Table 5.2 Parameter definitions for Gaussian distributed RVs

Parameters	Notations
Rice-Nakagami	$\rho = 0$, $\mathrm{Var}[\lvert U^L \rvert] = 0$
Log-normal	$\mathrm{Var}[\lvert U^L \rvert] = 0$, $g \to 0$
Gamma	$\rho = 0$, $g = 0$
Shadowed-Rician	$\mathrm{Var}[\lvert X_2 \rvert] = 0$
K distribution	$\Omega_o = 0$ and $\rho = 0$ or $\beta = 1$, $\Omega_o = 0$
HK (homodyned K) distribution	$\rho = 0$, $X_2 = g$, $\Omega_o = 0$
Exponential	$\rho = 0$, $\alpha \to \infty$
GG distribution	$\rho = 1$, then $g = 0$, $\Omega_o = 1$
Gamma-Rician	$\beta \to \infty$

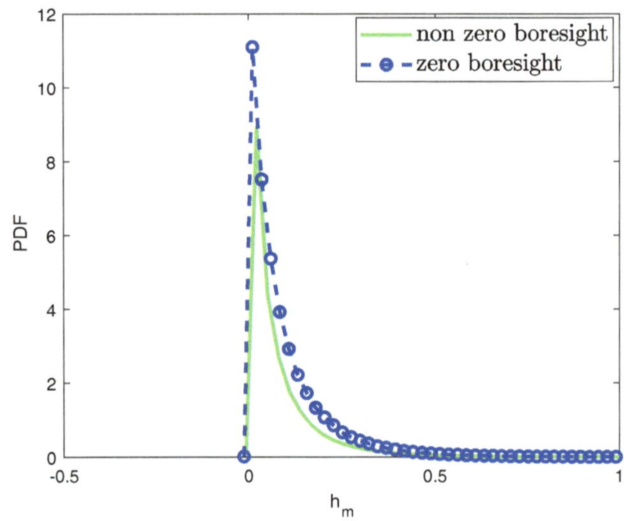

Fig. 5.3 PDF of zero and nonzero boresight-pointing errors

Now, by solving (5.38) using [16, eq.(07.34.21.0085.01)], the final channel PDF is obtained as

$$f_{h_m}(h_m) = \frac{\zeta^2 A}{2h_m} \sum_{p=1}^{\beta} a_p \left(\frac{\alpha\beta}{g\beta + \Omega_o'}\right)^{\frac{-(\alpha+p)}{2}} \times G_{1,3}^{3,0}\left(\frac{\alpha\beta h_m}{g\beta + \Omega_o'(h_l A_o)} \middle| \begin{matrix} 1+\zeta^2 \\ \zeta^2, \alpha, p \end{matrix}\right) \tag{5.39}$$

where $G_{\cdots}^{\cdots}(\cdot)$ is the Meijer G-function [16].

In Fig. 5.3, the channel distribution of the FSO system has been compared based on the PDF of the combined channel given by (5.39) for both zero and nonzero boresight-pointing errors while maintaining the same values for the pointing error standard deviations, specifically $\sigma_{txo} = \sigma_{tyo} = \sigma_{rxo} = \sigma_{ryo} = 4$ mrad. This figure clearly demonstrates the significant impact of boresight-pointing errors on FSO communications.

The comparison reveals that even when the pointing error standard deviations are identical, the presence of nonzero boresight-pointing errors cannot be ignored. The deviations caused by these pointing errors lead to noticeable changes

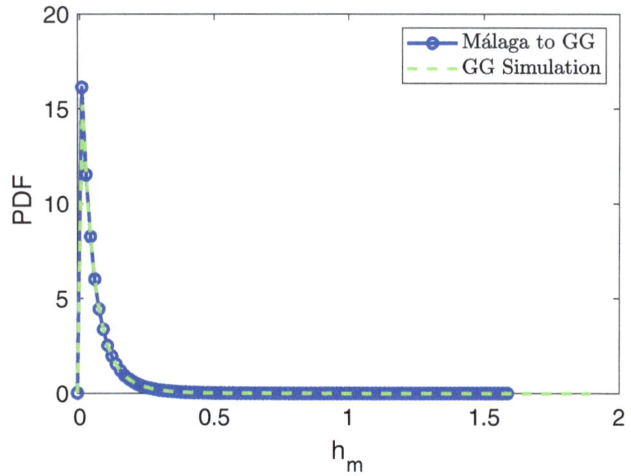

Fig. 5.4 Málaga distribution reduces to gamma-gamma distribution for specific parameter values

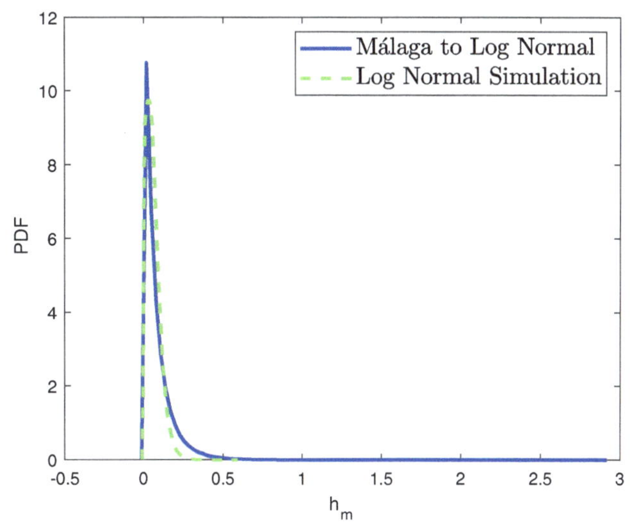

Fig. 5.5 Málaga distribution reduces to log-normal distribution for specific parameter values

in the channel distribution, which can adversely affect the performance of FSO communication systems. Therefore, it is crucial to consider the effects of boresight-pointing errors in the design and analysis of such communication systems to ensure reliable performance.

In Fig. 5.4, the combined channel PDF of the GG distribution is presented. Upon analyzing the expression for the PDF of Málaga distribution (5.39), it is observed that under specific parameter conditions, i.e., when the parameter $\rho \rightarrow 1$, $\Omega_o = 1$, and $b_0 \rightarrow 0$, the Málaga distribution approximates the GG distribution. This implies that under these parameter values, the Málaga distribution essentially transforms into the GG distribution.

The PDF plot in Fig. 5.5 visually confirms the approximation of the Málaga distribution to the log-normal distribution

Table 5.3 Parameters involved in the PDF of Malaga distribution

Symbol	Description
$\Omega_o^{'}$	$\Omega_o^{'} = \Omega_o + 2b_0\rho + 2\sqrt{2b_0\rho\Omega_o}cos(\theta_a - \theta_b)$
Ω_o	Mean power of the LOS signal
θ_a	Deterministic LOS phase
θ_b	Deterministic phase of the scatter component coupled to the LOS
g	$g = 2b_0(1 - \rho)$
$2b_0$	Mean power of all the scatter components
$0 < \rho < 1$	Amount of scattering power coupled to LOS component

for specific parameter values: $\rho \rightarrow 0$, $\Omega_o = 1$, and $b_0 \rightarrow 0$. Under these conditions, the Málaga distribution effectively transforms into the log-normal distribution. This transformation underscores the flexibility and adaptability of the Málaga distribution, making it a powerful tool for statistical modeling. It is capable of encompassing a wide range of distributions, including GG and log-normal distributions, under specific parameter conditions.

The instantaneous SNR of the FSO link under HD and IM/DD schemes is given by

$$\gamma_m = \frac{\overline{\gamma_m}|h_m|^u}{[k(g + \Omega_o^{'})A_o h_l]^u} \tag{5.40}$$

where $\overline{\gamma}_m$ denotes the average SNR of the FSO link and u indicates the unifying parameter ($u = 1$ implies HD technique, and $u = 2$ indicates IM/DD technique). Using the transformation of random variables technique, the unified expression for the PDF of the instantaneous SNR of the FSO link γ_m in terms of the Meijer G-function is given by

$$f_{\gamma_m}(\gamma_m) \approx \frac{\zeta^2 A}{2^u \gamma_m} \sum_{p=1}^{\beta} b_p \times G_{1,3}^{3,0}\left(B\left(\frac{\gamma_m}{\overline{\gamma}_m}\right)^{\frac{1}{u}} \middle| \begin{matrix} \zeta^2+1 \\ \zeta^2, \alpha, p \end{matrix}\right) \tag{5.41}$$

where $b_p = a_p\left(\frac{\alpha\beta}{g\beta+\Omega_o^{'}}\right)^{-\frac{\alpha+p}{2}}$ and $B = \frac{k(g+\Omega_o^{'})\alpha\beta}{g\beta+\Omega_o^{'}}$. The rest of the parameters are given in Table 5.3.

From the PDF, the unified CDF expression for the instantaneous SNR of the FSO link can be written as

$$F_{\gamma_m}(\gamma_m) \simeq D\sum_{p=1}^{\beta} c_p \times G_{u+1,3u+1}^{3u,1}\left(A_1 \middle| \begin{matrix} 1, K_1 \\ K_2, 0 \end{matrix}\right) \tag{5.42}$$

where $D = \frac{\zeta^2 A_M}{2^{2u}\pi^{u-1}}$, $c_p = b_p u^{\alpha+m-1}$, $A_1 = \frac{B^u \gamma_m}{u^{2u}\overline{\gamma}_m}$, $K_1 = \left[\frac{\zeta^2+1}{u}, \frac{\zeta^2+2}{u}\right]$, and $K_2 = \left[\frac{\zeta^2}{u}, \frac{\zeta^2+1}{u}, \frac{\alpha}{u}, \frac{\alpha+1}{u}, \frac{p}{u}, \frac{p+1}{u}\right]$.

5.3.3.2 Fisher-Snedecor $\mathcal{F}$ Distribution

In the literature, numerous irradiance PDF models have been proposed. For instance, under conditions of weak irradiance fluctuations where single scattering of the optical wave occurs, the log-normal distribution has been found to fit experimental data well. However, its accuracy diminishes as turbulence intensity increases [17]. Additionally, the log-normal distribution lacks a closed-form expression for important statistical metrics such as the moment-generating function, making the analytical performance analysis of digital communication over turbulent channel challenging.

To address this issue, the inverse Gaussian distribution was proposed in [18] as a mathematically tractable alternative that aligns well with experimental data. Another significant model is the log-normal-Rice (or Beckmann) distribution, derived from the product of a Rician and a log-normal RVs [19]. Despite its generally good fit, it suffers from the same mathematical intractability due to log-normal distribution. The GG PDF, introduced in [4], provides a mathematically feasible alternative to the log-normal-Rice distribution, fitting experimental data across a broad range of turbulence conditions. The Málaga distribution, proposed in [20], is derived from a combination of a shadowed-Rician and a gamma distribution. This universal model offers a good fit for both weak and strong irradiance fluctuations and encompasses several of the previously mentioned models as special or limiting cases. More recently, the inverse Gaussian gamma distribution model was introduced in [21], which fits experimental data well. However, its mathematical complexity presents significant challenges for modeling. This complexity makes it difficult to derive essential performance metrics of FSO systems, such as BER for modulation schemes and the outage probability in closed form.

$\mathcal{F}$ distribution offers an excellent fit to experimental data across all turbulence conditions while being mathematically simpler than the GG, inverse Gaussian gamma, and Màlaga distributions. Utilizing the doubly stochastic theory of scintillation, $\mathcal{F}$ distribution model postulates that small-scale irradiance variations of the propagating wave, represented by a gamma distribution, are modulated by large-scale irradiance fluctuations, described by an inverse gamma distribution. The resulting irradiance PDF corresponds to the $\mathcal{F}$ distribution for modeling wireless propagation in composite fading-shadowed channels [22].

Consider an electromagnetic wave propagating through a turbulent atmosphere with a random refractive index. By applying the modified Rytov theory [23], the optical field can be described mathematically as

$$U = U_o \exp(\Phi^S + \Phi^L), \tag{5.43}$$

where Φ^S and Φ^L are complex-valued, statistically independent components corresponding to the small-scale and large-scale atmospheric propagation effects, respectively. Using (5.43), the irradiance can be expressed as

$$h_{at,f} = |U|^2 = I^S I^L, \tag{5.44}$$

where I^S and I^L represent the small-scale and large-scale atmospheric effects, respectively. In this work, I^S is assumed to follow a gamma distribution with the PDF given by

$$f_{I^S}(x_f) = \frac{\alpha^\alpha x_f^{\alpha-1}}{\Gamma(\alpha)} \exp(-\alpha x_f), \tag{5.45}$$

where $\alpha > 0$ and is the distribution parameter.

Several experimental studies, such as [23], suggest that I^L follows a log-normal distribution. However, more recent works, e.g., [21,24], have approximated the distribution of I^L with an inverse gamma distribution for mathematical simplicity. As noted in [24], the inverse gamma distribution serves as an accurate approximation to the log-normal distribution because it fits empirical data well. Moreover, the inverse gamma distribution, related to the gamma distribution, exhibits semi-heavy-tailed characteristics that make it suitable for modeling large-scale fading [25]. The PDF of I^L is given by

$$f_{I^L}(y_f) = \frac{(\lambda-1)^\lambda y_f^{-\lambda-1}}{\Gamma(\lambda)} \exp\left(-\frac{\lambda-1}{y_f}\right), \tag{5.46}$$

where $\lambda = \alpha + 2$ and $\lambda > 1$ is the distribution parameter.

Given that I^S and I^L are independent, the PDF of $h_{at,f}$ can be expressed as

$$f_{h_{at,f}}(h_{at,f}) = \int_0^\infty x_f^{-1} f_{I^S}(x_f) f_{I^L}\left(\frac{h_{at,f}}{x_f}\right) dx_f. \tag{5.47}$$

By substituting equations (5.45) and (5.46) into equation (5.47) and using [26, eq. (3.326/2)], the PDF of $h_{at,f}$ can be written in closed form as

$$f_{h_{at,f}}(h_{at,f}) = \frac{\alpha^\alpha (\lambda-1)^\lambda h_{at,f}^{\alpha-1}}{B(\alpha,\lambda)(\alpha h_{at,f} + \lambda - 1)^{\alpha+\lambda}}, \tag{5.48}$$

where $B(\alpha, \lambda)$ is the beta function.

The CDF of $h_{at,f}$ can be derived in closed form using [26, eq. (3.194.1)] as

$$F_{h_{at,f}}(I) = \frac{\alpha^\alpha h_{at,f}^\alpha}{(\lambda-1)^\alpha B(\alpha,\lambda)} \, {}_2F_1\left(\alpha+\lambda, \alpha; \alpha+1; -\frac{\lambda h_{at,f}}{\lambda-1}\right), \tag{5.49}$$

where ${}_2F_1$ is the hypergeometric function and $B(\alpha, \lambda)$ is the beta function.

5.3.3.3 Doubly Inverted Gamma-Gamma Turbulence Model

For weak turbulence, where the turbulence correlation distance exceeds the wave coherence radius, the log-normal distribution has been found to align well with experimental data, particularly in the context of aperture averaging. Under stronger turbulence conditions, mathematical complexities necessitated the development of the extended Rytov theory, which models turbulence as a doubly stochastic process by combining the effects of small- and large-scale eddies. However, the gamma distribution poorly approximates the log-normal distribution, whereas inverse Gaussian and inverted gamma distributions provide better approximations while maintaining mathematical tractability. Consequently, the $\mathcal{F}$ distribution, proposed in the previous section, shows significant improvements over the GG distribution but is less consistent with empirical data. The PDF of GG distribution is given by

$$f_{Y_1}(I) = \frac{2(\alpha\beta)^{\frac{\alpha+\beta}{2}}}{\Gamma(\alpha)\Gamma(\beta)} I^{\frac{\alpha+\beta}{2}-1} K_{\alpha-\beta}\left(2\sqrt{\alpha\beta I}\right), \quad (5.50)$$

where $K_\nu(\cdot)$ denotes the modified Bessel function of the second kind with order ν.

The aforementioned distributions are based on the extended Rytov theory, which is itself an approximation of the two-scale hybrid method that neglects intermediate scales in moderate-to-strong turbulence conditions. Despite these advancements, a unified PDF model applicable across a broad range of turbulence conditions remains an open challenge. Additionally, the impact of multiple scattering on the PDF is often overlooked in many applications.

This section proposes an enhancement of the inverted gamma distribution by considering a doubly stochastic process that combines inverted gamma-gamma distributions to represent large-scale induced turbulence, while small-scale fading is modeled using the gamma distribution. This results in a three-parameter model with closed-form expressions for its PDF and CDF. Using strong scintillation theory, the model parameters are related to the variances induced by large- and small-scale turbulence values. Here, a recently introduced turbulence model, termed the doubly inverted gamma-gamma (IGGG) turbulence channel model is discussed. To evaluate its effectiveness and relevance, a comparative analysis is conducted against several existing models, namely, the GG, $\mathcal{F}$, and log-normal distributions.

To develop a PDF model that aligns more closely with empirical measurements than the $\mathcal{F}$ distribution, while maintaining its mathematical simplicity, a doubly stochastic process is proposed. This process uses a combination of inverted gamma-gamma distributions to model the fading induced by large-scale turbulence and it is given as

$$Y = Y_2 Y_3 \quad (5.51)$$

where Y_2 is inverted gamma distribution with parameter λ and is given by

$$p_{Y_2}(I_2) = \frac{(\lambda-1)^\lambda}{\Gamma(\lambda)} I_2^{-\lambda-1} e^{-\left(\frac{\lambda-1}{I_2}\right)}, \quad \lambda > 2 \quad (5.52)$$

Let Y_3 follow gamma distribution with parameter α. Similarly, small-scale induced fading X continues to be modeled by a gamma distribution with parameter β. The large-scale log-irradiance variance, $\delta^2_{\ln L}$, is assumed to be related to parameters α and λ as

$$\alpha = \lambda - 2 = \frac{1}{e^{1/2\delta^2_{\ln L}} - 1}. \quad (5.53)$$

The scintillation index (SI) is given by

$$\begin{aligned}\sigma_I^2 &= \frac{\langle (Y_3Y_2X)^2\rangle}{\langle Y_3Y_2X\rangle^2} - 1 \\ &= \left(1+\frac{1}{\alpha}\right)\left(1+\frac{1}{\lambda-2}\right)\left(1+\frac{1}{\beta}\right) - 1 \\ &= e^{\sigma^2_{\ln S}+\sigma^2_{\ln L}} - 1. \end{aligned} \quad (5.54)$$

By using the product of inverted gamma and GG distribution, the atmospheric turbulence following IGGG distribution can be written as

$$h_{at,g} = Y_1 Y_2 \quad (5.55)$$

Now, using (5.50) and (5.52) in (5.55) and applying the product of two independent random variables technique, the PDF of IGGG distribution $f_{h_{at,g}}(h_{at,g})$ is expressed as

$$f_{h_{at,g}}(h_{at,g}) = \int_0^\infty I^{-1} f_{Y_2}\left(\frac{h_{at,g}}{I}\right) f_{GG}(I) dI. \quad (5.56)$$

$$\begin{aligned} f_{h_{at,g}}(h_{at,g}) &= \frac{2(\beta\alpha)^{(\beta+\alpha)/2}(\lambda-1)^\lambda}{\Gamma(\beta)\Gamma(\alpha)\Gamma(\lambda-1)h_{at,g}^{\lambda+1}} \\ &\times \int_0^\infty I^{(\beta+\alpha)/2+\lambda-1} e^{-\frac{I(\lambda-1)}{h_{at,g}}} K_{\alpha-\beta}\left(2\sqrt{\beta\alpha I}\right) dI. \end{aligned} \quad (5.57)$$

The above integral can be solved using the following identity [26, Eq. (6.643.3), page 709]:

$$\begin{aligned} &\int_0^\infty x^{(\mu-1/2}e^{-\alpha x} K_{2\nu}\left(2\beta\sqrt{x}\right) dx \\ &= \frac{\Gamma(\nu+\mu+1/2)\Gamma(\mu-\nu+1/2)}{2\alpha^\nu\beta} \\ &\times e^{\beta^2/2\alpha} W_{-\mu,\nu}\left(\frac{\beta^2}{\alpha}\right), \end{aligned} \quad (5.58)$$

where $\mu = \Re(\mu+\nu+1/2) > 0$, $\Re$ denotes the real part, and the relation of Whittaker function $W_{\mu,\nu}(\cdot)$ is expressed in terms of the confluent hypergeometric function of the first kind, ${}_1F_1(\cdot)$, or the Kummer-M function.

The Whittaker function $W_{\mu,\nu}(z)$ is defined as

$$W_{\mu,\nu}(z) = \frac{\Gamma(-2\nu)}{\Gamma(-\mu-\nu+\frac{1}{2})} z^{\mu+\frac{1}{2}} e^{-z/2} {}_1F_1\left(\mu-\nu+\frac{1}{2}; 2\mu+1; z\right) + \frac{\Gamma(2\mu)}{\Gamma(-\mu-\nu+\frac{1}{2})} z^{-\mu+\frac{1}{2}} e^{-z/2} {}_1F_1\left(-\mu-\nu+\frac{1}{2}; -2\mu+1; z\right). \tag{5.59}$$

By solving (5.57) using (5.59) and (5.58), we get the PDF of IGGG channel as

$$f_{h_{at,g}}(h_{at,g}) = \left(\frac{\alpha\beta}{\lambda-1}\right)^{\alpha} \frac{\Gamma(\alpha+\lambda)\Gamma(\beta+\lambda)}{\Gamma(\beta)\Gamma(\alpha)\Gamma(\lambda)} h_{at,g}^{\alpha-1} \times U\left(\alpha+\lambda; 1+\alpha-\beta; \frac{\beta\alpha}{\lambda-1}I\right), \quad h_{at,g}>0, \tag{5.60}$$

where $U(\cdot;\cdot;\cdot)$ is the Kummer confluent hypergeometric function of the second kind and $U(\cdot)$ is given as

$$U(a;c;z) = \frac{\Gamma(1-c)}{\Gamma(a+1-c)} {}_1F_1(a;c;z) + \frac{\Gamma(c-1)}{\Gamma(a)} z^{1-c} {}_1F_1(a+1-c; 2-c; z). \tag{5.61}$$

It is worth noting that the IGGG distribution can be viewed as a generalization of both GG and $\mathcal{F}$ distributions. Specifically, as $\gamma \to \infty$, IGGG simplifies to GG as shown in Eq. (5.50). As $\alpha \to \infty$, Y in Eq. (5.51) has the same distribution as Y_2, and IGGG reduces to $\mathcal{F}$ distribution given by (5.48). Since IGGG encompasses GG and $\mathcal{F}$ distributions as special cases, it is anticipated to offer advantages over both distributions. By using [26, eq. (8.4.46.1)] and [26, eq. (8.2.2.15)], (5.60) can be expressed as

$$f_{h_{at,g}}(h_{at,g}) = \frac{1}{\Gamma(\alpha)\Gamma(\beta)\Gamma(\lambda)} \times G_{1,2}^{2,1}\left(\frac{\alpha\beta h_{at,g}}{\lambda-1}\middle| \begin{matrix} 1-\lambda \\ \alpha, \beta \end{matrix}\right), \quad h_{at,g} > 0. \tag{5.62}$$

The combined channel PDF of an FSO link is obtained as the product of path loss h_l, atmospheric turbulence considering IGGG distribution $h_{at,g}$, and pointing error h_P. The expression for $f_{h_g}(h_g)$ is given by

$$f_{h_g}(h_g) = \int_{\frac{h_g}{h_l A_o}}^{\infty} \frac{1}{h_l h_{at,g}} f_{h_p}\left(\frac{h_g}{h_l h_{at,g}}\right) f_{h_{at,g}}(h_{at,g}) dh_{at,g}. \tag{5.63}$$

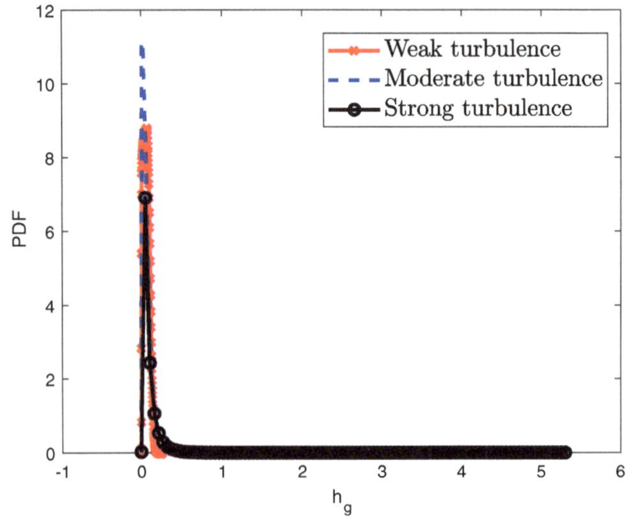

Fig. 5.6 PDF of IGGG distribution for different turbulence scenarios

By solving (5.63) using [16, eq.(07.34.21.0085.01)], we get the combined channel PDF as

$$f_{h_g}(h_g) \approx \frac{\zeta^2 A_{\gamma g}}{h_g} \times G_{3,1}^{2,3}\left(\frac{\alpha\beta h_g}{(\lambda-1)h_l A_o}\middle| \begin{matrix} 1-\lambda, 1+\zeta^2 \\ \zeta^2, \alpha, \beta \end{matrix}\right), \tag{5.64}$$

where $A_{\gamma g} = \frac{1}{\Gamma(\alpha)\Gamma(\beta)\Gamma(\lambda)}$.

In Fig. 5.6, the PDF of IGGG distribution based on the derived expression (5.64) has been presented under varying turbulence conditions. We consider $\omega_z = 3$ m with $C_n^2 = 1.7 \times 10^{-13}\,\text{m}^{-2/3}$ for strong turbulence, $C_n^2 = 5 \times 10^{-14}\,\text{m}^{-2/3}$ for moderate turbulence, and $C_n^2 = 4 \times 10^{-15}\,\text{m}^{-2/3}$ for weak turbulence. It is observed that the IGGG distribution models the PDF accurately under all three scenarios.

In Fig. 5.7, the PDF curves for IGGG, GG, and $\mathcal{F}$ turbulence channel models have been plotted. It is seen from the plots that the IGGG channel model can be approximated to GG and $\mathcal{F}$ models. According to doubly process, as $\lambda \to \infty$, IGGG reduces to $\mathcal{F}$ turbulence model, and when $\alpha \to \infty$, IGGG reduces to GG turbulence model.

For both IM/DD and HD systems, the instantaneous SNR γ_g of the FSO link is given by

$$\gamma_g = \frac{\overline{\gamma}_g (h_g)^u}{[h_{pl} K A_o]^u}, \tag{5.65}$$

where $K = \frac{\zeta^2}{1+\zeta^2}$ and $\overline{\gamma}_g$ represents the average SNR of the FSO link.

By utilizing equations (5.63) and (5.65) and employing the transformation of random variables, the unified PDF of the instantaneous SNR γ_g for the FSO link can be approximated as follows:

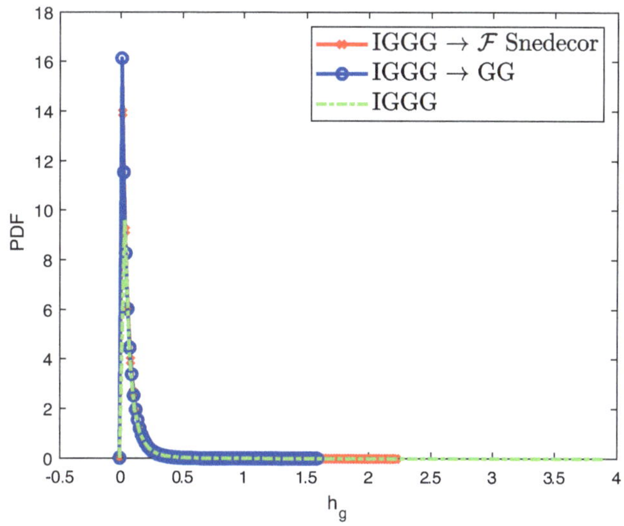

Fig. 5.7 IGGG reduces to GG and $\mathcal{F}$ Snedecor model

$$f_{\gamma_g}(\gamma_g) \approx \frac{\zeta^2 A_{\gamma_g}}{u\gamma} G_{2,3}^{3,1}\left(\frac{\alpha\beta K}{\lambda-1}\left(\frac{\gamma_g}{\overline{\gamma}_g}\right)^{\frac{1}{u}} \middle| \begin{matrix} 1-\lambda, \zeta^2+1 \\ \zeta^2, \alpha, \beta \end{matrix}\right). \tag{5.66}$$

5.4 Performance Analysis

This section provides a comprehensive analysis of the closed-form expressions for key performance metrics in FSO communication systems, including outage probability, average SER, and ergodic capacity. Additionally, asymptotic expressions for these metrics are derived to provide further insights.

The parameters listed in Table 5.4 are be used to obtain analytical and simulation results [8, 11]. Additionally, the orientation fluctuations caused by mounting FSO systems on tall buildings, which can sway due to earth vibrations or be affected by wind, have been accounted for. Consequently, a wide range of beamwidths ($w_z \in [1, 12\,\text{m}]$) is considered for analysis. The ground-based FSO systems typically assume a beamwidth of a few meters [8]. For consistency, a beamwidth of 3 m is used, unless stated otherwise.

FSO communication generally employs two types of receiver detection techniques, i.e., noncoherent (direct detection) and coherent (heterodyne) detection. Coherent detection offers superior sensitivity, data rates, and transmission distances compared to noncoherent detection. However, due to increased receiver complexity and implementation constraints, coherent detection is not widely adopted in FSO communication systems. Thus, the direct detection technique is primarily used to obtain most of the performance curves, with some plots comparing the performance of both direct and heterodyne detection techniques.

Table 5.4 Simulation parameters

Definition	Parameters
Turbulence parameters	$C_n^2 = 1.7 \times 10^{-13}\text{m}^{\frac{-2}{3}}$
System parameters	$\omega_z = 3\,m$; $r_a = 1$ cm λ_{FSO}=1550 nm
Optical source position SD Optical receiver position SD	$\sigma_{txp} = \sigma_{rxp} = 20$ cm $\sigma_{typ} = \sigma_{ryp} = 10$ cm
Boresight angle	$\theta'_{xt} = 1$ mrad ,$\theta'_{yt} = 2$ mrad, $\theta'_{xr} = 1$ mrad, $\theta'_{yr} = 2$ mrad
Boresight SD	$\sigma_{txo} = \sigma_{rxo} = 2$ mrad, $\sigma_{tyo} = \sigma_{ryo} = 3$ mrad

During the simulation process for computing outage, M_b bits (where $M_b = 10^6$) were randomly generated and used to create N_s MPSK symbols. Subsequently, a combined channel coefficient h_i was generated, and the transmitted signal x_s was subjected to this channel, resulting in a received signal $Y_D = (\eta P_{tp} h_i)^{u/2} x_s + n_o$, where n_o represents additive white Gaussian noise (AWGN). The received instantaneous SNR (γ_i) was then calculated based on the characteristics of the received signal and the channel conditions. The outage probability was determined by comparing γ_i with the threshold SNR γ_{th}. If $\gamma_i > \gamma_{th}$, the system was considered not to be in outage; otherwise, the system was considered to be in outage. The flowchart detailing the calculation of outage probability using Monte Carlo simulations is included in Fig. 5.8.

For the SER analysis of MPSK symbols, signal equalization was performed to obtain the equalized signal Y'_D. If the phase difference between Y'_D and x_s was 0 or a multiple of 2π, no error was detected; otherwise, an error was detected. A flowchart detailing the Monte Carlo simulation process for calculating the average SER is provided in Fig. 5.9.

The ergodic capacity was calculated based on the received instantaneous SNR γ_i using the expression $C = \log_2(1 + c_p \gamma_i)$. A flowchart illustrating the calculation of ergodic capacity is included in Fig. 5.10.

5.4.1 Málaga Distribution

Here, the statistical functions obtained for the Málaga distribution are utilized to obtain the closed-form expressions for various performance metrics. To validate the analytical findings, Monte Carlo simulations are conducted as outlined in Figs. 5.8, 5.9, and 5.10. The combined channel coefficient is generated using $h_m = h_{pl} h_p h_{at,m}$. For the analysis, the channel and Málaga distribution parameters are assumed as follows: $C_n^2 = 1.7 \times 10^{-13}\,\text{m}^{-2/3}$, indicative of strong turbulence, $\rho = 0.596$, $\Omega_o = 1.3265$, and $b_0 = 0.1079$, unless specified otherwise.

5.4.1.1 Outage Probability

An outage event occurs when the received instantaneous SNR falls below a predefined threshold γ_{th} value. The outage

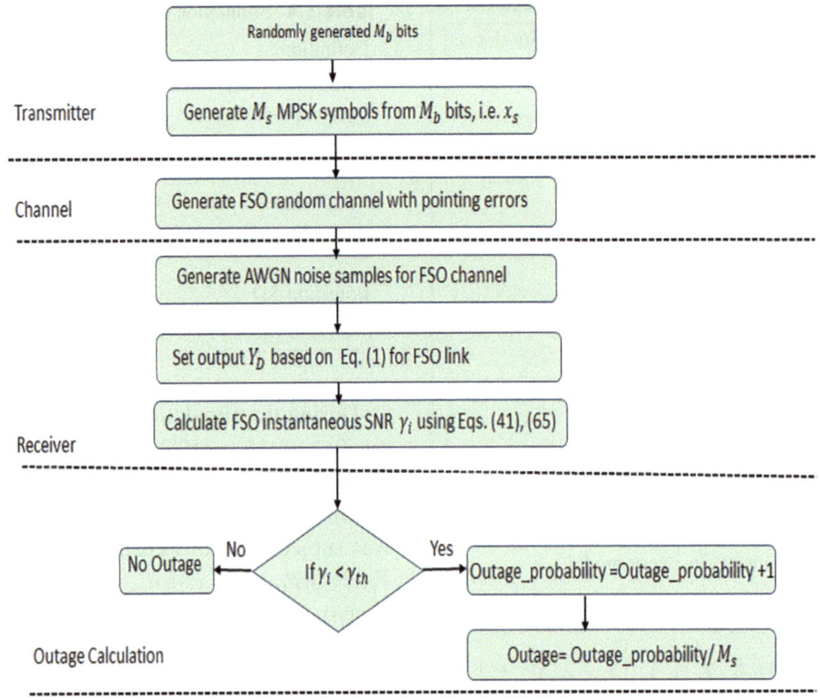

Fig. 5.8 Flowchart detailing the calculation of outage probability

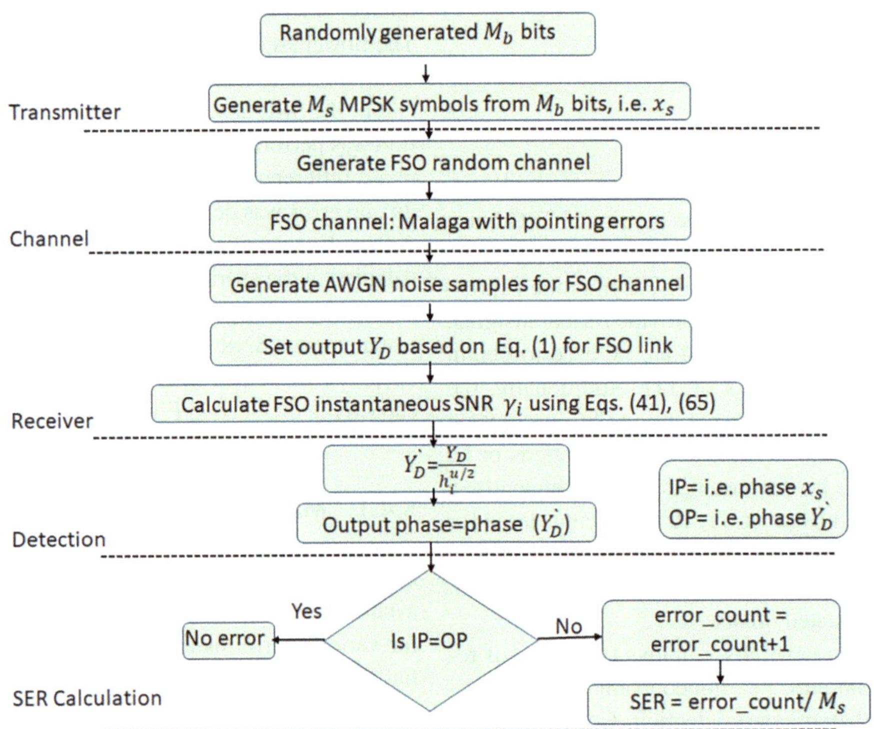

Fig. 5.9 Flowchart detailing the calculation of average SER

probability is defined as

$$P_{\text{out}}(\gamma_{\text{th}}) = P(\gamma_m < \gamma_{\text{th}}) = F_{\gamma_m}(\gamma_{\text{th}}), \tag{5.67}$$

where $F_{\gamma_m}(\gamma_{\text{th}})$ denotes the cumulative distribution function (CDF) of the instantaneous SNR of the FSO link, evaluated at γ_{th}, which is given by (5.42).

In Fig. 5.11, the outage performance of an FSO system is analyzed for three different cases listed in Table 5.1, over a

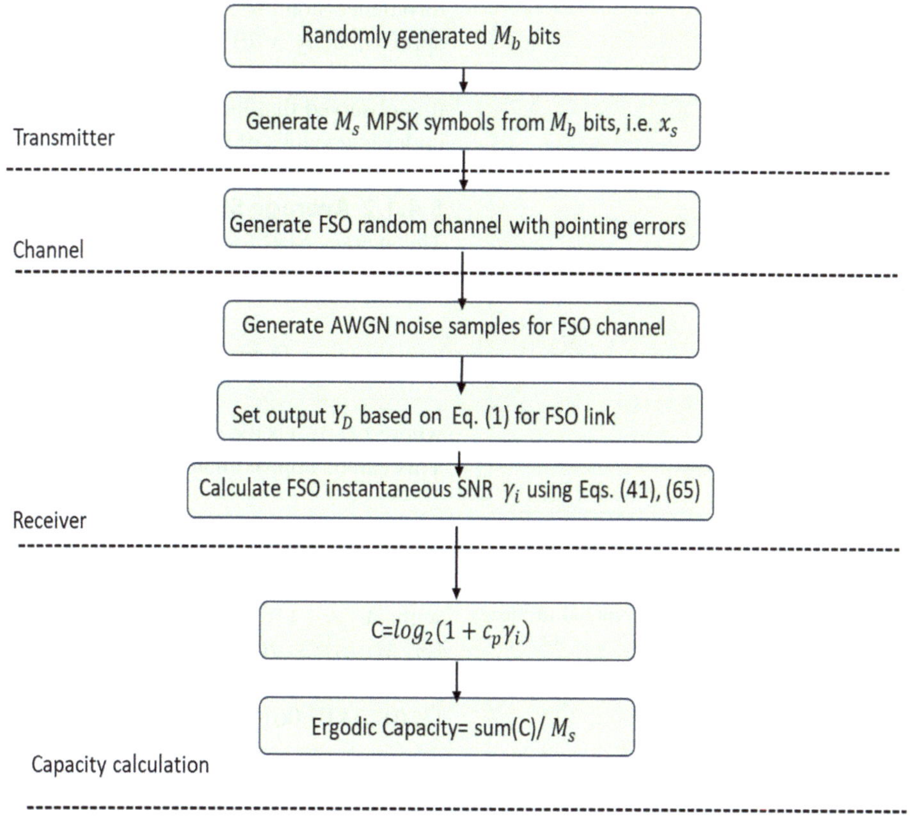

Fig. 5.10 Flowchart detailing the calculation of ergodic capacity

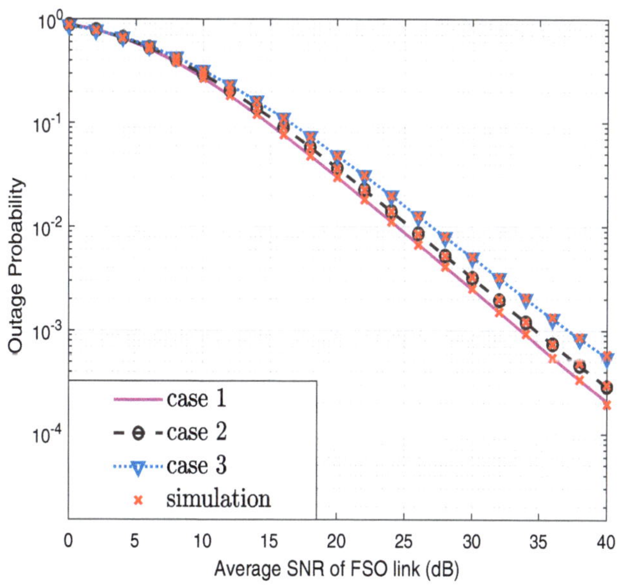

Fig. 5.11 Outage performance for different cases

fixed link distance of 300 m. The plots indicate that **Case 3** exhibits the highest outage probability (P_{out}), which is attributed to the orientation deviations of both the transmitter and receiver. In **Case 1**, where both the transmitter and receiver are fixed, the best performance is observed, as there are no orientation deviations. **Case 2**, where either the transmitter or receiver is hovering, performs better than **Case 3** but worse than **Case 1**. This analysis demonstrates that the presence of orientation deviations in FSO links significantly impacts the system's outage performance, with fixed transmitter and receiver setups yielding the most reliable results, as expected.

In Fig. 5.12, a detailed comparison of the outage probability performance across three different link distances, i.e., 250 m, 500 m, and 1000 m has been conducted. For a link distance $L = 250$ m, the atmospheric turbulence parameters are $\alpha = 7.37$ and $\beta = 5.86$. When the link distance is increased to $L = 500$ m, the parameters change to $\alpha = 4.14$ and $\beta = 2.17$. For the longest link distance of $L = 1000$ m, the parameters are $\alpha = 3.16$ and $\beta = 2.01$. It is observed that as the link distance increases, the outage probability P_{out} also increases for all three scenarios. This is because, with increasing distance, the scattering parameters α and β decrease, making the FSO system more susceptible to atmospheric turbulence. Thus, maintaining a stable FSO communication link over longer distances becomes more challenging due to the enhanced turbulence effects.

Figure 5.13 illustrates the impact of pointing errors on the outage performance of an FSO communication system. Pointing errors can occur due to misalignment between the transmitter and receiver, vibrations, or other environmental factors. Note that three different values for the pointing error

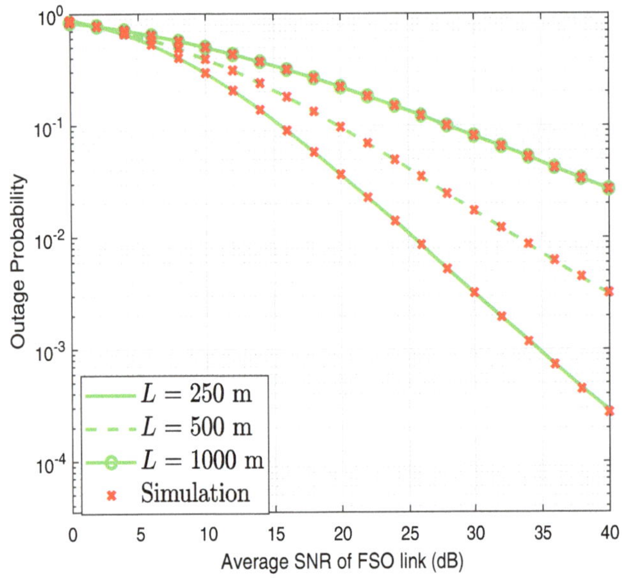

Fig. 5.12 Outage performance of FSO links for different link distances

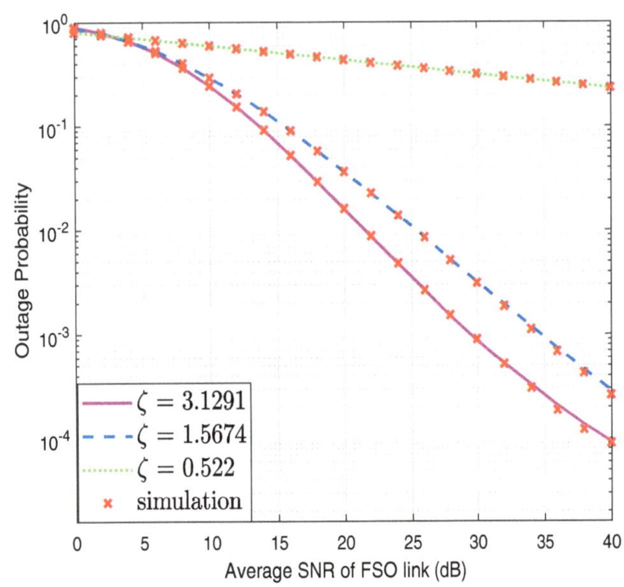

Fig. 5.13 Outage performance of FSO links for different values of pointing error severity

coefficient ζ are considered at a link distance of $L = 500$ m: (i) $\zeta = 3.1291$, (ii) $\zeta = 1.5674$, and (iii) $\zeta = 0.522$. The pointing error coefficient ζ represents the standard deviation of the pointing error displacement relative to the beam waist. A higher ζ value indicates a smaller pointing error, while a lower ζ value indicates a larger pointing error. It is important to note that the severity of pointing errors increases as the pointing error coefficient ζ decreases. Pointing errors can cause the received signal power to fluctuate, leading to higher outage probabilities. As observed from the plots, under strong turbulence conditions, at $P_{\text{out}} = 0.29$, an SNR gain of approximately 3 dB and 26 dB is achieved for $\zeta = 3.129$ compared to $\zeta = 1.567$ and $\zeta = 0.522$, respectively. Hence, it is observed that the degradation in performance is evident under low values of the pointing error coefficient.

5.4.1.2 Average SER

The average SER for the FSO communication system can be expressed as

$$P_e = \int_0^\infty p(e|\gamma_m) f_{\gamma_m}(\gamma_m) d\gamma_m, \tag{5.68}$$

where $p(e|\gamma_m)$ is the conditional SER for MPSK modulation. This can be approximated as

$$p(e|\gamma_m) \approx \frac{A}{2}\text{erfc}(B_1\sqrt{\gamma_m}), \tag{5.69}$$

with $A = 1$ for modulation order $M = 2$, $A = 2$ for $M > 2$, $B_1 = \sin(\pi/M)$, and erfc$(\cdot)$ denotes the complementary error function. Using the Meijer G-function [3, (07.34.03.0619.01)], (5.69) can be rewritten as

$$p(e|\gamma_m) = \frac{A}{2\sqrt{\pi}} G_{1,2}^{2,0}\left(B_1^2\gamma_m \middle| \begin{matrix} 1 \\ 0, 0.5 \end{matrix}\right). \tag{5.70}$$

Substituting (5.70) and (5.41) in (5.68) and solving the integral using [3, (07.34.21.0013.01)], the closed-form expression for SER is obtained as

$$P_e \approx C_2 \sum_{p=1}^{\beta} c_p G_{3u,3u+1}^{2u,3u+2}\left(A_2 \middle| \begin{matrix} 1, 0.5, K_1 \\ K_2, 0 \end{matrix}\right), \tag{5.71}$$

where $C_2 = \frac{A_M \zeta^2 A}{2^{2u}\pi^{u-\frac{1}{2}}}$ and $A_2 = \frac{B^2}{B_1^2 u^{2u}\overline{\gamma}_u}$.

In Fig. 5.14, the average SER is plotted for various MPSK modulation schemes. As anticipated, the system performance degrades with an increase in the modulation order M. Additionally, it is observed that the analytical SER curves precisely match the simulation curves for $M = 2$ and act as an upper bound for $M > 2$.

In Fig. 5.15, the average SER performance of an FSO system has been analyzed over a fixed link distance of 500 m for three different cases listed in Table 5.1. For these scenarios, we assume $C_n^2 = 4 \times 10^{-14}\,\text{m}^{-2/3}$ and $\omega_z = 3$ m. The plots indicate that **Case 3** exhibits the highest average SER, which is attributed to the orientation deviations of both the transmitter and receiver. This analysis demonstrates that the presence of orientation deviations in FSO links significantly impacts the system's average SER performance, with fixed transmitter and receiver setups yielding the most reliable results.

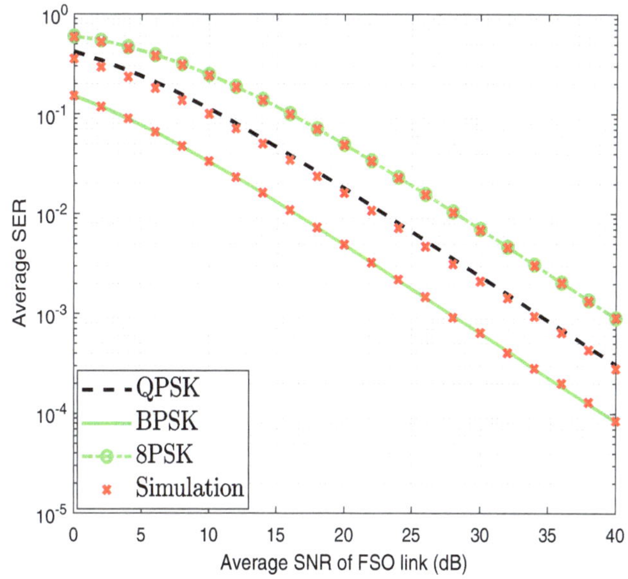

Fig. 5.14 Average SER performance of FSO link for different MPSK modulation scheme

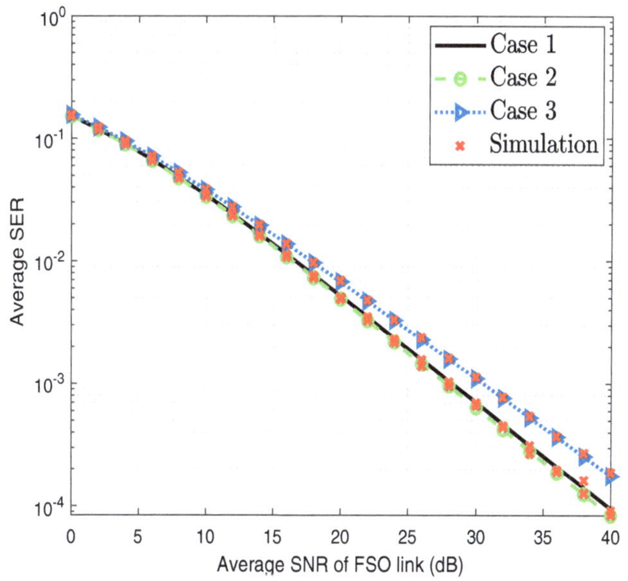

Fig. 5.15 Average SER performance comparison of different cases

5.4.1.3 Ergodic Capacity

It is important to note that atmospheric turbulence affecting FSO links is characterized as a slow fading process [27, 28]. This attribute stems from the fact that the coherence time of the FSO channel typically spans in milliseconds (ms), allowing the turbulence-induced fading to persist over extended duration encompassing numerous transmitted bits. Unlike traditional RF channels where interleaving can be employed to achieve independent fading samples in consecutive symbol intervals, implementing such a strategy in FSO channels is impractical. However, the presence of pointing errors and path loss introduces rapid signal fluctuations. These phenomena occur at high rates due to the dynamic nature of the atmosphere, influencing the received optical signal strength significantly. Therefore, when analyzing the ergodic capacity of FSO systems, it becomes imperative to account for these factors, assuming that the information symbols are sufficiently long to ensure the long-term statistical properties of the turbulence process.

Ergodic capacity analysis in the context of FSO systems has been extensively explored in existing literature [11, 27]. The normalized ergodic capacity of a FSO system can be mathematically expressed as [29]

$$\bar{C}_{\text{erg}} = \int_0^\infty \log_2\left(1 + c_u \gamma_m\right) f_{\gamma_m}(\gamma_m)\, d\gamma_m, \tag{5.72}$$

where c_u is a constant defined as $c_1 = 1$ for the HD scheme ($m = 1$) and $c_2 = \frac{e}{2\pi}$ for the DD scheme ($m = 2$). The logarithmic function in (5.72) is expressed in Meijer G-function form as

$$\log_2(1 + c_u\gamma_m) = \frac{1}{\ln(2)} G_{2,2}^{1,2}\left(c_u\gamma_m \middle| \begin{matrix}1, 1\\ 1, 0\end{matrix}\right) \tag{5.73}$$

After substituting (5.73) and (5.41) into (5.72), and solving using [16, eq. (07.34.21.0013.01)], the expression for normalized ergodic capacity $\bar{C}_{\text{erg}}$ is obtained as

$$\bar{C}_{\text{erg}} = C_3 \sum_{p=1}^{\beta} c_p G_{u+2,3u+2}^{3u+2,1}\left(A^3 \middle| \begin{matrix}0, 1, K_1,\\ K_2, 0, 0\end{matrix}\right) \tag{5.74}$$

where $C_3 = \frac{\zeta^2 A_M}{\ln(2)(\pi)^{-u+1} 2^{-2u-1}}$ and $A^3 = \frac{B^u}{c_u u^{2u} \bar{\gamma}_m}$.

Figure 5.16 illustrates the ergodic capacity performance for three different link distance values, i.e., 200 m, 500 m, and 1000 m. It is to be noted that $L = 200$ m results in the atmospheric turbulence parameter values as $\alpha = 8.37$ and $\beta = 6.86$. Similarly, $L = 500$ m and $L = 1000$ m result in $\alpha = 4.14$, $\beta = 2.17$ and $\alpha = 3.16$, $\beta = 2.27$, respectively. It is observed that as the link distance increases, $\bar{C}_{\text{erg}}$ decreases for all three cases. This is because, with increasing distance, the scattering parameters α and β decrease, making the FSO system more susceptible to turbulence. Therefore, with an increase in link distance value, ergodic capacity decreases.

In Fig. 5.17, the normalized ergodic capacity performance is shown under varying turbulence conditions. It is assumed that $\omega_z = 3\,\text{m}$ with $C_n^2 = 1.7 \times 10^{-13}\,\text{m}^{-2/3}$ for strong turbulence, $C_n^2 = 5 \times 10^{-14}\,\text{m}^{-2/3}$ for moderate turbulence, and $C_n^2 = 4 \times 10^{-15}\,\text{m}^{-2/3}$ for weak turbulence. The results show that ergodic capacity decreases as turbulence conditions deteriorate. This trend is due to a decrease in atmospheric scattering parameter values α and β, which enhance the FSO system's sensitivity to turbulence.

In Fig. 5.18, a link distance of $Z = 500$ m and two different beamwidth values of $w_z = 0.8$ m and $w_z = 3$

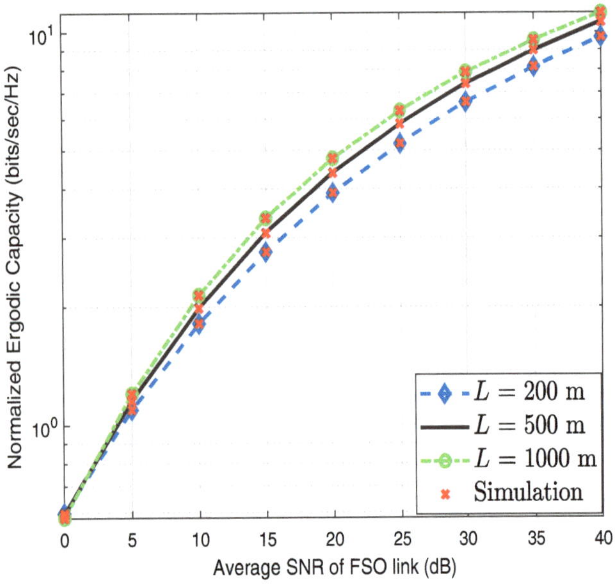

Fig. 5.16 Normalized ergodic capacity performance for different link distance

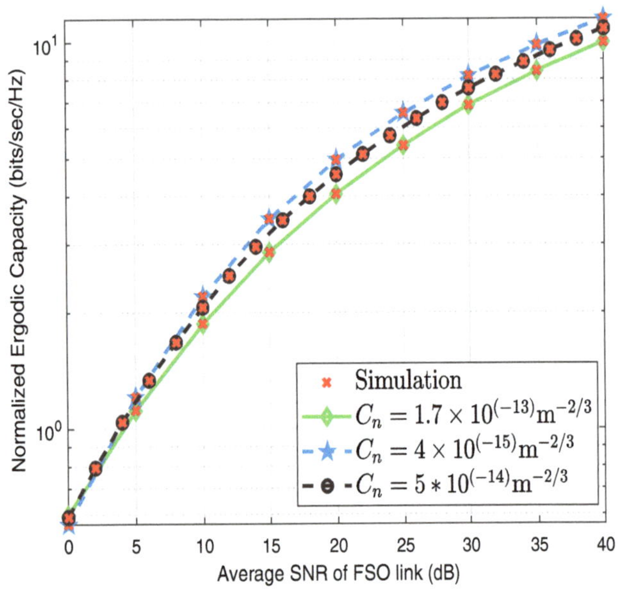

Fig. 5.17 Normalized ergodic capacity performance under different turbulence conditions

m have been considered. The plots demonstrate that ergodic capacity increases with the increase in beamwidth. For $\omega_z = 0.8$ m case, to achieve an ergodic capacity of 2.14 bits/sec/Hz, the average SNR must be at least 3 dB higher compared to the case when $\omega_z = 3$ m. The plots indicate that increasing ω_z leads to an increase in ergodic capacity. This is because a wider beamwidth mitigates pointing error losses and enhances the received signal strength.

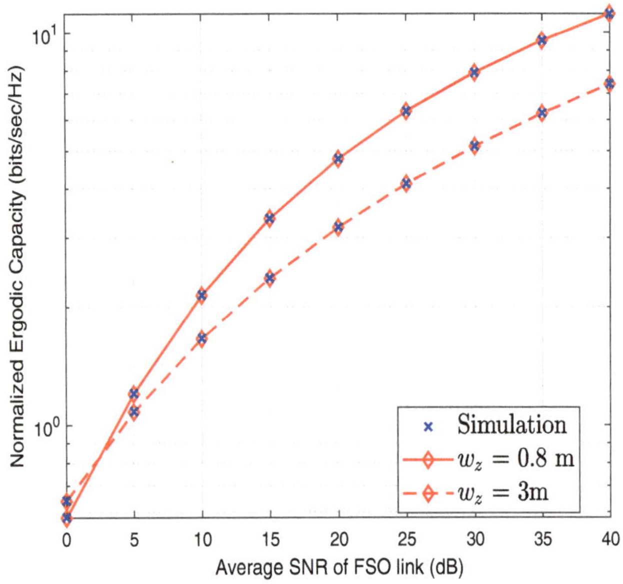

Fig. 5.18 Normalized ergodic capacity performance for different values of incoming optical beamwidth

5.4.1.4 Asymptotic Analysis

In this section, the closed-form asymptotic expressions for outage probability, average SER, and ergodic capacity under high SNR conditions (i.e., $\overline{\gamma}_m \to \infty$) for Málaga distribution are derived. By analyzing these asymptotic expressions, we gain valuable insights into the system's behavior at high SNR levels. This is particularly important for understanding the limits of system performance and for designing robust communication systems that can operate effectively even in adverse conditions. Furthermore, from the derived asymptotic expressions, we determine the diversity gain, denoted as G^d, which is a critical parameter that reflects the system's ability to mitigate the effects of fading. A higher diversity gain indicates better performance in combating channel impairments. The diversity gain is extracted from the high SNR asymptotic analysis, providing a deeper understanding of how different system parameters influence the overall reliability and efficiency of the communication link.

The findings presented in this section set the stage for further discussion and analysis in the subsequent section, where the implications of the derived expressions are explored in greater depth, along with their practical applications.

5.4.1.5 Outage Probability

The asymptotic expression for outage probability is derived assuming that the average SNR of the FSO link tends to infinity. The asymptotic outage probability of the FSO communication system is given by

$$P_{\infty,\mathrm{out}}(\gamma_{\mathrm{th}}) = F_{\infty,\gamma_m}(\gamma_{\mathrm{th}}) \tag{5.75}$$

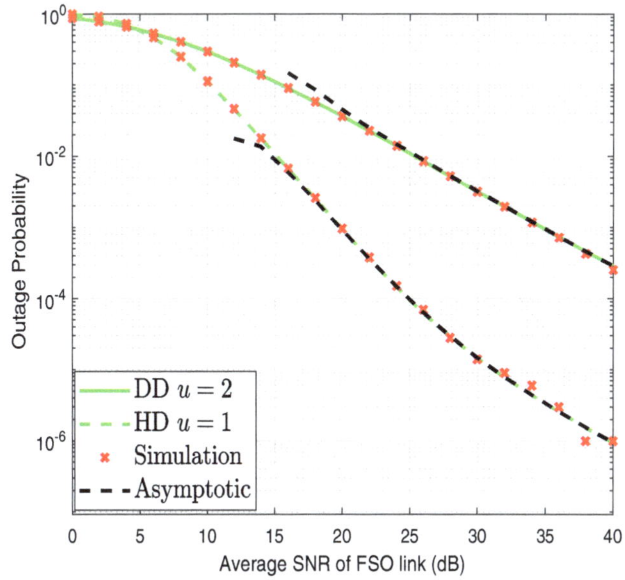

Fig. 5.19 Outage probability comparison of different detection scheme

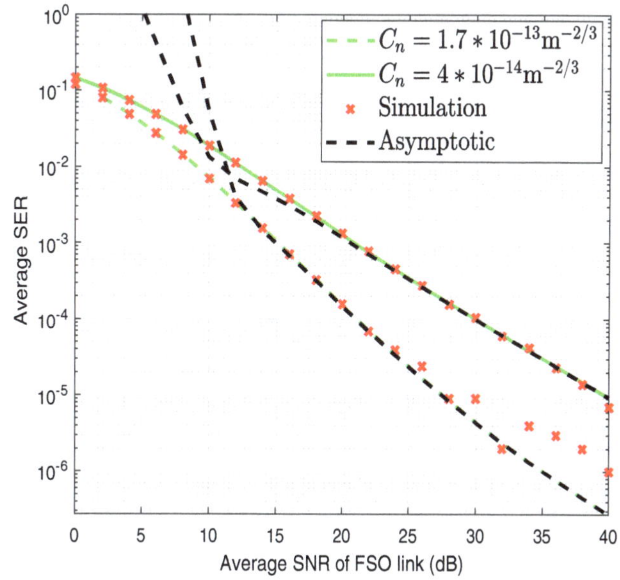

Fig. 5.20 Asymptotic SER comparison for different turbulence parameter

where $F_{\infty,\gamma_m}(\gamma_{\text{th}})$ is the asymptotic CDF of the FSO link. By applying [16, eq. (07.34.06.0040.01)] on (5.67), $F_{\infty,\gamma_m}(\gamma_{\text{th}})$ can be expressed as

$$F_{\infty,\gamma_m}(\gamma_{\text{th}}) = D \sum_{p=1}^{\beta} c_p \sum_{r=1}^{3u} A_2^{K_{2,r}} \frac{\Lambda_1}{K_{2,r}} \tag{5.76}$$

where $\Lambda_1 = \frac{\prod_{s=1;s\neq r}^{3u} \Gamma(K_{2,s}-K_{2,r})}{\prod_{s=1}^{u} \Gamma(K_{1,s}-K_{2,r})}$.

It is observed from (5.76) that $P_{\infty,\text{out}} \propto \overline{\gamma}_m^{-G^d}$, where G^d indicates the diversity gain, given by $G^d = \min(\frac{\zeta^2}{u}, \frac{\alpha}{u}, \frac{1}{u})$.

In Fig. 5.19, the comparison of outage probability for different detection schemes assuming $L = 700$ m is depicted. The asymptotic outage curves are also included in the plot. As shown in the plot, the DD technique exhibits the highest outage probability than that of HD technique, since the detection in DD is noncoherent in nature. Additionally, it is observed that the outage probability for DD scheme at an SNR of 20 dB is 9×10^{-4}, and at 26 dB, it is 7×10^{-4}. Using these values, a diversity gain of 0.54 is obtained, which aligns with the diversity gain $G^d = \min(\zeta^2/u, \alpha/u, 1/u)$ obtained using the asymptotic expression. This confirms the accuracy of the derived diversity gain. Finally, the asymptotic curves converge with the exact outage curves in the high SNR region, validating the derived asymptotic expression.

5.4.1.6 SER Analysis

The asymptotic expression for the average SER of FSO system is obtained by applying [16, (07.34.06.0040.01)] on (5.71) and is given as

$$P_{\infty,\text{e}} = C_2 \sum_{p=1}^{\beta} c_p \sum_{r=1}^{3u} A_2^{K_{2,r}} \frac{\Gamma(K_{2,r}+\frac{1}{2})\Lambda_1}{K_{2,r}} \tag{5.77}$$

Similar to asymptotic outage probability expression, it is observed from (5.77) that $P_{\infty,\text{e}} \propto \overline{\gamma}_m^{-G^d}$, where G^d indicates the diversity gain, given by $G^d = \min(\frac{\zeta^2}{u}, \frac{\alpha}{u}, \frac{1}{u})$.

In Fig. 5.20, the average SER performance across two different turbulence scenarios has been compared assuming $C_n^2 = 1.7 \times 10^{-13}\,\text{m}^{-2/3}$ for the strong turbulence scenario and $C_n^2 = 5 \times 10^{-14}\,\text{m}^{-2/3}$ for the moderate turbulence scenario. It is evident that the asymptotic curves closely match the slope of the performance curves and converge with the exact SER curve in the high SNR region. From Fig. 5.20, it is also noticed that an average SER value of 1.5×10^{-4} is achieved at an SNR of 20 dB and 3.3×10^{-5} at 34 dB. Using these data points, a diversity gain of 1.61 is determined. This result aligns with the diversity gain $G^d = \min(\zeta^2/u, \alpha/u, 1/u)$, confirming the accuracy of the derived diversity gain.

In Fig. 5.21, the average SER performance for different detection schemes is compared for a link distance of 300 m. It can be observed that the asymptotic curves accurately follow the slope of the performance curves and coincide with the exact SER curves in the high SNR region. Also, it is noted that an SER value of 9.9×10^{-5} is achieved at an SNR of 30 dB and 2.3×10^{-5} at 35 dB. Using these values, a diversity gain of 1.26 is determined, which is consistent with the diversity gain $G^d = \min(\zeta^2/u, \alpha/u, 1/u)$ derived in (5.77). This confirms the accuracy of the derived diversity gain.

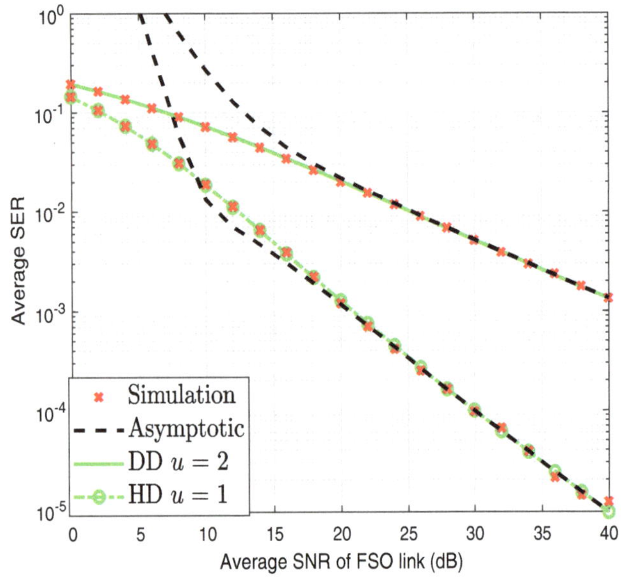

Fig. 5.21 Asymptotic SER comparison for different detection scheme

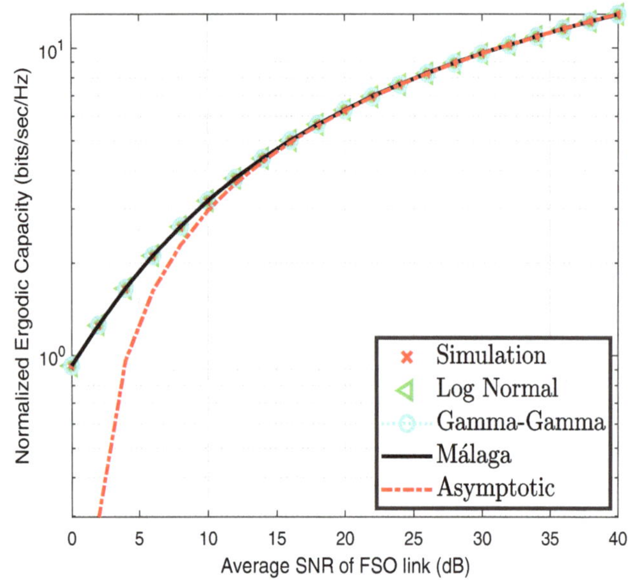

Fig. 5.22 Asymptotic ergodic capacity FSO link versus average SNR of FSO link for different distributions

5.4.1.7 Capacity Analysis

The asymptotic capacity of the FSO system is obtained using the method of moments. To calculate it, the first-order derivative of the nth-order moment of $f_{\gamma_m}(\gamma_m)$ at $n = 0$ needs to be obtained. Firstly, the nth-order moment of $f_{\gamma_m}(\gamma_m)$ is given by

$$E[\gamma_m^n] = \int_0^\infty \gamma_m^n f_{\gamma_m}(\gamma) d\gamma_m$$
$$= \frac{u\zeta^2 A_m \Gamma(\alpha + un)}{2^u(\zeta^2 + un)} \left(\frac{c_u \bar{\gamma}_m}{B^u}\right)^n \sum_{p=1}^{\beta} b_p \Gamma(p + un) \tag{5.78}$$

Taking the first-order derivative of (5.78) at $n = 0$, the asymptotic capacity $\bar{C}_{\infty,\gamma_m}$ of the FSO link is given as

$$\bar{C}_{\infty,\gamma_m} \approx \frac{1}{\log_e(2)} \frac{u\Gamma(\alpha)A_M}{2^u} \sum_{p=1}^{\beta} b_p \Gamma(p)$$
$$\times \left\{ b\left[\psi(\alpha) + \psi(p) - \frac{1}{\zeta}\right] + \log_2\left(\frac{c_u \bar{\gamma}_m}{B^u}\right)\right\} \tag{5.79}$$

where $\psi(\cdot)$ denotes the psi function [26].

Figure 5.22 illustrates the normalized ergodic capacity performance of the FSO link under various atmospheric turbulence distributions. The Málaga distribution transforms into GG distribution when $\rho = 1$, $\Omega_p' = 1$, and $g = 0$. Similarly, the Málaga distribution approximates the log-normal distribution as $\rho \to 0$ and $g \to 0$. These relationships show how the Málaga distribution simplifies to established turbulence models based on specific parameter values.

5.4.2 IGGG Distribution

In this section, the performance analysis and the simulation results of the IGGG distribution are presented. To validate our analytical findings, we conduct Monte Carlo simulations by generating 10^6 message symbols and independent channel coefficients. The combined channel coefficient is generated using $h_g = h_{pl} h_p h_{at,g}$. For the analysis, it is assumed that $C_n^2 = 1.7 \times 10^{-13}\,\text{m}^{-2/3}$, indicative of strong turbulence. The values presented in Table 5.4 are used to generate the analytical and simulation results.

5.4.2.1 Outage Probability

An outage will occur when the received instantaneous SNR (γ_g) of FSO link falls below the threshold SNR (γ_{th}). So, the outage probability is given by

$$P_{out} = F_{\gamma_g}(\gamma_{th}) = \int_0^{\gamma_{th}} f_{\gamma_g}(\gamma_g) d\gamma_g \tag{5.80}$$

The unified expression for CDF is obtained by substituting (5.66) in (5.80) and can be calculated in terms of Meijer's G-function using [16, eq. (07.34.21.0084.01)] as

$$F_{\gamma_g}(\gamma_{th}) \approx \frac{g^2 A_{\gamma_g} u^{(\alpha+\beta+\lambda-2.5)}}{(2\pi)^{(u-1)1.5}}$$

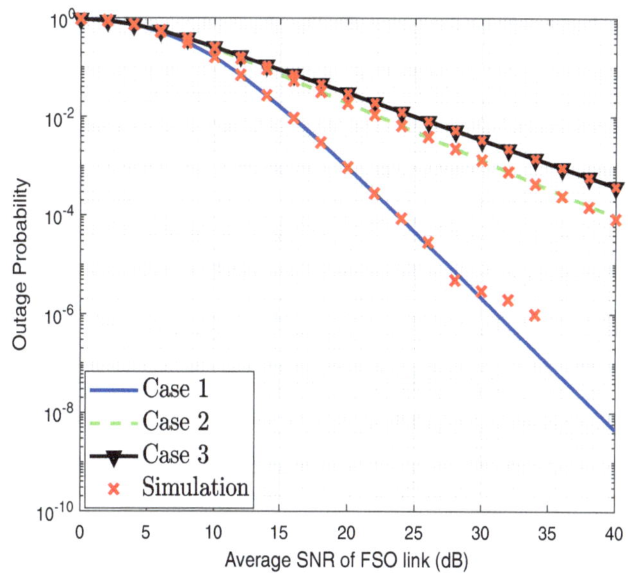

Fig. 5.23 Outage performance of FSO link using IGGG turbulence model for different cases

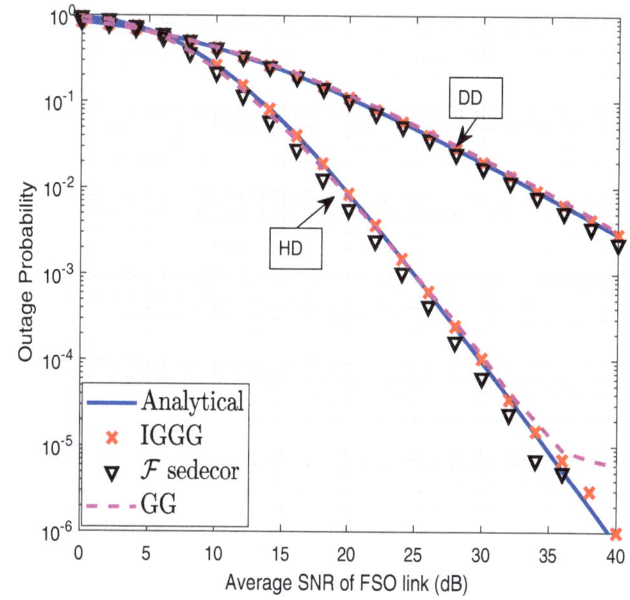

Fig. 5.24 Outage probability of IGGG, GG, and Fisher-Snedecor $\mathcal{F}$ channel models

$$G^{3u,u+1}_{2u+1,u+1}\left(B^u\left(\frac{\gamma_g}{\bar{\gamma}_g}\right)\middle|\begin{matrix}1,\frac{(1-\lambda)}{u},\frac{(u-\lambda)}{u},\frac{\zeta^2+1}{u},\frac{\zeta^2+u}{u}\\ \frac{\zeta^2}{u},\frac{\zeta^2+u-1}{u},\frac{\alpha}{u},\frac{\alpha+u-1}{u},\frac{\beta}{u},\frac{\beta+u-1}{u}\end{matrix}\right), \tag{5.81}$$

where $B=\frac{\alpha\beta K}{(\lambda-1)u}$.

In Fig. 5.23, the outage probability curves of the FSO system considering IGGG turbulence distribution under three distinct cases are presented in Table 5.1. The scenario considers a distance of $L = 500$ meters. Under these conditions, **Case 1** demonstrates an SNR gain of 10 dB and 15 dB compared to **Case 2** and **Case 3**, respectively, to achieve an outage probability of 9.6×10^{-4}.

The analysis highlights the significant impact of position and orientation deviations on the system's outage performance. In the first case, where both the transmitter and receiver are fixed, the system shows the lowest outage probability, indicating the best performance. This is because fixed positions eliminate orientation deviations, ensuring a stable and reliable communication link. In the second case, where either the transmitter or receiver is hovering, the performance degrades relative to the first scenario. The hovering component introduces orientation deviations, leading to potential misalignments and a reduction in received signal power, which in turn increases the outage probability. The third case, with both the transmitter and receiver hovering, results in the highest outage probability. The combined orientation deviations of both components exacerbate alignment issues, causing significant performance degradation among the three scenarios.

This comprehensive analysis underscores the crucial role of maintaining orientation and position stability in FSO systems. Fixed transmitter and receiver setups significantly enhance system reliability and reduce outage probability, while hovering components introduce deviations that can negatively impact the performance. These insights are vital for optimizing FSO communication systems, especially in dynamic environments.

In Fig. 5.24, the outage probability curves are plotted for IGGG, GG, and $\mathcal{F}$ turbulence models. It is seen from the plots that the IGGG channel model can be approximated to GG and $\mathcal{F}$ turbulence models. According to doubly process, as $\lambda \to \infty$, IGGG reduces to $\mathcal{F}$ turbulence model, and when $\alpha \to \infty$, IGGG reduces to GG turbulence model.

Figure 5.25 illustrates the influence of pointing errors on the outage performance of the FSO communication system. The performance is evaluated at a link distance of 300 meters, considering three different values of the pointing error coefficient ζ, i.e., (i) 0,818, (ii) 1.632, and (iii) 3.277. The pointing error coefficient ζ reflects the severity of misalignment between the transmitter and receiver. Lower values of ζ correspond to more severe pointing errors, which can significantly affect the reliability of the communication link.

In the first scenario with $\zeta = 3.1291$, the system experiences the least impact from pointing errors. This is due to the higher value of the coefficient, indicating a more stable alignment between the transmitter and receiver. Consequently, the outage probability is relatively low, showcasing better performance. The second scenario, where $\zeta = 1.5674$, represents a moderate level of pointing errors. Here, the alignment stability decreases compared to the first scenario, leading to a noticeable increase in outage probability. This moderate pointing error coefficient reflects a situation where

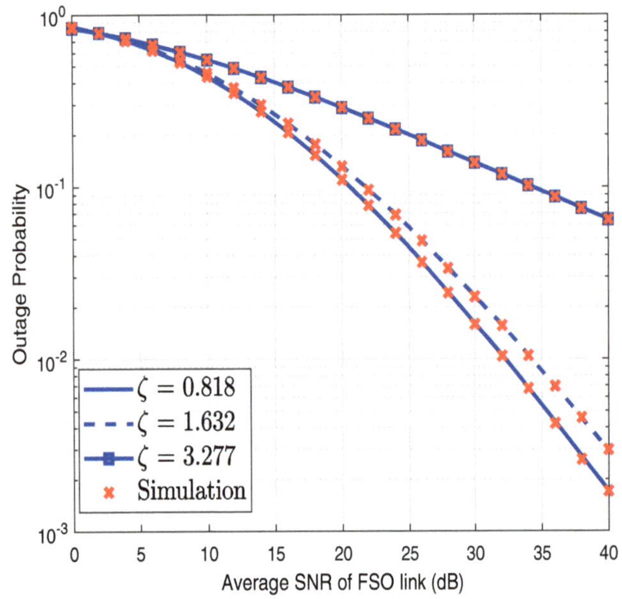

Fig. 5.25 Outage probability of IGGG distribution for different pointing error severity

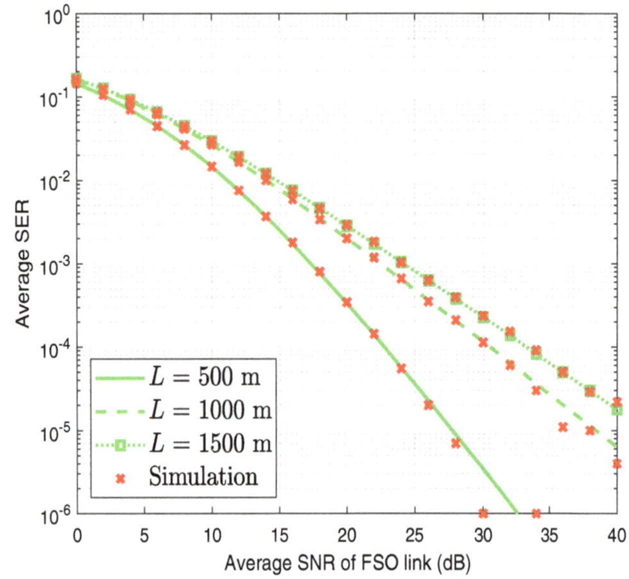

Fig. 5.26 Average SER plot for different link distance values

misalignment is more frequent, reducing the received signal power and thus degrading the system's performance. In the third scenario, with $\zeta = 0.522$, the system encounters the most severe pointing errors. The low value of the pointing error coefficient indicates significant misalignment issues, which drastically increase the outage probability. The severe misalignment causes substantial power loss and signal degradation, making reliable communication more challenging.

Overall, this analysis emphasizes the critical impact of pointing errors on the performance of FSO communication systems. Maintaining precise alignment between the transmitter and receiver is crucial, especially under strong turbulence conditions. As the severity of pointing errors increases, the system's outage probability increases.

5.4.2.2 Average SER

The average SER of the FSO system for the MPSK modulation scheme is given by

$$P_{ser} = \int_0^\infty p(e_1/\gamma_g) f_{\gamma_g}(\gamma_g) d\gamma_g, \tag{5.82}$$

where the conditional SER is given by

$$p(e_1/\gamma_g) \approx \frac{A}{2}\text{erfc}(D\sqrt{\gamma_g}), \tag{5.83}$$

Now, using [16, eq. (07.34.03.0619.01)], the above equation eq.(5.83) can be written in terms of Meijer's G-function as

$$p(e_1/\gamma_g) = \frac{A}{2\sqrt{\pi}} G_{1,2}^{2,0}\left(D^2\gamma_g \middle| \begin{matrix} 1 \\ 0, 0.5 \end{matrix}\right) \tag{5.84}$$

By substituting (5.66) and (5.84) in (5.82) and solving it using [16, eq.(0.7.34.21.0013.01)], the closed-form expression for average SER can be written as

$$P_{ser} = \frac{g^2 A_{\gamma_g} u^{(\alpha+\beta+\lambda-2.5)}}{2\sqrt{\pi}(2\pi)^{(u-1)2.5}} G_{2u+2,3u+1}^{3u,u+2} \left(B^u\left(\frac{1}{\overline{\gamma}_g}D^2\right) \middle| \begin{matrix} K_1, 1, 0.5, K_2 \\ K_3, 0 \end{matrix}\right), \tag{5.85}$$

where $K_1 = \left[\frac{1-\lambda}{u}, \ldots, \frac{1-\lambda+u-1}{u}\right]$, $K_2 = \left[\frac{\zeta^2+1}{u}, \ldots, \frac{1+u^2+u-1}{u}\right]$ and $K_3 = \left[\frac{\zeta^2}{u}, \ldots, \frac{\zeta^2+u-1}{u}, \frac{\alpha}{u}, \ldots, \frac{\alpha+u-1}{u}, \frac{\beta}{u}, \ldots, \frac{\beta+u-1}{u}\right]$.

In Fig. 5.26, we compare the average SER performance for three different link distances: 500 m, 1000 m, and 1500 m. For 500 m link distance, the atmospheric turbulence parameters are $\alpha = 4.463$, $\beta = 2.658$, and $\lambda = 6.658$. Similarly, for 1000 m link distance, the parameters are $\alpha = 4.2002$, $\beta = 1.3770$, and $\lambda = 6.2002$. Finally, for 1500 m link distance, the parameters are $\alpha = 3.2002$, $\beta = 1.16$, and $\lambda = 5.2002$.

The analysis shows that as the link distance increases, the severity of atmospheric turbulence also increases, which results in a higher average SER. The increased turbulence causes greater signal distortion and attenuation, leading to a degradation in the SER performance of the FSO system. This highlights the importance of considering link distance and atmospheric conditions when evaluating the performance of FSO communication systems.

Figure 5.27 illustrates the average SER for various MPSK modulation schemes. For a fixed link distance of $L = 500$

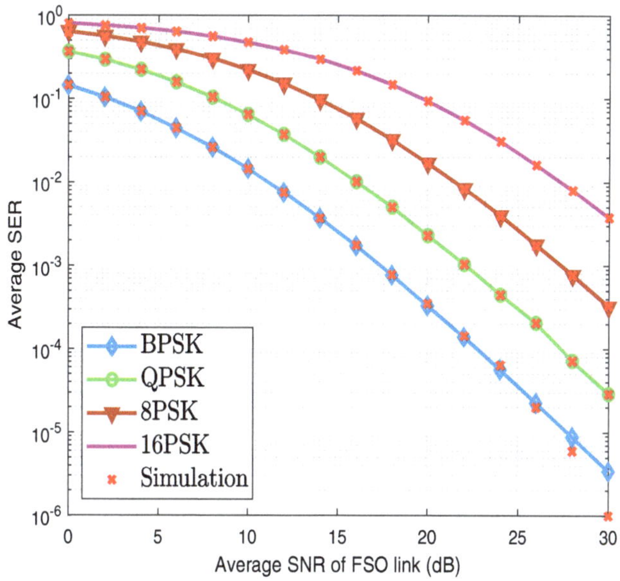

Fig. 5.27 Average SER of MPSK signal of FSO communication system using IGGG channel model

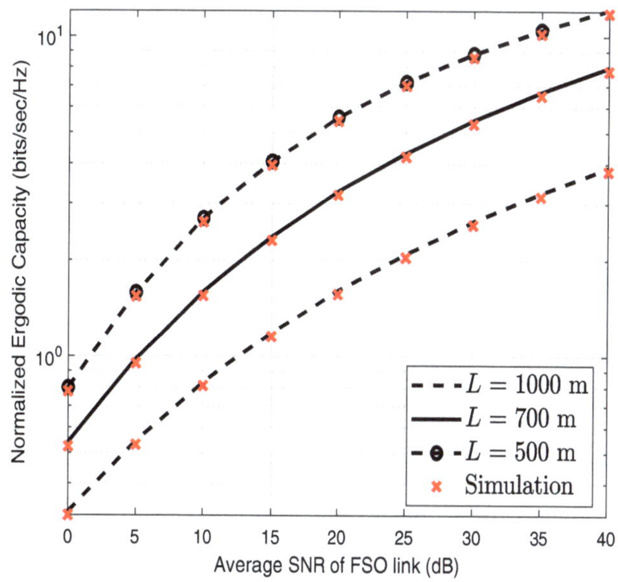

Fig. 5.28 Ergodic capacity for various link distances

m, it is observed that to achieve an average SER of 0.05, BPSK requires an SNR gain of 6.5 dB compared to QPSK, 11 dB compared to 8-PSK, and 15 dB compared to 16-PSK. This trend indicates that as the modulation order increases, the average SER also increases. Higher-order modulation schemes, such as 8-PSK and 16-PSK, are more susceptible to noise and other channel impairments compared to lower-order schemes like BPSK and QPSK. Consequently, higher-order modulations require a higher SNR to achieve the same SER performance. This behavior is expected because higher-order modulation schemes transmit more bits per symbol, which makes them more efficient in terms of spectral efficiency but also more prone to errors due to the reduced Euclidean distance between constellation points. Therefore, while they offer higher data rates, they demand better channel conditions and higher SNR to maintain an acceptable SER.

5.4.2.3 Ergodic Capacity

Capacity is the highest achievable data rate that can be transmitted over the combined channel. The normalized ergodic capacity of the FSO system has been calculated under the influence of pointing error fluctuations using the following integral:

$$\overline{C}_{erg} = \int_0^\infty \zeta_2(1 + c_u\gamma_g) f_{\gamma_g}(\gamma_g) d\gamma_g, \tag{5.86}$$

where c_u is a constant and takes the value as $c_1 = 1$ for HD and $c_2 = \frac{e}{2\pi}$ for IM/DD. By using [16, eq. (07.34.03.0830.01)], the logarithmic term in (5.86) can be written in terms of Meijer's G-function as

$$\log_2(1 + c_u\gamma_g) = \frac{1}{\ln 2} G_{2,2}^{1,2}\left(c_u\gamma_g \middle| \begin{matrix} 1, 1 \\ 1, 0 \end{matrix}\right). \tag{5.87}$$

The closed-form expression for ergodic capacity is obtained by substituting (5.66) and (5.87) in (5.86), and solving using [16, eq. (07.34.21.0013.01)], as

$$\overline{C}_{erg} = \frac{\zeta^2 A_{\gamma_g} u^{(\alpha+\beta+\lambda-2.5)}}{\ln(2)(2\pi)^{(u-1)1.5}} G_{2u+2,3u+2}^{3u+2,u+1}\left(\frac{B^u}{\bar{\gamma}_g c_u} \middle| \begin{matrix} K_1, 0, 1, K_2 \\ K_3, 0, 0 \end{matrix}\right). \tag{5.88}$$

In Fig. 5.28, the normalized ergodic capacity is illustrated for three distinct link distances, i.e., 500 m, 700 m, and 1000 m. The plot clearly demonstrates a decline in ergodic capacity as the link distance increases. This inverse relationship between link distance and ergodic capacity can be attributed to several factors inherent in FSO communication systems. As the distance between the transmitter and receiver increases, the optical beam undergoes greater attenuation due to atmospheric absorption and scattering. Additionally, the likelihood of signal degradation due to turbulence and misalignment increases with longer distances. These effects collectively reduce the received signal strength, thereby decreasing the ergodic capacity.

Thus, while FSO systems offer high data rates and flexibility, maintaining optimal performance over longer distances remains challenging due to these physical and environmental factors. The plot underscores the necessity for careful design and optimization of FSO links, especially in scenarios involving significant link distances.

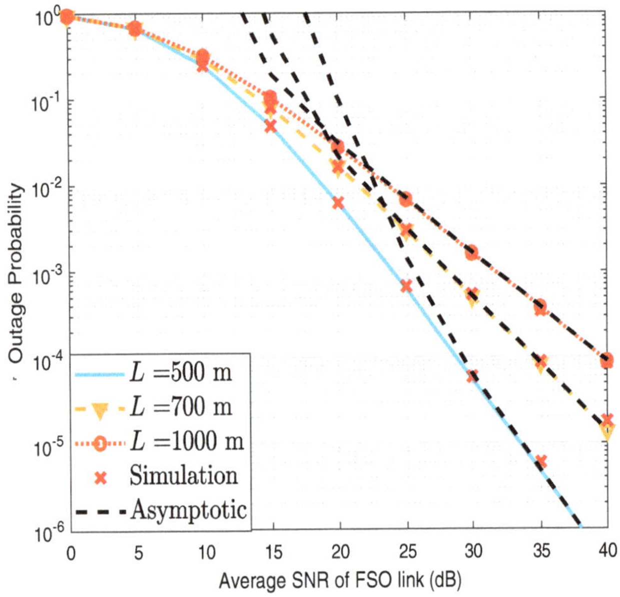

Fig. 5.29 Asymptotic outage analysis of IGGG distribution for different link distance

5.4.3 Asymptotic Analysis

In this section, the closed-form asymptotic expressions for outage probability, average SER, and ergodic capacity under high SNR conditions (i.e., $\overline{\gamma}_g \to \infty$) for IGGG distribution are derived.

5.4.3.1 Outage Probability

The asymptotic expression for the outage probability of the FSO system has been derived under the assumption that the average SNR of the FSO system approaches infinity. The asymptotic CDF $F^{\infty}_{\gamma_g}(\gamma_{th})$ is calculated by assuming $\gamma_g \to \infty$, and after some mathematical manipulation using [16, eq.(07.34.06.0040.01)], it can be written as

$$F^{\infty}_{\gamma_g}(\gamma_{th}) = \frac{\zeta^2 A_{\gamma_g} u^{(\alpha+\beta+\lambda-2.5)}}{(2\pi)^{(u-1)1.5}} \sum_{k=1}^{3u} \frac{\prod_{\substack{j=1\\ j\neq k}}^{3u} \Gamma(b_{1,j}-b_{1,k}) \prod_{j=1}^{u+1} \Gamma(1-a_{1,j}+b_{1,k})}{\prod_{j=u+1}^{2u+1} \Gamma(a_{1,j}-b_{1,k}) \Gamma(1+b_{1,k})} B_1^{b_{1,k}}, \tag{5.89}$$

where $a_{1,j} = \{1, K_1, K_2\}$, $b_{1,j} = \{K_{3,j}, 0\}$, and $B_1 = B^u\left(\frac{\gamma_{th}}{\overline{\gamma}_g}\right)$.

In Fig. 5.29, the outage probability of FSO system using IGGG distribution is obtained by varying link distance. The asymptotic outage curves are also included in the plot. The analysis indicates that as the link distance increases, the severity of atmospheric turbulence also increases, leading to an increase in outage probability. The heightened turbulence results in greater signal distortion and attenuation, which in turn degrades the outage performance of the FSO system. By performing the diversity gain calculation in (5.89), it is observed that $P_{out,\infty} \propto \overline{\gamma}_g^{-G^d}$, where $G^d = \min\left\{\frac{\zeta^2}{u}, \frac{\alpha}{u}, \frac{\beta}{u}\right\}$ indicates the diversity gain. It is evident that G_d mainly depends on the type of detection scheme u, pointing error coefficient (ζ), and scattering parameters (α, β).

Additionally, it is observed that for a link distance of 500 m and the SNR of 28 dB, the outage probability is noticed as 6.26×10^{-5} and at 33 dB, it is 8×10^{-5}. Using these values, a diversity gain of 0.4 is obtained, which aligns with the diversity gain $G^d = \min(\zeta^2/u, \alpha/u, \beta/u)$. This confirms the accuracy of the derived diversity gain for the FSO system over IGGG distribution. Finally, the asymptotic curves converge with the exact outage curves in the high SNR region, validating the derived asymptotic expression.

5.4.3.2 Average SER

The asymptotic SER calculation can be computed similarly to the asymptotic outage probability expression. By employing [16, eq.(07.34.06.0040.01)] in (5.85), asymptotic SER for FSO link is computed as

$$P^{\infty}_{ser}(e) = \frac{K}{2} + K(1-a_i)\frac{\zeta^2 A_{\gamma_g} u^{(\alpha+\beta+\lambda-2.5)}}{2\sqrt{\pi}(2\pi)^{(u-1)2.5}} \sum_{k=1}^{3u} \frac{\prod_{\substack{j=1\\ j\neq k}}^{3u} \Gamma(b_{2,j}-b_{2,k}) \prod_{j=1}^{u+2} \Gamma(1-a_{2,j}+b_{2,k})}{\prod_{j=u+2}^{2u+2} \Gamma(a_{2,j}-b_{2,k}) \Gamma(1+b_{2,k})} B_2^{b_{2,k}}, \tag{5.90}$$

where $a_{2,j} = \{K_1, 1, 0.5, K_2\}$, $b_{2,j} = \{K_{3,j}, 0\}$, and $B_2 = B^u\left(\frac{1}{\overline{\gamma}_g}\right)$.

In Fig. 5.30, we compare the average SER performance across different turbulence models, i.e., $C_n^2 = 1.7 \times 10^{-13}\text{m}^{-2/3}$ is assumed for strong turbulence parameter, $C_n^2 = 5 \times 10^{-14}\text{m}^{-2/3}$ is assumed for moderate turbulence parameter, and $C_n^2 = 4 \times 10^{-15}\text{m}^{-2/3}$ is assumed for the weak turbulence parameter. It is evident that the asymptotic curves closely match the slope of the performance curves and converge with the exact SER curve in the high SNR region.

This analysis highlights the consistency and reliability of the derived theoretical models. The close agreement between the asymptotic and exact SER curves, particularly in the high SNR region, underscores the robustness of the theoretical framework used to predict system performance under various turbulence conditions. This validation through diversity gain calculation further reinforces the precision of the derived models in characterizing the performance of FSO systems.

It is evident that G^d mainly depends on the type of detection scheme u, pointing error coefficient (ζ), and scattering parameters (α, β).

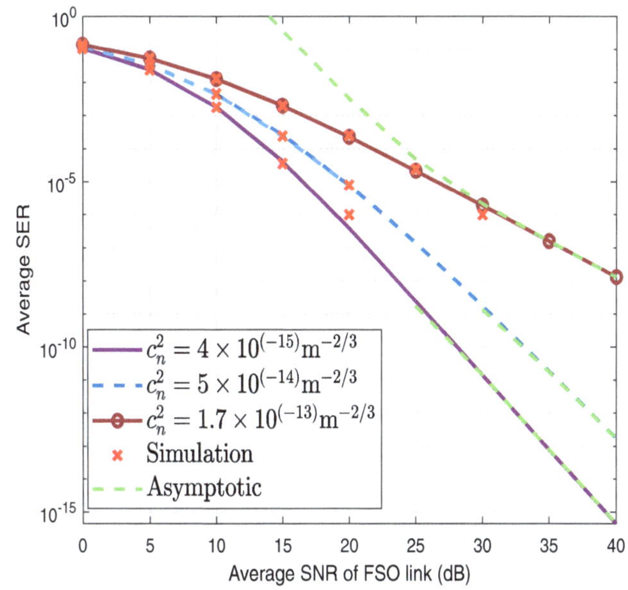

Fig. 5.30 Asymptotic SER comparison of various turbulence models

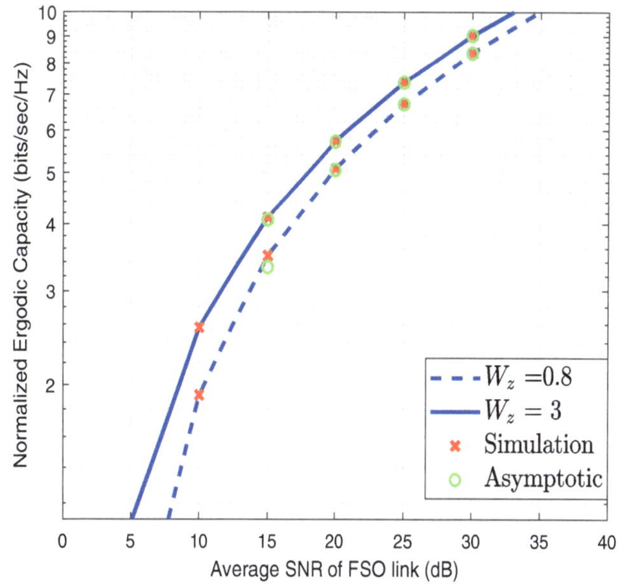

Fig. 5.31 Asymptotic ergodic capacity comparison of FSO system by varying incoming optical beamwidth

5.4.3.3 Ergodic Capacity

The expression for the asymptotic capacity of an FSO system is calculated similarly to that for asymptotic outage and SER calculations. By employing [16, eq.(07.34.06.0040.01)] in (5.85), the asymptotic capacity for FSO link is computed as

$$C_{fso}^{\infty} = \frac{\zeta^2 A_\gamma u^{(\alpha+\beta+\lambda-2.5)}}{\ln(2)(2\pi)^{(u-1)1.5}} \sum_{k=1}^{3u+2} \frac{\prod_{\substack{j=1 \\ j\neq k}}^{3u+2} \Gamma(b_{3,j}-b_{3,k}) \prod_{j=1}^{u+1} \Gamma(1-a_{3,j}+b_{3,k})}{\prod_{j=u+1}^{2u+2} \Gamma(a_{3,j}-b_{3,k})\Gamma(1+b_{3,k})} B_3^{b_{3,k}} \tag{5.91}$$

where $a_{3,j} = \{K_1, 0, 1, K_2\}$, $b_{3,j} = \{K_{3,j}, 0, 0\}$, and $B_3 = B^u\left(\frac{1}{\overline{\gamma}_{gC_u}}\right)$.

In Fig. 5.31, the ergodic capacity curve is depicted with variations in the optical beamwidth of the incoming signal (ω_z). The results indicate that the severity of pointing errors is significantly influenced by ω_z. Consequently, a reduction in ω_z leads to a noticeable decline in ergodic capacity performance. The asymptotic curves accurately follow the slope of the performance curves and align with the exact capacity curve in the high SNR region validating the derived expression.

5.5 Summary

In this chapter, an extensive performance analysis of the FSO communication system was conducted. The analysis considers various factors including atmospheric turbulence, nonzero boresight-pointing errors, and atmospheric attenuation. The unified closed-form expressions for key performance metrics such as outage probability, average SER, and ergodic capacity were derived by utilizing the generalized Málaga distribution and the IGGG distribution to model atmospheric turbulence-induced fading while also considering nonzero boresight-pointing errors and path loss. Additionally, the log-normal and gamma-gamma distributions are obtained as special cases of the Málaga distribution. Furthermore, it was shown that the IGGG channel model can effectively approximates the existing Fisher-Snedecor $\mathcal{F}$ and gamma-gamma turbulence models. Furthermore, the asymptotic closed-form expressions for all the performance metrics were derived, and the diversity gain of the FSO system under both Málaga and IGGG turbulence distributions was obtained. The numerical results revealed that the performance metrics are significantly influenced by parameters such as beamwidth (ω_z) and the severity of pointing errors. Moreover, the effects of various system and link parameters were also examined, including link lengths, atmospheric turbulence conditions, weather conditions, and the positional and orientation deviations caused by building sways and wind. The findings demonstrated that the system achieves optimal performance when both the transmitter and receiver are static, whereas the performance deteriorates when both are hovering under different scenarios.

Homework Problems

1. What are the primary advantages of free-space optics (FSO) compared to traditional fiber-optic cables?
2. What causes fluctuations in the intensity and phase of optical waves in FSO systems?

3. Which statistical models are the most notable for characterizing irradiance fluctuations in optical communication, and what are their respective strengths?
4. Derive the expressions for the outage capacity of Malaga and IGGG distributions based on the obtained CDF expressions.
5. **Given:**
   - Visibility range $V_r = 10$ km
   - Wavelength $\lambda_{fso} = 1550$ nm (or 0.00155 mm)
   - Size of scattering particles $r = 0.01$ mm

   Find the path loss h_l.
6. **Given:**
   - Radius of detection aperture $r_a = 0.005$ m
   - Standard deviation of pointing error $\sigma_p = 0.001$ m

   Find the power reduction at the receiver due to pointing error misalignment?
7. Derive the expressions for average SER and capacity of Malaga and IGGG distributions using CDF-based approach?
8. How does the performance of an FSO system differ when both the optical source and receiver are fixed versus when one is hovering ?
9. Derive the expressions for average SNR of Malaga and IGGG distributions?
10. How do BPSK and MPSK modulation schemes compare in terms of phase separation, bit error rate (BER), energy efficiency, bandwidth efficiency, and complexity of modulation and demodulation ?
11. How is diversity gain calculated, and what insights about the system can be drawn from the obtained diversity gain value?

References

1. Zhu, X., Kahn, J.M.: Free-space optical communications through atmospheric turbulence channels. IEEE Trans. Commun. **50**(8), 1293–1300 (2002)
2. Ghassemlooy, Z., Popoola, W.: Terrestrial Free-Space Optical Communications, Mobile and Wireless Communications: Network Layer and Circuit Level Design. InTech, Rijeka, Croatia (2010)
3. Andrews, L.C., Phillips, R.L.: Laser Beam Propagation Through Random Media. Bellingham, Washington (1998)
4. Al-Habash, M.A., Andrews, L.C., Phillips, R.L.: Mathematical model for the irradiance probability density function of a laser beam propagating through turbulent media. Opt. Eng. **40**(8), 1554–1562 (2001)
5. Jurado-Navas, A., Garrido-Balsells, J.M., Paris, J.F., Puerta-Notario, A.: A unifying statistical model for atmospheric optical scintillation. Weblink: http://arxiv.org/abs/1102.1915
6. Peppas, K., Alexandropoulos, G.C., Xenos, E.D., Maras, A.: The Fisher-Snedecor F-distribution model for turbulence-induced fading in free-space optical systems. J. Lightw. Technol. **38**(6), 1286–1295 (2020)
7. Shishter, Y.M., Young, R., Ali, F.H.: Irradiance probability density function for turbulence-induced fading in free space optics. J. Opt. Soc. Am. A. **39**(7), 1267–1274 (2022)
8. Singh, D., Swaminathan, R.: Comprehensive performance analysis of hovering UAV-based FSO communication system. IEEE Photon. J. **14**(5), 1–13 (2022)
9. Singh, D., Reddy, C.S.S., Swaminathan, R.: Hovering UAV-based FSO communications with DF relaying: A performance analysis. In: Proceedings of IEEE National Conference on Communications, Guwahati, India, 2023, pp. 1–6
10. Yahia, O.B., Erdogan, E., Kurt, G.K., Altunbas, I., Yanikomeroglu, H.: HAPS selection for hybrid RF/FSO 480 satellite networks. IEEE Trans. Aerosp. Electron. Syst. **58**, 2855–2867 (2022)
11. Ansari, I.S., Alouini, M.-S., Cheng, J.: Ergodic capacity analysis of free-space optical links with nonzero boresight pointing errors. IEEE Trans. Wireless Commun. **14**(8), 4248–4264 (2015)
12. Abdi, A., Lau, W.C., Alouini, M.S., Kaveh, M.A.: A new simple model for land mobile satellite channels: first- and second-order statistics. IEEE Trans. Wireless Commun. **2**(3), 519–528 (2003)
13. Uniyal, S., Vishwakarma, N., Swaminathan, R.: Multihop IRS-assisted free space optics communication with DF relaying: a performance analysis. Appl. Optics **62**(18), 4716–4726 (2023)
14. Andrews, L., Phillips, R.: Laser Beam Propagation Through Random Media, 2nd edn. SPIE (2005)
15. Sharma, S., Madhukumar, A.S., Swaminathan, R.: Effect of pointing errors on the performance of hybrid FSO/RF networks. IEEE Access **7**, 131418–131434 (2019)
16. Wolfram Research Inc.: Mathematica edition: version 8 Champaign, IL, USA (2010). Weblink: https://functions.wolfram.com/HypergeometricFunctions/MeijerG/
17. Andrews, L.C., Al-Habash, M.A., Hopen, C.Y., Phillips, R.L.: Theory of optical scintillation: Gaussian beam wave model. Waves Random Media **11**, 271–291 (2001)
18. Chatzidiamantis, N.D., Sandalidis, H.G., Karagiannidis, G.K., Matthaiou, M.: Inverse Gaussian modeling of turbulence-induced fading in free-space optical systems. J. Lightw. Technol. **29**(10), 1590–1596 (2011)
19. Wang, T., Strohbehn, J.: Perturbed log-normal distribution of irradiance fluctuations. J. Opt. Soc. Amer. A. **64**(7), 994–999 (1974)
20. Jurado-Navas, A., Garrido-Balsells, J.M., Paris, J.F., Puerta-Notario, A.: A Unifying Statistical Model for Atmospheric Optical Scintillation Numerical Simulations of Physical and Engineering Processes, pp. 181–206. InTech, Rijeka, Croatia (2011)
21. Cheng, M., Guo, Y., Li, J., Zheng, X., Guo, L.: Inverse gaussian gamma distribution model for turbulence-induced fading in free-space optical communication. Appl. Optics **57**(12), 3031–3037 (2018)
22. Yoo, S.K., Cotton, S.L., Sofotasios, P.C., Matthaiou, M., Valkama, M., Karagiannidis, G.K.: The Fisher—Snedecor F distribution: A simple and accurate composite fading model. IEEE Commun. Lett. **21**(7), 1661–1664 (2017)
23. Andrews, L., Phillips, R., Hopen, C., Al-Habash, M.: Theory of optical scintillation. J. Opt. Soc. Amer. A. **16**(6), 1417–1429 (1999)
24. Churnside, J.H., Clifford, S.F.: Log-normal Rician probability density function of optical scintillations in the turbulent atmosphere. J. Opt. Soc. Amer. **4**(10), 1923–1930 (1987)
25. Ramirez-Espinosa, P., Lopez-Martinez, F.J.: On the Utility of the Inverse Gamma Distribution in Modeling Composite Fading Channels. In: Proceedings of IEEE GLOBECOM, Waikoloa, HI, USA, 2019, pp. 1–6
26. Gradshteyn, I., Ryzhik, I.M.: Tables of Integrals, Series, and Products, 6th edn. Academic, New York, NY, USA (2000)
27. Ansari, I.S., Yilmaz, F., Alouini, M.: Performance analysis of free space optical links over Málaga (M) turbulence channels with pointing errors. IEEE Trans. Wireless Commun. **15**(1), 91–102 (2016)
28. Farid, A.A., Hranilovic, S.: Outage capacity optimization for free-space optical links with pointing errors. J. Lightw. Technol. **25**(7), 1702–1710 (2007)
29. Vishwakarma, N., Swaminathan, R.: On the capacity performance of hybrid FSO/RF system with adaptive combining over-generalized distributions. IEEE Photon. J. **14**(1), 1–12 (2022)

6 Visible Light Communication

6.1 Introduction

The amount of traffic transported by telecommunication networks has dramatically increased during the last ten years, as observed worldwide. The growing need for cloud-based applications, video calls, high-definition television, and high-speed Internet services computers has supported the need for additional innovation, study, and development in new, developing technologies that can provide end users with extremely high data rates. Spectrum congestion is a result of the current radio frequency (RF) wireless spectrum shortage, which requires immediate care. This is presently driving the so-called "tragedy of the commons" paradigm, whereby all users will contribute to the depletion of a common resource—in this case, available spectrum—without having any apparent goal of doing so. These kinds of circumstances occur in high-density settings like sporting arenas, music halls, airports, emergency rooms, etc., where strong user demand may result in a dire situation of restricted access. Particularly in densely populated areas, multipath propagation effects negatively impact the performance and availability of current RF-based communications. Relatively few high-definition channels can be supported in a given area due to the limited bandwidth and spectrum congestion of these RF systems [1]. This issue is especially problematic for indoor applications, where a high number of users compete for the same channel capacity, resulting in insufficient bandwidth to distribute among them. It is estimated that over 70% of wireless traffic takes place indoors, in locations such as homes and offices. Therefore, seamless indoor wireless communication requires low-cost and highly reliable technology. The only two viable solutions are to increase the bandwidth of the RF technology or to adopt alternatives such as optical technologies.

Future generations of wireless systems are expected to incorporate visible light technology, an environmentally friendly optical technology that has emerged as a promising solution for wireless communication. VLC operates by modulating the intensity of light emitted from LEDs to transmit data. This data is then received by photodiodes or other light-sensitive devices. Since the visible light spectrum (ranging from 380 nm to 750 nm wavelengths) is unregulated and abundant, VLC can potentially alleviate the "WiFi spectrum crunch," a problem caused by the overcrowding of the electromagnetic spectrum allocated to WiFi [2]. With VLC, data transmission rates can theoretically reach the order of terabytes per second, far exceeding the capabilities of current RF-based systems. Furthermore, in VLC, the channel characteristics change due to the movement of people or any obstruction, and thus, when the link is temporarily blocked or shadowed, the link experiences a much-reduced signal-to-noise ratio at the receiver. Hence, the time-varying effect of the channel cannot be entirely ignored in modeling the VLC system. In order to enhance user mobility, one must investigate the effect of user dynamics, such as varying pause time and velocity, on the received power. In an indoor VLC system, the received optical power depends on various factors, such as the location of the emitting LEDs, the desired user's location, and the different types of obstacles present in the room [3]. Other users in a multiuser indoor environment can also act as an obstruction to the intended user. Moreover, these other users might occasionally pause in between movements for a fixed period of time (referred to as a pause time). Pauses of moving users can impede the transmission of LoS and Non-line-of-sight (NLoS) signals, causing an abrupt drop in received power. The blockage height and radius significantly influence this abrupt decrease in the received power [4, 5].

A significant milestone in the history of VLC was reached in 2011, when Harald Haas introduced the concept of light fidelity (LiFi). LiFi, a type of VLC, enables high-speed, bidirectional, and fully networked wireless communication. Haas's demonstration generated substantial interest in VLC, spurring a surge in both research and commercial interest in the technology [6].

V. A. Bohara, Swaminathan R, *Optical and Wireless Access Networks*, Textbooks in Telecommunication Engineering, https://doi.org/10.1007/978-3-031-89696-5_6

6.1.1 Applications of VLC

The unique characteristics of VLC make it especially suitable for a wide range of applications:

- **Indoor Positioning:** VLC can be used for highly accurate indoor positioning systems. By leveraging the existing lighting infrastructure, VLC-based positioning systems can provide location-based services in environments where GPS signals are weak or unavailable, such as shopping malls, airports, and hospitals [7].
- **Vehicular Communications:** VLC can facilitate communication between vehicles and traffic infrastructure, enhancing road safety and enabling autonomous driving. LED headlights and traffic lights can be used as transmitters, while photodiodes in vehicles can serve as receivers [8].
- **Internet of Things (IoT):** VLC offers a promising solution for the IoT ecosystem, where a vast number of devices require connectivity. VLC can be integrated into smart lighting systems to enable data transmission between IoT devices, reducing the dependence on RF-based communication and alleviating spectrum congestion [9].
- **Medical Applications:** In healthcare settings, VLC can be utilized for the secure transmission of sensitive data. Since visible light does not interfere with medical equipment, VLC can be employed in hospitals to transmit patient data and monitor medical devices without causing electromagnetic interference [10].
- **Underwater Communication:** Unlike RF waves, visible light can penetrate water more effectively, making VLC a suitable technology for underwater communication. This can be particularly useful for underwater exploration, environmental monitoring, and communication between divers [11].

The wide range of applications of VLC systems are illustrated in Figs. 6.1, 6.2, 6.3, and 6.4.

Fig. 6.1 Vehicular communication

Fig. 6.2 Internet-of-Things

6.2 System Model of VLC

In the considered VLC system, LED-based transmitter acts as a source, and user equipment acts as the receivers in an indoor communication scenario. It is assumed that the LED transmitter is fixed and mounted above the user equipment, where LoS communication exists between the transmitter and receiver as shown in Fig. 6.5.

The relationship between the angles of irradiance (ϕ_k) and incidence (ψ_k) is given as [12]

$$\cos(\psi_k) = \cos(\phi_k) = \frac{L}{d_k}, \tag{6.1}$$

where L is the vertical distance between the transmitter and receiver and d_k is the Euclidean distance between them, which is given by

$$d_k = \sqrt{r_k^2 + L^2}, \tag{6.2}$$

where r_k denotes the horizontal link distance from the center of the illuminated area to the kth actuator.

The channel gain h_k between the transmitter and the kth receiver is expressed as

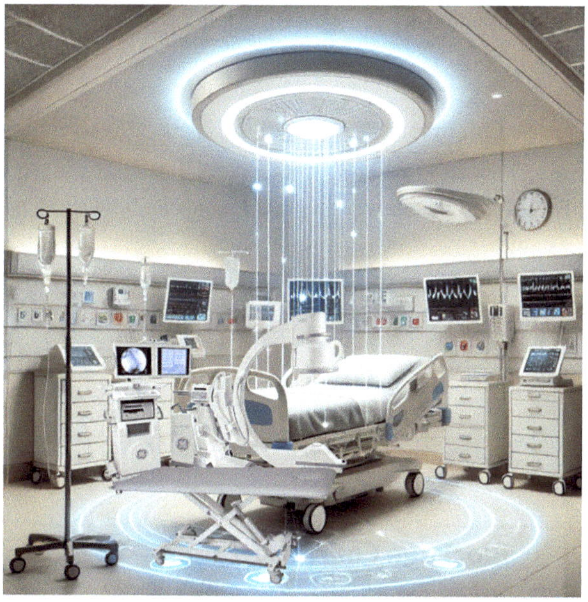

Fig. 6.3 Medical applications

Fig. 6.4 Underwater communication

$$h_k = \frac{(m+1)\cdot A_p \cdot R_p}{2\pi \cdot d_k^2} \cdot \cos^m(\phi_k) \cdot T(\psi_k) \cdot g(\psi_k) \cdot \cos(\psi_k), \tag{6.3}$$

where

- A_p is the physical area of the photodiode
- R_p is the responsivity of the photodiode
- $T(\psi_k)$ is the gain of the optical filter
- $g(\psi_k)$ is the gain of the optical concentrator
- m is the Lambertian order of emission determined by the semi-angle at the half luminance $\phi_{1/2}$

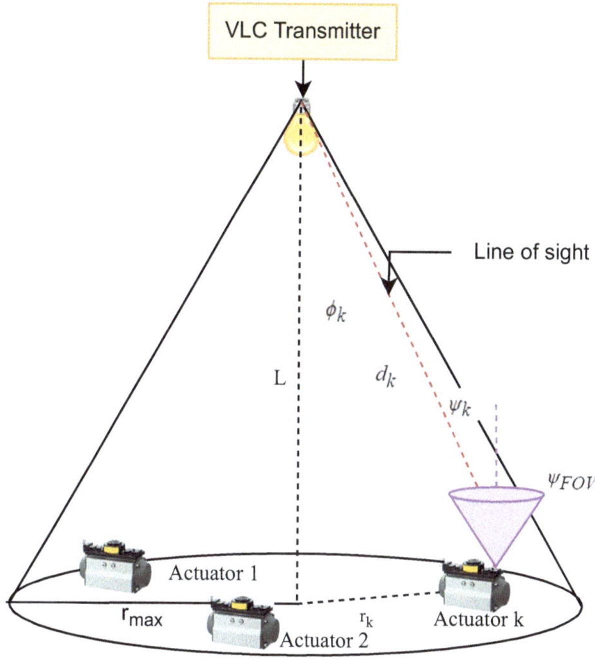

Fig. 6.5 VLC system model

The gain of the optical concentrator $g(\psi_k)$ is given as [12]

$$g(\psi_k) = \frac{n^2}{\sin^2(\psi_{\text{FOV}})}, \tag{6.4}$$

where n is the refractive index of the concentrator and ψ_{FOV} is the field of view of the receiver.

From Eqs. (6.1) and (6.2), the following relation is written as

$$(\cos(\phi_k))^m \cos(\psi_k) = \left(\frac{L}{d_k}\right)^{m+1} = \left(\frac{L}{\sqrt{r_k^2 + L^2}}\right)^{m+1}. \tag{6.5}$$

The channel gain h_k in the VLC system is represented in a concise manner as follows:

$$h_k = \chi \cdot \left(r_k^2 + L^2\right)^{-\frac{m+3}{2}}, \tag{6.6}$$

where χ is written as

$$\chi = \frac{(m+1)\cdot L^{m+1} \cdot T(\psi_k) \cdot G(\psi_k) \cdot R_p \cdot A_p}{2\pi}. \tag{6.7}$$

The PDF of the channel gain h_k is derived using the relation between h_k and the PDF of r_k and by using the transformation of random variables, which is given as [13]

$$f_{h_k}(h) = \left|\frac{d}{dh}u^{-1}(h)\right| \cdot f_{r_k}(u^{-1}(h)), \tag{6.8}$$

where $u(r) = \chi \cdot (r^2 + L^2)^{-\frac{m+3}{2}}$. By taking the inverse of this function and differentiating it with respect to h_k leads to the PDF of channel gain h_k and is given as

$$f_{h_k}(h) = \frac{2}{r_{\max}^2} \cdot \frac{1}{m+3} \cdot \left(\chi \cdot L^{m+1}\right)^{\frac{2}{m+3}} \cdot h^{-\frac{2}{m+3}-1} . \quad (6.9)$$

The PDF of h_k^2 is calculated using the PDF of channel gain h_k. Given that $h_k^2 = t$, the PDF of h_k^2, denoted by $f_{h_k^2}(t)$, is derived using the transformation of random variable technique and the same is obtained as

$$f_{h_k^2}(t) = \frac{1}{r_{\max}^2} \cdot \frac{1}{m+3} \cdot \left(\chi \cdot L^{m+1}\right)^{\frac{2}{m+3}} \cdot t^{-\frac{1}{m+3}-1} . \quad (6.10)$$

Now, the PDF of instantaneous SNR is derived using the PDF of h_k^2, since the instantaneous SNR γ is given as

$$\gamma = \overline{\gamma} \cdot h_k^2 , \quad (6.11)$$

where $\overline{\gamma}$ is the average SNR and h_k^2 is the squared VLC channel gain. The PDF of the instantaneous SNR γ is given by

$$f_\gamma(\gamma) = \frac{1}{\overline{\gamma}} \cdot f_{h_k^2}\left(\frac{\gamma}{\overline{\gamma}}\right) . \quad (6.12)$$

Substituting (6.10) in (6.12), the PDF of the instantaneous SNR can be simplified as

$$f_\gamma(\gamma) = \frac{1}{r_{\max}^2} \cdot \frac{1}{m+3} \cdot \left(\chi \cdot L^{m+1}\right)^{\frac{2}{m+3}} \cdot \overline{\gamma}^{\frac{1}{m+3}} \cdot \gamma^{-\frac{1}{m+3}-1} . \quad (6.13)$$

Now, to derive the cumulative distribution function (CDF) of instantaneous SNR γ, the PDF of instantaneous SNR will be integrated over the minimum and maximum bounds of the instantaneous SNR. The CDF of γ, denoted as $F_\gamma(\gamma)$, is given by

$$F_\gamma(\gamma) = \int_{\gamma_{\min}}^{\gamma_{\max}} f_\gamma(\gamma) d\gamma , \quad (6.14)$$

where $\gamma_{\min}$ and $\gamma_{\max}$ define the operational range of the VLC system, $\gamma_{\max}$ represents ideal conditions for optimal communication performance, and $\gamma_{\min}$ sets the threshold below which the system experiences outages or degraded performance in terms of BER, capacity, and overall link reliability. The expressions for $\gamma_{\min}$ and $\gamma_{\max}$ are given by [13]

$$\gamma_{\min} = \kappa_{\min} \cdot \overline{\gamma} \quad \text{and} \quad \gamma_{\max} = \kappa_{\max} \cdot \overline{\gamma} , \quad (6.15)$$

where

$$\kappa_{\min} = \frac{(m+1) \cdot L^{m+1}}{(r_{\max}^2 + L^2)^{\frac{m+3}{2}}} \quad \text{and} \quad \kappa_{\max} = \frac{(m+1) \cdot L^{m+1}}{L^{m+3}}$$

After substituting (6.13) in (6.14), the CDF can be written as

$$F_\gamma(\gamma) = \int_{\gamma_{\min}}^{\gamma_{\max}} \frac{1}{r_{\max}^2} \cdot \frac{1}{m+3} \cdot \left(\chi \cdot L^{m+1}\right)^{\frac{2}{m+3}} \cdot \overline{\gamma}^{\frac{1}{m+3}} \cdot \gamma^{-\frac{1}{m+3}-1} d\gamma . \quad (6.16)$$

By solving the above integral, $F_\gamma(\gamma)$ can be written as

$$F_\gamma(\gamma) = \omega - \Upsilon \cdot \gamma^{-\frac{1}{m+3}} , \quad (6.17)$$

where

$$\omega = \frac{r_{\max}^2 + L^2}{r_{\max}^2} \quad \text{and} \quad \Upsilon = \frac{\left(\chi^2 \cdot \overline{\gamma}\right)^{\frac{1}{m+3}}}{r_{\max}^2}$$

6.3 Blockage Characterization

The received optical power in the indoor VLC system varies with respect to the distance from the respective transmitting LEDs to the intended user inside the room. When there are multiple users inside the room, these users and other static obstacles like furniture, etc. act as a blockage to the intended user and create shadowing. This shadowing results in a sudden fall in received optical power as it can block both the LoS and the non-line-of-sight (NLoS) signals from the LED to the intended user. The amount of power reduction will depend on the height and width of the obstacles [4]. In this section, the effect of shadowing with static and dynamic blockages are considered. The optimum VLC parameters, such as LED semi-angle and the receiver FOV, under the effect of shadowing are also discussed in this section.

In order to analyze the impact of static and dynamic human blockages in the indoor VLC system performance, generally two models are utilized to characterize the blockages. The first model, with Matern hard-core point process (MHCP), realizes the static blockages wherein two blockages are forbidden to an absolute minimum distance. Further, to include the effects of dynamic human blockages moving with uniform velocity inside the room, the random waypoint (RWP) model is employed.

6.3.1 Characterization of Static Human Blockages Using MHCP

This section discusses the realization of the locations of the stationary human blockages inside the room using MHCP. In MHCP, two or more points are forbidden to be closer than a certain minimum distance. The MHCP is also used to plan the deployment of base stations in cellular networks and modeling of blockages and interference calculation in wireless networks [4]. One way to achieve such a minimum distance between points is to start with a parent point process with no such restriction and then remove points that violate the above condition. For the downlink communication (from AP to users), a target number of blockages (N_B) is considered. To model the location of the blockages, a modified version of type-II MHCP process is used, wherein the points are a result of random thinning of a parent PPP Φ_P with intensity λ_P. In particular, for each realization of Φ_P, a random mark [0,1] is assigned to each x_i in Φ_P. Then, x_i is deleted if there exists an x_j, $j \neq i$, such that the mark assigned to x_j is lower than the mark assigned to x_i. Let the resulting MHCP so created be denoted by Φ_B with intensity λ_B. Finally, the condition $N_B = \Phi_B(A)$ is checked, and the downlink analysis of the network is performed for only such realizations in which $N_B = \Phi_B(A)$ holds. The above MHCP is used to characterize the location of stationary homogeneous (same radii $r_1 \quad or \quad r_2$) and heterogeneous (different radii) blockages, where $r_1 < r_2$.

6.3.2 Characterization of Dynamic Blockages Using RWP

In this section, the dynamic effect of human blockages moving within the room with the uniform velocity has been analyzed. The RWP mobility model is a simple and straightforward stochastic model that describes a human blockage's movement behavior in given room size. In this model, a blockage randomly chooses a destination point ("waypoint") in the room and moves with uniform velocity on a straight line to the next destination point. After waiting for a specific pause time, it chooses a new destination and moves with uniform velocity to the destination and so on. Further, it may be noted the below analysis can be extended to the cases of nonuniform velocity distribution; however, for ease of analysis and analytical tractability, this section only considers uniform velocity distribution [4]. As discussed earlier, in a VLC system, the achievable user's data rate is related to the user-AP distance; therefore, including the spatial user distribution becomes essential.

Furthermore, note that human blockages are not necessarily distributed uniformly in typical indoor environments across the network area. On the contrary, their location distribution is characterized by a higher probability of being located near the center of the room than near the walls. Incidentally, the stationary distribution of the RWP model captures such a spatial configuration. For example, as discussed in [4], stationary distribution of users without pause time in an RWP model can be approximated by an elliptic-paraboloid function.

In the analysis, RWP model is considered in a rectangular area of size $R = a \times a$; a point (waypoint) S_1 is selected randomly in R. A pause time at point S_1 is chosen randomly in an interval $[t_{\min}, t_{\max}]$, upon the termination of pause time, a new waypoint S_2 is chosen uniformly at random in R, and the node starts moving from S_1 to S_2 along a straight line trajectory with velocity chosen uniformly at random in an interval $[v_{\min}, v_{\max}]$. The pause and movement process is repeated as above when the node arrives at the destination. For the sake of simplicity and without loss of generality, the RWP model in the below analysis has been used without any pause time. As the users continue to move, the distribution of the position of users approaches a stationary distribution. In particular, for a square area of size, $a \times a$, the stationary distribution of the user locations is

$$f_X(x) = f_{XY}(x, y) = \frac{36}{a^6}(x^2 - \frac{a^2}{4})(y^2 - \frac{a^2}{4}), \qquad (6.18)$$

where $\frac{-a}{2} \leq x \leq \frac{a}{2}$, $\frac{-a}{2} \leq y \leq \frac{a}{2}$ shows the coordinate of each node in the square area and $\mathbb{X} = [x \quad y]$ is the position vector. This stationary distribution is also represented in Fig. 6.6, which confirms the concentration of the users near the center of the room.

6.3.3 Effect of Blockage on SNR Calculation

In this section, the received SNR is calculated for the stationary blockage process using MHCP. It is assumed when the PD is in blockage, it receives no signal. The optical signal transmitted by the ith LED is given by

$$s_i(t) = P_{t_i}[1 + M_I x_i(t)], \qquad (6.19)$$

where P_{t_i} is the transmit power at the ith LED, x_i is the corresponding modulating OOK signal, and M_I is the modulation index [4]. Here, the first term (P_{t_i}) in 6.19 takes care of illumination, while the second term $P_{t_i} M_I x_i$ is for communication. After photodetection, assuming that the DC component of the detected electrical signal is filtered out at the receiver, the received signal at photodetector j is given by

$$y_j = \mathbf{R} P_{r_j} + n_j, \qquad (6.20)$$

where $\mathbf{R}$ is responsivity of photodiode, n_j is AWGN with $n_j = \mathbb{N}(0, {\sigma_j}^2)$, and P_{r_j} is expressed as

$$P_{r_j} = \sum_{i=1}^{N} H_{ij} P_{t_i} M_I x_i, \qquad (6.21)$$

where N is the number of LEDs and H_{ij} is the VLC channel coefficient between ith LED and jth PD, the channel impulse response can be expressed as

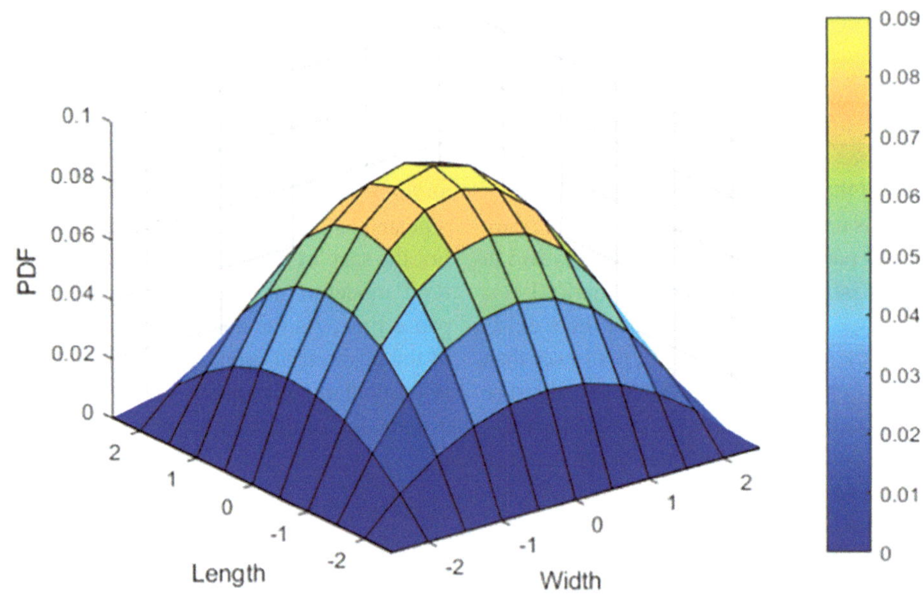

Fig. 6.6 Stationary RWP model location PDF

$$H_{ij} = \frac{(m+1)A_{PD}}{2\pi d_{ij}^2} \cos^m(\theta_{ij}) \cos(\phi_{ij})\, g_c(\phi_{ij})\, g_f, \tag{6.22}$$

where d_{ij} is the distance between the ith LED and the jth PD. The irradiance angle is denoted as θ_{ij}, while the incidence angle is represented by ϕ_{ij}. The physical area of the photodetector is designated as A_{PD}. g_f and g_c are denoted as the optical filter gain and concentrator gain, respectively. Further, the parameter m to indicate the Lambertian order. If there is a human blockage between the LED and the PD at distance d with probability $P_B(d)$ between transmitting LED and the PD, the VLC channel gain with blockage can be expressed as

$$H_{ij}^B = H_{ij}[1 - P(B)], \tag{6.23}$$

where $P(B)$ denotes the probability of encountering blockages. To effectively model human blockages, Matérn hard-core point process (MHCP) is a well-established approach in the literature [4].

The signal-to-noise ratio (SNR) at the receiver in the presence of blockages can be expressed as

$$SNR_B = \frac{(\mathbf{R} \sum_{i=1}^{N} H_{ij}^B P_{t_i} M_I x_i)^2}{\sigma_j^2} \tag{6.24}$$

The noise variance σ_j^2 at jth PD is the total noise comprising of shot noise and thermal noise.

6.3.4 Received SNR Profile with Human Blockage Analysis

In this section, the VLC system with human blockages is considered inside a standard room size of 5 m × 5 m × 3 m. The two transmitter configurations of 4 and 8 LEDs in a rectangular geometry are considered. The locations and the orientations of the VLC transmitters and the receiver are considered according to [4]. Both LoS and NLoS (with one-point reflection because of higher data rate threshold) [4] signals are considered in the analysis. The received SNR profile with human blockages inside the room has been plotted. Figures 6.7 and 6.8 show the received SNR profile with 4 LEDs having 6 and 8 human blockages of the same radius inside the room. It can be seen that there is a sudden fall in the received SNR whenever there is a blockage between the transmitting LED and the user inside the room. Based upon the distance of blockage from the transmitting LED, its shadow region has been calculated. As the distance from the LEDs to the blockage increases, it can be shown that the shadow region also decreases and vice versa. This results in different size areas of received SNR profile in shadowing due to the blockage and can also be seen in Figs. 6.7 and 6.8.

6.4 Behavior of Dynamic Human Blockages in a VLC Model

This section describes the behavior of dynamic human blockages characterized by the RWP mobility model in a VLC

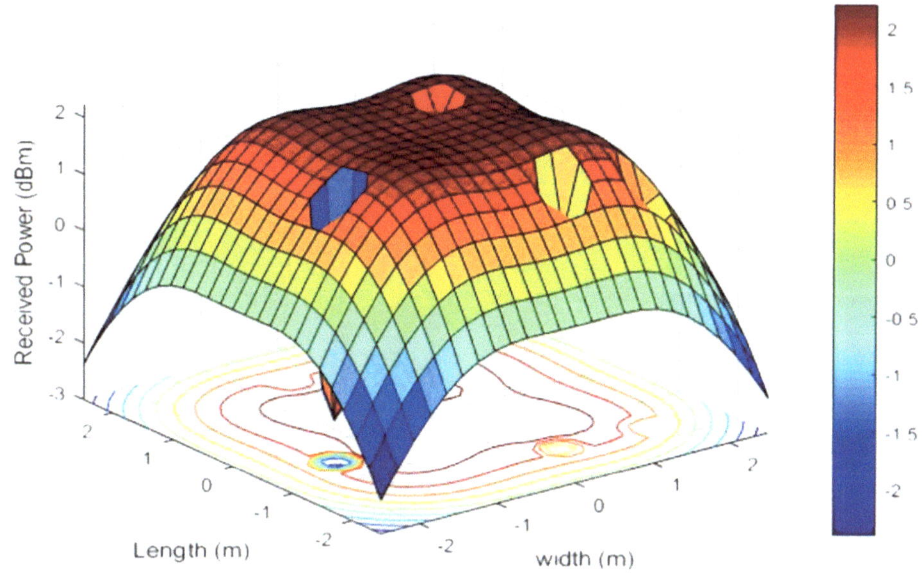

Fig. 6.7 $\lambda_B = 6$

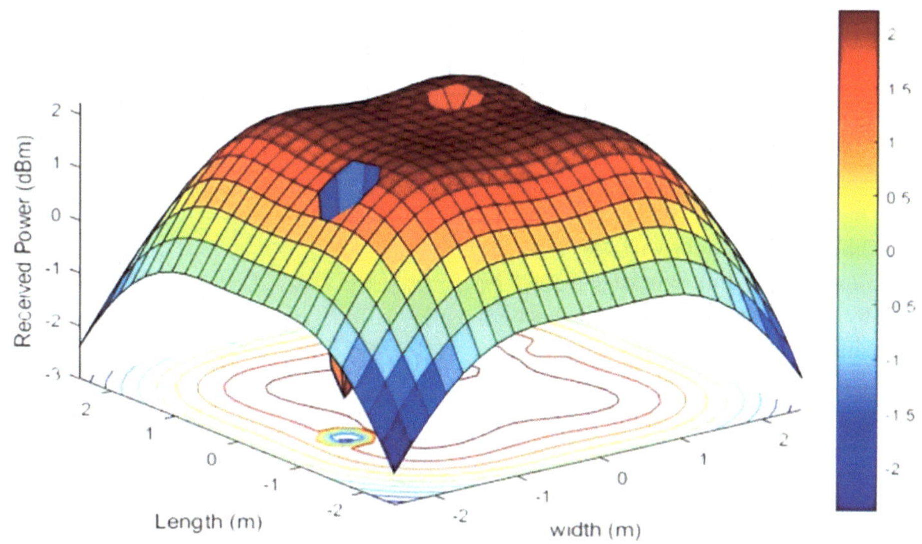

Fig. 6.8 $\lambda_B = 8$

channel model. The obstacles are considered to be cylindrical, radius of each blockage is r_B, and the height is h_B. All the transmitters (Tx) are located at 3 m above the ground, and the receiver (Rx) is at a height of 0.85 m (receiver plane height).

6.4.1 Shadowing Effect

This subsection discusses the shadowing effect for dynamic blockages characterized using the RWPMM. Therefore, a shadow of that blockage will be created each time a blockage pauses in the LoS path between any Tx and the Rx. Since the blockages are cylindrical, the shadow is supposed to be rectangular. There may be a communication link interruption between that specific Tx and Rx if the shadow is long enough to cover the receiver. This is known as the shadowing effect. A blockage requires to be in a certain region (known as the shadow region) for it to disrupt the Tx-Rx relationship. The condition of blockage to block the users has been derived later in this subsection.

However, it should be noted that if a blockage during its movement passes through a point that lies in the LoS path between any Tx and the Rx, the shadow formed (if any) on the Rx will be for a short and insignificant duration. It will have a negligible effect on the communication channel. Thus, we consider the shadowing effect of influencing the channel only when a blockage rests for a particular pause time in the shadow region between any Tx and the Rx before choosing a new destination point.

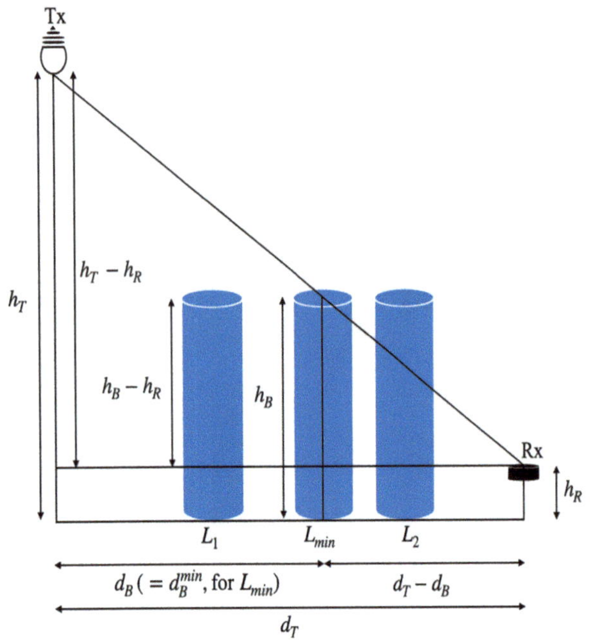

Fig. 6.9 Schematic for calculation of minimum blockage distance

As shown in Fig. 6.9, the distance of the blockage and the Rx from the base of the Tx are d_B and d_T, respectively. The shadow whose length is $d_T - d_B$ will fall on the Rx if the distance between the blockage and the base of the Tx is greater than d_B^{min}. Otherwise, the light will pass over the blockage, and no communication link blockage would be observed (as depicted by the blockage present at the location L_1 in Fig. 6.9). Using simple trigonometry, from Fig. 6.9

$$\frac{h_T - h_R}{d_T} = \frac{h_B - h_R}{d_T - d_B} \implies d_B = d_T.\left(1 - \frac{h_B - h_R}{h_T - h_R}\right)$$

$$\implies d_B = d_T.\left(\frac{h_T - h_B}{h_T - h_R}\right), \quad (6.25)$$

It is clear from Fig. 6.9 that if an obstacle pauses at a location $L1$, i.e., less than the distance $d_B^{\min}$, the shadowing effect will not occur, and consequently, the blockage will not affect the VLC link during that rest time.

6.4.2 Blockage Probability

The blockage probability $P_B(d_B)$ is calculated in this subsection. The object location distribution obtained finds the blockage probability, i.e., the probability that the center of an object (human) falls in the shadow region ABCD (Fig. 6.10) at a distance of d_B from the transmitter's base.

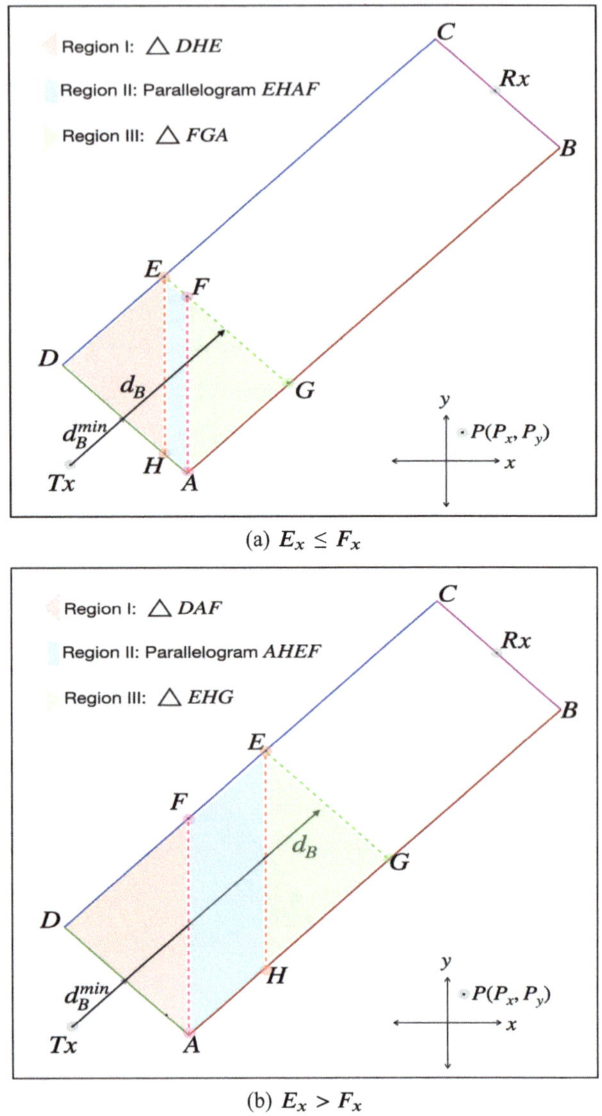

Fig. 6.10 Model to calculate blockage probability

Consider a Tx to be at any arbitrary point (T_x, T_y), and the Rx is positioned at (R_x, R_y) in the space $[0, a]^2$. The obstacle should at least be at a distance of $d_B^{\min}$ in order to block the link. So, to find the blockage probability, to contemplate the cases whenever a shadow is formed by any obstacle present at a distance of $d_B^{\min}$ or more from the base of Tx, i.e., it should be presented somewhere in the area ABCD (the shadow region) as shown in Fig. 6.10.

Let the line joining the base of the Tx and Rx (LoS) be given as $y = mx + c$, where

$$m = \frac{R_y - T_y}{R_x - T_x}, \quad (6.26)$$

and

$$c = -\frac{T_x(R_y - T_y)}{R_x - T_x} + T_y, \tag{6.27}$$

It has to be taken into account the entire shadow region (ABCD) instead of just the LoS. Since the obstacle is cylindrical, it must consider the LoS region with a buffer of $\pm\, r_B$ on both sides. So, a rectangular region ABCD described above, with width $2 \times r_B$, is referred to as the shadow region. This results in a shadow falling on the Rx if the center of the blockage is within this shadow region. Since the shadow region is rectangular, $\overline{AB} \parallel \text{LoS} \parallel \overline{CD}$, and thus, $\overline{AB}$ and $\overline{CD}$ will also have a slope of m:

$$\begin{aligned} \overline{AB}:\ & y = mx + c_2, \\ & \text{where } c_2 = -r_B\sqrt{1+m^2} + c \end{aligned} \tag{6.28}$$

and

$$\begin{aligned} \overline{CD}:\ & y = mx + c_1, \\ & \text{where } c_1 = r_B\sqrt{1+m^2} + c \end{aligned} \tag{6.29}$$

Let a line perpendicular to the LoS at a distance of d_B from the base of the LED intersect $\overline{AB}$ and $\overline{CD}$ at points G and E, respectively. Thus, we get

$$\begin{aligned} \overline{EG}:\ & y = -\frac{1}{m}x + \frac{T_x}{m} + T_y + d, \\ & \text{where } d = \frac{|d_B|}{\sqrt{1+\left(-\frac{1}{m}\right)^2}},\ m \neq 0 \end{aligned} \tag{6.30}$$

To find the blockage probability $P_B(d_B)$ as a function of d_B, the area AGED is calculated under the distribution curve obtained, where it substitutes $p_s = 0$ to analyze the effect of dynamic blockages with pause time solely.

Since the rectangle AGED is not aligned with the X-Y axes in the Cartesian plane, it is not integrable, hence intractable. Therefore, we divide the rectangle AGED into three regions. Let P_x and P_y denote any arbitrary point P's x and y coordinates. Before dividing the rectangle into three regions, two cases should be taken into account. The first one being, if $E_x \leqslant F_x$, and the second one, $E_x > F_x$. These conditions will alter the limits of the integration.

Examining the first case, i.e., $E_x \leqslant F_x$; if it extends a line parallel to $x = 0$ (Y-axis), passing from A, intersecting $\overline{EG}$ at F, and another line (parallel to Y-axis) from E intersecting $\overline{AD}$ at H. There are three regions : i) $\triangle$DHE (region I), ii) $\|^{gm}$EHAF (region II), and iii) $\triangle$FGA (region III).By using (6.28), (6.29), and (6.30) as the limits for the integration of the distribution to find the (blockage) probability of an obstacle being in the region AGED

$$\begin{aligned} P_B(d_B) &= \int_{D_x}^{E_x}\int_{AD}^{CD} f_{X,Y}(x,y)\,dy\,dx \\ &+ \int_{E_x}^{F_x}\int_{AD}^{EG} f_{X,Y}(x,y)\,dy\,dx \\ &+ \int_{F_x}^{G_x}\int_{AB}^{EG} f_{X,Y}(x,y)\,dy\,dx \\ &= \int_{D_x}^{E_x}\int_{-\frac{1}{m}x+\frac{x_T}{m}+y_T+d^{\min}}^{mx+c_1} P_{X,Y}(x,y)\,dy\,dx \\ &+ \int_{E_x}^{F_x}\int_{-\frac{1}{m}x+\frac{x_T}{m}+y_T+d^{\min}}^{-\frac{1}{m}x+\frac{x_T}{m}+y_T+d} P_{X,Y}(x,y)\,dy\,dx \\ &+ \int_{F_x}^{G_x}\int_{mx+c_2}^{-\frac{1}{m}x+\frac{x_T}{m}+y_T+d} P_{X,Y}(x,y)\,dy\,dx, \end{aligned} \tag{6.31}$$

$\forall\, D_x < E_x \leqslant F_x,\ A_x = F_x$ and $E_x = H_x$

Now, in the second case, i.e., $E_x > F_x$; if it extends a line parallel to $x = 0$ (Y-axis), passing from A, intersecting $\overline{CD}$ at F, and another line (parallel to Y-axis) from E intersecting $\overline{AB}$ at H. The three regions are obtained as shown in Fig. 6.10: i) $\triangle$DAF (region I), ii) $\|^{gm}$AHEF (region II), and iii) $\triangle$HGE (region III)

$$\begin{aligned} P_B(d_B) &= \int_{D_x}^{A_x}\int_{\overline{AD}}^{\overline{CD}} P_{X,Y}(x,y)\,dy\,dx \\ &+ \int_{A_x}^{E_x}\int_{\overline{AB}}^{\overline{CD}} P_{X,Y}(x,y)\,dy\,dx \\ &+ \int_{E_x}^{G_x}\int_{\overline{AB}}^{\overline{EG}} P_{X,Y}(x,y)\,dy\,dx \\ &= \int_{D_x}^{A_x}\int_{-\frac{1}{m}x+\frac{x_T}{m}+y_T+d^{\min}}^{mx+c_1} P_{X,Y}(x,y)\,dy\,dx \\ &+ \int_{A_x}^{E_x}\int_{mx+c_2}^{mx+c_1} P_{X,Y}(x,y)\,dy\,dx \\ &+ \int_{E_x}^{G_x}\int_{mx+c_2}^{-\frac{1}{m}x+\frac{x_T}{m}+y_T+d} P_{X,Y}(x,y)\,dy\,dx, \end{aligned} \tag{6.32}$$

$\forall\, E_x \geqslant F_x,\ A_x = F_x$ and $E_x = H_x$.

Hence, using either (6.31) or (6.32), the blockage probability of an obstacle (characterized using RWPMM) could be calculated depending on its distance from the base of the transmitter. No closed form of the integrals in the above equations was available. The simulation and analytical results of the blockage probability are analyzed in the results and discussion section.

6.4.3 Received SNR Evaluation

In this section, the received SNR is calculated of the VLC channel in the presence of dynamic blockages characterized by RWPMM.

To calculate the blockage probability $P_B(d_B)$, it can use either (6.31) or (6.32) depending on the location of the blockage. It is assumed that no signal is received whenever the PD is blocked by the obstacle. The optical signal $s_i(t)$ transmitted by the ith LED is as follows:

$$s_i(t) = P_T^i[1 + M_I x_i(t)], \quad (6.33)$$

where P_T^i is the ith LED's the transmit power, M_I is the modulating index, and $x_i(t)$ is the corresponding OOK modulated signal [14]. The first term in (6.33) (P_T^i) accounts for the illumination whereas the second term ($P_T^i M_I x_i(t)$) for the communication part. It is assumed that the DC component of the detected electric signal is filtered out at the Rx after photodetection. The y_j is the received signal at the photodetector j, and is expressed as

$$y_j = \mathcal{R}P_{r_j} + n_j, \quad (6.34)$$

where $\mathcal{R}$ is the responsivity of the PD and n_j is the additive white Gaussian noise (AWGN) with zero mean and σ_j^2 variance. Thus, we can write the AWGN as $n_j = \mathcal{N}(0, \sigma_j^2)$. The receiver plane is divided into a 25×25 grids which essentially represents the number of users' locations and the received power is integrated over all values of j in order to cover the whole room for analysis. Further, received power at the photodetector due to the ith LED, P_r^i is given by

$$P_r^i = P_T^i H_{LoS}^i + \sum_{w=1}^{N_w} P_T^i H_{NLOS^w}^i, \quad (6.35)$$

where $i = 1, 2, \ldots, N$ and N denotes the number of LEDs. If there is a blockage in between the ith LED and the PD, at a distance of $d_{B,i}$ from the base of the ith LED, the received power due to the ith LED, in the presence of blockage, can be expressed as

$$P_r^{B,i} = P_r^i(1 - P_B(d_{B,i})), \quad (6.36)$$

Using (6.34), (6.35), and (6.36), the total SNR due to all the LEDs at the receiver in the presence of blockages is given as follows:

$$SNR_B = \sum_{j=1}^{M} \frac{(\mathcal{R}\sum_{i=1}^{N} P_{r_j}^{B,i})^2}{\sigma_j^2}. \quad (6.37)$$

Table 6.1 System model parameters for dynamic blockages characterized by RWPMM

Parameter	Value
Room size	5 m × 5 m × 3 m
Receiver plane height h_R	0.85 m
Receiver active area A_{PD}	0.1 cm^2
Receiver elevation θ	90°
Field of views (FOVs) of receiver ψ_c	60°
Height of the objects h_B	1.80 m
Velocity v of the blockages	1 and 2 m/s
Transmitter height h_T	3 m
LED transmitted power P_T	200 mW
LED irradiance angle $\phi_{\frac{1}{2}}$	60°
Responsivity $\mathcal{R}$	0.55 A/W
Pause time t_p	10, 50 and 100 seconds
Optical filter gain T_s	1

where M is the number of receiver locations inside the room (Table 6.1). Thus, it can use the (6.35) and (6.37) to calculate the instantaneous received power and SNR to determine the receiver's performance (Fig. 6.11).

6.4.4 Result Analysis

Analytical and simulation plots for the blockage probability have been presented for two Rx locations: (2.5, 2.5) and (2.8, 3.2). We considered only 1 blockage moving with a uniform velocity of 1 m/s in the system, with two different pause times 10 and 100 seconds.

It can be observed in both cases that as the distance from the base of the LED increases, the blockage probability also increases. This is quite intuitive, as an increase in d_B results in the expansion of the area of rectangle AGED (Fig. 6.10) and, consequently, an increase in $P_B(d_B)$.

For the Rx at the center of the room, i.e., (2.5, 2.5), the blockage probability plots are identical for all 4 LEDs due to symmetry. However, for the Rx at (2.8, 3.2), $P_B(d_B)$ will vary accordingly, depending on the Tx's location. The minimum and maximum range of values of $P_B(d_B)$ is for the LED at (3.75, 3.75) and (1.25, 1.25), respectively. This is because the Tx3 (1.25, 1.25) is farthest from the location of Rx (2.8, 3.2), and hence, the area of blockage (AGED) would be maximum in this case. Following this, the probability of the communication link getting blocked is also higher. Similarly, the Tx2 (3.75, 3.75) is nearest to the Rx, and hence, the probability of getting blocked is the least in this case. A similar justification could be provided to understand the behavior of $P_B(d_B)$ for other LED locations.

The blockage probability increases with the increase in pause time. Values of $P_B(d_B)$ for 100 seconds are greater than the corresponding values of $P_B(d_B)$ for 10 seconds. With the

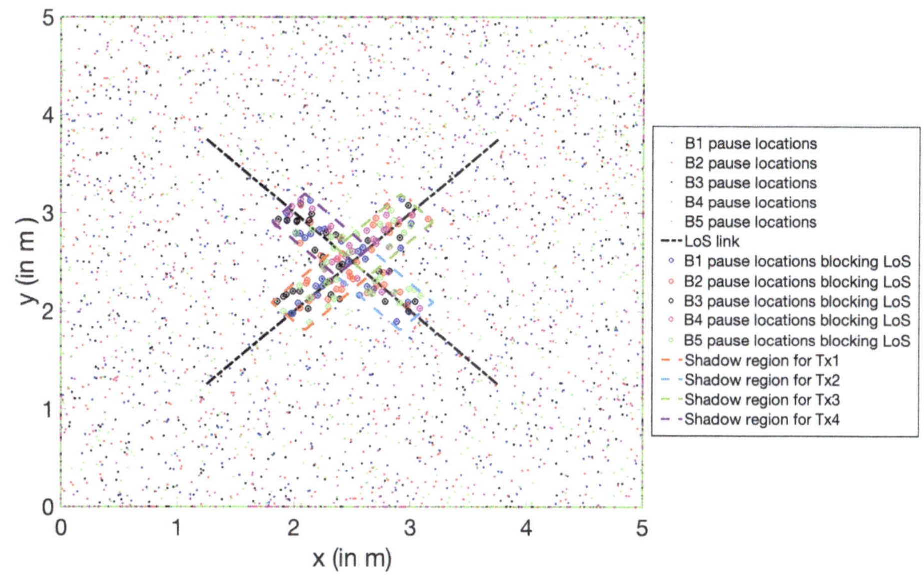

Fig. 6.11 Simulation system model

increase in pause time, the distribution values ($f_{X,Y}(x, y)$) of the blockage also increase, which is justified by the plots. Thus, for identical limits in (6.31) and (6.32), if the value of the distribution is more, $P_B(d_B)$ will be more. This explains the nature of $P_B(d_B)$ for different pause times.

6.4.5 Outage Probability Analysis

Outage probability is a critical performance metric, particularly in VLC systems, as it quantifies the likelihood that the received SNR will be insufficient to maintain a reliable communication link. A VLC system is said to be in outage, when the received instantaneous SNR falls below a predetermined threshold, denoted as γ_{th}. The outage probability P_{out} in terms of CDF of instantaneous SNR is written as

$$P_{out}(\gamma_{th}) = P(\gamma < \gamma_{th}) = F_\gamma(\gamma_{th}) \tag{6.38}$$

The outage probability will be obtained by substituting (6.17) in (6.38) and replacing γ with γ_{th}.

The system parameters assumed in the outage performance evaluation are listed in Table 6.2.

Table 6.2 System parameters for outage analysis

Notation	Description	Value
$T(\psi_k)$	Gain of optical filter	8
$G(\psi_k)$	Gain of optical concentrator	8
γ_{th_dB}	Threshold SNR (dB)	5
R_p	Responsivity of the photodiode (A/W)	0.84
A_p	Area of the photodiode (cm^2)	0.2
L	Vertical distance between the transmitter and receiver (m)	1.8
$\phi_{1/2}$	Semi-angle of the LED (degrees)	60
m	Lambertian order of the LED	$-1/\log_2(\phi_{1/2})$
r_{max}	Maximum radial distance (m)	2

6.4.6 Outage Capacity Analysis

The outage capacity C_{out} is an essential performance metric in VLC systems, representing the probability that the system's capacity falls below a certain threshold C_{th}. The normalized capacity in terms of bits/sec/Hz of a VLC system is given by $C_{norm} = \log_2(1+\gamma)$. Now, the outage capacity is defined as [15]

$$C_{out} = \Pr\left(\log_2(1+\gamma) \le C_{th}\right) \tag{6.39}$$

$$= \Pr\left(1+\gamma \le 2^{C_{th}}\right) \tag{6.40}$$

$$= \Pr\left(\gamma \le 2^{C_{th}} - 1\right) \tag{6.41}$$

$$C_{out} = F_\gamma\left(2^{C_{th}-1}\right) \tag{6.42}$$

In order to evaluate the outage capacity, both simulation and analytical studies are conducted. The system parameters assumed for outage capacity evaluation are listed in Table 6.3.

6.4.7 Average BER Analysis for BPSK Modulation

The average BER in VLC system refers to the probability that a transmitted symbol will be incorrectly detected at the receiver. In other words, it is the rate at which errors occur in the received symbols compared to what was originally transmitted. BER is an important performance metric

Table 6.3 System parameters for outage capacity analysis

Notation	Description	Value
$T(\psi_k)$	Gain of optical filter	8
$G(\psi_k)$	Gain of optical concentrator	8
$C_{\mathrm{th\_dB}}$	Threshold capacity (dB)	− 3
R_p	Responsivity of the photodiode (A/W)	0.8
A_p	Area of the photodiode (cm^2)	0.2
L	Vertical distance between the transmitter and receiver (m)	1.8
$\phi_{1/2}$	Semi-angle of the LED (degrees)	40
m	Lambertian order of the LED	$-1/\log_2(\phi_{1/2})$
$r_{\max}$	Maximum radial distance (m)	2

because it directly affects the quality and reliability of the communication system. It is influenced by various factors, such as SNR, modulation scheme, channel conditions, and error correction techniques.

The mathematical framework to derive the average BER has been formulated using the CDF-based approach [16]. The average BER expressions considering a binary modulation scheme in terms of CDF are written as

$$\mathrm{BER} = \frac{1}{2\sqrt{\pi}} \int_0^{\infty} \frac{F_\gamma(\gamma)}{\sqrt{\gamma}} e^{-\gamma} d\gamma \tag{6.43}$$

By substituting Eq. (6.17) in (6.43) and by restricting the limits of integration from $\gamma_{\min}$ to $\gamma_{\max}$, the average BER expression is rewritten as [12]

$$\mathrm{BER} = \frac{1}{2\sqrt{\pi}} \int_{\gamma_{\min}}^{\gamma_{\max}} \frac{\left(\omega - \Upsilon \cdot \gamma^{-\frac{1}{m+3}}\right)}{\sqrt{\gamma}} e^{-\gamma} d\gamma \tag{6.44}$$

After solving the above integral using the upper incompete gamma function [13] as shown below

$$\Gamma(\alpha, x) = \int_x^{\infty} e^{-t} t^{\alpha-1}\, dt\,,$$

the closed-form expression for average BER is written as

$$\mathrm{BER} \tag{6.45}$$

$$= \frac{1}{2\sqrt{\pi}} \times \left\{ w \times \Gamma(0.5) \times [\Gamma(0.5, \gamma_{\min}) - \Gamma(0.5, \gamma_{\max})] \right\} \tag{6.46}$$

$$+ \left\{ \Upsilon \times \Gamma\left(\frac{m+1}{2m+6}\right) \times \left[\Gamma\left(\frac{m+1}{2m+6}, \gamma_{\max}\right) \right.\right. \tag{6.47}$$

$$\left.\left. - \Gamma\left(\frac{m+1}{2m+6}, \gamma_{\min}\right) \right] \right\} \tag{6.48}$$

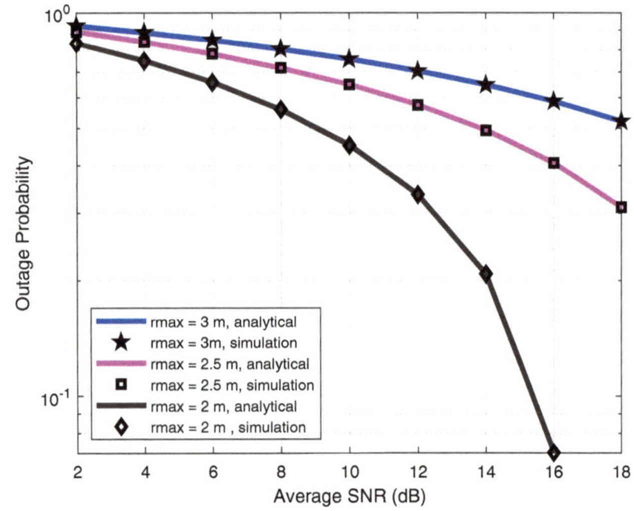

Fig. 6.12 Outage probability for VLC system with different maximum radial distances

6.5 Simulation Results and Discussions

In Fig. 6.12, the outage probability performance of the VLC system for varying maximum cell radii, specifically $r_{\max} = 2$ m, $r_{\max} = 2.5$ m, and $r_{\max} = 3$ m, is presented. The maximum cell radius defines the radial distance from the center of the illuminated area to its outer boundary, with the LED transmitter positioned above the cell center. The results show that increasing the maximum cell radius leads to a higher outage probability. For instance, at $r_{\max} = 3$ m, the outage probability at an average SNR of 18 dB is approximately 0.52. In contrast, for $r_{\max} = 2$ m, a similar outage probability is observed at an average SNR of 9 dB, indicating an SNR gain of 9 dB for the smaller cell radius.

Figure 6.13 illustrates the BER performance for varying maximum cell radii, specifically $r_{\max} = 1.8$ m, $r_{\max} = 2$ m, and $r_{\max} = 2.2$ m, with respect to the average SNR in a BPSK modulated VLC system. The results show that as $r_{\max}$ increases, the BER increases for the same average SNR values. For instance, at $r_{\max} = 2.2$ m, the BER at an average SNR of 20 dB is approximately 2.2×10^{-3}, while at $r_{\max} = 1.8$ m, a similar BER is achieved at an average SNR of 15 dB. This corresponds to an SNR gain of approximately 5 dB for the smaller cell radius.

This behavior is attributed to the fact that larger cell radii spread the light over a wider area, reducing signal intensity at the edges and increasing the likelihood of signal outages. In contrast, smaller cell radii focus the light within a more confined area, leading to stronger signal reception and lower outage and BER values. These findings emphasize the trade-off between expanding coverage area and maintaining communication reliability, underscoring the importance of

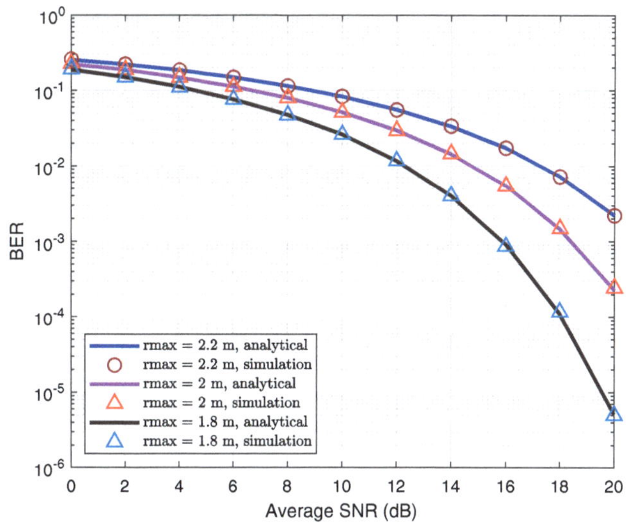

Fig. 6.13 Variation of maximum cell radius on BER in BPSK modulation

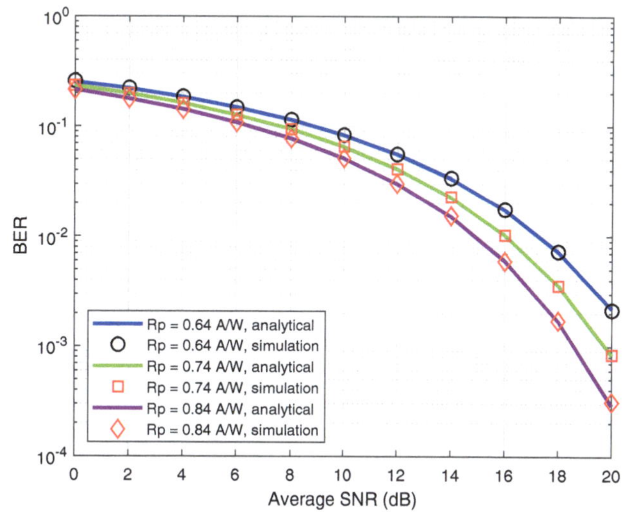

Fig. 6.15 Variation of responsivity on BER in BPSK modulation

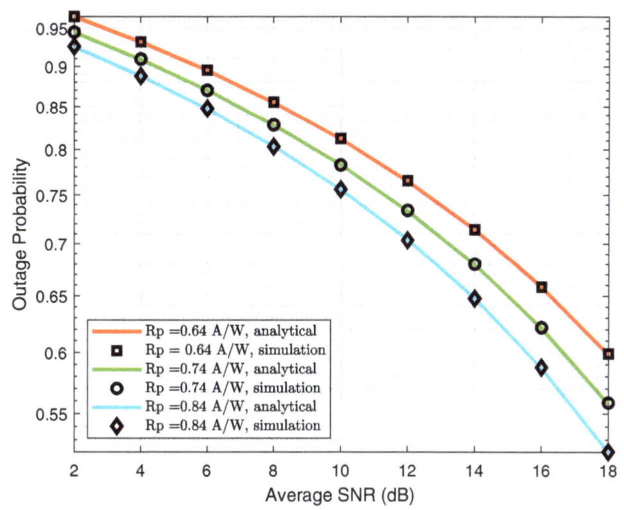

Fig. 6.14 Outage probability for VLC system with different photodiode responsivity

optimizing the maximum cell radius to ensure robust VLC system performance.

Figure 6.14 shows the outage probability performance of the VLC system for varying photodiode responsivity values, specifically $R_p = 0.64$ A/W, $R_p = 0.74$ A/W, and $R_p = 0.84$ A/W, as a function of the average SNR. The photodiode responsivity, R_p, represents the efficiency of the photodiode in converting the received light into an electrical signal. Higher responsivity results in better signal reception and thus improved system performance. The results demonstrate that as the responsivity increases, the outage probability decreases for the same average SNR values. For example, at $R_p = 0.84$ A/W, the outage probability at an average SNR of 18 dB is approximately 0.599. In contrast, for $R_p = 0.64$ A/W, the same outage probability is achieved at an average SNR of 15.5 dB, representing an SNR gain of approximately 2.5 dB for the higher responsivity.

Figure 6.15 illustrates the BER performance for varying photodiode responsivity values, specifically $R_p = 0.64$ A/W, $R_p = 0.74$ A/W, and $R_p = 0.84$ A/W, as a function of the average SNR in a BPSK modulated VLC system. The results indicate that increasing the photodiode responsivity reduces the BER for the same average SNR values. For example, at $R_p = 0.84$ A/W, the BER at an average SNR of 20 dB is approximately 0.0022. In comparison, for $R_p = 0.64$ A/W, a similar BER is achieved at an average SNR of 17.5 dB, corresponding to an SNR gain of 2.5 dB for the higher responsivity.

This behavior is explained by the fact that higher photodiode responsivity enables more efficient conversion of light into electrical signals, resulting in a stronger received signal and a lower likelihood of outages and BER. As responsivity increases, the system becomes more robust, maintaining reliable communication at lower SNR levels. These findings highlight the importance of optimizing photodiode responsivity to improve VLC system performance, particularly in scenarios where minimizing outage probability and BER is crucial.

Figure 6.16 presents the outage probability performance for different photodetector areas, specifically $A_p = 0.1\,\text{cm}^2$, $A_p = 0.15\,\text{cm}^2$, and $A_p = 0.2\,\text{cm}^2$, as a function of the average SNR in a VLC system. The photodetector area (A_p) plays a significant role in determining how much light is captured, directly influencing the strength of the signal and thus the system's reliability. The results indicate that as the photodetector area increases, the outage probability decreases for the same average SNR values. For instance, at $A_p = 0.2\,\text{cm}^2$, the outage probability at an average SNR of 20 dB is approximately 0.648, while for $A_p = 0.1\,\text{cm}^2$,

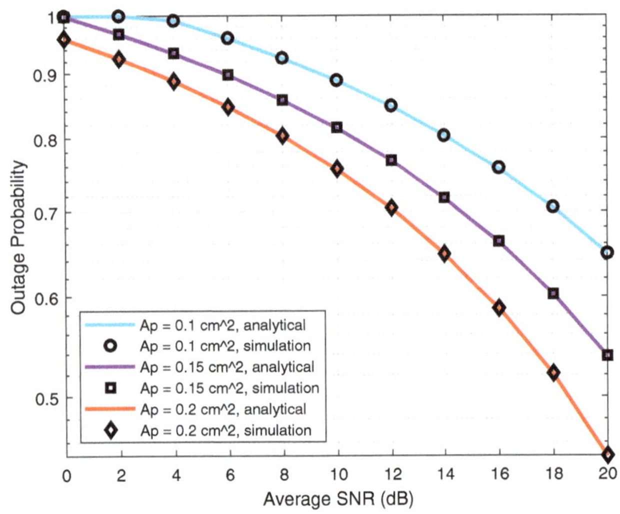

Fig. 6.16 Outage probability for VLC system with different photodiode areas

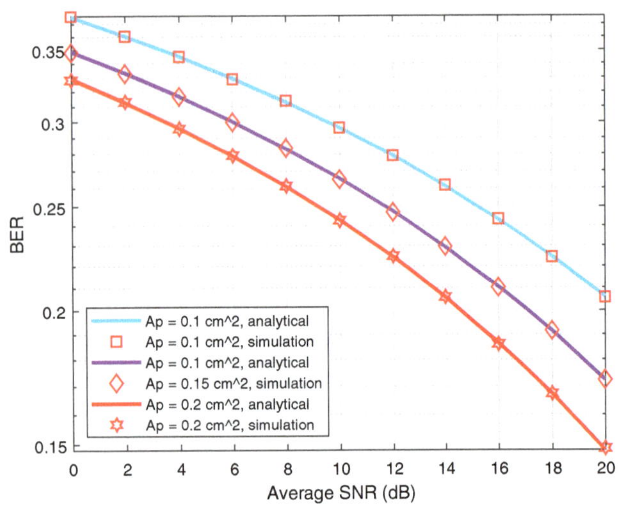

Fig. 6.17 Variation of photodiode area on BER in BPSK modulation

a similar outage probability is achieved at an average SNR of 14 dB, yielding an SNR gain of 6 dB for the larger photodetector area.

Figure 6.17 illustrates the BER performance for varying photodetector areas, specifically $A_p = 0.1\,\text{cm}^2$, $A_p = 0.15\,\text{cm}^2$, and $A_p = 0.2\,\text{cm}^2$, as a function of the average SNR in a BPSK modulated VLC system. The results show that increasing the photodetector area significantly reduces the BER for the same average SNR values. For instance, at $A_p = 0.2\,\text{cm}^2$, the BER at an average SNR of 20 dB is approximately 0.206. In contrast, for $A_p = 0.1\,\text{cm}^2$, the same BER is achieved at a lower average SNR of 14 dB, resulting in an SNR gain of 6 dB for the larger photodetector area.

This reduction in outage probability and BER with increasing photodetector area can be attributed to the enhanced ability of larger detectors to collect more light, which improves overall signal quality, reduces the likelihood of the signal falling below the threshold SNR, and decreases the possibility of receiving bits with errors. The findings indicate that optimizing the photodetector area is vital for minimizing outage probability and BER, thereby improving the robustness of VLC systems, especially in environments where reliable communication is critical.

Figure 6.18 illustrates the comparison of the outage capacity (C_{out}) for varying capacity thresholds, specifically $C_{\text{th}} = -3\,\text{dB}$, $C_{\text{th}} = 0\,\text{dB}$, and $C_{\text{th}} = 3\,\text{dB}$, as a function of the average SNR in a VLC system. The capacity threshold (C_{th}) defines the minimum capacity required to maintain reliable communication, and increasing this threshold generally increases the probability of outage due to higher capacity requirements. At $C_{\text{th}} = -3\,\text{dB}$, an outage capacity of approximately 0.7879 is observed at an average SNR of 20 dB, while the same outage capacity is achieved at an average SNR of 14 dB for $C_{\text{th}} = 3\,\text{dB}$, indicating an SNR gain of 6 dB. This behavior is explained by the fact that lower threshold values allow for more lenient capacity requirements, reducing the likelihood of the instantaneous capacity falling below the threshold.

As the capacity threshold increases, the probability of outage increases because the system requires a higher instantaneous capacity to maintain communication. However, once the threshold reaches higher values, the outage capacity begins to saturate, as the system cannot sustain such high capacity requirements, leading to frequent outages. These findings demonstrate that optimizing the capacity threshold is essential for balancing system performance, with lower thresholds yielding better outage performance.

6.5.1 Monte Carlo Simulation Details

To validate the analytical findings, Monte Carlo simulations are performed for each performance metric. For outage probability, the simulation procedure is carried out by generating 10^6 samples of cell radius values considering users are uniformly distributed in the cell area and same number of independent random channel coefficients as well. The random channel coefficient is generated using (6.6). After that, the calculation of $\gamma_{\min}$ is carried out using (6.14). Further, the calculation of instantaneous SNR is carried out using (6.11). The VLC system is said to be in outage if the instantaneous SNR falls below the system designer threshold SNR. For better clarity and understanding, the simulation steps are depicted in Fig. 6.19.

Now, for the simulation of outage capacity, a procedure similar to outage probability is adopted in the terms of generating cell radius values and random channel coefficient. After that, the instantaneous SNR is also calculated in the similar manner. The system capacity is said to be in outage if the

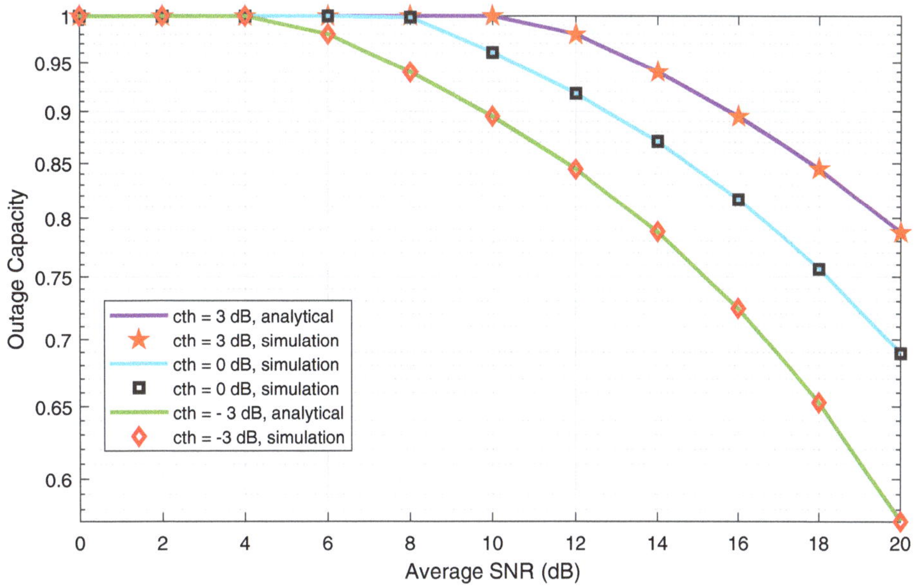

Fig. 6.18 Outage capacity comparison for VLC system

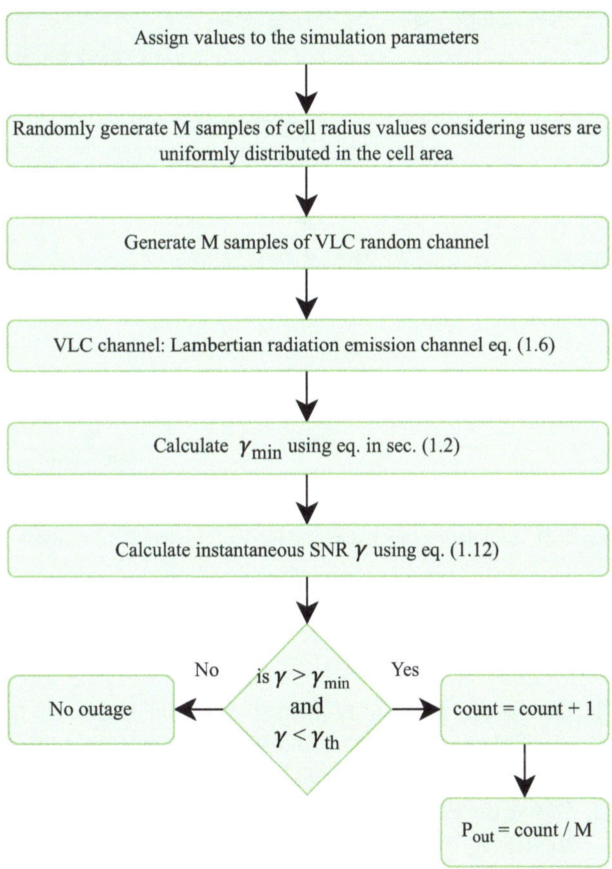

Fig. 6.19 Outage probability simulation steps for VLC system

calculated capacity is below the threshold value. The capacity is calculated using the following equation, $C = \log_2(1+\gamma)$. For each instantaneous SNR, capacity is calculated and the condition for outage is checked. The complete simulation process is depicted in Fig. 6.20

For BER simulation, random symbols corresponding to BPSK modulation scheme are generated. Afterward, the symbols are sent through the VLC channel modeled using (6.6). From the received symbol, maximum-likelihood (ML) detection is employed to retrieve the transmitted symbols. Finally, symbol error rate (SER) is computed by comparing the received symbols with the transmitted symbols. For BPSK modulation, it is to be noted that the SER is equal to BER. The detailed steps involved in Monte Carlo simulations for computing BER are given in Fig. 6.21.

6.6 Summary

In this chapter, an extensive performance analysis framework for a VLC communication system and the analysis of dynamic human blockages for indoor VLC are discussed. The analytical expressions for various performance metrics such as outage probability, outage capacity, and average BER considering BPSK modulation scheme are obtained. The impact of various system parameters such as responsivity, area of photo detector, maximum cell radius, threshold capacity, etc.

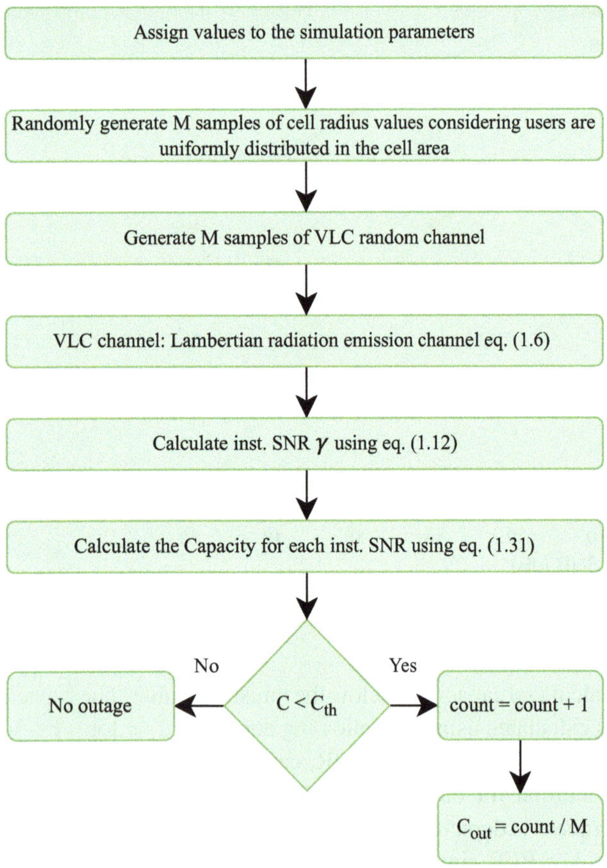

Fig. 6.20 Simulation block diagram for VLC system outage capacity analysis

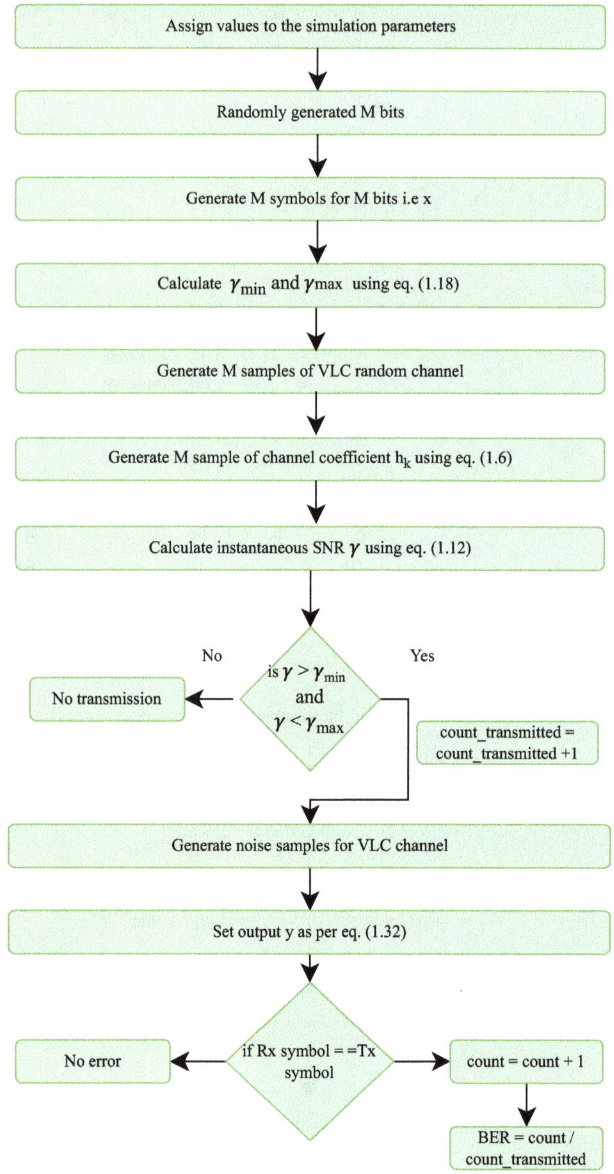

Fig. 6.21 Simulation block diagram for VLC system BER analysis

on the system performance is discussed in great detail. Furthermore, the effect of blockages was analyzed for an indoor VLC by utilizing the random waypoint mobility model with pause time to realize the dynamic human blockages. The analytical expression of the received SNR was calculated for varying sizes of human blockages using RWP. The analytical results were in close agreement with the simulation results, which validates the analytical framework. Additionally, the analysis was extended for multiple blockage scenarios with different pause times. The system performance of this time-varying channel was investigated in terms of instantaneous power. The received power profiles also show the effect of shadowing due to blockages.

These findings underscore the importance of optimizing vital system parameters to ensure reliable communication links in VLC systems. By carefully tuning these parameters, the performance of the VLC system can be significantly improved, resulting in more robust and efficient communication with effect of blockages. This comprehensive analysis provides valuable insights into the design and optimization of VLC systems, ensuring robust and efficient communication in various application scenarios.

Homework Problems

1. Derive the probability density function (PDF) of the squared channel gain h_k^2 in a VLC system, starting from the PDF of channel gain h_k.
2. Derive the PDF of the instantaneous signal-to-noise ratio (SNR), γ_k, of a VLC system.
3. Derive the cumulative distribution function (CDF) of the instantaneous SNR, γ_k, for a VLC system using the PDF of γ_k. Discuss the significance of this CDF in analyzing system performance.
4. Define outage probability and outage capacity in the context of a VLC system. Explain the significance of the

parameters r_{max} (maximum radial distance), A_p (area of the photodiode), and R_p (responsivity of the photodiode) on the outage probability and outage capacity of the system.
5. Define BER in the context of a VLC system. Discuss the significance of the parameters r_{max} (maximum radial distance), A_p (area of the photodiode), and R_p (responsivity of the photodiode) in optimizing the BER of the system.
6. Calculate the values of γ_{min} and γ_{max} for a VLC system under the following conditions: $r_{max} = 2\,\text{m}$, $L = 1.8\,\text{m}$, and Lambertian order $m = 1.44$. Compute these thresholds for the following average SNR values ($\overline{\gamma}$): (a) 5 dB, (b) 10 dB, and (c) 15 dB.
7. Find the VLC channel gain with blockage probability of 0.3 if the channel gain without blockage given as 4.97×10^{-11}.
8. Define outage capacity. Prove the outage capacity, $C_{out} = F_\gamma\left(2^{C_{th}-1}\right)$.
9. Find the distance of the blockage from the base of the Tx (d_B) for height of transmitter and receiver of 3.5m and 0.8m, respectively. Given, $d_T = 0.5\,\text{m}$ and height of the blockage is 1 m.
10. Explain shadowing effect with its disadvantages in VLC.

References

1. Jokanovic, D., Josipovic, M.: RF spectrum congestion: Resolving an interference case. In: Proceedings of IEEE International Conference on Microwaves, Communications, Antennas and Electronic Systems (COMCAS), Tel Aviv, Israel, 2011, pp. 1–4
2. Singh, S., Kakamanshadi, G., Gupta, S.: Visible light communication-an emerging wireless communication technology. In: Proceedings of 2nd International Conference on Recent Advances in Engineering & Computational Sciences (RAECS), Chandigarh, India, 2015, pp. 1–3
3. Singh, A., Srivastava, A., Bohara, V.A., Jagadeesan, A.K.: Performance of indoor VLC system under random placement of LEDs with nonimaging and imaging receiver. IEEE Syst. J. **16**(1), 868–879 (2022)
4. Singh, A., Ghatak, G., Srivastava, A., Bohara, V.A., Jagadeesan, A.K.: Performance analysis of indoor communication system using off-the-shelf leds with human blockages. IEEE Open J. Commun. Soc. **2**, 187–198 (2021)
5. Singh, A., Srivastava, A., Bohara, V.A.: Optimum LED semiangle and the receiver FOV selection for indoor VLC system with human blockages. In: Proceedings of IEEE 95th Vehicular Technology Conference (VTC2022-Spring), 2022, pp. 1–7
6. Haas, H.: Visible light communication. In: Proceedings of Optical Fiber Communications Conference and Exhibition (OFC), Los Angeles, CA, USA, 2015, pp. 1–72
7. Kabiri, A., Tatababa, F.S.: Large scale indoor VLC positioning using image sensor with limited field of view. In: The Proceedings of 29th Iranian Conference on Electrical Engineering (ICEE), Tehran, Iran, Islamic Republic, 2021, pp. 859–863
8. Matus, V. et al.: Implementation of a low-cost vehicular VLC system and can bus interface. In: Proceedings of 11th International Symposium on Communication Systems, Networks & Digital Signal Processing (CSNDSP), Budapest, Hungary, 2018, pp. 1–5
9. Kumari, M., Sharma, N., Chauhan, R., Joshi, K., Kumar, A.: Integrated PON-VLC system for IoT applications. In: Proceedings of the third International Conference on Distributed Computing and Electrical Circuits and Electronics (ICDCECE), Ballari, India, 2024, pp. 1–6
10. Caputo, S., Borghini, G., Jayousi, S., Rashid, A., Mucchi, L.: Visible light communications for healthcare applications: opportunities and challenges. In: Proceedings of IEEE 17th International Symposium on Medical Information and Communication Technology (ISMICT), Lincoln, NE, USA, 2023, pp. 1–6
11. Arfan, M., Lakshminarayana, C.: VLC for underwater operations: Li-Fi solution for underwater short range communication. In: Proceedings of International Conference on Electrical, Electronics, Communication, Computer, and Optimization Techniques (ICEECCOT), Msyuru, India, 2018, pp. 638–642
12. Yang, L., Yan, X., Benevides da Costa, D., Tsiftsis, T.A., Yang, H.-C., Alouini, M.-S.: Indoor mixed dual-hop VLC/RF systems through reconfigurable intelligent surfaces. IEEE Wireless Commun. Lett. **9**(11), 1995–1999 (2020)
13. Yin, L., Wu, X., Haas, H.: On the performance of non-orthogonal multiple access in visible light communication. In: Proceedings of IEEE 26th Annual International Symposium on Personal, Indoor, and Mobile Radio Communications (PIMRC), 2015 pp. 1354–1359
14. Neokosmidis, I., Kamalakis, T., Walewski, J.W., Inan, B., Sphicopoulos, T.: Impact of nonlinear LED transfer function on discrete multitone modulation: analytical approach. J. Lightwave Technol. **27**(22), 4970–4978 (2009)
15. Singh, D., Swaminathan, R., Pham, A.T.: Multiple HAPS-based space-air-ground networks with FSO communication: a performance analysis. OPtica (OSA) Appl. Optics **63**(9), 2362–2375 (2024)
16. Uniyal, S., Vishwakarma, N., Singh, D., Swaminathan, R.: IRS-aided hybrid FSO/RF communication system with selection combining. In: Proceedings of IEEE GLOBECOM Workshops, Kuala Lumpur, Malaysia, 2023, pp. 1–6

7 Optical Wireless Communication for Vehicular Networks

7.1 Introduction

The previous chapter discussed the channel modeling and performance analysis for fiber-less optical networks. There is a need to look into measures through which the requirements of high mobility traffic, such as vehicular networks, could be met. In this chapter, we focus on how optical wireless networks could be integrated with vehicular networks to meet the critical requirements in high-mobility scenarios. This chapter is further subdivided into hybrid RF-VLC Vehicle-to-everything (V2X) models and integrated FiWi-based vehicular networks. In the first part of this chapter, we discuss how VLC can complement and strengthen RF communication in V2X network. The second part of this chapter develops a framework that integrated PON with wireless fronthaul network to provide the services to the vehicular users. Further, various types of vehicular data traffic standards and their corresponding QoS requirements are also explained in this chapter.

7.2 Hybrid RF-VLC V2X System

Next-generation technologies are expected to support improved road safety and driving comfort in the future Intelligent transportation system (ITS). To fully support Connected autonomous vehicle (CAV), next-generation vehicles will be equipped with a wide range of sensors, and thus, there is a strong demand for reliable near-real-time exchange of sensing and control data. Such a demand will be filled by V2X communication, which comprises a wide range of communication technologies such as vehicle-to-vehicle (V2V) and vehicle-to-infrastructure (V2I) communications [1]. The most salient V2X communication technologies are Dedicated short-range communication (DSRC)-aided V2X and Cellular-V2X (C-V2X). While DSRC-V2X represents a mature cost-efficient V2X technology, C-V2X has attracted much attention in recent years due to its significantly improved coverage, throughput, and latency. Thanks to sophisticated cellular infrastructure, C-V2X outperforms due to centralized resource allocation and enhanced communication/sensing capabilities. Several Third Generation Partnership Project (3GPP) V2X initiatives (such as long-term evolution-vehicle-to-everything (LTE-V2X) and Fifth generation (5G) New Radio (NR)-V2X) have contributed to the prominence of C-V2X. With an abundance of both advanced sensors and communication devices, however, new challenges arise for the emerging next-generation V2X networks [2]. More explicitly, stringent system reliability ($\sim$99.999%), end-to-end latency (<5 ms), coverage-quality, spectral efficiency, energy rating, networking, and privacy/security specifications should be met to support various C-V2X use case requirements [3]. Although the current C-V2X technology (such as 5G-NR-V2X) offers substantial performance gains over its predecessor, the improved performance is achieved at the cost of requiring additional spectral/hardware resources, while utilizing cellular-based system architectures and mechanisms. Thus, V2X networks based on cellular standards may not be able to meet the aforementioned rigorous requirements and use cases of the emerging intelligent autonomous vehicles. A paradigm shift from conventional communication networks in favor of more flexible and diversified approaches is necessary. In fact, this transformation is beginning to take shape with the intensifying research into 6G wireless communication networks aiming for incorporating disruptive concepts [4]. In addition to intelligent and ubiquitous V2X systems, 6G is expected to provide significant data rate increases (e.g., up to Tbps), extremely fast wireless access (e.g., in the range of sub-milliseconds), and massive increase in wireless connections (e.g., billions of connected devices) as well as more extensive, more energy-efficient, and more environmentally friendly 3D communications.

Pure RF links may suffer from excessive RF interference in scenarios of high road traffic density, which increases the communication latency owing to packet delivery failures

V. A. Bohara, Swaminathan R, *Optical and Wireless Access Networks*, Textbooks in Telecommunication Engineering, https://doi.org/10.1007/978-3-031-89696-5_7

and aggressive ARQ retransmission attempts. This, in turn, increases the spectrum congestion in dynamic vehicular environments. As a possible solution, the non-interfering unlicensed VLC band may be harnessed in unison with the RF band for improved V2X communications, while supporting enhanced security [5]. Furthermore, a VLC-enabled V2X system will need minimal setup costs as VLC-based V2X can use LEDs or LDs that is already present in the vehicular head- and tail-lights or in street/traffic lights. Despite the aforementioned benefits, stand-alone VLC networks also have their drawbacks, including their limited coverage distance, sensitivity to background light, and line-of-sight (LOS) blockage. These impediments are conveniently circumvented by the classic RF wireless networks, which exhibit wider coverage and higher transmission integrity in the absence of LOS. 6G-V2X is envisioned to play a pivotal role in attaining the ambitious goals for the Internet of vehicles by satisfying the more rigorous Key performance indicators (KPIs) that were partially fulfilled by 5G for vehicle communications. Indeed, it is expected that 5G use case categories will evolve to Further enhanced Mobile Broadband (FeMBB), mobile broadband and low-latency (MBBLL), ultramassive machine-type communication (UMMTC), and massive Low-Latency Machine-Type communication (mLLMT) with extreme requirements such as data rates over 1 Tbps, end-to-end delays lower than 1 ms, network availability and reliability beyond 99.999% [6], extreme connection density of over 10^7 devices/km^2, and spectrum efficiency over five times that of 5G while supporting extreme mobility [6]. By intelligently combining, it is shown that the integration of VLC and RF improves the overall system performance to meet the stringent KPI requirements of 6G-V2X networks.

There are two major V2X streams: dedicated short-range communication (DSRC)-aided V2X and cellular-V2X (C-V2X) [7, 8]. While DSRC represents a mature cost-efficient V2X technology, CV2X has attracted much attention in recent years because of its significantly improved coverage, throughput, and latency. The latter is due to the improved physical layer, centralized resource allocation, as well as sophisticated cellular infrastructure.

7.2.1 Channel Modeling

As shown in Fig. 7.1, there are five primary scenarios in which VLC can complement and strengthen RF communication in V2X networks: (1) V2V communications via front lights or back lights, (2) U2V (UAV-to-Vehicle) communication, (3) V2V communication via Reconfigurable Intelligent Surfaces (RIS), (4) V2X communications via traffic lights, and (5) V2X communications via street lights. The latter may be viewed as a second layer of ubiquitous small-cell VLC BSs. In addition to increasing data rates, VLC has the potential to address some of the limitations of traditional V2X communication based on RF. For example, in the left bottom corner of Fig. 9.2, the RF-based V2V communication of two cars separated by a large bus may suffer from severe packet loss due to the shadowing effect. In this case, the transmitting car may use VLC to communicate with the bus; subsequently, the bus could forward the messages to the receiving car in the shadowed region. Moreover, the data packets can also be relayed using traffic/street lights at urban intersections, allowing vehicles to interact across perpendicular streets, where traditional RF-based solution is often plagued by severe packet loss (Table 7.1).

For a Vehicular VLC (V-VLC) system, the channel DC gain between k-th user and RSU can be modeled using Lambertian emission model as follows [9]:

$$h_k = \frac{(m+1)A_R}{2\pi d_k^2} \cos^m(\phi_k)\cos(\Psi_k)T_s(\Psi_k)G(\Psi_k), \quad (7.1)$$

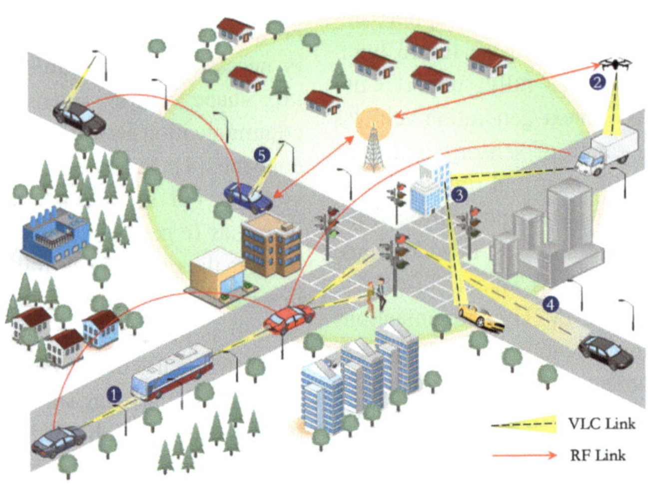

Fig. 7.1 Illustration of a generic hybrid RF-VLC communication in a vehicular network

Table 7.1 Comparison of DSRC/IEEE 802.11p, C-V2X, VLC, LTE, and 5GNR V2X

Features	DSRC/IEEE 802.11p	C-V2X	VLC	LTE	5G NR
Standard	IEEE 802.11p	3GPP Rel-14& 15	IEEE 802.15.7	3GPPRel-10/11/12	3GPPRel-15/16
Frequency Band(s)	5.86–5.92 GHz	2.4 GHz, 5.2 GHz	300–800 THz	450 MHz–4.99 GHz	700 MHz–100 GHz
Channel width	10 MHz	20 MHz	NA	Upto 100 MHz	NA
Bit Rate	3–27 Mbps	6.54 Mbps	11.67 Kbps–96 Mbps THz	Upto 1 Gbps	Upto 20 Gbps
Range	Upto 1 km	100–500 m	<200 m	Tens of meters to 30 km	Ubiquitous
Mobility support	Upto 60 km/h	Low	Low	Upto 350 km/h	Upto 500 km/h
V2V support	Yes:Ad hoc	Yes:Ad hoc	Yes	Yes:via D2D	Yes
V2I support	Yes	Yes	Yes	Yes	Yes
Deployment	RSU	Hotspot, access Point	Available road lights	May use the available eNodes B	NSA and SA mode
Market penetration	Low	High	Low	Potentially high	NA

where, A_R, d_k, ϕ_k, and Ψ_k denote the area of PD, the Euclidean distance between the kth vehicle and RSU, the angle of irradiance, and the angle of arrival (AoA) respectively. $T_s(\Psi_k)$ denotes the gain of the optical filter at the receiver; in aforementioned, m is the order of the Lambertian model, which is given by $m = -\frac{\ln(2)}{\ln\left(\cos\left(\phi_{\frac{1}{2}}\right)\right)}$. $G(\Psi_k)$ denotes the gain of optical concentrator at the receiver front end, which is given as

$$G(\Psi_k) = \begin{cases} \frac{n^2}{\sin^2 \Psi_{FOV}}; & \text{if } 0 \leq \phi_k \leq \Psi_{\text{FOV}}, \\ 0; & \text{if } \phi_k > \Psi_{\text{FOV}}, \end{cases} \tag{7.2}$$

where n denotes the refractive index of the optical concentrator. As given in [6], $\cos(\phi_k) = \cos(\psi_k) = \frac{x_k}{\sqrt{(h^2+x_k^2)}}$ where, h denotes the height of RSU. (5.1) can be rewritten as:

$$h_k = \frac{(m+1)A_R}{2\pi} \frac{x_k^{(m+1)}}{(h^2+x_k^2)^{\frac{(m+3)}{2}}} T_s(\Psi_k)G(\Psi_k), \tag{7.3}$$

The channel power can be obtained as: $Z_k^{\text{V-VLC}} = (h_k)^2$. In the case of Vehicular RF (V-RF) communication, the received signal amplitude in RF RF-based V2V channel is assumed to follow the Rayleigh probability distribution function (PDF). Since Rayleigh fading is considered, fading gain (h_x) is an exponential random variable of unit mean.

7.2.2 Performance Evaluation

This section presents the performance analysis for the outage probability, throughput, and data-oriented characterization for a hybrid RF-VLC V2X system.

7.2.2.1 Outage Probability

In this subsection, the performance of the V-VLC in the presence of the aggregate interference and noise variance, σ^2 in terms of outage probability as a performance metric using the moment generating functional (MGF) approach, is characterized. In interference-limited channels, an outage occurs when the SINR falls below a given SINR threshold, β [10]. Mathematically,

$$\begin{aligned} P_{\text{out,VLC}}(\beta) &= \mathbb{P}(\text{SINR} < \beta), \\ &= \mathbb{P}\left(\frac{S}{I_{\text{VLC}} + \sigma^2} < \beta\right). \end{aligned} \tag{7.4}$$

The desired electrical signal power S and interference I can be formulated as

$$\begin{aligned} S &= \mathcal{R}^2 Z_o P_{\text{VLC}}, \\ I_{\text{VLC}} &= \sum_{x_k \in \Phi_{\text{PPP}}} \mathcal{R}^2 Z_k P_{\text{VLC}}. \end{aligned} \tag{7.5}$$

In aforementioned, P_{VLC} and $\mathcal{R}$ denote the transmission power for V-VLC and the responsivity of PD respectively. The electrical SINR can be represented as:

$$\text{SINR} = \frac{1}{\frac{I}{\mathcal{R}^2 Z_o P_{\text{VLC}}} + \frac{1}{\alpha_0}}, \tag{7.6}$$

where $\alpha_0 = \frac{\mathcal{R}^2 Z_o P_{\text{VLC}}}{\sigma^2}$. From (5.4), the outage probability for V-VLC assuming that desired vehicle is transmitting can be given as:

$$P_{\text{out,VLC}}(\beta) = \mathbb{P}\left(\frac{I}{\mathcal{R}^2 Z_o P_{\text{VLC}}} + \frac{1}{\alpha_0} > \frac{1}{\beta}\right). \tag{7.7}$$

A random variable W is defined as

$$W = \frac{I}{\mathcal{R}^2 Z_o P_{\text{VLC}}} + \frac{1}{\alpha_0}, \tag{7.8}$$

Hence, Eq. (5.7) can be rewritten as

$$P_{\text{out,VLC}}(\beta) = \mathbb{P}(W > \beta^{-1}) = 1 - \mathcal{F}_W(\beta^{-1}). \tag{7.9}$$

In general, it is quite difficult to obtain a closed-form solution for $\mathcal{F}_W(\beta^{-1})$. Hence, it is essential to make use of numerical inversion of Laplace transform to find CDF, $\mathcal{F}_W(\beta^{-1})$. The CDF of a random variable W is related to the Laplace transform of $\mathcal{F}_W(w)$ as

$$\mathcal{F}_W(w) = \frac{1}{2\pi j} \int_{c-j\infty}^{c+j\infty} \mathcal{L}_{\mathcal{F}_{W(w)}} \exp(sw) ds. \quad (7.10)$$

where j denotes imaginary unit ($\sqrt{-1}$). The aforementioned integral can be discretized to get a series using the trapezoid rule, and then the infinite series can be truncated to get a finite sum using the Euler summation [11]. Also, $\mathcal{L}_{\mathcal{F}_{W(w)}}(s) = \frac{\mathcal{L}_{W(s)}}{s}$. Moreover, (5.11) can be approximated as

$$P_{\text{out,VLC}}(\beta) \approx 1 - \frac{2^{-B}\exp(\frac{A}{2})}{\beta^{-1}} \sum_{b=0}^{B} \binom{B}{b} \times \sum_{c=0}^{C+b} \frac{(-1)^c}{D_c} \Re\left\{\frac{\mathcal{L}_W(s)}{s}\right\}. \quad (7.11)$$

where $D_c = 2$ (if $c = 0$) and $D_c = 1$ (if $c = 1, 2, 3, ...$) and $s = \frac{(A+j2\pi c)}{2\beta^{-1}}$. The estimation error is controlled by three parameters A, B, and C. Using the well-established result given in [11] and [12], in order to achieve an estimation accuracy of $10^{-\eta}$ (i.e., having the $(\eta-1)$th decimal correct), A, B, and C have to be at least equal to $\eta\ln 10$, $1.243\eta - 1$, and 1.467η, respectively. Setting $A = 8\ln 10$, $B = 11$, $C = 14$ achieves stable numerical inversion with an estimation error of 10^{-8}.

Using the definition of the Laplace transform of the probability distribution of a random variable,

$$\begin{aligned}\mathcal{L}_W(s) &= \mathbb{E}_I\left[\exp\left(-s\left(\frac{I}{\mathcal{R}^2 Z_o P_{\text{VLC}}} + \frac{1}{\alpha_0}\right)\right)\right], \\ &= \mathbb{E}_I\left[\exp\left(-\frac{s}{\alpha_0}\right)\exp\left(-\frac{sI}{\mathcal{R}^2 Z_o P_{\text{VLC}}}\right)\right], \\ &= \exp\left(-\frac{s}{\alpha_0}\right)\mathbb{E}_{\phi_{\text{PPP}}} \\ &\times \left[\prod_{x\in\phi_{\text{PPP}}} \exp\left(-\frac{sk'x^{2(m+1)}}{Z_o(h^2+x^2)^{(m+3)}}\right)\right].\end{aligned} \quad (7.12)$$

Here $k' := \left(\frac{(m+1)A_R}{2\pi} T_s(\psi)G(\psi)\right)^2$. The expectation in (5.13) can be solved using probability generating functional Laplace (PGFL)[1] defined for a homogeneous Poisson point process [10, Th 4.9].

$$\begin{aligned}&\mathbb{E}_{\phi_{\text{PPP}}}\left[\prod_{x\in\phi_{\text{PPP}}} \exp\left(-\frac{sk'x^{2(m+1)}}{Z_o(h^2+x^2)^{(m+3)}}\right)\right] \\ &= \exp\left[-\lambda_{MAC}\int_{-\infty}^{-R}(1 - \exp \right. \\ &\left.\times\left(-\frac{sk'x^{2(m+1)}}{Z_o(h^2+x^2)^{(m+3)}}\right)\right) dx\right].0 \le \phi_k \le \Psi_{\text{FOV}}\end{aligned} \quad (7.13)$$

Equation (5.13) can be determined numerically using standard software tools like `MATLAB` or `MATHEMATICA`.

Assuming free space path loss propagation model, the interference experienced at RSU for RF-based V2V communication can be expressed as sum of RF power received from all the interferers as

$$I_{\text{RF}} = \sum_{x\in\Phi_{\text{PPP}}} P_{\text{RF}} G_t G_r \ell h_x (h^2 + x^2)^{-\frac{\alpha}{2}}, \quad (7.14)$$

where $\ell = \frac{c^2}{(4\pi)^2 f_0^2}$. In aforementioned expression, P_{RF}, α, G_t, and G_r are the transmission power for V-RF, the path loss exponent, the antenna gains for transmitter and receiver, respectively [13]. In order to calculate outage probability $P_{\text{out,RF}}(\zeta)$, it is more convenient to express it as a function of probability of successful transmission, $P_s(\zeta)$, where $P_{\text{out,RF}}(\zeta)$ can be expressed as

$$P_{\text{out,RF}}(\zeta) = 1 - P_s(\zeta) \quad (7.15)$$

The probability of successful transmission, $P_s(\zeta)$ can be calculated as

[1] The PGFL can be envisioned as an equivalent for point process of the characteristic function (CF) or moment generating function (that provides an alternative description of random variables). It enables to compute the Laplace transform (LT) of random variables of the form $F = \sum_{X_i\in\Psi_{\text{PPP}}} g(X_i)$. In mathematical form, LT of such function can be given as:

$$\mathcal{L}(s) = \mathbb{E}[e^{-sF}] = \mathbb{E}\left[\exp(-s\sum_{X_i\in\Psi} g(X_i))\right] = \mathbb{E}\left[\prod_{X_i\in\Psi} e^{-sg(X_i)}\right].$$

$$
\begin{aligned}
P_s(\zeta) &= \mathbb{P}(\mathrm{SINR} > \zeta), \\
&= \mathbb{P}\left(\frac{P_{\mathrm{RF}} G_t G_r \ell h_x D^{-\alpha}}{I_H + I_V + \mathcal{N}_0} > \zeta\right), \\
&= \mathbb{E}_{I_H+I_V}\left[\mathbb{P}\left(h_x > \frac{\zeta}{P_{\mathrm{RF}} G_t G_r \ell D^{-\alpha}}\right.\right. \\
&\quad \times (I_H + I_V + \mathcal{N}_0))\Big], \\
&= \exp\left(-\frac{\zeta \mathcal{N}_0}{P_{\mathrm{RF}} G_t G_r \ell D^{-\alpha}}\right) \\
&\quad \times \mathbb{E}_{I_H+I_V}\left[\exp(-\frac{\zeta(I_H + I_V)}{P_{\mathrm{RF}} G_t G_r \ell D^{-\alpha}})\right], \\
&= \underbrace{\mathcal{L}_{I_H}\left(\frac{\zeta}{P_{\mathrm{RF}} G_t G_r \ell D^{-\alpha}}\right)}_{1^{st}} \underbrace{\mathcal{L}_{I_V}\left(\frac{\zeta}{P_{\mathrm{RF}} G_t G_r \ell D^{-\alpha}}\right)}_{2^{nd}} \\
&\quad \underbrace{\times \exp\left(-\frac{\zeta \sigma^2}{P_{\mathrm{RF}} G_t G_r \ell D^{-\alpha}}\right)}_{3^{rd}}.
\end{aligned} \tag{7.16}
$$

where $\mathcal{L}(.)$, I_H and I_V denote the Laplace transform, the interference from H-road and V-road, respectively. Kindly note that there are three factors in (5.16), which can be interpreted as follows: the first and second factors represent the reduction in probability of successful transmission taking into consideration the impact of interference from the interferers from H-road and V-road, respectively; the third factor denotes the probability of successful transmission for interference-free scenario. For a one-dimensional PPP, the Laplace transform of the aggregate interference originating from the interferers can be given as

$$
\begin{aligned}
\mathcal{L}_{I_V}(s) &= \mathbb{E}[\exp(-s I_{\mathrm{RF}})], \\
&= \mathbb{E}\left[\prod_{x\in\phi} \exp(-s P_{\mathrm{RF}} G_t G_r \ell h_x (h^2 + x^2)^{-\frac{\alpha}{2}})\right], \\
&\overset{(a)}{=} \mathbb{E}_x\left[\prod_{x\in\phi} \mathbb{E}_{h_x}\{\exp(-s P_{\mathrm{RF}} G_t G_r \ell h_x (h^2+x^2)^{-\frac{\alpha}{2}})\}\right], \\
&= \mathbb{E}_x\left[\prod_{x\in\phi} \frac{1}{1 + s P_{\mathrm{RF}} G_t G_r \ell (h^2 + x^2)^{\frac{-\alpha}{2}}}\right], \\
&\overset{(b)}{=} \exp\left(-2\lambda_{\mathrm{MAC}} \int_0^{\infty} \frac{1}{1+(h^2+x^2)^{\frac{\alpha}{2}}/s P_{\mathrm{RF}} G_t G_r \ell} dx\right),
\end{aligned} \tag{7.17}
$$

Here (a) holds due to independence of fading coefficients h_x, (b) uses the definition of PGFL for PPP.

Special Case: The step (b) can be further expressed in simplified form by setting the path loss exponent, $\alpha = 2$ as

$$
\mathbb{E}[\exp(-s I_{\mathrm{RF}})] = \exp\left(-\lambda_{\mathrm{MAC}} \frac{\pi (s P_{\mathrm{RF}} G_t G_r \ell)}{\sqrt{h^2 + s P_{\mathrm{RF}} G_t G_r \ell}}\right), \tag{7.18}
$$

Substituting $s = \frac{\zeta}{P_{\mathrm{RF}} G_t G_r \ell D^{-\alpha}}$ yields the result as

$$
\mathcal{L}_{I_V}\left(\frac{\zeta}{P_{\mathrm{RF}} G_t G_r \ell D^{-\alpha}}\right) = \exp\left(-\lambda_{\mathrm{MAC}} \frac{\pi \zeta D^{\alpha}}{\sqrt{h^2 + \zeta D^{\alpha}}}\right). \tag{7.19}
$$

Corollary: When the height of RSU is very low, the step (b) can also be expressed in alternative form as:

$$
\mathbb{E}[\exp(-s I_{\mathrm{RF}})] = \exp\left(-2\lambda_{\mathrm{MAC}} (s P_{\mathrm{RF}} G_t G_r \ell)^{\frac{1}{\alpha}} \frac{\pi}{\alpha} \csc(\frac{\pi}{\alpha})\right), \tag{7.20}
$$

Substituting $s = \frac{\zeta}{P_{\mathrm{RF}} G_t G_r \ell D^{-\alpha}}$ yields the result as

$$
\begin{aligned}
&\mathcal{L}_{I_V}\left(\frac{\zeta}{P_{\mathrm{RF}} G_t G_r \ell D^{-\alpha}}\right) \\
&\quad = \exp\left(-2\lambda_{\mathrm{MAC}} (\zeta)^{\frac{1}{\alpha}} D \frac{\pi}{\alpha} \csc\left(\frac{\pi}{\alpha}\right)\right).
\end{aligned} \tag{7.21}
$$

$$
\begin{aligned}
&\mathcal{L}_{I_H}\left(s = \frac{\zeta}{P_{\mathrm{RF}} G_t G_r \ell D^{-\alpha}}\right) \\
&= \exp\Big(-\lambda_{\mathrm{MAC}}\Big[\int_{-\infty}^{-R} \frac{1}{1 + (h^2 + x^2)^{\frac{\alpha}{2}}/s\zeta D^{\alpha}} dx \\
&\quad + \int_0^{\infty} \frac{1}{1 + (h^2 + x^2)^{\frac{\alpha}{2}}/s\zeta D^{\alpha}} dx\Big]\Big).
\end{aligned} \tag{7.22}
$$

Here, the expression for outage probability of a hard-switching-based hybrid V-VLC/V-RF network is derived. Hard-switching-based hybrid V-VLC/V-RF network considers only single link to be operational. For simple receiver implementation, it is assumed that only one of the links is active at a certain point. The outage probability of such hybrid network configuration can given as[2]

$$
P_{\mathrm{out},hyb}(\beta, \zeta) \geq P_{\mathrm{out,VLC}}(\beta) P_{\mathrm{out,RF}}(\zeta). \tag{7.23}
$$

7.2.2.2 Throughput

From a system design perspective, the outage probability cannot be considered as sufficient metric to characterize the performance, since an MAC that allows few concurrent transmissions leads to high packet reception probabilities but low throughput as well. Hence, to suitably compare the impact of different MAC protocols, it is necessary to characterize the throughput for the intersection scenario for different network configurations. For the case with the desired transmitter located at x_{tx}, the throughput for a specific network configuration with system bandwidth, B_s and SINR

[2] Note that the equality holds when V-VLC and V-RF links are considered to be independent.

threshold, β can be expressed as [14]

$$\mathcal{T} = p_A(x_{tx})(1 - P_{\text{out}})\log_2(1+\beta)B_s. \quad (7.24)$$

where $p_A(x_{tx})$ denotes the access probability of a transmitter located at x_{tx} [14].

7.2.2.3 Data-Oriented Characterization: DOR and IOR

Many safety/warning-related messages are so critical that a large latency is intolerable, especially during accident-prone scenarios. DOR and IOR can be of strong interest to vehicular communication and can be used to compare the performances of stand-alone V-VLC and V-RF based on offering high reliability as well as low latency. Any vehicular transmission systems should be able to efficiently support the large amount of data traffics with tolerable latency. As such, the data-oriented metric, minimum transmission time (MTT) is required, which is defined as the minimum time duration required to transmit a certain amount of data, H over a channel with bandwidth, B. Then, the DOR can be defined as the probability that MTT required for transmitting a certain amount of data is greater than a predefined delay threshold duration [15]. Mathematically,

$$\text{DOR}^{\text{MAC}} = \mathbb{P}(MTT > T_{th}), \quad (7.25)$$

where T_{th} denotes the delay threshold duration and can be related to the delay requirement of the data to be transmitted. As a matter of fact, DOR may serve as an statistical measure for the Quality-of-service (QoS) experienced by individual data transmission for stringent delay requirement by a particular network. Irrespective of multiple access scheme, the DOR for small data transmission within a given coherence time can be defined as

$$\text{DOR}^{\text{MAC}} = \mathbb{P}(\text{SINR} < 2^{\frac{H}{BT_{th}}} - 1). \quad (7.26)$$

Apart from DOR analysis, the characterization of the amount of data that can be successfully transmitted over a specific spectral–temporal resource block also needs to be taken into account for any vehicular transmission system. Here, another data-oriented metric is utilized, maximum entropy throughput (MET), which is defined as the maximum amount of information that can be transmitted over a certain time duration, T, and system channel bandwidth B. Then, the IOR can be defined as the probability that MET over a certain time duration is less than a threshold entropy value, represented by H_{th} [15]. Mathematically,

$$\text{IOR}^{\text{MAC}} = \mathbb{P}(\text{SINR} < 2^{\frac{H_{th}}{BT}} - 1). \quad (7.27)$$

Table 7.2 V-VLC and V-RF system parameters

Parameter	Symbol	Value
Lambertian Order	m	1
PD active detection area	A_d	1 cm^2
Half angle of PD's field of view (FOV)	Ψ_{FOV}	60°
LED semi-angle	$\Phi_{\frac{1}{2}}$	70°
Transmission power for V-VLC	P_{VLC}	36.5 dBm [16]
Transmission power for V-RF	P_{RF}	23 dBm [17]
Noise PSD in VLC	σ^2	10^{-22} W/Hz
Noise PSD in RF	N_0	−174 dBm/Hz
RF path loss exponent	α	4
Absolute temperature	T_k	298° K
System Bandwidth	B_s	20 MHz
Optical filter gain	$T_s(\Psi_k)$	1
Carrier frequency for V-RF	f_c	5.9 GHz
Refractive index	n	1.5
Height of LED traffic light	h	8 m
Access probability	ρ	0.01–0.9
Traffic Intensity	λ_R	0.01
Carrier frequency for V-VLC	f_c	450 THz

7.2.2.4 Numerical Results and Discussion

This section presents numerical results that substantiate the theoretical findings. The system model parameters are adopted in accordance with practical vehicular scenarios and listed in Table 7.2. Specifically, the potential benefit of stand-alone V-VLC/V-RF links or hybrid V-VLC/V-RF links is presented. In order to validate the accuracy of the theoretical findings, Monte Carlo simulations are performed by averaging over 10,000 realizations of PPPs and fading channel parameters. A worst-case scenario where interference from interferers arise from infinite road segment ($\mathcal{B} = \mathbb{R}^1$) is also shown.

Figure 7.2 gives comparison of outage performance for standalone V-VLC links, V-RF links, and hybrid V-VLC/V-RF links with slotted ALOHA protocol. Here, it can be observed that when the desired transmitter's location from RSU increases, the outage probability increases. Also, the outage performance of stand-alone V-VLC links is comparatively better than stand-alone V-RF for low communication range. For instance, when access probability, $\rho = 0.01$, the outage performance of stand-alone V-VLC links is better compared to V-RF links when desired transmitter's location is upto 80 m. However, standalone V-RF is reliable option for long distance communication. It can also be noted that there exits limitation of stand-alone V-VLC over V-RF when distance between the desired vehicle and RSU is less than 5 m which comes at the cost of restricted receiver's FOV. It is worth mentioning here that irrespective of distance between the desired vehicle and RSU, a hybrid V-VLC/V-RF link outperforms stand-alone V-VLC or V-RF link as expected. Let us recall that hybrid outage is a joint event of V-VLC and V-RF links wherein the outage probability of hybrid V-

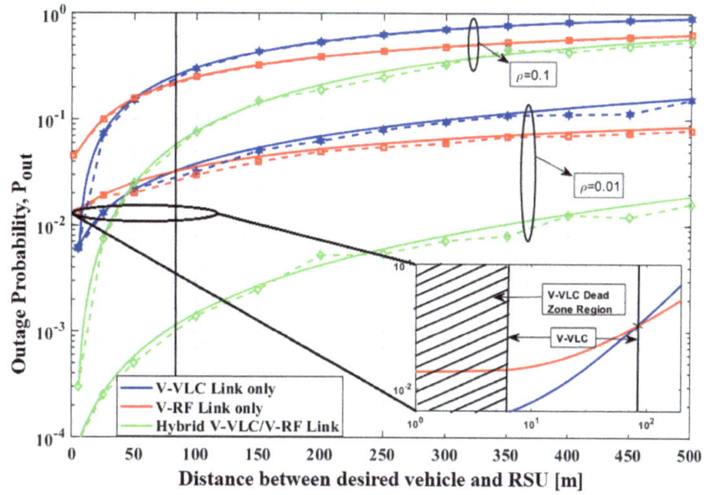

Fig. 7.2 Comparison of analytical (solid line) and simulation (dashed line) results for outage probability, P_{out} versus distance for V-VLC links only (blue), V-RF links only (red), and hybrid V-VLC/V-RF Links (green) with slotted ALOHA protocol

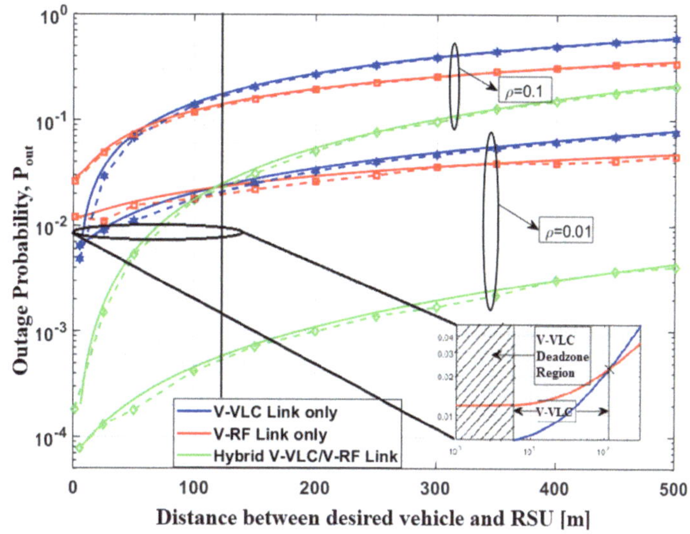

Fig. 7.3 Comparison of analytical (solid line) and simulation (dashed line) results for outage probability, P_{out} versus distance for V-VLC link only (blue), V-RF link only (red), and hybrid V-VLC/V-RF link (green) with CSMA CA protocol

VLC/V-RF will be less than the outage probability associated with either standalone V-VLC or V-RF links. However, this comes at the expense of extra resources required during practical implementation of a hybrid V-VLC/V-RF link.

The performance gain achieved in proposed scenario by using CSMA over slotted ALOHA is further studied. The CSMA contention radius is set to $\delta = 10\,\text{m}$. Figure 7.3 illustrates that for access probability, ρ=0.01, the outage performance of stand-alone V-VLC links is better compared to V-RF links when the desired transmitter's location, $R < 120\,\text{m}$. However, stand-alone V-RF is reliable option when desired transmitter's location, $R > 120\,\text{m}$. As before, hybrid V-VLC/V-RF links outperform stand-alone V-VLC and V-RF links. Interestingly, a low access probability reduces the outage probability. For ease of validation and visualization, the outage probability, P_{out} is plotted as a function of access probability, ρ as can be seen in Fig. 7.3.

Figure 7.4a shows the ALOHA case, with an increase in access probability, ϱ outage probability increases due to the availability of more interferers. Also, the throughput first increases (due to more active transmitters) with increase in access probability and then decreases (due to extensive amount of interference), leading to an optimal value of access probability, ϱ. Nevertheless, in order to ensure a certain QoS, one must also consider a guarantee on the outage performance as well. With an outage probability below 20% when distance between the desired vehicle and RSU, $R = 50\,\text{m}$, the optimal value of access probability for standalone V-VLC link, $\rho = 0.12$, results in a throughput of about 60 Kbps as can be seen in Fig. 7.4a. In addition, the outage performance of

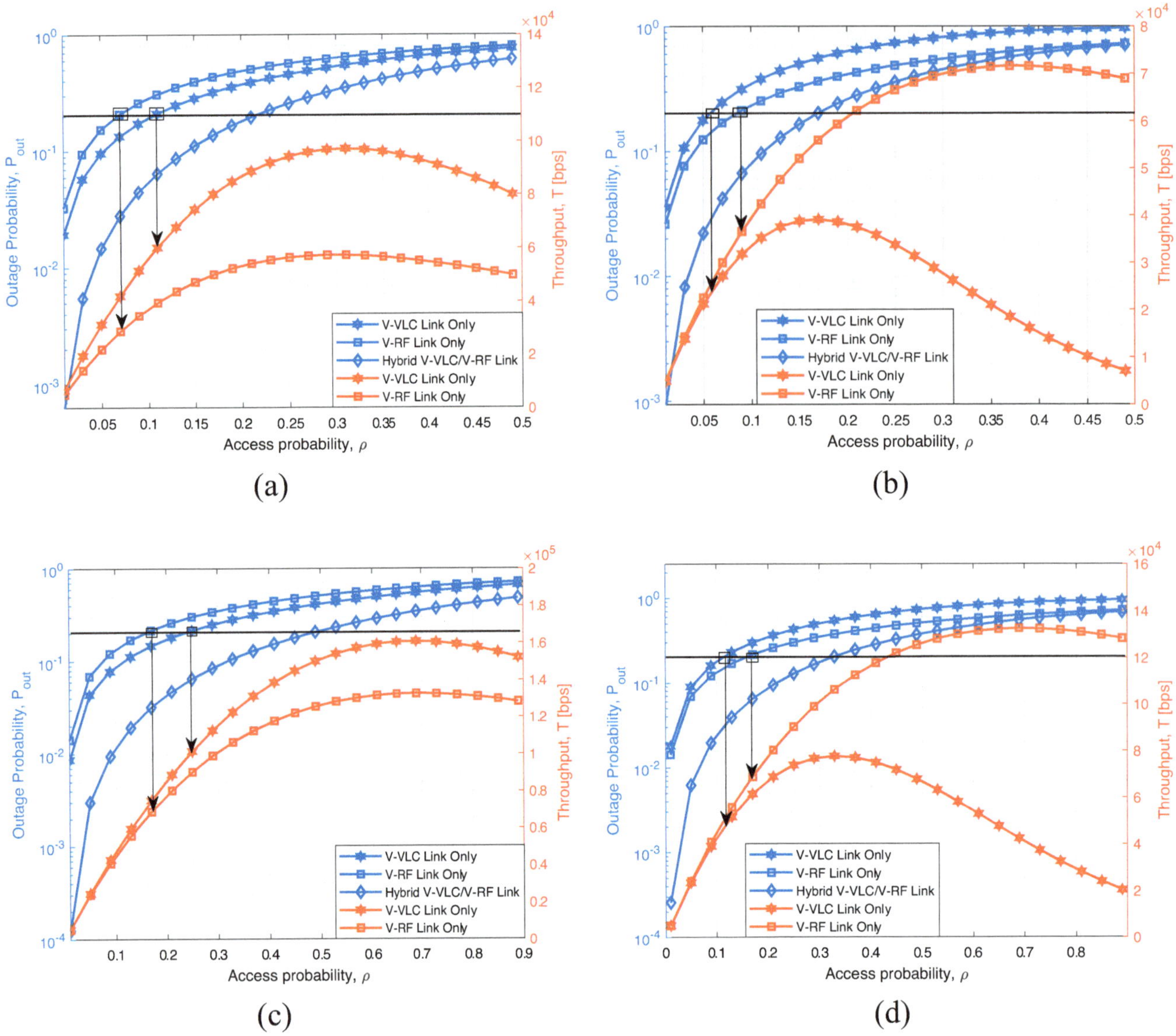

Fig. 7.4 Outage probability, Pout, and throughput, $\mathcal{T}$ as a function of access probability, ϱ for stand-alone V-VLC link, V-RF link, and hybrid V-VLC/VRF link for two different distance between the desired vehicle and RSU, $\mathcal{R}$. (**a**) ALOHA case, $R = 50$ m; (**b**) ALOHA case, $R = 150$ m; (**c**) CSMA case, $R = 50$ m; (**d**) CSMA case, $R = 150$ m

standalone V-VLC is better than stand-alone V-RF. However, increasing the distance, R, the opposite trend can be observed in Fig. 7.4b. As expected, hybrid V-VLC/V-RF outperforms stand-alone V-VLC or V-RF in terms of outage performance. Kindly note that the throughput of hybrid V-VLC/V-RF system depends on maximum throughput offered by either stand-alone V-VLC or stand-alone V-RF system.

For CSMA (Fig. 7.4c), large contention region (i.e., low access probability) reduces the outage probability. Analogous to ALOHA, the throughput first increases with increase in access probability, ρ, and then decreases. To ensure an outage probability below 20% when distance between the desired vehicle and RSU, $R = 50$ m, the optimal value of access probability for stand-alone V-VLC link, $\rho = 0.25$, results in a throughput of about 100 Kbps. Hence, in this scenario, the use of CSMA instead of ALOHA leads to more increase in the throughput for the same transmitter location. It can be inferred from aforementioned comparison that implementation of hybrid V-VLC/V-RF link leads to considerable improvement in throughput and outage probability improvement, compared to scenarios utilizing either V-VLC link or V-RF link separately.

Figure 7.5a compares DOR performance comparison for stand-alone V-VLC link, V-RF link and hybrid V-VLC/V-RF link with ALOHA protocol. In particular, DOR of both strategies as function of the delay threshold is plotted, T_{th} for different data amount $H \in \{50 \text{ KB}, 100 \text{ KB}\}$. It can be observed that for both the choices of H, there exists

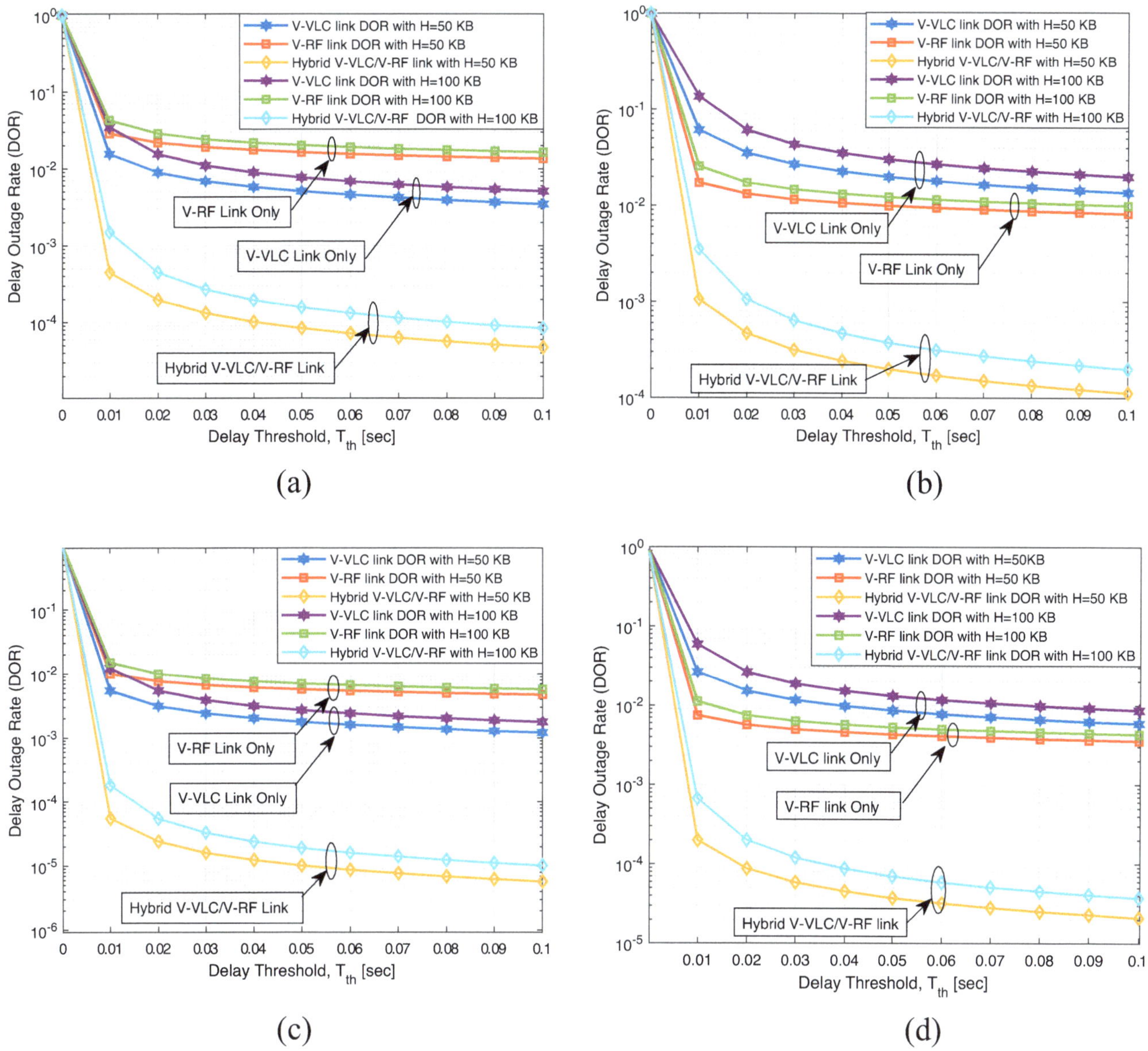

Fig. 7.5 Delay outage performance of V-VLC Link only, V-RF Link only, and hybrid V-VLC/V-RF Link for CSMA case for two different distances between the desired vehicle and RSU. Here, access probability, ϱ is 0.01.: (**a**) ALOHA case, $R = 50$ m; (**b**) ALOHA case, $R = 150$ m; (**c**) CSMA case, $R = 50$ m; and, (**d**) CSMA case, $R = 150$ m

complementary behavior in the DOR performance of stand-alone V-VLC and V-RF network depending on the distance between the desired vehicle and RSU, R. Specifically, with access probability, $\varrho = 0.01$ and transmitter's location, $R = 50$ m, DOR performance of stand-alone V-VLC is comparatively better than V-RF as can be seen from Fig. 7.5a. The opposite is true when R is 150 m as shown in Fig. 7.5b. As before, DOR performance is observed to improve by employing CSMA for both stand-alone V-VLC as well as V-RF networks as depicted in Fig. 7.5c and d. The reason is as follows: DOR typically depends on SINR, which is better for proposed scenario with CSMA/CA (due to low interference). It is interesting to note that for data traffic with stringent delay requirement, hybrid V-VLC/V-RF ensures minimum delay in transmitting given amount of information between the desired vehicle and RSU as compared to stand-alone V-VLC or V-RF network.

The performance of IOR for stand-alone V-VLC, V-RF, and hybrid V-VLC/V-RF network configuration for two different values on the transmitter's location, $R \in \{50 \text{ m}, 150 \text{ m}\}$ is compared. In Fig. 7.6, the IOR for all network configurations with ALOHA and CSMA as a function of threshold entropy, H_{th} for two different time durations, $T \in \{60 \text{ ms}, 30 \text{ ms}\}$ is shown. In particular, Fig. 7.6a shows that for a given IOR performance with slotted ALOHA, a stand-alone V-VLC Link is a reliable

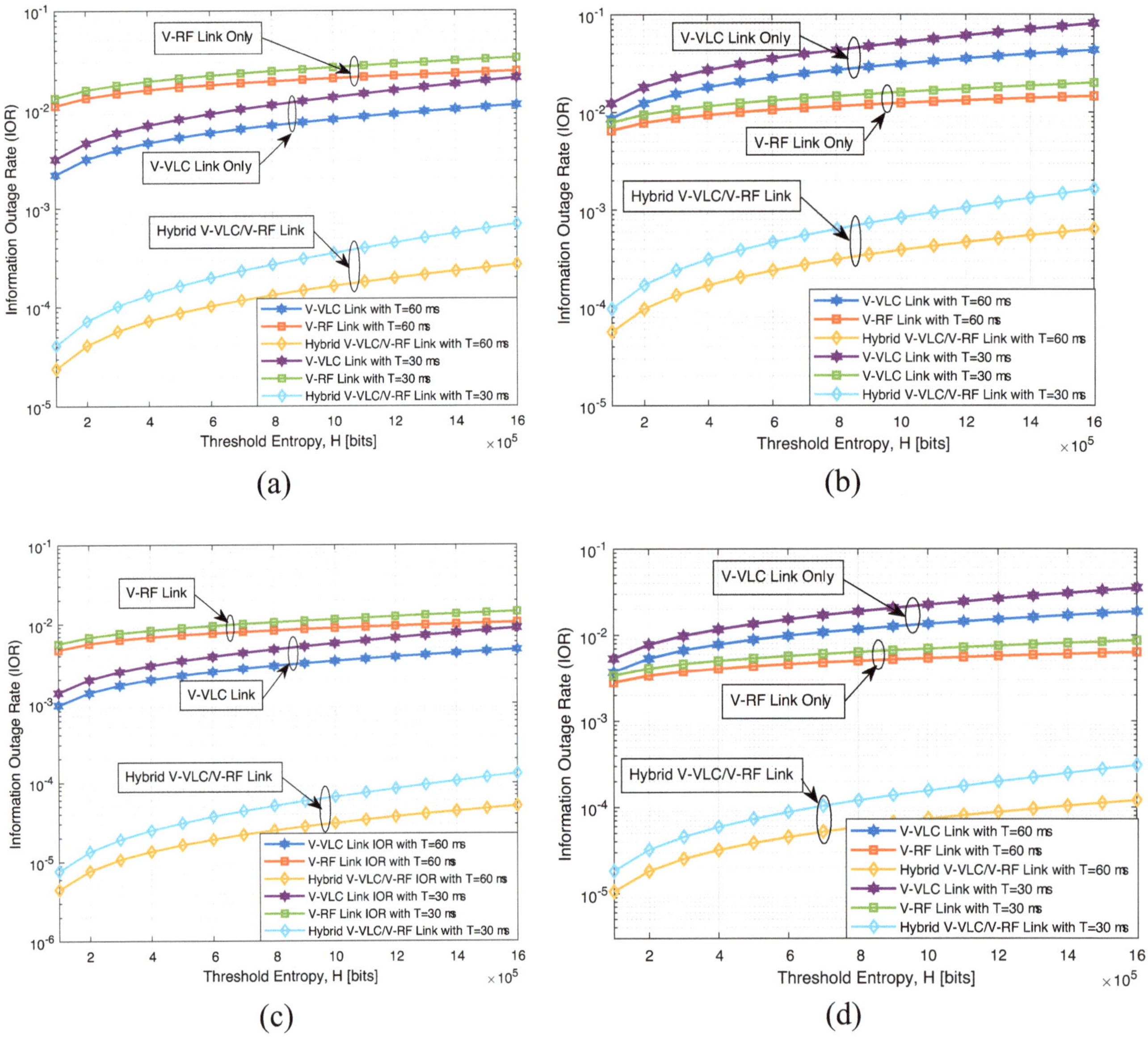

Fig. 7.6 Information Outage Rate of V-VLC Link only, V-RF Link only and hybrid V-VLC/V-RF Link for ALOHA case for two different distances between the desired vehicle and RSU. Here, access probability, ϱ is 0.01: (**a**) ALOHA case, $R = 50\,\text{m}$; (**b**) ALOHA case, $R = 150\,\text{m}$; (**c**) CSMA case, $R = 50\,\text{m}$; and, (**d**) CSMA case, $R = 150\,\text{m}$

option over a stand-alone V-RF link when the distance between the desired vehicle and RSU, R is 50 m. However, the complementary insights stand true when D is 150 m, as evident from Fig. 7.6b.

In Fig. 7.6c, the improvement in the IOR performance of all the network configurations is again observed by using CSMA over ALOHA. The reason is as follows: IOR typically depends on SINR, which is better for the proposed scenario with CSMA/CA (due to low interference). Again, the performance of stand-alone V-VLC or V-RF link depends on the distance between the legitimate vehicle and RSU, R as evident from Fig. 7.6c and d. It is worth mentioning that irrespective of distance, R, a hybrid V-VLC/V-RF system always guarantees the maximum amount of information flow between the desired vehicle and RSU as compared to a stand-alone V-VLC or V-RF network in the given time duration.

7.3 Integrated Fiber-Wireless Access Network for Vehicular Network

The previous section deals with the VLC-based modeling of V2I and V2V scenarios. However, the section is limited to the fronthaul networks as it does not talk about the backhaul access network. The previous architecture could utilize the optical fiber at the backend to enhance the performance of wireless vehicular networks. Thus, this section explores the integration of optical fiber network at the backhaul with wireless network at the fronthaul for the vehicular networks. In particular, FiWi-based V2I network can facilitate some of the stringent requirements of 6G vehicular networks, including higher capacity, lower delay, and ubiquitous connectivity. FiWi-based V2I network integrates the NG-PON2 with IEEE 802.11p-based V2I network. In this section, the various kinds of vehicular data traffic and their desired KPI, namely throughput, delay, and reliability, are reviewed. Depending on the KPI requirements, the V2I traffic is classified into four classes and assigned to different Transmission containers (T-CONTs) of the ONU. Further, in order to minimize the delay of the network, an Machine learning (ML)-based T-CONT priority assignment wavelength allocation algorithm is utilized to minimize the number of wavelength switching instances in the PON. The performance of the ML-based wavelength allocation algorithm [18] is compared with the other approaches, namely random and equal T-CONT-based wavelength allocation algorithms. Simulation results demonstrate the efficiency of the proposed algorithm vis-a-vis other approaches in terms of End-to-end (E2E) delay, throughput, and reliability.

This section is intended to improve QoS for different vehicular users in an FiWi-assisted V2I network.[3] The vehicular users are assumed to generate different kinds of vehicular data traffic, namely data related to advanced driving, platooning, remote driving, and extended sensors [19]. The vehicular data traffic is thus classified into four classes depending on its throughput, delay, and reliability requirements, also termed key performance indicators (KPIs). The ONUs and Road side units (RSUs) are collocated and are collectively called ONU-RSUs. The received data at ONU are the same as that received at RSU. At ONU-RSUs, the vehicular data traffic is assigned to the T-CONTs based on the priority of the vehicular data traffic. The different T-CONTs at the ONU, namely T-CONT$_1$, T-CONT$_2$, T-CONT$_3$, T-CONT$_4$, and T-CONT$_5$ are allocated services in the optical network based on the priority of the traffic. In order to provide services to the vehicular data traffic based on the KPIs, the vehicular data traffic is classified into four classes as discussed in the later sections.

[3]This section is specific to vehicular data traffic. However, the algorithm proposed in the section is agnostic to the type of traffic used. However, the vehicular network has critical safety and security requirements; thus, ensuring that the QoSs is met for vehicular traffic is a critical challenge for the network. Therefore, in this section, vehicular networks is considered.

7.3.1 Channel Modeling

In this section, the system model for the considered FiWi-based V2I network is presented. In addition, a detailed classification for vehicular traffic is also presented in this section.

7.3.2 FiWi-Based Vehicle-to-Infrastructure Network

As shown in Fig. 7.7, an FiWi network consisting of an NG-PON2 as the backhaul and IEEE 802.11p-based V2I network as the fronthaul is considered. The OLT of NG-PON2 is located at the central office. The ONUs are connected to the OLT via a passive power splitter. All spectral components are available at each ONU, where the wavelength is selected. Moreover, each ONU is capable of handling adding/dropping of wavelengths and there are three configurations for the demultiplexer selection node— (a) demux using wavelength selective switches (WSS), (b) fixed AWG and fiber cross-connect, and (c) power splitter with coherent receiver or tunable filter. The ONUs are collocated with RSUs and are collectively called ONU-RSUs. As ONU and RSU are collocated, the amount of vehicular data traffic received at the RSU is directly sent to the ONU [20]. There are T-CONTs available at the ONU that allocates the different network services to the vehicular data traffic based on the priority of the network traffic.

Among the five T-CONTs at the ONU, T-CONT$_1$ has the highest priority and is generally reserved for background traffic. Similarly, T-CONT$_5$ has the least priority and is assigned the data traffic with low priority. The T-CONT mapping is shown in Fig. 7.7 for T-CONT$_2$, T-CONT$_3$, T-CONT$_4$, and T-CONT$_5$ that shows the type of traffic load at each ONU-RSU. The T-CONT mapping at the ONU-RSU suggests the amount of high-priority vehicular data traffic available at the ONU-RSU. In this section, the priority of the data traffic is decided on the basis of delay, throughput, and reliability of the data traffic. The ONU-RSUs are assumed to be collocated with the RSUs. The RSUs serve the vehicles through a wireless medium, and each vehicle can generate four classes of traffic that are assigned to the T-CONTs based on their priority. The vehicular traffic classification and assignment of traffic to different T-CONTs are defined in detail in later sections.

7.3.3 Vehicular Data Traffic Use Cases

In this subsection, the classification of vehicular data traffic and its use cases are discussed in brief.

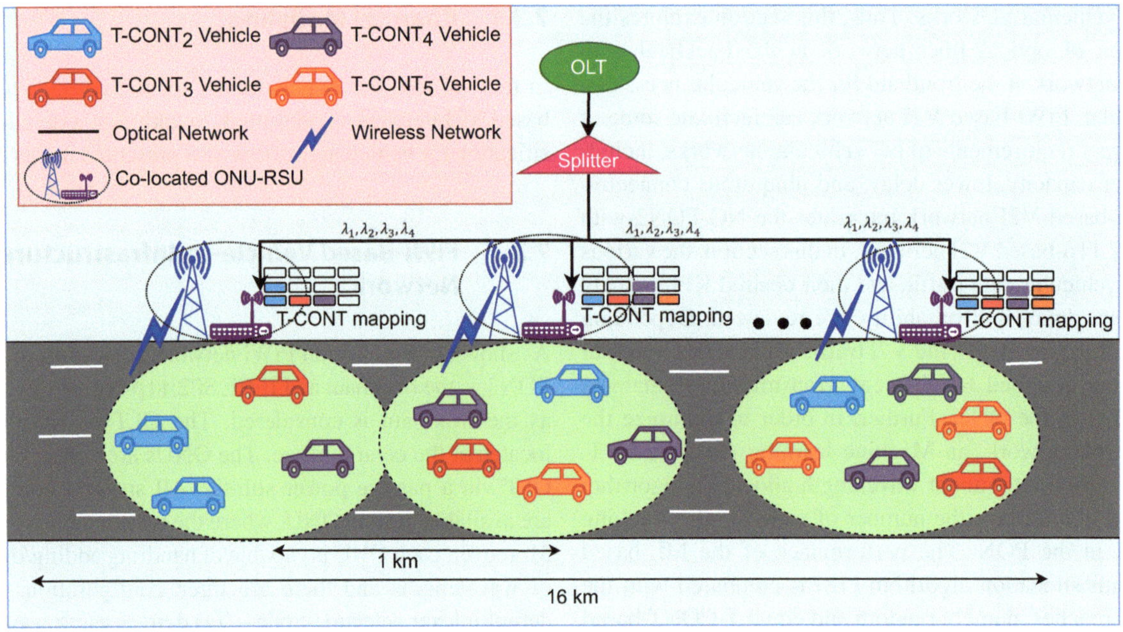

Fig. 7.7 FiWi based V2I network

7.3.3.1 Vehicle Platooning

Vehicle platooning minimizes the distance between the vehicles and is used to accelerate a group of vehicles together. It includes the following use cases [21, 22]:

- Cooperative driving: It includes the exchange of platooning information among multiple vehicles. The E2E delay requirement of these services is in the range of 10–25 msec, and the throughput required is in the range of 50–65 Mbps. These services require reliability of 90–99% and have minimum communication range of 80–350 m.
- Reporting: These services include the reporting information exchange between vehicles and RSUs. The E2E delay requirement for these services is 500 msec.
- Information sharing: The information exchange between vehicles and RSUs require an E2E delay of 20 msec, the throughput required is 50 Mbps and minimum communication range required is 180–350 m.

7.3.3.2 Advanced Driving

The services for automated driving include collision avoidance, automated driving, trajectory alignment. The KPIs for the use cases are defined as follows [19]:

- Cooperative collision avoidance: These services include the collision avoidance information exchange among V2X users. The maximum tolerable E2E delay for these services is 10 msec, and with maximum throughput of 10 Mbps. These services require high reliability of 99.99%.
- Automated driving (among vehicles): It requires information sharing among the vehicles. The maximum tolerable E2E delay for these services is 100 msec, with maximum throughput of 53 Mbps. These services require minimum communication range of 360–700 m.
- Automated driving (between vehicles and RSUs): These services require information transmission between vehicles and RSU. The minimum requirement for E2E delay, throughput, and communication range is 100 msec, 50 Mbps, and 360–700 m, respectively.
- Emergency trajectory alignment: It includes trajectory alignment among the vehicles. The minimum E2E delay for such services is 3 msec. These services require high reliability of 99.99%. The minimum communication range and throughput requirement for these applications are 500 m and 30 Mbps, respectively.
- Intersection safety information: These include the safety information among RSU and vehicles regarding intersections. The downlink and uplink throughput required for these services are 50 and 25 Mbps, respectively.
- Cooperative lane change: These include the transmission among vehicles to improve safety and change lanes effectively. These services require E2E delay in the range of 10–25 msec and reliability of 90–99%.
- Video sharing: These include video sharing between a vehicle and a V2X application server. These services require uplink throughput of 10 Mbps.

7.3.3.3 Extended Sensors

In order to ensure the enhanced perception of vehicular environment beyond what their own sensors can detect and get the holistic view of the on-road vehicular traffic, these data services include exchange of data from the local sensors or live video data among the vehicles. Some of the use cases for extended sensors are as follows [21, 22]:

- Sensor information: These include sharing of sensor information among the vehicles. The range of E2E delay requirement for such services is 3–100 msec, reliability in the range of 95–99.99%, the minimum communication range for these applications is 50–1000 m, and the range of throughput requirement is 20–1000 Mbps.
- Video sharing: To ensure the holistic view of the road, these include live video sharing among vehicles. The range for E2E delay requirement for these services is 10–50 msec, throughput requirement is in the range of 10–700 Mbps, the range of reliability requirement is 90–99%, and minimum communication range required is 100–400 m.

7.3.3.4 Remote Driving

These include information exchange between vehicle and V2X application server for remotely operating the vehicle. The downlink throughput required for these services is 25 Mbps, and uplink throughput requirement is 1 Mbps with maximum E2E delay of 5 msec and high reliability of 99.99% [19].

According to T-CONT to which these traffics are assigned, a priority is assigned to these traffic in order to provide relevant services to the end users. The assignment of traffic to different T-CONTs is defined in detail in the later sections.

7.3.4 T-CONT Assignment for V2X Traffic

In general, the traffic is assigned to the T-CONTs based on the desired service requirement of the users. For instance, in [20, 23], the T-CONTs are assigned to different kinds of network traffic, namely voice, video, best-effort, and background traffic, based on their desired delay requirement. The background traffic has the highest priority; therefore, it was assigned T-CONT$_1$ of the ONU. The delay for voice traffic is the lowest and, therefore, is assigned T-CONT$_2$. This was followed by video and best-effort traffic, which were assigned T-CONT$_3$ and T-CONT$_4$, respectively. Similarly, here, four classes of vehicular traffic depending on their KPI requirements are assigned. Thus, a class of traffic with high throughput, high reliability, and low delay requirement is prioritized more than the other classes. Remote driving has the maximum reliability requirement of 99.99%, while the other types of vehicular data traffic have lower reliability requirements. Thus, extended sensors are given the highest priority and assigned to T-CONT$_2$. Extended sensors and advanced driving have a minimum delay requirement of 3 msec; however, the throughput requirement for extended sensors is higher and, therefore, is given the second highest priority. Further, the minimum delay requirement for platooning is higher than the other vehicular data traffic and thus, is given the lowest priority and assigned to T-CONT$_5$. The detailed traffic assignment to different T-CONTs is given in Table 7.3.

7.3.5 Problem Formulation

The fundamental problem under consideration is then cast as an optical delay minimization problem, which is dependent on the minimization of the number of wavelength switching instances at the ONU-RSU and is subject to the following set of constraints:

$$(P1) \min_{N_S} \sum_{t=0}^{T_H-1} \sum_{k=2}^{5} D_k^{\mathrm{O}}(t), \tag{7.28a}$$

$$s.t. \quad \sum_{i=1}^{N_{\text{ONU-RSU}}} U_{v,i} \leq 1, \tag{7.28b}$$

$$D_k^{\text{T-CONT}}(t) \leq D_k^{\mathrm{th}},\ \forall\, t \in [0, T_H - 1],\ k \in [2, 5], \tag{7.28c}$$

$$\sum_{t=0}^{T_H-1} \sum_{k=2}^{5} D_k^{\text{T-CONT}}(t) = \sum_{t=0}^{T_H-1} \sum_{k=2}^{5} D_k^{\mathrm{O}}(t) + D_k^{\mathrm{W}}(t), \tag{7.28d}$$

$$D_2^{\text{T-CONT}}(t) < D_3^{\text{T-CONT}}(t) < D_4^{\text{T-CONT}}(t) < D_5^{\text{T-CONT}}(t),\quad \forall\, t \in [0, T_H - 1], \tag{7.28e}$$

where $D_k^{\mathrm{O}}(t)$ is the optical delay of kth T-CONT traffic at time t, N_S is the number of wavelength switching instances at the ONU-RSU, T_H is the number of time slots, $N_{\text{ONU-RSU}}$ is the number of RSUs in the network, $U_{v,i}$ denotes the association of vehicle v with the ith RSU, $D_k^{\text{T-CONT}}(t)$ is the delay of the kth T-CONT at time t, where $k \in [2, 5]$, D_k^{th} is the delay threshold of kth T-CONT, $D_k^{\mathrm{O}}(t)$, and $D_k^{\mathrm{W}}(t)$ is the optical and wireless delay of kth T-CONT traffic at time t, respectively. In problem $(P1)$, the first constraint (7.28b) assures the association of the vehicle with the RSU such that each vehicle is connected with only one RSU. Constraint (7.28c) ensures that the end-to-end delay requirement of each T-CONT is maintained within the ITS standard. In constraint (7.28d) the e2e delay of FiWi-based V2I network is presented. This e2e delay is a combination of optical link delay and wireless link delay. Further, constraint (7.28e) ensures that the priority of the T-CONTs is maintained. $T_H = 24$ hours to minimize the number of hourly wavelength switching instances during an entire day. The overhead delay is taken as 16 μs [24]. The delay thresholds for different T-

Table 7.3 KPIs for V2X use cases [19]

Use case	Delay (msec)	Reliability (%)	Throughput (Mbps)	T-CONT Assigned
Extended Sensors	3–100	90–99.99	10–1000	3
Advance driving	3–100	90–99.99	10–50	4
Remote driving	5	99.99	25	2
Platooning	10–500	90–99.99	50–65	5

CONTs are considered as: $D_2^{th} = 50\,\text{msec}$, $D_3^{th} = 10\,\text{msec}$, $D_4^{th} = 10\,\text{msec}$, and $D_5^{th} = 25\,\text{msec}$ [19].

The aforementioned problem is solved using the following wavelength allocation algorithms: 1) equal T-CONT-based wavelength allocation algorithm, 2) random wavelength allocation algorithm, and 3) proposed ML-based T-CONT priority assignment wavelength allocation algorithm. The ML-based T-CONT priority assignment wavelength allocation algorithm utilizes the prediction of the number of vehicles on the road. The distribution of the traffic to the RSUs and type of traffic generated by the vehicles are of known distribution. Thus, the prediction of the number of vehicles on the road would quantify the amount and type of hourly vehicular data traffic available at each ONU-RSU. It has been shown later that the proposed ML-based T-CONT priority assignment algorithm achieves a good compromise between delay minimization, throughput maximization, and reliability enhancement.

7.3.6 Delay Analysis

This section presents the performance analysis for FiWi-assisted V2I network. The e2e delay for the FiWi network is calculated by adding the delay for the optical and V2I network. The delay at the jth ONU at tth hour is calculated as [24]:

$$D_{\text{ONU}_j}(t) = (T_{oh} + T_{\text{tuning}}(t))/(1 - \Lambda_{\text{ONU-RSU}_j}(t)), \quad (7.29)$$

where T_{oh} is the delay due to overheads at the ONU-RSU, T_{tuning} is the tuning time from one wavelength to the next. According to ITU-T G.989.2, there are following three classes of wavelength tuning delay: (a) slow (25 ms to 1 s), (b) middle (10 μs to 25 ms), and (c) fast (<10 μs) [25]. Consequently, we have considered a middle tuning delay of 50 μs [25–27]. Thus, the wavelength tuning delay is considered as follows:

$$T_{\text{tuning}} = \begin{cases} 50\,\mu\text{s}, & \text{if wavelength switching} \\ 0\,\mu\text{s}, & \text{otherwise,} \end{cases} \quad (7.30)$$

i.e., $T_{\text{tuning}}(t) = 0$ if there is no switching of wavelength at the hour t, else $T_{\text{tuning}}(t) = 50\,\mu\text{s}$ [24]. An hourly wavelength switching design is considered wherein the wavelength can be switched at each hour depending on the priority of traffic at the ONU-RSUs. $\Lambda_{\text{ONU-RSU}_j}(t)$ is the normalized load of ONU-RSU$_j$ on a wavelength at time t. The overhead delay is taken as 16 μs [24].

Table 7.4 Parameters for simulation

Parameter	Definition	Value
$D_{o,s}$	OLT-splitter distance	2.2 km [28]
t_T	Transport delay/Propagation delay	5 μs [28]
t_Q	Queuing delay	60 μs [28]
t_P	Processing delay at OLT	200 μs [28]
$N_{\text{ONU-RSU}}$	Number of ONU-RSUs	16 [20]

The delay for the optical link, i.e., between jth ONU and OLT, is a combination of delay at the ONU, propagation delay, which is given by $t_T(D_{o,s} + D_{s,\text{ONU}})$, queuing delay, and processing delay. Here, $D_{o,s}$ and $D_{s,\text{ONU}}$ are the OLT-splitter distance and splitter ONU distance, respectively [28].

$$D_{O_j} = D_{\text{ONU}_j} + D_{tr}(D_{o,s} + D_{s,\text{ONU}}) + D_Q + D_P. \quad (7.31)$$

where D_Q is the queuing delay, and D_P is the processing delay. The queuing delay, $D_Q = [0, 1000]\,\mu\text{s}$ and processing delay $D_P = 200\,\mu\text{s}$ [28].

For wireless network, the delay at vehicle v for a predefined throughput $R_k(\Gamma_v, L_v, \epsilon)$ can be calculated as [29]:

$$D_w = \left[\frac{\sqrt{B}Q^{-1}(\epsilon)\log_2(e)\sqrt{1-(1+\Gamma_v)^{-2}}}{B\log_2(1+\Gamma_v) - R_k(\Gamma_v, L_v, \epsilon)} \right]^2, \quad (7.32)$$

where B is the bandwidth at each ONU-RSU, ϵ is the reliability, $R_k(\Gamma_v, L_v, \epsilon)$ is the maximum achievable ergodic rate of the vehicle v given in [29], Γ_v is SINR at vth vehicle given by $p_v\Psi_v$ for zero forcing receiver, Ψ_v is given $\frac{\chi_v\beta_v(M-K)}{p_B\beta_v(1-\chi_v)+M\sigma^2}$, M is the number of antennas at RSU, χ_v is the estimation accuracy, K is the number of vehicles to which RSU is connected [29]. The value for each parameter is mentioned in Table 7.4. Further, the e2e delay of vehicle-to-infrastructure link is calculated by using the delay of optical and wireless link given in (7.29) and (7.32), respectively.

Similarly, for a given delay requirement, L_v, the achieved V2I throughput for Ultra-reliable low latency communication (URLLC) system is given as [29]:

$$R_k(\Gamma_v, L_v, \epsilon) = B\log_2(1+\Gamma_v) - \sqrt{\frac{B}{L_k}}\log_2(e)Q^{-1}(\epsilon)\sqrt{1-(1+\Gamma)^{-2}}. \quad (7.33)$$

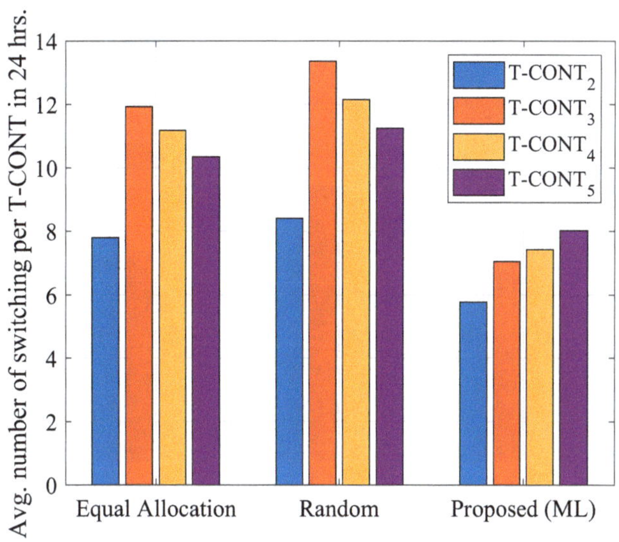

Fig. 7.8 Average number of wavelength switching instances per T-CONT for each ONU-RSU in 24 hours for equal allocation, random, proposed (ML) [18] algorithm

It is assumed that the optical fiber cable has negligible packet loss [30]. Thus, the e2e throughput of the vehicular network is assumed to be bounded by the V2I network throughput.

7.3.7 Performance Analysis

A 16000 m $\times$ 200 m road segment of IS-94 highway between Minneapolis and Saint Paul is considered for performance evaluation. The hourly on-road vehicular traffic data was taken from the Minnesota Department of Transportation (MnDOT) dataset [31]. The hourly vehicular traffic is distributed to the ONU-RSUs using Poisson distribution. Each vehicle can generate four kinds of traffic as mentioned in Table 7.3. These vehicles are connected to the RSUs using wireless links. The ONUs are collocated with the RSUs and are collectively called ONU-RSUs. The considered ONUs are tunable ONUs, i.e., each ONU can be tuned to any of the available wavelengths. The number of wavelengths available at the ONUs is considered to be four. The distance between two RSU units is considered to be 1 km [32]. The distance between OLT and splitter is considered to be 2.2 km [28]. The transportation, queuing, and processing delay of PON are considered to be 5 µs, 60 µs, and 200 µs, respectively. For simulations, the queuing delay is fixed to 60 µs; however, the queuing delay is generally of the order of micro seconds [28] and therefore, would not significantly affect the e2e delay. The impact of queuing delay on the e2e delay has further been explained in detail later. Table 7.4 summarizes the parameters used for the simulation of FiWi-assisted V2I network.

The average number of wavelength switching instances for each T-CONT in T_H = 24 hours is shown in Fig. 7.8. It can be observed that for random and equal T-CONT-based wavelength allocation algorithms, the number of wavelength switches of different T-CONTs is proportional to the traffic at each T-CONT. However, for the ML-based algorithm in [18], the number of wavelength switching is proportional to the priority of each T-CONT. Further, the number of wavelength switches is highest for the random wavelength allocation algorithm followed by an equal T-CONT-based wavelength allocation algorithm and is the least for the ML-based algorithm in [18] wavelength allocation algorithm. This impacts the e2e delay of the network as shown in Fig. 7.10.

In accordance to ITU-T, the following three classes as the wavelength tuning delay are considered: (a) slow (25 ms to 1 s), (b) middle (10 µs to 25 ms), and (c) fast (<10 µs) [25]. Thus, the performance of the different wavelength allocation schemes for different wavelength tuning delays is compared in Fig. 7.9. It can be observed that for slow wavelength switching, the QoS requirement of the users is not met for all the T-CONTs. This is because the wavelength tuning delay is very high, which increases the end-to-end delay of the vehicular network. However, for middle and fast wavelength tuning delays, the QoS of the vehicular traffic is met.

The e2e delay for FiWi-based V2I network has been depicted in Fig. 7.10. It can be observed that the delay for the ML-based wavelength allocation algorithm [18] is the least followed by equal T-CONT allocation and random wavelength allocation algorithm. Further, it can be seen that for random and equal T-CONT-based wavelength allocation algorithms, the delay of the different T-CONTs is proportional to the traffic at each T-CONT. However, for the ML-based T-CONT prioritization algorithm, the e2e delay follows the same trend as the priority of the T-CONTs, i.e., T-CONT$_2$ has the highest priority, i.e., least delay followed by T-CONT$_3$, T-CONT$_4$, and T-CONT$_5$.

Figure 7.11 shows the variation of e2e delay of the V2I network with respect to queuing delay for high priority T-CONT, i.e., T-CONT$_2$ and low priority T-CONT, i.e., T-CONT$_4$. It can be seen that the e2e delay performance is linearly varying with respect to queuing delay. It is interesting to note that the queuing delay is usually very low, i.e., of the order of microseconds due to which it does not affect the e2e performance of the V2I network. It can be seen that for random wavelength allocation, the e2e delay is the highest, followed by equal T-CONT-based wavelength allocation followed by the the ML-based wavelength allocation algorithm. Further, it is interesting to note from Fig. 7.11 that even for the worst-case queuing delay of 1 msec, the performance for the FiWi network is within the delay requirement of 5 msec for T-CONT$_2$ using the ML-based algorithm in [18].

Figure 7.12 shows the variation of achievable throughput with respect to reliability of the V2I link. It is well known that as the reliability of the network increases, the packets can reach the destination more reliably and the overall throughput

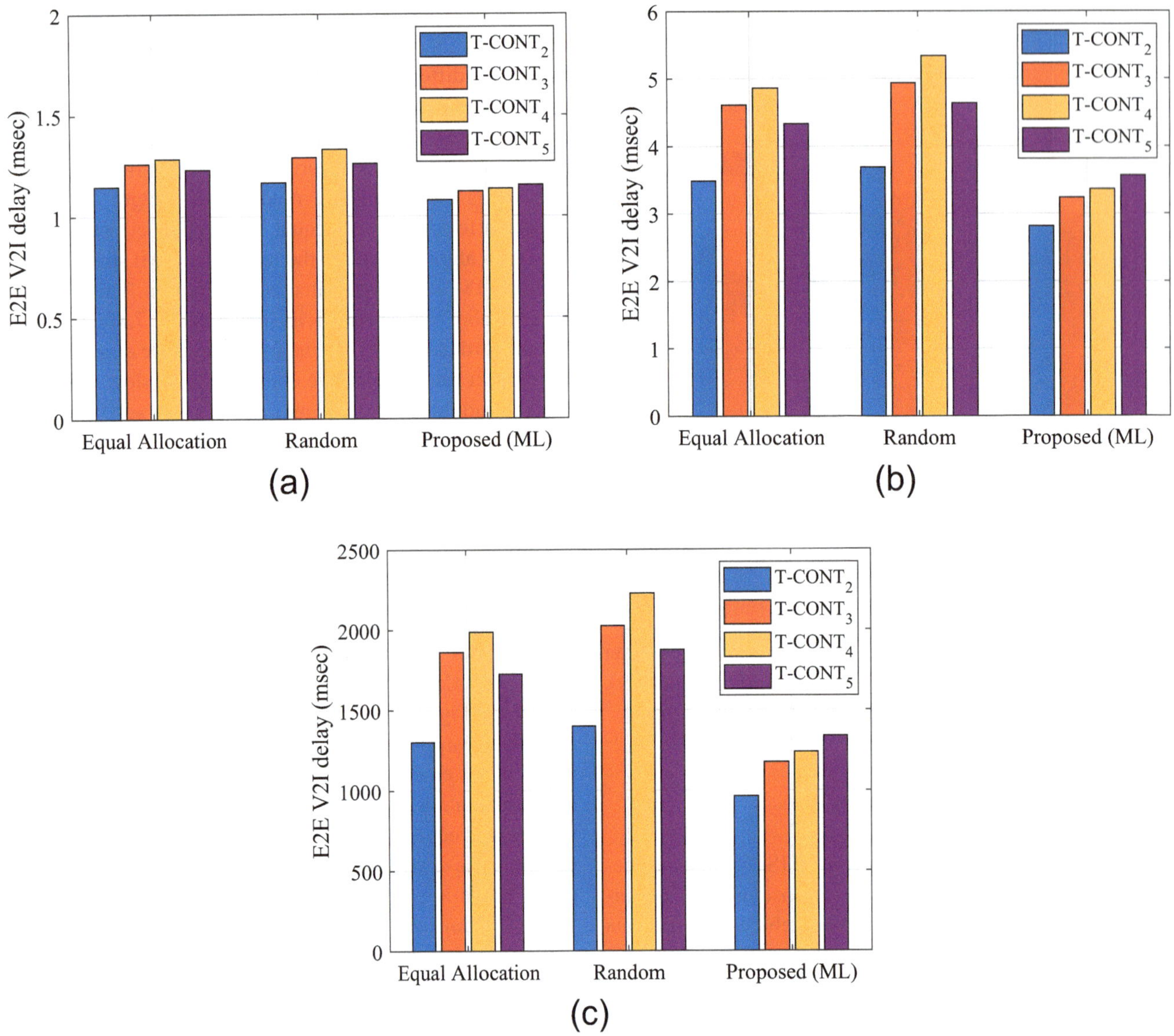

Fig. 7.9 E2E delay for different wavelength tuning delay classes: (a) Fast wavelength tuning (5 μs), (b) Middle wavelength tuning (50 μs), and (c) Slow wavelength tuning (25 ms) for equal allocation, random, proposed (ML) [18] algorithm

of the V2I network increases. This can be observed from Fig. 7.12, specifically, for stringent delay requirement of the V2I link, i.e., for $D_{V2I} = 0.01$ msec, there is an increase in throughput from 38 to 44 Mbps as reliability increases from 10^{-6} to 10^{-1}. Further, as the e2e delay requirement of the user increases, the vehicle can accommodate more traffic within the same e2e delay requirement. This variation in throughput decreases as the delay requirement of the network is relaxed from $D_{V2I} = 0.01$ msec to $D = 10$ msec. Furthermore, with the increase in delay requirement, the achieved throughput is close to the ergodic rate. In addition, it is interesting to note that there is only a slight increase in achievable network throughput with the change in delay from 1 to 10 msec, i.e., the delay requirement of T-CONTs in the V2I network, thus, the achievable throughput of the T-CONTs is not much affected much with the change in network load as can be observed form Fig. 7.13.

The achievable throughput per vehicle is shown in Fig. 7.13. The bandwidth to each T-CONT is assigned in proportion to the network load, thus, the throughput per user for each T-CONT is similar. It can be observed that using the ML-based algorithm in [18], the required performance can be met for up to 100% loaded network for T-CONT$_2$, T-CONT$_3$ and T-CONT$_4$. However, the throughput requirement of T-CONT$_5$ is met only for 65% of the network load. This is because T-CONT$_5$ has the least priority and, therefore, is not able to meet the throughput requirement for high network load. Further, it is interesting to note that there is no impact of delay requirements of the different T-CONTs on the throughput of the V2I network. It is due to the fact that

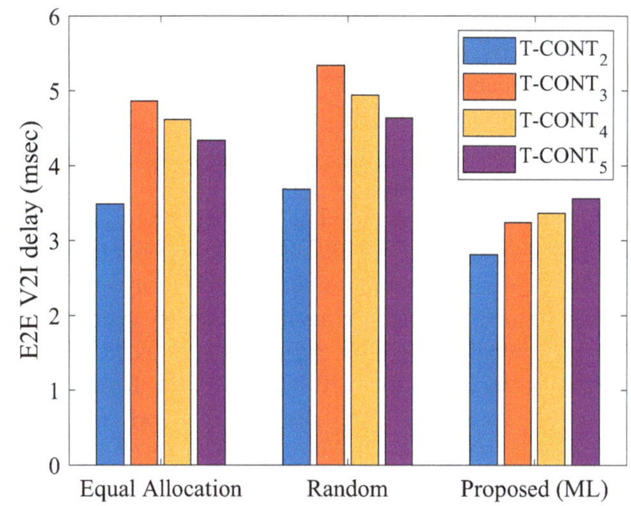

Fig. 7.10 E2E delay for V2I network for equal allocation, random, proposed (ML) [18] algorithm

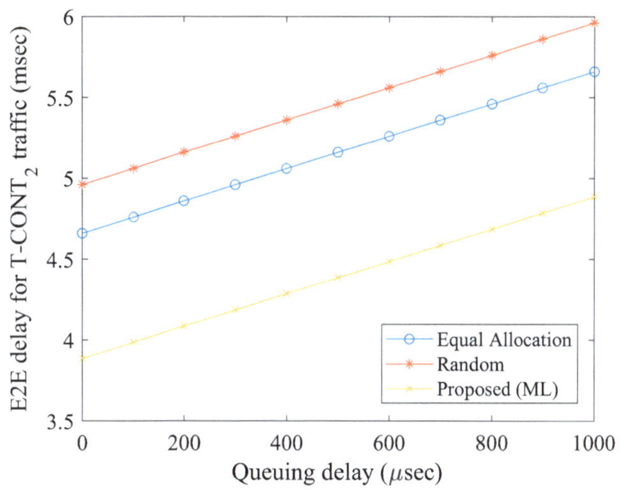

Fig. 7.11 E2E delay for T-CONT$_2$ with respect to queuing delay

the e2e delay requirement for the different T-CONTs of V2I network has a very slight impact on the throughput achieved by the V2I network as can be seen from Fig. 7.12.

The e2e link reliability for all the T-CONTs of the FiWi-based V2I network is calculated as follows:

$$P_T^S = P_{V2I}^S \cdot P_V^S \cdot P_{\text{RSU}}^S \cdot P_{\text{ONU-OLT}}^S, \tag{7.34}$$

where P_{V2I}^S, P_V^S, P_{RSU}^S is the V2I link reliability, vehicle equipment reliability, and the reliability of RSU, respectively. $P_{\text{ONU-OLT}}^S$ is the probability of successful transmission from ONU-RSU to OLT. The failure of V2I transmission can be denoted as:

$$\begin{aligned} P_T^F &= 1 - P_T^S = 1 - (1 - P_{\text{V2I}}^F)(1 - P_V^F) \\ &\times (1 - P_{\text{RSU}}^F)(1 - P_{\text{ONU-OLT}}^F), \end{aligned} \tag{7.35}$$

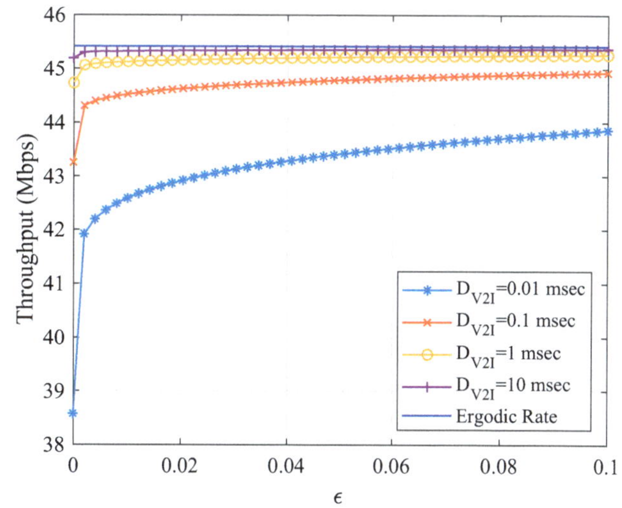

Fig. 7.12 Throughput with respect to delay and reliability for V2I link for 70% network load

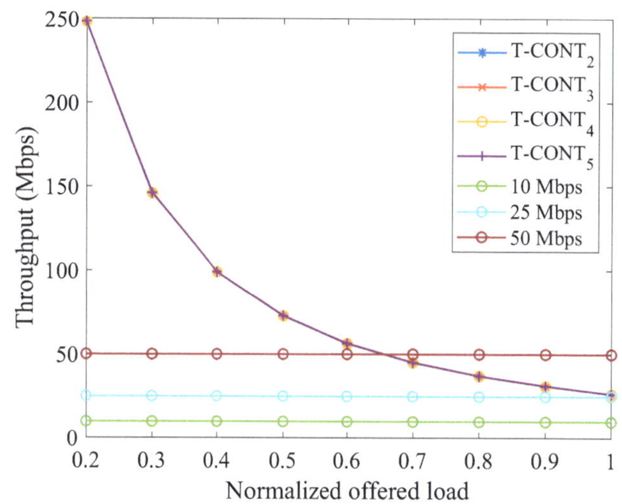

Fig. 7.13 Throughput at each vehicle for the ML-based T-CONT priority assignment algorithm

where P_{V2I}^F, P_V^F, and P_{RSU}^F are the probability of failure of V2I link, vehicle equipment, and RSU, respectively. $P_{\text{ONU-OLT}}^F$ is the probability of failure of transmission from ONU to OLT, which is given by:

$$P_{\text{ONU-OLT}}^F = 1 - (1 - P_{\text{ONU}}^F)(1 - P_C^F)(1 - P_{\text{cong}}^F), \tag{7.36}$$

where P_{ONU}^F denotes the probability of failure of ONU, P_{cong}^F is the probablity of failure of transmission due to optical network congestion, and P_C^F is the probability of fiber cut. Thus, reliability for V2I transmission is given as:

$$\begin{aligned} P_T^S &= 1 - P_T^F \\ &= (1 - P_{V2I}^F)(1 - P_V^F)(1 - P_{\text{RSU}}^F)(1 - P_{\text{ONU}}^F) \\ &\times (1 - P_C^F)(1 - P_{\text{cong}}^F). \end{aligned} \tag{7.37}$$

It can be observed from Fig. 7.14 that with the increase in wireless link failure, the reliability of the network decreases. Further, it can also be seen that as the probability of fiber cut (P_f) increases, the reliability of the network decreases. Moreover, for a high probability of fiber cut failure, i.e., $P_f = 10^{-1}$, the network is not able to meet the reliability threshold of 99%, i.e., it is not suitable for high priority T-CONT$_2$ traffic; however, for other T-CONTs, the reliability requirement of the network is met for V2I link failure probability of 10^{-3}. However, as the fiber cut failure decreases to 10^{-2}, the network is able to meet the reliability requirement of T-CONT$_2$ only for wireless failure probability of less than 10^{-3}. Similarly, if the probability of fiber cut decreases to 10^{-3}, the network is able to meet the reliability requirements of T-CONT$_2$ for wireless failure probability less than 10^{-2}. For the other T-CONTs, the reliability requirement is met for higher V2I link failure probabilities as well. Thus, if fiber cut probability is high, then the network may not be able to meet the reliability requirement of T-CONT$_2$ for high V2I link failure probability. While for the other T-CONTs, the reliability requirement of the network could be met even for high fiber cut probability.

Figure 7.15 compares the link reliability for V2I link for $P^F_{\text{cong}} = 10^{-3}$ and $P^F_{\text{cong}} = 10^{-2}$. It can be observed that as the network congestion increases, the link reliability of the V2I link decreases. Specifically, the link relaibility decreases by 1% for a probability of fiber cut failure 10−2 as the network congestion increases from 10^{-6} to 10^{-2}. Furthermore, for $P_f = 10^{-3}$ it can be observed that as the network congestion increases to 10^{-2}, the reliability requirement of T-CONT$_2$ traffic of the V2I link is not met even at 10^{-4} V2I link failure probabiliy. Similarly, for $P_f = 10^{-1}$, the relaibility requirement cannot be met for any of the T-CONTs.

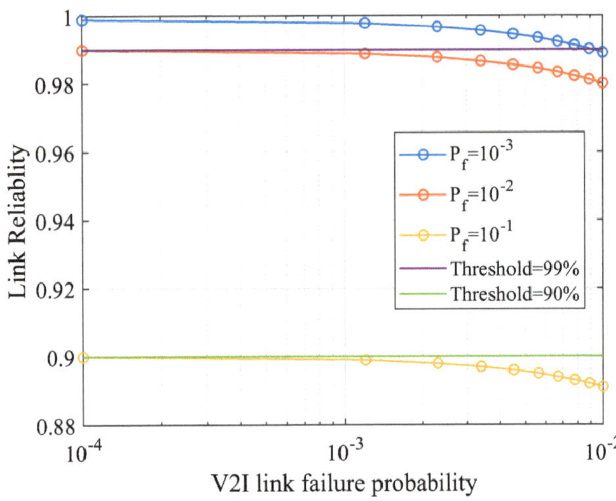

Fig. 7.14 Link Reliability for V2I link for $P^F_V = P^F_{\text{RSU}} = P^F_{\text{ONU}} = P^F_{\text{cong}} = 10^{-6}$

7.4 Summary

In this chapter, outdoor optical wireless systems are discussed. Specifically, this chapter focuses on providing a detailed analysis of two use cases for outdoor vehicular scenarios, namely, (a) hybrid RF-VLC V2X system and (b) wavelength allocation in the integrated fiber-wireless access network. The potential benefit of employing hybrid V-VLC/V-RF over stand-alone V-VLC or V-RF network configuration for BSMs dissemination at road intersections, ensuring high reliability and low latency, is shown. Depending on the transmitter's location, it is found that the stand-alone V-VLC and V-RF exhibit complementary roles in terms of outage probability, throughput, DOR, and IOR. The presented framework also shows the limitation of stand-alone V-VLC over V-RF communication when the distance be-

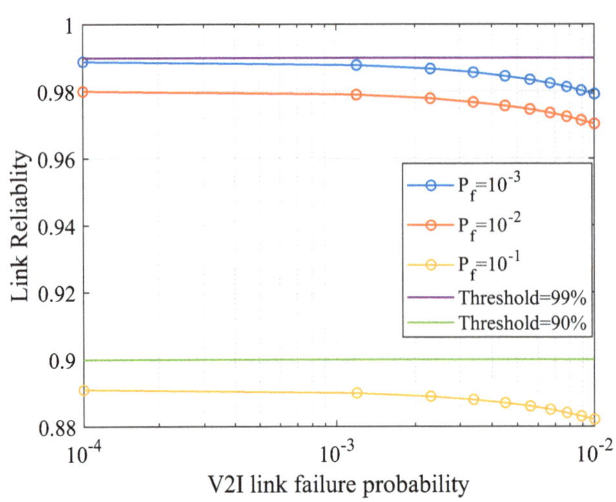

Fig. 7.15 Link Reliability for V2I link for $P^F_V = P^F_{\text{RSU}} = P^F_{\text{ONU}} = 10^{-6}$: **(a)** $P^F_{\text{cong}} = 10^{-3}$ and **(b)** $P^F_{\text{cong}} = 10^{-2}$

tween the desired vehicle and RSU is less than the critical distance defined for the V-VLC system. It is also found that hybrid V-VLC/V-RF outperforms stand-alone V-VLC or V-RF networks, thus serving as a better alternative option to meet diverse application needs for future intelligent transportation systems. It is expected that this comparative analysis may stimulate more innovations for hybrid VLC/RF-based vehicular networks. Further, the performance of ML-based T-CONT priority assignment wavelength allocation algorithm to meet the stringent requirements of 6G vehicular traffic is added. In order to facilitate the aforementioned, a mapping between the various kinds of vehicular data traffics and their 6G KPIs, namely throughput, delay, and reliability, is studied. Thereafter, a classification of the V2I traffic among four classes based on their KPI requirements is shown. Each class of traffic is assigned to a T-CONT for wavelength allocation at each ONU-RSU. By exploiting vehicular traffic predictions through ML, a framework is proposed to allocate the wavelength to the ONU-RSUs. The efficiency of the ML-based T-CONT priority assignment wavelength allocation algorithm is compared with the other approaches, namely equal T-CONT allocation and random wavelength allocation algorithm in terms of e2e delay, throughput, and reliability.

Homework Problems

1. Define the different optical access technologies in fiber-to-the-everything (FTTX) network.
2. What are the advantages of VLC links in comparison to RF links in vehicle-to-everything system?
3. What are the disadvantages of standalone VLC links in comparison to RF links in vehicle-to-everything system?
4. Explain the key differences between DSRC/IEEE 802.11p, C-V2X, VLC, LTE, and 5GNR for vehicle-to-everything system?
5. Derive the expression for outage probability for vehicular VLC system.
6. Discuss the KPI requirements for V2X use cases.
7. Describe the different configurations for the demultiplexer selection node at ONU.
8. Consider an FiWi-assisted V2I network where tuning time at a particular time slot is 50 μs, and the delay due to overheads is 0.5 msec. Consider the normalized load at the ONU is 70%. Find the delay at the ONU.
9. Consider an FiWi-based V2I transmission network where the probability of failure of the V2I link is 10^{-4}, and the probability of failure of vehicle equipment, ONU, and RSU is 10^{-6}. For the 10^{-3} probability of failure of transmission due to optical network congestion and the 10^{-4} probability of fiber cut, find the reliability for V2I transmission.

References

1. Chen, S., Hu, J., Shi, Y., Peng, Y., Fang, J., Zhao, R., Zhao, L.: Vehicle-to-everything (V2X) services supported by LTE-based systems and 5G. IEEE Commun. Standards Mag. **1**(2), 70–76 (2017)
2. Noor-A-Rahim, M., Liu, Z., Lee, H., Khyam, M.O., He, J., Pesch, D., Moessner, K., Saad, W., Poor, H.V.: 6G for vehicle-to-everything (V2X) communications: Enabling technologies, challenges, and opportunities. Proc IEEE **110**(6), 712–734 (2022)
3. Moya Osorio, D.P., Ahmad, I., Sánchez, J.D.V., Gurtov, A., Scholliers, J., Kutila, M., Porambage, P.: Towards 6G-enabled internet of vehicles: security and privacy. IEEE Open J. Commun. Soc. **3**, 82–105 (2022)
4. Calvanese Strinati, E., Barbarossa, S., Gonzalez-Jimenez, J.L., Ktenas, D., Cassiau, N., Maret, L., Dehos, C.: 6G: the next frontier: From holographic messaging to artificial intelligence using subterahertz and visible light communication. IEEE Veh. Tech. Mag. **14**(3), 42–50 (2019)
5. Memedi, A., Dressler, F.: Vehicular visible light communications: a survey. IEEE Commun. Surveys Tut. **23**(1), 161–181 (2021)
6. Singh, G., Pal, T., Bohara, V.A., Srivastava, A.: Optical ris empowered hybrid rf-vlc v2x communication systems: a promising new frontier for smart transportation applications. In: 2023 IEEE International Conference on Advanced Networks and Telecommunications Systems (ANTS), pp. 120–125 (2023)
7. Bazzi, A., Masini, B.M., Zanella, A., Thibault, I.: On the performance of IEEE 802.11 p and LTE-V2V for the cooperative awareness of connected vehicles. IEEE Trans. Veh. Tech. **66**(11), 10419–10432 (2017)
8. Hu, Y., Li, H., Chang, Z., Han, Z.: End-to-end backlog and delay bound analysis for multi-hop vehicular ad hoc networks. IEEE Trans. Wirel. Commun. **16**(10), 6808–6821 (2017)
9. Komine, T., Nakagawa, M.: Fundamental analysis for visible-light communication system using LED lights. IEEE Trans. Consumer Electron **50**(1), 100–107 (2004)
10. Haenggi, M.: Stochastic Geometry for Wireless Networks. Cambridge University Press, Cambridge (2012)
11. O'cinneide, C.A.: Euler summation for fourier series and laplace transform inversion. Stochast Models **13**(2), 315–337 (1997)
12. Abate, J., Whitt, W.: Numerical inversion of laplace transforms of probability distributions. ORSA J. Comp. **7**(1), 36–43 (1995)
13. Goldsmith, A.: Wireless Communications. Cambridge University Press, Cambridge (2005)
14. Steinmetz, E., Wildemeersch, M., Quek, T.Q.S., Wymeersch, H.: Packet reception probabilities in vehicular communications close to intersections. IEEE Trans. Intell. Transp. Syst. **22**, 1–11 (2020)
15. Yang, H.C., Alouini, M.S.: Wireless transmission of big data: Data-oriented performance limits and their applications. arXiv preprint arXiv:1805.09923 (2018)
16. Singh, G., Srivastava, A., Bohara, V.A.: Stochastic geometry-based interference characterization for RF and VLC-based vehicular communication system. IEEE Syst. J. **15**, 1–11 (2020)
17. Choi, J., Marojevic, V., Nealy, R., Reed, J.H., Dietrich, C.B.: DSRC and IEEE 802.11 ac adjacent channel interference assessment for the 5.9 GHz band. In: 2019 IEEE 89th Vehicular Technology Conference (VTC2019-Spring), pp. 1–5 (2019)
18. Gupta, A., Singh, A.P., Srivastava, A., Bohara, V.A., Srivastava, A., Maier, M.: Traffic prediction assisted wavelength allocation in vehicle-to-infrastructure communication: a fiber-wireless network based framework. Vehic Commun **45**, 100713 (2024). https://doi.org/10.1016/j.vehcom.2023.100713. https://www.sciencedirect.com/science/article/pii/S2214209623001432

19. Service Requirements for Enhanced V2X Scenarios (3GPP TS 22.186 Version 15.3.0 Release 15) (2018). https://www.etsi.org/deliver/etsi_ts/122100_122199/122186/15.03.00_60/ts_122186v150300p.pdf
20. Kaur, R., Gupta, A., Srivastava, A., Chatterjee, B.C., Mitra, A., Ramamurthy, B., Bohara, V.A.: Resource allocation and QoS guarantees for real world IP traffic in integrated XG-PON and IEEE802.11e EDCA networks. IEEE Access **8**, 124883–124893 (2020)
21. Maaloul, S., Aniss, H., Kassab, M., Berbineau, M.: Classification of C-ITS services in vehicular environments. IEEE Access **9**, 117868–117879 (2021)
22. Service Requirements for Enhanced V2X Scenarios (3GPP TS 22.186 Version 15.3.0 Release 15) (2018). https://www.etsi.org/deliver/etsi_ts/122100_122199/122186/15.03.00_60/ts_122186v150300p.pdf
23. Gupta, A., Goel, H., Bohara, V.A., Srivastava, A.: Performance evaluation of integrated XG-PON and IEEE 802.11ac based EDCA networks. In: 2020 IEEE International Conference on Advanced Networks and Telecommunications Systems (ANTS), pp. 1–6 (2020)
24. Dixit, A., Lannoo, B., Colle, D., Pickavet, M., Demeester, P.: Dynamic bandwidth allocation with optimal wavelength switching in TWDM-PONs. In: 2013 15th International Conference on Transparent Optical Networks (ICTON), pp. 1–4 (2013)
25. Series G: Transmission systems and media, digital systems and networks access networks – optical line systems for local and access networks: Multiple-wavelength passive optical networks (MW-PONs) (2023). https://www.itu.int/rec/dologin_pub.asp?lang=e&id=T-REC-G.989.3-202306-IAmd1PDF-E&type=items
26. Series G: Transmission systems and media, digital systems and networks: Digital sections and digital line system—optical line systems for local and access networks- 40-gigabit-capable passive optical networks 2 (NG-PON2): Physical media dependent (PMD) layer specification (2019). https://www.itu.int/rec/dologin_pub.asp?lang=s&id=T-REC-G.989.2-201902-IPDF-E&type=items
27. Garg, S., Dixit, A.: Novel dynamic bandwidth and wavelength allocation algorithm for energy efficiency in TWDM-PON. In: 2019 21st International Conference on Transparent Optical Networks (ICTON), pp. 1–4 (2019)
28. Yu, Y., Ranaweera, C., Lim, C., Guo, L., Liu, Y., Nirmalathas, A., Wong, E.: Hybrid fiber-wireless network: an optimization framework for survivable deployment. IEEE/OSA IEEE/OSA J. Opt. Commun. Netw. **9**(6), 466–478 (2017)
29. Yang, H., Zheng, K., Zhao, L., Hanzo, L.: Twin-timescale radio resource management for ultra-reliable and low-latency vehicular networks. IEEE Trans. Veh. Technol. **69**(1), 1023–1036 (2020)
30. Tsagkaris, K., Tselikas, N.D., Pleros, N.: A handover scheme based on moving extended cells for 60 ghz radio-over-fiber networks. In: 2009 IEEE International Conference on Communications, pp. 1–5 (2009)
31. Metro Interstate Traffic Volume Data Set (2018). https://archive.ics.uci.edu/ml/datasets/Metro+Interstate+Traffic+Volumel
32. Salari, M., Kattan, L., Gentili, M.: Optimal roadside units location for path flow reconstruction in a connected vehicle environment. Transport Res Part C Emerg Technol **138**, 103625 (2022). https://www.sciencedirect.com/science/article/pii/S0968090X22000705

8 Indoor Optical Wireless Communication

8.1 Introduction

Full use of the capacity provided by the fiber delivered to homes will necessitate the use of broadband links, including indoor wireless access technologies capable of operating at Gbits/s. In recent years, optical wireless (OW) has emerged as a strong candidate for high-speed indoor communications, as a complementary scheme to RF systems. The main advantages of optical wireless communications are unregulated and unlicensed electromagnetic spectrum, high-quality data transmission, immunity to electromagnetic interference, and highly secured communications. Indoor OW communications include two main technologies: visible light and infrared communications (VLC and IRC). Concerns about energy saving in general lighting are leading to the replacement of incandescent and fluorescent lamps with more energy-efficient solid-state lighting (SSL) devices. These SSL visible light sources can provide communications as well lighting, and this emerging technology has recently drawn great interest in research and practical applications. IRC technology is relatively well developed, and there are various applications and devices that conform to the Infrared Data Association (IrDA) standards. However, through the integration of VLC-based LiFi access network into current wireless infrastructures, the opportunity arises to ease the burden on RF systems and alleviate challenges posed by spectrum limitation. LiFi technology has emerged as a viable alternative to wireless fidelity (WiFi), offering a multitude of benefits. The concept of VLC can be extended to LiFi in order to avail high data rate, reliable, bidirectional, and fully networked indoor wireless communications. The concept of VLC can be extended to LiFi in order to achieve indoor wireless communications that have high data rates, are secure, and are fully networked and bidirectional in nature. LiFi offers various advantages over its RF counterpart (WiFi). First, because of the transmission properties of the light, LiFi offers inherent security. Second, as it operates in the visible light spectrum, it offers a huge unlicensed spectrum. Third, it does not suffer from electromagnetic interference from the large number of devices that operate in the GHz range, and hence, it is suitable for airplanes and hospitals environment. Moreover, LiFi does not cause health hazards [1].

The ubiquitous accessibility of LEDs, unlicensed spectrum, with data rate in the order of Gbps make LiFi an appealing choice for indoor communications [1]. However, despite of its numerous advantages, there is one major drawback in LiFi, which is restricting its global adaptation. LiFi can only deliver high data rates in Line-of-sight (LoS) links; in case of non-LoS or blockages, the throughput drops quickly, leading to coverage holes [1]. In contrast, the RF counterpart (WiFi) supports moderate data rates with nearly omnipresent connectivity. As LiFi and WiFi operate in the non-overlapping spectrum, it is possible for these technologies to coexist together and complement one another [1]. A network in which LiFi and WiFi coexist together and user is allowed to receive data from either LiFi or WiFi access point (AP) is termed as a hybrid LiFi-WiFi network.

Furthermore, multiple access (MA) is among the core physical layer technologies of OW communications, which enables wireless APs to identify a large number of different terminal users and serve them simultaneously. Current systems choose to use orthogonal multiple access method, i.e., users are orthogonal to each other in at least one radio resource dimension (e.g., frequency, time, and code). To support multiple access in VLC, conventional orthogonal frequency division multiple access (OFDMA) may not be employed as VLC has the requirement of real and positive signal transmission. In order to overcome the aforementioned drawback, Direct current-biased optical OFDMA (DCO-OFDMA) and Asymmetrically clipped optical OFDMA (ACO-OFDMA) have been studied. However, DCO-OFDMA suffers from deteriorated Bit error rate (BER), and ACO-OFDMA suffers from reduced spectral efficiency [2]. This has given rise to the concept of NOMA—a novel paradigm to guarantee high data rate and energy-

V. A. Bohara, Swaminathan R, *Optical and Wireless Access Networks*, Textbooks in Telecommunication Engineering, https://doi.org/10.1007/978-3-031-89696-5_8

efficient MA services. NOMA involves supporting a large number of devices by non-orthogonally overloading resource elements (REs). NOMA can be combined with visible light communication (VLC) to increase the spectral efficiency because of the following reasons [3]:

- The dominant LoS component and the short distance between light-emitting diode (LED) and photodetector (PD) give rise to a high signal-to-noise ratio (SNR) at VLC receiver, which helps improve the performance of NOMA.
- The limited modulation bandwidth of the LEDs can be efficiently utilized by NOMA to achieve high spectral efficiency.

NOMA is categorized into code-domain NOMA (CD-NOMA) and power-domain NOMA (PD-NOMA). PD-NOMA assigns different power levels to users based on their channel conditions and spatial locations. In CD-NOMA, each user is assigned a unique codebook (CB) or signature sequence. Sparse Code Multiple Access (SCMA), a prominent CD-NOMA scheme, utilizes multidimensional CBs to improve multi-user capacity and connectivity. Several research works incorporating SCMA in VLC (SCMA-VLC) systems have been presented in the literature. In [4, 5], experimental demonstration of SCMA-VLC system with overloading factor of 1.5 was reported.[1] In [6], a hybrid power-domain SCMA was investigated for VLC, showing improved data rate at the cost of reduced transmission distance. In [7], color-domain SCMA was proposed for VLC for a large throughput gain over conventional white light SCMA system. All the aforementioned works used the predefined RF complex SCMA CBs and made suitable changes (such as Hermitian symmetry and DC-level shifting) so that they can be applied to the VLC system. The authors in [8] focused on SCMA constellation design for VLC system in the presence of thermal noise with six users and eight REs. In [9] and [10], SCMA CB has been designed for three users with four REs considering the impact of varying shot noise in VLC systems.

In the VLC system, the optical receiver (i.e., PD) measures the intensity of the incident optical signal and converts it into electrical signal. The interaction of the incoming photons with the matter of PD is a statistical process [11]. The fluctuations in the number of photons detected result in fluctuating current, which causes shot noise at the optical receiver. The generated shot noise depends on the incident optical signal itself [12–14]. Further, the thermal motion of the electronic carriers results in fluctuating voltage known as thermal noise. In case of strong ambient light, the impact of thermal noise is more and the shot noise becomes very low. Thus, the noises at the VLC system are assumed to be independent of the received signal [15–17]. However, in case of indoor scenario, the distance between transmitter and receiver is less and the dominant signal is LoS [18]; thus, the received power is large. However, in practical VLC systems, typical illumination and communication scenarios offer very high SNR [19, 20]. In case of indoor scenario, since the received power is large, the shot noise is more dominant and cannot be neglected. The variance of shot noise is given as

$$\sigma_{sh}^2 = 2qx_p\Re \mathrm{B} + 2qI_B I_2 \mathrm{B}, \tag{8.1}$$

where q is the electronic charge, x_p is the intensity of incident optical signal, $\Re$ is responsivity, I_B is the background noise current, and I_2 is the noise-bandwidth factor, and B is the effective bandwidth of the optical receiver.

The thermal noise is assumed to be AWGN, which is independent of the intensity of the incident signal [13]. The shot noise affected optical intensity channel can be distinguished as an input-dependent Gaussian noise (IDGN), where variance of IDGN can be used to characterize the physical properties of the VLC system [13,21]. Consequently, the received signal y can be expressed as

$$y = x_p + \sqrt{x_p}Z_1 + Z_0, \qquad y \in \mathbb{R},\ x_p \geq 0, \tag{8.2}$$

where $Z_1 \sim \mathcal{N}_{\mathbb{R}}[0, \varsigma^2\sigma^2]$ denotes a real Gaussian RV with zero mean and $\varsigma^2\sigma^2$ variance describing IDGN (shot noise), and $Z_0 \sim \mathcal{N}_{\mathbb{R}}[0, \sigma^2]$ describes the thermal noise. Here, Z_1 and Z_0 are assumed to be independent, and $\varsigma^2 \in [1, 10]$ denotes the strength of shot noise with respect to thermal noise. In an indoor scenario, as the transmitter and the receiver parameters change, the received power changes, and thus, the shot noise varies. Also, the variation in shot noise can be depicted by considering different shot noise factor ς^2 values.

8.2 SCMA Data Transmission

Consider a downlink indoor LoS VLC system where multiple users are being served, each equipped with a single receiving device. All U users are grouped together for SCMA, and each group is supported by K REs (e.g., time or frequency slots), providing the load factor $\lambda = U/K$. In Fig. 8.1, the data transmission and reception of six users ($U = 6$) on four REs ($K = 4$) with $\lambda = 1.5$ is shown.

At the transmitter, $b = \log_2 M$ bits are mapped to K-dimensional real and positive codewords, with M number of codewords in a CB. These codewords are sparse vectors with N nonzero elements, $K > N$. Each user has a dedicated CB allotted to it. The CB of uth user can be expressed as

$$\mathbb{A}_u = \mathbf{V}_u\mathbf{C}_u, \qquad \text{for } j = 1, \ldots, J, \tag{8.3}$$

where $\mathbf{V}_u \in \mathbb{B}^{K\times N}$ and $\mathbf{C}_u \in \mathbb{R}_{>0}^{N\times M}$ denote the mapping matrix and constellation matrix of the uth user, respectively. The constellation matrix $\mathbf{C}_u = [\mathbf{c}_u^1, \ldots, \mathbf{c}_u^M]$, where $\mathbf{c}_u^m \in$

[1] The overloading factor of 1.5 means that the number of users is 1.5 times the number of resource elements.

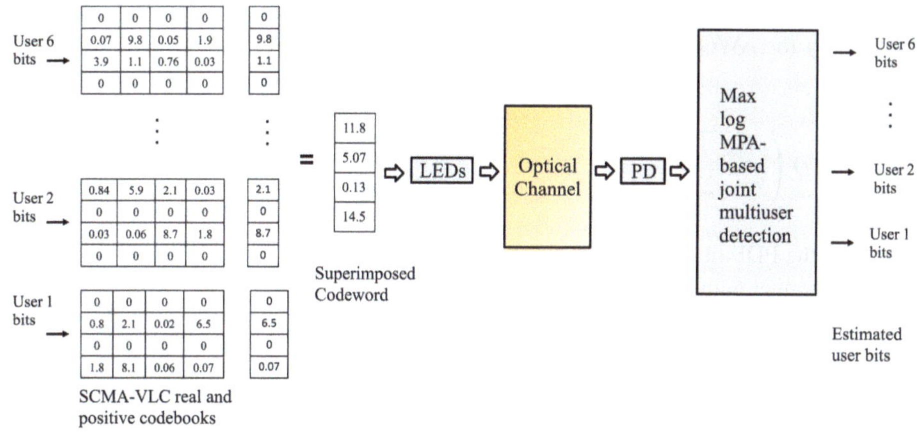

Fig. 8.1 The system model diagram of 4 × 6 SCMA with the CBs at the transmitter in VLC channel

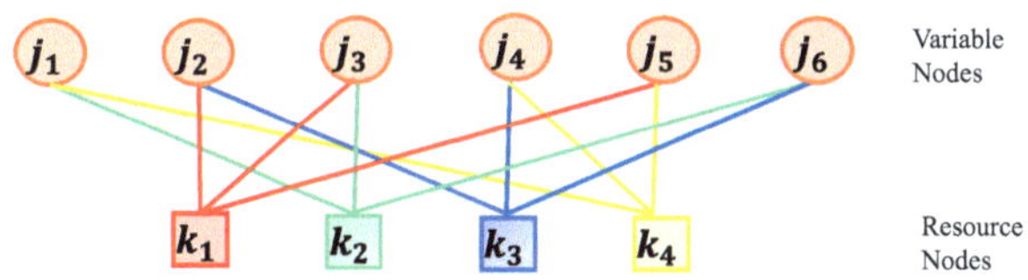

Fig. 8.2 Factor graph representation of 4 × 6 SCMA system with $d_f = 3$ and $N = 2$

$\mathbb{R}^{N\times 1}_{>0}$ denotes the mth constellation point of $\mathbf{C}_u$. The whole CB structure of SCMA can be represented by the factor graph matrix, $\mathbf{F} = (\mathbf{f}_1, \mathbf{f}_2, \ldots, \mathbf{f}_u)$, where $\mathbf{f}_u = \text{diag}(\mathbf{V}_u\mathbf{V}_u^T)$. The user u is being served by resource node (RN) k, if $\mathbf{F}_{kj} = 1$ $(1 \leq k \leq K, 1 \leq j \leq J)$. For a 4 × 6 SCMA block having $K = 4$ REs and $U = 6$ users, one example of the factor graph matrix of size 4 × 6 is given as

$$\mathbf{F}_{4\times 6} = \begin{bmatrix} 0\,1\,1\,0\,1\,0 \\ 1\,0\,1\,0\,0\,1 \\ 0\,1\,0\,1\,0\,1 \\ 1\,0\,0\,1\,1\,0 \end{bmatrix}. \tag{8.4}$$

Figure 8.2 shows an example of factor graph representation of six users (or Variable Nodes (VNs)) with four RNs, N=2 and data of d_f users are being superimposed on each RN ($d_f = 3$ for $\mathbf{F}_{4\times 6}$ shown in (8.4)). The binary mapping matrices corresponding to the six users are given as follows:

$$\mathbf{V}_1 = \begin{bmatrix} 0\,0 \\ 1\,0 \\ 0\,0 \\ 0\,1 \end{bmatrix}, \mathbf{V}_2 = \begin{bmatrix} 1\,0 \\ 0\,0 \\ 0\,1 \\ 0\,0 \end{bmatrix}, \mathbf{V}_3 = \begin{bmatrix} 1\,0 \\ 0\,1 \\ 0\,0 \\ 0\,0 \end{bmatrix}, \\ \mathbf{V}_4 = \begin{bmatrix} 0\,0 \\ 0\,0 \\ 1\,0 \\ 0\,1 \end{bmatrix}, \mathbf{V}_5 = \begin{bmatrix} 1\,0 \\ 0\,0 \\ 0\,0 \\ 0\,1 \end{bmatrix}, \mathbf{V}_6 = \begin{bmatrix} 0\,0 \\ 1\,0 \\ 0\,1 \\ 0\,0 \end{bmatrix}. \tag{8.5}$$

The set of CBs for U users is dependent on the constellation matrices $\mathbf{C}_u,\ j = 1, \ldots, J$ as the mapping matrices remain fixed as shown in (8.5). Assuming synchronous multiplexing between users, the received signal can be expressed as

$$\begin{aligned} \mathbf{y} &= \sum_{j=1}^{J} \text{diag}(\mathbf{h}_j)\mathbf{x}_j + \text{diag}\left(\sum_{j=1}^{J} \text{diag}(\mathbf{h}_j)\mathbf{x}_j\right)^{\frac{1}{2}} \mathbf{n}_{sh} + \mathbf{n}_{th}, \\ &= \sum_{j=1}^{J} \text{diag}(\mathbf{h}_j)\mathbf{V}_j\mathbf{c}_u \\ &\quad + \text{diag}\left(\sum_{j=1}^{J} \text{diag}(\mathbf{h}_j)\mathbf{V}_j\mathbf{c}_u\right)^{\frac{1}{2}} \mathbf{n}_{sh} + \mathbf{n}_{th}, \end{aligned} \tag{8.6}$$

where $\mathbf{x}_u \in \mathbb{R}^{K\times 1}_{>0}$ is a codeword of uth user, diag$(\mathbf{h}_u)$ denotes the channel matrix of uth user, the second term represents the IDGN with $\mathbf{n}_{sh} \sim \mathcal{N}_{\mathbb{R}}[\mathbf{0}, \varsigma^2\sigma^2\mathbf{I}]$, and $\mathbf{n}_{th} \sim \mathcal{N}_{\mathbb{R}}[\mathbf{0}, \sigma^2\mathbf{I}]$. The channels are assumed to be unit vectors for simplicity, and the extension to the general VLC values is straightforward.

8.3 SCMA CB Design

For SCMA CB design and optimization, most researchers aim for maximization of the minimum Euclidean distance (MED) between the superimposed codewords [22,23]. In the AWGN channel, MED is taken as one of the key performance indicators for determining the performance of the designed SCMA CB. Let $\mathbf{s}_i$ be a superimposed codeword from the combined constellation matrix $\mathbb{M}_v \in \mathbb{R}^{K\times M^J}_{>0}$. Then, the MED between $\mathbf{s}_i \in \mathbb{M}_v$ and $\mathbf{s}_u \in \mathbb{M}_v$ is given as

$$d'_{iu} = (\mathbf{s}_i - \mathbf{s}_u)^T(\mathbf{s}_i - \mathbf{s}_u), \qquad \forall\, i < u \in [1, M^U]. \tag{8.7}$$

The MED emerges as a critical parameter because of the pair-wise error probability (PEP) for AWGN channel, which is given as

$$P'(\mathbf{s}_i \to \mathbf{s}_u) = Q\left(\sqrt{\frac{|\mathbf{s}_i - \mathbf{s}_u|^2}{2\sigma^2}}\right). \tag{8.8}$$

However, in case of IDGN, the PDF of the received signal $\mathbf{y}$ is given as (assuming unit channel gain)

$$f(\mathbf{y}|\mathbf{s}_i) = \frac{1}{(2\pi)^{K/2}}\frac{1}{|\Sigma_i|^{\frac{1}{2}}}\exp\{-\frac{1}{2}(\mathbf{y}-\mathbf{s}_i)^T\Sigma_i^{-1}(\mathbf{y}-\mathbf{s}_i)\}, \tag{8.9}$$

where $\Sigma_i = \text{diag}\{\varsigma^2\sigma^2\mathbf{s}_i + \sigma^2\mathbf{1}\}$ denotes the covariance matrix for the $\mathbf{s}_i$th superimposed codeword. It can be seen from (8.9) that the variance is now dependent on the superimposed codeword.

In the case of only thermal noise, MED is taken as a critical parameter for designing CB where variance is independent of the superimposed codeword. However, in the case of shot noise, when variance is no longer independent, the conventional MED cannot serve as an optimal performance indicator. With the effect of shot noise, the shape of constellation points gets distorted. To mitigate the effect of IDGN, the authors in [10] introduced an improved optimization objective based on rotated MED for designing constellation matrices. The rotated Euclidean distance (RED) between superimposed codewords $\mathbf{s}_i$ and $\mathbf{s}_u$ can be expressed as

$$d_{iu} = (\mathbf{s}_i - \mathbf{s}_u)^T\mathbf{G}_i^{-\frac{1}{2}}\mathbf{G}_u^{-\frac{1}{2}}(\mathbf{s}_i - \mathbf{s}_u),\ \forall i < u \in [1, M^U], \tag{8.10}$$

where $\mathbf{G}_i = \varsigma^2\text{diag}\{\mathbf{s}_i\} + \mathbf{I}$, $\mathbf{G}_u = \varsigma^2\text{diag}\{\mathbf{s}_u\} + \mathbf{I}$ and $\mathbf{I} \in \mathbb{B}^{K\times K}$ represents the Identity matrix. It is to be noted from (8.10) that the new parameter RED incorporates the shot noise effect while computing the Euclidean distance between superimposed codewords. The authors in [10] aimed at maximizing the rotated minimum Euclidean distance (R-MED) subject to the power constraints. The objective function proposed in [10] for SCMA CB design is given as

$$\max_{\mathbf{L}\in\mathbb{R}_{>0}^{NMU\times 1}}\left\{\min_{1\le i\le u\le M^U} d_{iu}(\mathbf{L})\right\}, \tag{8.11}$$

where $\mathbf{L} = \text{vec}(\mathbf{C})$. This was further simplified as

$$\max_{\mathbf{L}\in\mathbb{R}_{>0}^{NMU\times 1},\ d_{\min}>0} d_{\min}, \tag{8.12}$$

where $d_{\min}$ denotes the minimum value of RED. It has been verified in [10] that R-MED-based CB outperforms the MED-based CB for VLC systems affected by shot noise. The results in [10] have shown that for $U = 3, K = 4, M = 4, \varsigma^2 = 5$ and $\sigma^2 = 0.01$, BER of 10^{-3} is achieved at $P_e = 10$ and $P_e = 7$ for MED and R-MED-based CBs, where P_e denotes the maximum electrical power allotted to a user. Since from the literature we can understand that RED works as an important parameter for designing CB than MED, we have included RED in our optimization problem. Further, the authors in [9] introduced the concept of distance range (DR) based on the maximum and the minimum values of the RED. They have mentioned that for the low-to-medium SNR region, the BER is no longer dependent on just R-MED. The optimization problem proposed in [9] for CB designing is

$$\max_{d_{\min}>0, d_{\max}>0}\{d_{\min} - \alpha d_{\max}\} \tag{8.13a}$$

$$\text{s.t.}\quad L_i > 0, \qquad \forall i \in [1, NMU], \tag{8.13b}$$

$$\frac{1}{M}\text{Tr}(\mathbf{C}_u^T\mathbf{C}_u) \le P_e, \quad \forall u \in [1, U], \tag{8.13c}$$

$$d_{iu} \ge d_{\min}, \quad \forall i < u, \tag{8.13d}$$

$$d_{iu} \le d_{\max}, \quad \forall i < u, \tag{8.13e}$$

where $d_{\max}$ denotes the maximum value of RED, and $\alpha > 0$ denotes the weighting factor. The objective function shown in (8.13a) is dependent on $d_{\min}$ and $d_{\max}$. The real and positive optical signal constraint is considered in (8.13b) and maximum electrical power constraint is considered in (8.13c). All the REDs will lie between $d_{\min}$ and $d_{\max}$ (8.13d, 8.13e). Also, it is to be noted that, lesser the overlapping between the constellation points of a user, better the performance of the designed CBs.

The objective function mentioned in (8.11) considers all the REDs while designing and optimizing the CBs to get optimum results. However, considering all the REDs makes the objective function non-convex and non-differentiable [24]. To solve (8.11), authors in [17] have proposed the objective function as shown in (8.13). It should be noted that the numerical methods used for solving the nonlinear maximin problem aim at approximating the $F(x) = \min_i f_i(x)$ by close enough smooth functions, and the closeness of approximation decides the nearness to optimal solutions. Considering $\sum_i \phi(f_i(x))$, the approximation to $F(x)$ becomes better when $\phi(.)$ is a strong convex function (i.e., large positive second derivative). One approach to solve this non-convex problem is to use well-known "log sum-exponentials" continuous approximation [25]. With this approximation, the objective function in (8.11) transforms to [26]

$$\min_{\mathbf{L}\in\mathbb{R}_{>0}^{NMU\times 1}}\frac{1}{\beta}\ln\left\{\sum_{i\neq u} e^{-\beta\, d_{iu}(\mathbf{L})}\right\}, \tag{8.14}$$

where $\beta > 0$ is a tuning parameter, which results in the smoothness of (8.14) and affects the accuracy of the model.

- Example 1: Let $\beta = 1$, and there are two distances $d_1 \gg d_2$, then
  $- \ln(e^{-d_1} + e^{-d_2})$
  $\approx -\ln(e^{-d_2}) = d_2 = \min\{d_1, d_2\}$.

The function in (8.14) is differentiable but still non-convex; however, optimization algorithms based on the derivation of the Hessian of the objective function can find the local optimal solution [27]. Also, if the objective function is optimized number of times with initialization from small set of random numbers and choosing the β parameter wisely, the likelihood of obtaining the global optimal solution increases [28]. Thus, the optimization problem is formulated as

$$\min_{\mathbf{L}\in\mathbb{R}_{>0}^{NMU\times 1}} \quad \frac{1}{\beta}\ln\left\{\sum_{i\neq u} e^{-\beta\, d_{iu}(\mathbf{L})}\right\} \tag{8.15a}$$

$$\text{s.t.} \quad L_i > 0, \qquad \forall\, i \in [1, NMU], \tag{8.15b}$$

$$\frac{1}{M}\mathrm{Tr}(\mathbf{C}_u^T\mathbf{C}_u) \leq P_e, \quad \forall\, u \in [1, U], \tag{8.15c}$$

where (8.15b) denotes the real positive intensity constraint of the transmitted codeword elements, and (8.15c) denotes that the average electrical power of the generated codewords should be less than the maximum electrical power (P_e) allotted per user. The optimization problem in (8.15) can be solved using commercially available solvers like MATLAB's *fmincon*.

8.3.1 Performance Analysis

For the initial simulation, consider the case where three users share four REs with $M = 4$ and the number of nonzero elements in the codeword is $N = 2$. For three users, the CBs (DR-based and Log-sum-exp-based) are generated by solving the optimization problems (8.13, 8.15). Figure 8.3 shows the two-dimensional (2D) scatter plots of the generated DR-based CBs (8.13) and the Log-sum-exp-based CBs (8.15) for $\varsigma^2 = 10$ and $P_e = 30$. It can be noted that the constellation points under DR criterion (8.13) are substantially overlapping as shown in Fig. 8.3a–c. However, the overlapping between Log-sum-exp-based constellation points for all three users is negligible as shown in Fig. 8.3d–f. As the overlapping increases between different constellation

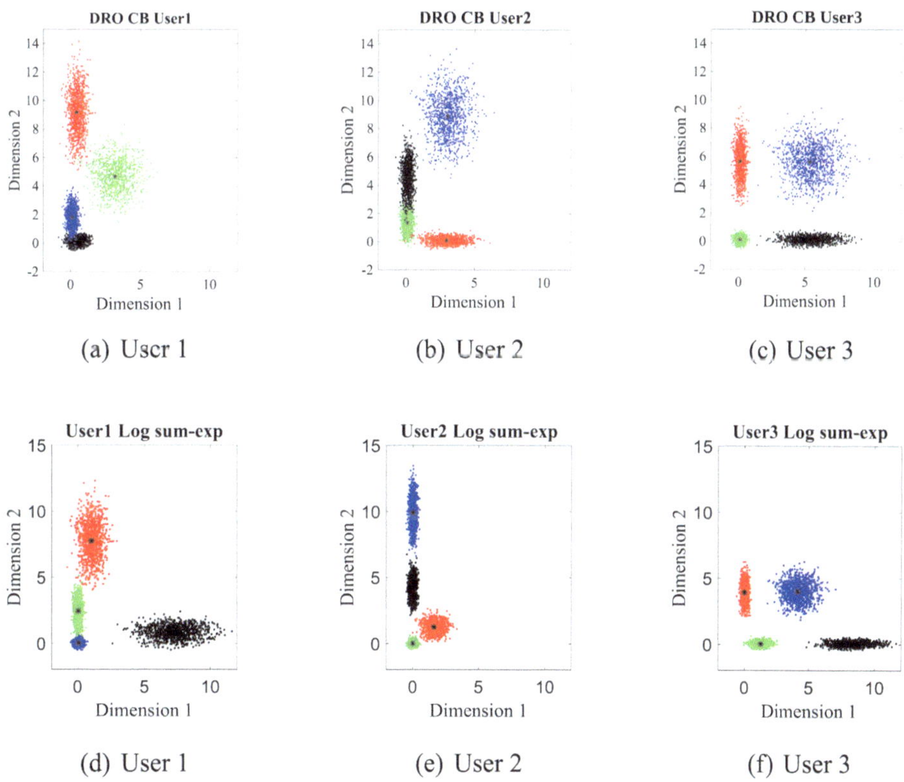

Fig. 8.3 SCMA constellations scatter plots with $U = 3$, $K = 4$, $P_e = 30$, $\varsigma^2 = 10$ (a)–(c): DR-based constellation sets, (d)–(f): Log-sum-exp-based constellation sets

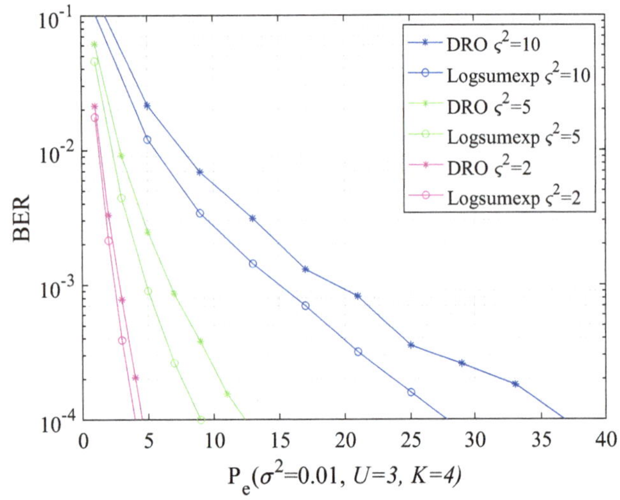

Fig. 8.4 BER performance comparison for $U = 3$ between DR-based CBs and Log-sum-exp CBs for $\sigma^2 = 0.01$, $M = 4$

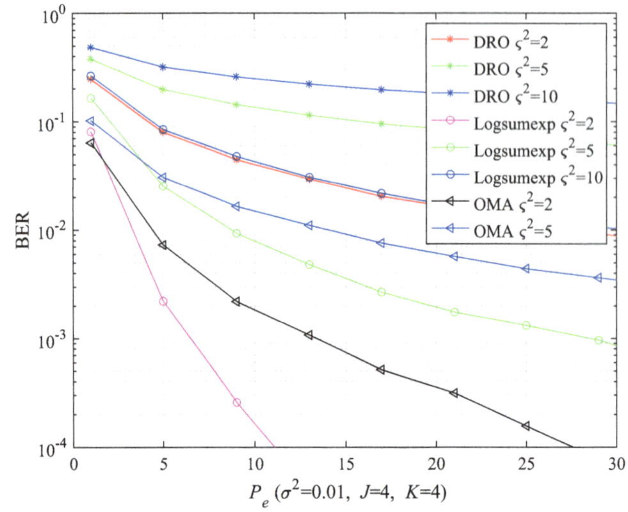

Fig. 8.5 BER performance comparison for $U = 4$ between SCMA DR-based CBs, Log-sum-exp CBs, and OMA scheme for $\sigma^2 = 0.01$, $M = 4$

points, it becomes difficult for the receiver to detect the symbols correctly, and thus, the overall performance deteriorates.

Figure 8.4 shows the bit error rate (BER) performance for three users of the Log-sum-exp-based (8.15) SCMA CBs and the DR-based CBs (8.13), respectively. The negligible overlapping in the case of log-sum-exp CBs as compared to DR CBs (in Fig. 8.3) reasons the performance gain obtained in Fig. 8.4. It can be noticed that for the low value of ς^2 (e.g., $\varsigma^2 = 2$), BER performance of both CBs is close, and for high values of ς^2 (e.g., $\varsigma^2 = 5, 10$), the Log-sum-exp-based CBs yield better performance gain as compared to DR-based CBs. It is to be noted that log-sum-exp CBs achieve the BER of 10^{-3} at $P_e = 15$ while DR-based CBs achieve it at $P_e \approx 20$ for $\varsigma^2 = 10$. This suggests that for large values of ς^2, the advantage of the Log-sum-exp-based CB is more significant than DR-based CB.

Consider the case where four users share four REs. Figure 8.5 shows the BER performance of four users where Log-sum-exp-based CB outperforms DR-based CBs for different values of ς^2. The reason behind the performance difference can be understood from the level of overlapping between the constellation points for both CBs in Fig. 8.6. It can be observed in Fig. 8.6 that the DR-based 2D constellation points suffer from significant overlapping, especially for user 1 and user 4, while Log-sum-exp-based 2D constellation points are barely overlapping. This results in significant performance improvement for Log-sum-exp-based CBs. Also in Fig. 8.5, the comparison with pulse amplitude modulation orthogonal multiple access (4 PAM-OMA) scheme is shown. It is noted that the SCMA outperforms the OMA scheme for different values of ς^2. This is expected as the OMA scheme offers no frequency diversity in contrast to the SCMA scheme.

8.4 Hybrid LiFi/WiFi for VLC Systems

The rapid increase in massive mobile gadgets such as smartphones, tablets, and laptops has triggered an insatiable craving for wireless data access in indoor communication. Despite wireless providers deploying additional access infrastructure like the high-throughput Wi-Fi APs, the primary limitation arises from the overuse of the current RF spectrum. This leads to contention, interference, increased latency, and reduced network throughput, commonly known as the "spectrum crunch." To tackle this challenge, innovative strategies are essential to unleash the greater capacity potential of the wireless link.

Optical communication technologies, including VLC, have surfaced as a promising contender for this purpose. VLC harnesses the visible light spectrum to facilitate wireless data transmission, offering advantages such as broader spectrum availability and the potential for higher data rates compared to RF-based communication. Through the integration of VLC-based Li-Fi network into current wireless infrastructures, the opportunity arises to ease the burden on RF systems and alleviate challenges posed by the spectrum limitation.

However, it is essential to acknowledge that Li-Fi is still an emerging technology as mentioned before, it has challenges such as limited coverage and requirement of LoS communication link [29]. However, Li-Fi has a huge potential to enhance wireless capacity and meeting the escalating demand for data access in wireless networks.

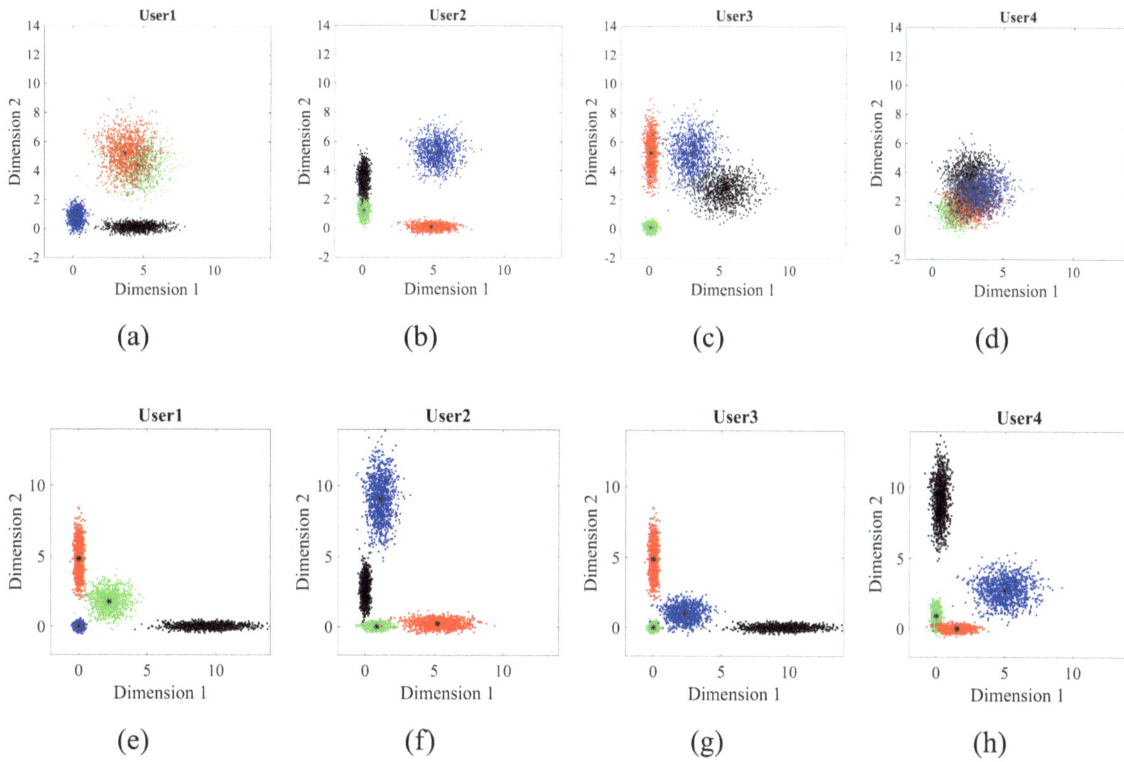

Fig. 8.6 SCMA constellations with $U = 4, K = 4, P_e = 30, \varsigma^2 = 10$ (a)–(d): DR based constellation sets, (e)–(f): Log-sum-exp based constellation sets

The Li-Fi AP incorporates a panel of LEDs for transmitting data and a PD for receiving transmitted data. Likewise, the user utilizes an Li-Fi dongle to receive data via light signals. The Li-Fi dongle is equipped with IR LEDs for transmitting data in the uplink direction and a photodiode for receiving data in the downlink direction through light. In indoor wireless communication, Wi-Fi APs as well as users utilize RF signals. However, by integrating Li-Fi and Wi-Fi technologies, the drawbacks of limited coverage in Li-Fi and lower throughput in Wi-Fi can be alleviated. As a result, a hybrid Li-Fi/Wi-Fi network [29] provides a wide coverage area and enhanced throughput, surpassing stand-alone Wi-Fi-only and Li-Fi-only networks.

8.4.1 Li-Fi Channel Modeling

The impulse response of Li-Fi channel is modeled between the Li-Fi AP i and user u as [29]:

$$H_{u,i}^{\text{Li-Fi}} = \frac{(m+1)A_{PD}}{2\pi d_{u,i}^2} \cos^m\left(\theta_{u,i}\right) \cos\left(\phi_{u,i}\right) g_c\left(\phi_{u,i}\right) g_f \tag{8.16}$$

where, $d_{u,i}$ represents the Euclidean distance between the AP i and user u. The irradiance angle is denoted as $\theta_{u,i}$, while the incidence angle is represented by $\phi_{u,i}$. The physical area of the photo-detector is designated as A_{PD}. We use g_f to denote the optical filter gain, and concentrator gain as g_c. Further, the parameter m to indicate the Lambertian order, as defined in [29].

Further, the received signal-to-interference ratio (SINR) at PD from Li-Fi AP is expressed as [29]:

$$\gamma_{u,i}^{\text{Li-Fi}} = \frac{\left(R_{PD} H_{u,i}^{\text{Li-Fi}} P_{\text{opt}}/\mathbb{C}_{\mho}\right)^2}{N_{\text{Li-Fi}} B_{\text{Li-Fi}} + \sum_{\beta \neq i} \left(R_{\text{PD}} H_{\text{u,i}}^{\text{Li-Fi}} P_{\text{opt}}/\mathbb{C}_{\mho}\right)^2}, \tag{8.17}$$

where R_{PD} represents the PD responsivity, and $\mathbb{C}_{\mho}$ signifies optical-to-electrical power conversion coefficient. Furthermore, P_{opt} denotes the transmitted power from Li-Fi AP. We define $N_{\text{Li-Fi}}$ as the power spectral density (PSD) of noise, while $B_{\text{Li-Fi}}$ stands for the channel bandwidth (BW) associated with each Li-Fi AP.

8.4.2 Wi-Fi Channel Modeling

For indoor RF propagation, the path loss model has a slope of 2 and 3.5 upto and after the breaking-point distance, respectively [29]. The path loss can be written as

$$L(d) = \begin{cases} L_{fs}(d) + X_s, & d \leq d_{BP} \\ L_{fs}(d) + 35\log_{10}(d/d_{BP}) + X_s, & d > d_{BP}. \end{cases} \tag{8.18}$$

Here, d_{BP} is the breaking-point distance, X_s denotes a Gaussian random variable with zero mean and standard deviation of σ. The free space path loss, denoted as $L_{fs}(.)$, is expressed as follows:

$$L_{fs}(d) = 20\log_{10}(d) + 20\log_{10}(f_c) - 147.5, \quad (8.19)$$

here f_c signifies the carrier frequency. The Wi-Fi channel multipath propagation is modeled as follows:

$$H_{u,i}^{\text{Wi-Fi}} = \sqrt{\frac{\mathbb{K}}{\mathbb{K}+1}}e^{j\Theta} + \sqrt{\frac{\mathbb{K}}{\mathbb{K}+1}}X_g, \quad (8.20)$$

where X_g signifies a random variable of complex Gaussian distribution with a mean of zero and variance of 1, and Θ represents the arrival or departure angle of the LOS component. In (8.20), the parameter $\mathbb{K}$ denotes the Ricean factor, with $\mathbb{K}$=1 before and K =0 after d_{BP}. The gain of Wi-Fi channel concerning user u is modeled as described in [29]:

$$G_{u,i}^{Wi-Fi} = \left|H_{u,i}^{\text{Wi-Fi}}\right|^2 10^{-\frac{PL\left(d_{u,i}\right)}{10}}, \quad (8.21)$$

where $PL(d_{u,i})$ represents pathloss according to the Wi-Fi multipath propagation model. The SNR between user u and Wi-Fi AP i is computed as [30]:

$$\gamma_{u,i}^{\text{Wi-Fi}} = \frac{G_{u,i}^{\text{Wi-Fi}}P_{\text{Wi-Fi}}}{N_{\text{Wi-Fi}}B_{\text{Wi-Fi}}}. \quad (8.22)$$

Where $P_{\text{Wi-Fi}}$ represents the transmit power, $N_{\text{Wi-Fi}}$ and $B_{\text{Wi-Fi}}$ denote noise PSD and BW of Wi-Fi AP, respectively.

8.4.3 Data Transmission Protocol

LiFi currently works in the following way. The LiFi AP schedules the channel access for multiple users using Media access control (MAC) protocol for uplink and downlink communication [31]. The MAC protocol for LiFi is standardized by the IEEE 802.15.7 and IEEE 802.11bb standards [31]. Carrier sense multiple access with collision aggregation (CSMA/CA) is used as the MAC protocol to schedule the packets and provide access to the channel in these standards.

LiFi's RF counterpart, Wireless fidelity (Wi-Fi), i.e., IEEE 802.11 uses distributed coordination function (DCF) MAC mechanism to provide access to the channel [31]. This mechanism employs CSMA/CA MAC protocol with binary exponential backoff algorithm. In this case, multiple users contend with each other to access the channel. If a user senses that the channel is busy, then it stops attempts to access the channel for a random period of time before trying to access the channel again.

Although widely used in practice, the exponential backoff algorithm suffers from a major drawback. The key drawback, known as hidden node problem, is that users not in the vicinity of the listening range of the channel may still interfere with other users during their channel access. To resolve this problem, prior notification is optionally sent to other users called Request to Send (RTS). A Clear to Send (CTS) request is sent by the access point (AP) to indicate that other stations should not send packets. This RTS/CTS mechanism mitigates the hidden node problem in WiFi [31] by avoiding collisions of the data packets at the cost of additional overhead.

The key challenge of Li-Fi networks is that it has an even more severe hidden node problem than Wi-Fi while accessing the channel. This is because the users cannot sense each other [31], since the uplink and downlink links use different spectrums. Therefore, RTS/CTS-based CSMA/CA MAC mechanism is extensively used to resolve the hidden node problem in this case. However, this mechanism significantly reduces the number of packets transmitted due to overheads incurred by the RTS/CTS frames [31].

8.4.4 Performance Analysis

The performance of the hybrid LiFi/WiFi is analyzed by considering a room of dimensions $5 \times 5 \times 3$ m^3, as depicted in Fig. 8.7. In this setup, four Li-Fi APs are strategically placed at the center of each of the four quadrants of the room's ceiling. Furthermore, a single Wi-Fi AP is centrally mounted on the ceiling. To model user locations, a uniform random distribution is employed. Moreover, for the effective deployment of hybrid Li-Fi/Wi-Fi Network (HLWN), a central controller (CC) [31] or network controller unit is taken into account.

In the context of the HLWN, N_{L} is denoted as the total count of Li-Fi APs and N_{W} as the total count of Wi-Fi APs. Furthermore, the total number of users in this hybrid network is denoted as N, which is represented as a set $\mathbb{U} = u \mid u \in [1, N]$.

The data rate received by the user u from Wi-Fi AP is expressed as:

$$R_{\text{Wi-Fi}} = B_{\text{Wi-Fi}}\log_2(1 + \gamma_{i,w}^{\text{Wi-Fi}}), \quad (8.23)$$

where $\gamma_{i,w}^{\text{Wi-Fi}}$ is SNR of ith user connected to Wi-Fi AP w.

The capacity of the users connected to Li-Fi AP is calculated as follows [1]:

$$R_{\text{Li-Fi}} = \frac{B_{\text{Li-Fi}}}{2}\log_2(1 + \frac{6}{\pi e}\gamma_{i,l}^{\text{Li-Fi}}), \quad (8.24)$$

where $B_{\text{Li-Fi}}$ denotes the bandwidth of each Li-Fi AP. $\gamma_{i,l}^{\text{Li-Fi}}$ is SINR between user i and Li-Fi AP l.

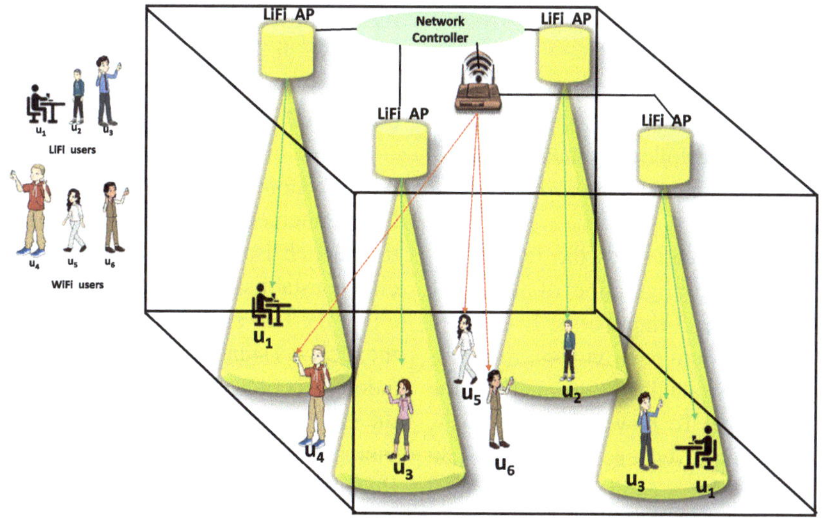

Fig. 8.7 System model for simulation setup of hybrid Li-Fi/Wi-Fi network

Table 8.1 Parameters of Wi-Fi channel model [31]

Channel parameter	Symbol	Value
Central carrier frequency	f_c	2.4 GHz
Transmit Power of Wi-Fi AP	$P_{\text{Wi-Fi}}$	20 dBm
Bandwidth of Wi-Fi AP	$B_{\text{Wi-Fi}}$	20 MHz
Wi-Fi noise PSD	$N_{\text{Wi-Fi}}$	−75 dBm/MHz

Table 8.2 Parameters of Li-Fi channel Model [31]

Channel parameter	Symbol	Value
Height of the AP from user level	h	2.15 m
Area of PD	A_{PD}	1 cm^2
Gain of optical filter	g_{filter}	1
Half intensity radiation angle	$\theta_{1/2}$	30°
FOV of PD	ϕ_{max}	60°
Optical to electric conversion efficiency	k	1
PD's responsivity	R_{PD}	0.53 A/W
Li-Fi AP's optical power	P_{opt}	3 Watt
Bandwidth of Li-Fi AP	$B_{\text{Li-Fi}}$	40 MHz
Li-Fi noise PSD	$N_{\text{Li-Fi}}$	−210 dBm/MHz

8.4.5 Performance Evaluations

The user's connectivity to a specific AP is contingent upon SNR or SINR value at the user's location. It is evident that the SNR exhibits a peak at the center of each quadrant, which corresponds to the AP location. Additionally, with increase in the distance from the AP, the SNR peak gradually diminishes. Consequently, as the users move out of the range of the Li-Fi AP, they seamlessly transition to using the Wi-Fi AP in HLWN. Average throughput is evaluated and compared for variable number of users of hybrid network with stand-alone Li-Fi and Wi-Fi networks. The evaluation has been done based on the parameters outlined in Tables 8.1 and 8.2.

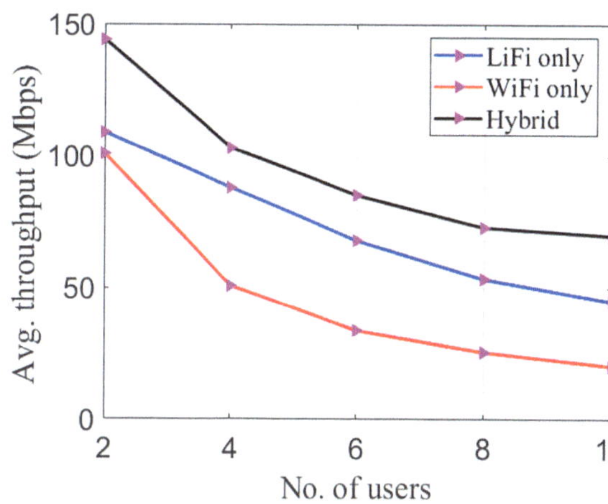

Fig. 8.8 Average throughput for simulation setup for varying the users

8.4.6 Simulation-Based Evaluation

8.4.6.1 Average Throughput

The performance of the average throughput in a hybrid Li-Fi/Wi-Fi network, considering varying numbers of users, is presented in Fig. 8.8. It is evident from the result that the average throughput has a decreasing trend as an increase in the number of users, which aligns with the expected trend. These simulation results provide valuable insights into the throughput behavior of the hybrid Li-Fi/Wi-Fi network. Similarly, the throughput performance for the experimental setup involving two and three users has also been evaluated. Figure 8.9 showcases the throughput performance obtained from the simulation setup for the case where $N = 3$. It can be observed that average throughput of the hybrid network is higher as compared to the Li-Fi only and Wi-Fi only network.

8.4.7 Experimental Evaluation

The performance of the hybrid Li-Fi/Wi-Fi network testbed is evaluated for real-time scenarios, encompassing both static and dynamic user conditions. The following subsections illustrate the experimental setup, along with the corresponding results and analyses.

The experimentation and evaluation of the hybrid Li-Fi/Wi-Fi network involved the configuration of $M = 1$ (number of Wi-Fi AP) and $N = 3$ (number of Li-Fi AP). Consequently, within an office room environment, three Li-Fi APs and one Wi-Fi AP are strategically installed on the ceiling, as depicted in Fig. 8.10. To power the Li-Fi APs, regulated power supplies are employed, each AP requires 36 volts for operation. Each Li-Fi AP is equipped with a panel containing 24 visible light LEDs for downlink communication and an IR LED for uplink communication, as illustrated in Fig. 8.10. When a user enters the field of view (FOV) range of a Li-Fi AP, the sync LED initiates blinking, indicating the start of communication between the user device and the AP. The user's device establishes communication with the Li-Fi AP through a dongle, which includes a transceiver of an IR LED for uplink communication and a photodiode for downlink communication.

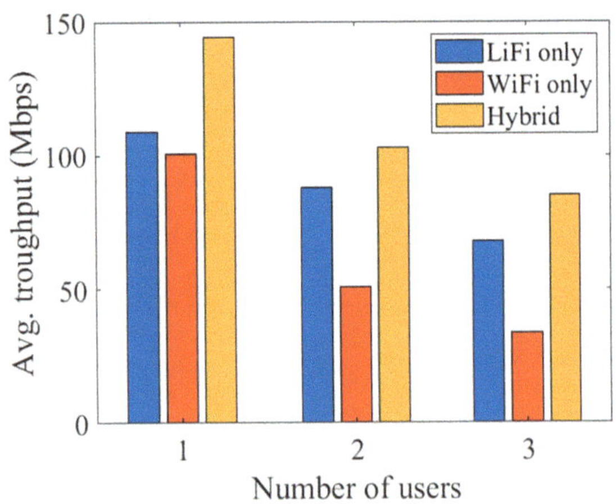

Fig. 8.9 Average throughput for simulation setup for varying the users

On the contrary, the Wi-Fi AP is powered by a 12-volt power adapter. All four APs, including both Li-Fi and Wi-Fi, are interconnected through a Gigabit Ethernet switch, and this switch is, in turn, linked to a central server. Users or clients establish wireless connections with the server using either the Li-Fi or Wi-Fi AP, depending on their location or the received signal strength (RSS). For Li-Fi communication, a dongle is powered by a USB power supply and connects to the client or user device via an Ethernet cable. Serial communication is facilitated through a Universal Asynchronous Receiver Transmitter (UART), while the controller makes decisions regarding seamless handover, such as transitioning between the Li-Fi and Wi-Fi APs or vice versa.

8.4.7.1 Observations

In the provided experimental configuration, data transmission from the server to the client occurs either through the Li-Fi AP or Wi-Fi AP, contingent on the client's location and movement dynamics as outlined by [29]. Both the server and the APs are interlinked via a network switch. Upon receiving data from the server, the Li-Fi AP transmits the information to the client using light as the communication medium. However, when the client is in motion, it receives data through the Li-Fi AP up to a defined signal strength threshold. Subsequently, if the signal strength drops below this threshold, the client seamlessly switches to the Wi-Fi AP from LiFi AP. This ensures continuous data reception from the server, as depicted in Figs. 8.11 and 8.12 with handover.

The observations are conducted through the live streaming of a 4K video from the server to the client's device. Simultaneously, the server initiated ping requests to the client's device while the client moved within the room. The outcomes of these observations are outlined below:

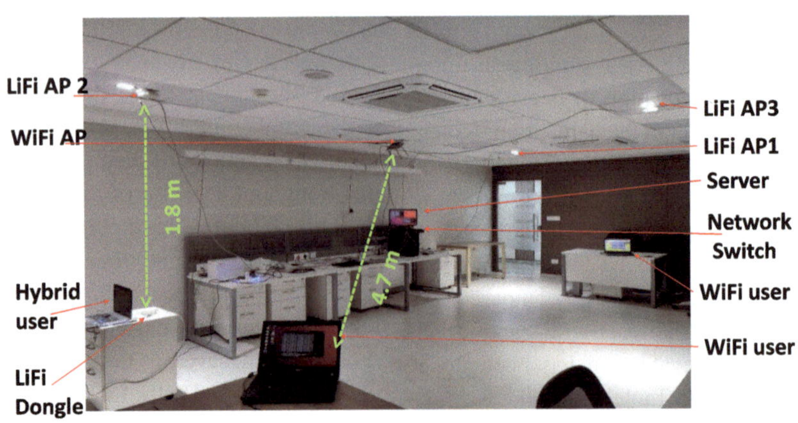

Fig. 8.10 Experimental setup of hybrid Li-Fi/Wi-Fi network

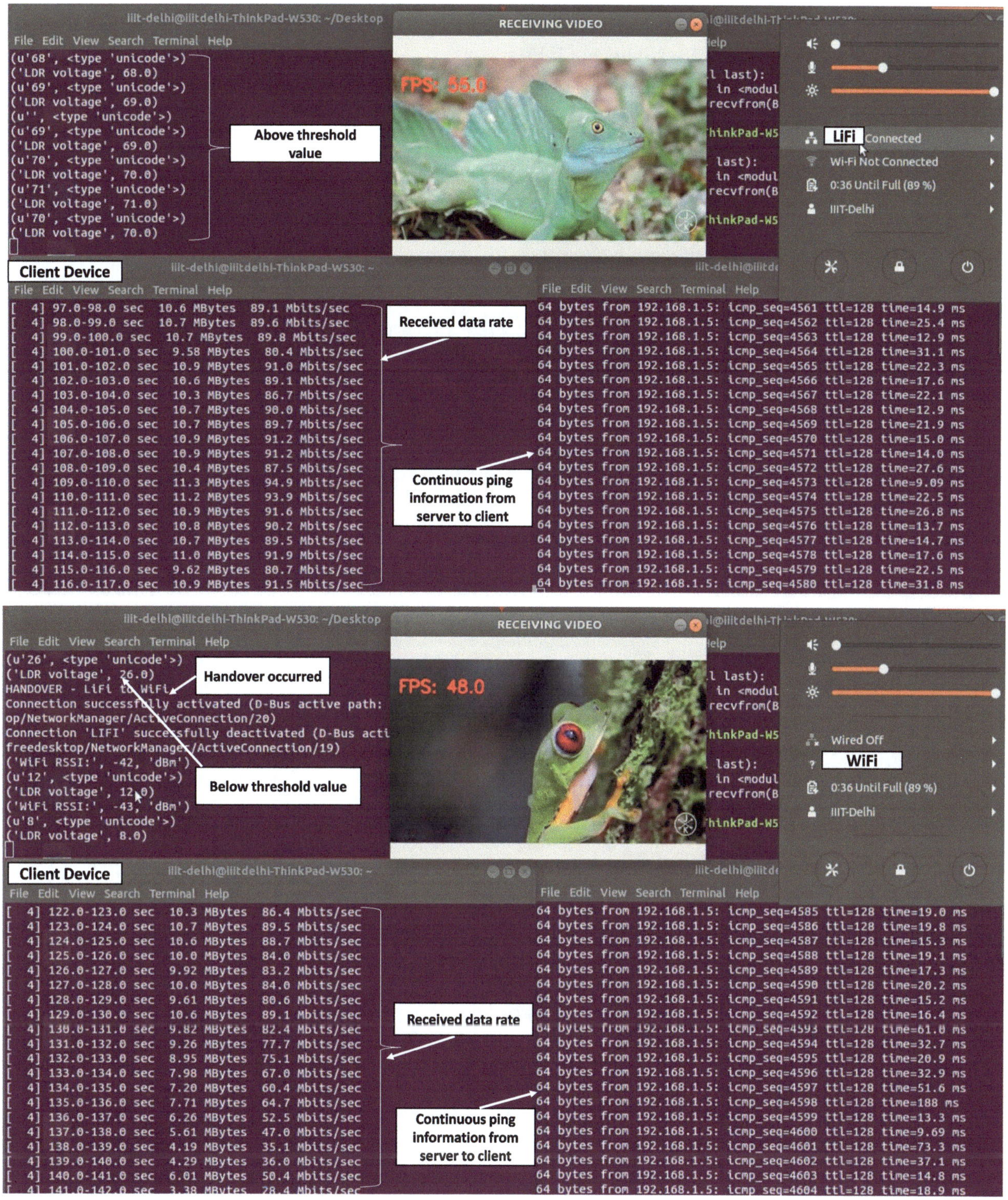

Fig. 8.11 Handover at client side experienced from LiFi to WiFi AP during live video streaming

- Initially, the client device establishes a connection with the Li-Fi AP. As the client moves, if the received voltage at the light detector (LDR) drops below a certain threshold value, a handover occurs. This entails the client device switching its connection from the Li-Fi AP to the Wi-Fi AP. Importantly, the video streaming process remains uninterrupted on the client device. This continuity is confirmed by the absence of timeouts between the server and client device, as evident from the ping information illustrated in Fig. 8.11.

Fig. 8.12 Server to client live video streaming in a hybrid LiFi/WiFi network

Table 8.3 Testing scenarios for user connection and AP switching status

Scenarios	User connection	Switching status of AP	Mobility status of the user
Hybrid Li-Fi/Wi-Fi network	One user connected to Li-Fi AP and one user connected to Wi-Fi AP	3 Li-Fi APs = ON, Wi-Fi AP = ON	Hybrid user—mobile Wi-Fi user—static
	One user connected to Li-Fi AP and two users connected to Wi-Fi AP		Hybrid user—Mobile Wi-Fi user—static
Stand-alone Wi-Fi network	Two users are connected to Wi-Fi AP	3 Li-Fi APs = OFF, Wi-Fi AP = ON	1 Wi-Fi user—mobile 1 Wi-Fi user—static
	Three users are connected to WiFi AP		1 Wi-Fi user—Mobile 2 Wi-Fi user—Static

- Additionally, as depicted in Fig. 8.12, it becomes evident that the client device transitions to a WiFi AP connection when the LDR voltage falls below the specified threshold. Subsequently, as the client device changes position and the LDR voltage surpasses the threshold once again, the client device seamlessly switches back from WiFi to the LiFi AP connection. Remarkably, throughout these transitions, the video streaming and ping activities persist, maintaining uninterrupted communication, as illustrated in the figure.

8.4.7.2 Experimental Results and Analysis

The performance of the hybrid Li-Fi/Wi-Fi networks is evaluated experimentally for two scenarios. In scenario 1, there are two users present in the room. One user remains stationary and maintains a constant connection to the Wi-Fi AP, while the second user stands beneath Li-Fi AP 1 and then proceeds to walk toward Li-Fi AP 2. Consequently, the second user qualifies as a hybrid user, as the connection dynamically shifts between Li-Fi and Wi-Fi, depending on the user's movement, as outlined in Table 8.3.

8.4.7.3 Average Throughput

The average throughput of the network has been measured in two scenarios as illustrated in Table 8.3: (a) using only the Wi-Fi network, and (b) utilizing the hybrid Li-Fi/Wi-Fi network, as illustrated in Fig. 8.13. The observations indicate that when there are two or three users in the network, the hybrid Li-Fi/Wi-Fi network demonstrates an improvement in throughput performance by 23.44% and 36.13% respectively, compared to the Wi-Fi-only network. It is evident that as the increase in number of users, the average throughput of the network decreases due to bandwidth sharing in the Wi-Fi-only network, which has also been observed in section III. In contrast, in the hybrid Li-Fi/Wi-Fi network, users can effectively utilize either Li-Fi or Wi-Fi network bandwidth based on their connectivity and position. This dynamic allocation of resources contributes to the enhanced performance and throughput observed in the hybrid Li-Fi/Wi-Fi network.

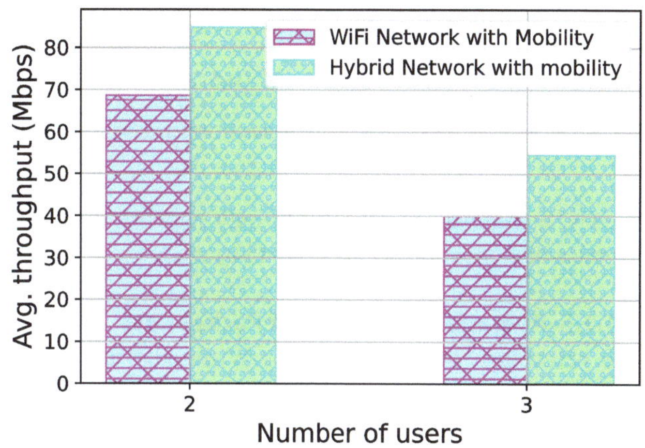

Fig. 8.13 Average throughput performance in experimental setup

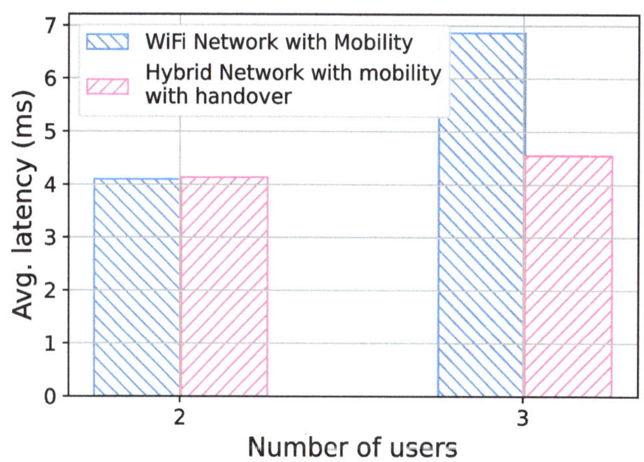

Fig. 8.14 Average latency performance in experimental setup

8.4.7.4 Average Latency

The average latency of both Wi-Fi-only and hybrid Li-Fi/Wi-Fi networks has been measured, as depicted in Fig. 8.14, for scenarios illustrated in Table 8.3. For the two-user case, the average latency of the hybrid Li-Fi/Wi-Fi network is observed to be the same as the Wi-Fi-only network. Conversely, in the three-user scenario, the average latency of the hybrid Li-Fi/Wi-Fi network is found to be 33.33% lower than that of the Wi-Fi-only network. It is important to note that in the three-user case, two users remain fixed and connected to the Wi-Fi AP at all times, while the other user is mobile and switches between Li-Fi and Wi-Fi AP based on their position. Consequently, the introduction of latency caused by the two fixed Wi-Fi users within the hybrid Li-Fi/Wi-Fi network contributes to the observed decrease in average

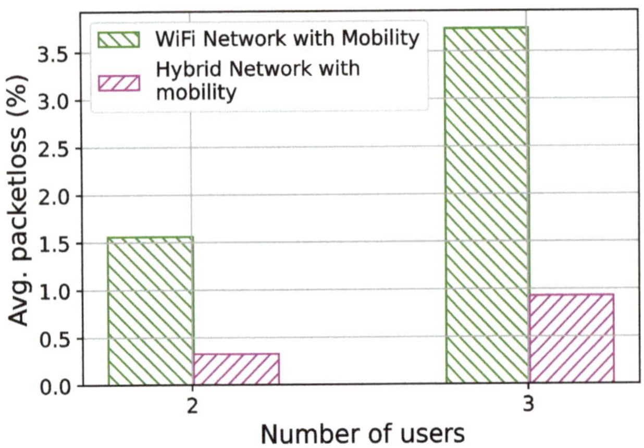

Fig. 8.15 Average packet loss performance in experimental setup

latency as compared to the Wi-Fi-only network due to an increase in contention between Wi-Fi users.

8.4.7.5 Average Percentage of Packet Loss

The percentage of packet loss for Wi-Fi-only and hybrid Li-Fi/Wi-Fi networks is measured for varying numbers of users, as depicted in Fig. 8.15. The measurements indicate that the Wi-Fi-only network experiences a higher percentage of packet loss compared to the hybrid Li-Fi/Wi-Fi network, specifically in scenarios illustrated in Table 8.3. In these cases, the hybrid network exhibits a substantial reduction of 73.51 and 74.32% in packet loss compared to the Wi-Fi-only network, respectively. These observations can be attributed to the heightened contention among users as well as the collision of packets, which intensifies with an increasing number of users.

8.5 Link Aggregation Enabled Hybrid Li-Fi/Wi-Fi Network

Since the LiFi and WiFi operates on different spectrums; a user is able to receive data concurrently from LiFi AP as well as WiFi AP, and such network is referred as link aggregation (LA)-enabled heterogeneous LiFi-WiFi network (HLWN). This LA can further improve the capacity, reliability, delay, and reduce the number of handovers. This section provides the performance analysis of LA-enabled HLWN framework against the hybrid LiFi-WiFi, stand-alone WiFi, and stand-alone LiFi networks in terms of average network data rate, QoS, and the coverage probability for predefined QoS requirement. An algorithm, henceforth known as, LA-SINR, is shown to further increase the system performance of LA-enabled HLWN in terms of the average data rate and coverage probability. Moreover, another algorithm (LA-EQoS) is proposed in [1], which is shown to further increase the system performance of LA-enabled HLWN in terms of the average data rate and coverage probability. A more realistic framework with LA overhead and adaptive modulation has been considered in this section.

The final user data rate for each user μ from a particular AP α is [30]:

$$R_{\mu} = \sum_{\alpha \in \mathbb{AP}} g_{\mu,\alpha} \rho_{\mu,\alpha} \eta_{\mu,\alpha} B_{\alpha} \tag{8.25}$$

where, B_α is the bandwidth of AP α, whereas, $g_{\mu,\alpha}$ is a binary association variable. The value of $g_{\mu,\alpha}$ denotes the set of users connected to a particular AP α_{Li} gets an equal proportion of time resource from the AP. The $\eta_{\mu,\alpha}$ denotes the spectral efficiency of the user; that is determined using the adaptive modulation and coding (AMC) scheme, which changes as per the SINR value [Table III]. $\rho_{\mu,\alpha}$ is resource allocation (RA) indicator denoting the amount of time resource AP α assigns to the user μ connected to it. In this work, equal resource allocation has been considered, and hence, the RA indicator can be calculated as follows:

$$\begin{aligned} &\rho_{\mu,\alpha} = 1/N_{\alpha} \\ &\text{s.t. } 0 \le \rho_{\mu,\alpha} \le 1 \text{ and } \sum_{\mu \in \boldsymbol{U}_{\alpha}} \rho_{\mu,\alpha} \le 1 \\ &\forall \mu \in \mathbb{U} \text{ and } \forall \alpha \in \mathbb{AP} \end{aligned} \tag{8.26}$$

where, $\boldsymbol{U}_\alpha$ represents the set of N_α users connected to AP α.

8.5.1 LA-SINR Algorithm

In this algorithm, every user receives data from both Li-Fi and Wi-Fi AP simultaneously. The user association with Li-Fi AP is based on highest SINR, hence this algorithm is termed as LA-SINR. The details of LA-SINR are explained in Algorithm 1. The binary association variable for LA-SINR is different for Wi-Fi and LiFi APs and can be calculated using (8.22) and (8.17) respectively. With the inclusion of LA overhead (β_{ov}) for a more practical realization, using (8.23) and (8.24), the final data rate for each user for the LA-SINR algorithm can be calculated as:

$$R_{\mu,\text{LA-SINR}} = \beta_{ov} \times \left\{ R_{\mu,\text{LiFi}} + R_{\mu,\text{WiFi}} \right\}, \tag{8.27}$$

where, β_{ov} is the LA overhead. Based on the values considered in literature [32], an average link aggregation overhead value of 80 %, i.e., $\beta_{ov} = 0.8$ has been considered.

8.5.2 LA-EQoS Algorithm

LA-EQoS maximizes the QoS of users in terms of data rate. The primary objective is to supplement the users that have

Algorithm 2 LA-SINR algorithm

Initialisation :
N - Total number of users
for all user μ
Assign $g_{\mu,\alpha}$.
end
Final rate calculation using (8.27)

low QoS from Li-Fi AP with an additional Wi-Fi connection, thereby increasing the overall data rate of individual users. LA-EQoS will reduce the load on Wi-Fi AP, as only those users that have low QoS from Li-Fi AP will connect with the WiFi AP.

Let N_{LA} be the number of users from set $\mathbb{U}_{\mathbb{LA}}$ that undergo LA. The remaining users will receive data only from Li-Fi AP assigned to them. The final rate for the users in the case of LA-EQoS algorithm is:

$$R_{\mu,\text{LA-EQoS}} = \begin{cases} \beta_{ov} \times \left\{ R_{\mu,\text{LiFi}} + R_{\mu,\text{WiFi}} \right\}, & \text{If } \mu \in \mathbb{U}_{\mathbb{LA}} \\ R_{\mu,\text{LiFi}}, & \text{Otherwise} \end{cases} \tag{8.28}$$

$g_{\mu,\alpha}$: binary association variable
$\eta_{\mu,\alpha}$: spectral efficiency of the user
$\rho_{\mu,\alpha}$: RA indicator

The optimal value of N_{LA} is determined using exhaustive search to optimize the overall data rate meanwhile ensuring the QoS for all the users.

Algorithm 3 LA-EQoS algorithm

Initialisation :
N - Total number of users
N_{LA} - Total number of LA users
for all user μ
Assign $g_{\mu,\alpha}$ using.
end
Rate calculation using (8.24)
Determine N_{LA} users having lowest QoS.
Assign these users to set $\mathbb{U}_{\mathbb{LA}}$.
for all user μ
if $\mu \in \mathbb{U}_{LA}$ **then**
$g_{\mu,\alpha_w} = 1$. Calculate final rate using (8.28)
end
end

8.5.2.1 Average Data Rate

Figure 8.16 illustrates the average data rate with respect to number of users. It is evident from the plot that LA-EQoS algorithm outperforms the LA-SINR algorithm and the hybrid network for different number of users. Furthermore, as the number of users increases, the average data rate reduces because of higher load and the data rate for hybrid network converges with the LA-SINR when $N = 10$. This is because each of the N users in LA-SINR is always connected to Li-Fi AP, which results in a smaller fraction of time resource allocation to all the users. In addition to this, the LA overhead associated with LA users further decreases the overall aggregated data rates from Li-Fi and Wi-Fi. Overall, the proposed LA-EQoS outperforms the LA-SINR and hybrid network.

8.5.3 Coverage Probability Analysis

In Fig. 8.17, the coverage probability vs. N has been plotted for a predefined QoS requirement. The QoS requirement of $R_0 = 20$ Mbps has been considered. It can be observed from Fig. 8.17, the LA-EQoS outperforms other network configurations for the given QoS requirement. The proposed LA-EQoS algorithm has almost 100% coverage for $N = 10$ users while for the LA-SINR algorithm, 100% coverage is limited to $N = 4$ users. The hybrid and stand-alone WiFi networks provides 100% coverage up to $N = 5, 3$ respectively. On the other hand, the stand-alone LiFi does not provide 100% coverage for any number of users. Additionally, more than 80% coverage can be observed by LA-EQoS algorithm up to $N = 14$ while for LA-SINR algorithm, it is up to $N = 8$. All the remaining networks provide less than 80% coverage when number of users is greater than 6. Thus, we can conclude that LA-EQoS performs better in terms of coverage probability as compared to other networks.

8.6 Summary

This chapter discusses the system models for indoor optical wireless communication. This chapter discusses the MA schemes for integrated OW communications. Specifically, this chapter focuses on SCMA codebook design and its performance analysis for VLC-based networks. Further, another system model for a VLC-based integrated hybrid Li-Fi/Wi-Fi network is presented in this chapter. This hybrid Li-Fi/Wi-Fi network explores the coexistence of LiFi and WiFi together, where the user is allowed to receive data from either Li-Fi or WiFi access points (APs). In addition to the simulated hybrid Li-Fi/Wi-Fi network, this chapter also presents an experimental evaluation of the network to understand the key challenges and performance of the system in practical deployment setup.

Homework Problems

1. Explain how NOMA can be combined with VLC to improve spectral efficiency. What are the key characteristics of VLC that enhance the performance of NOMA?

Fig. 8.16 Average data rate vs N

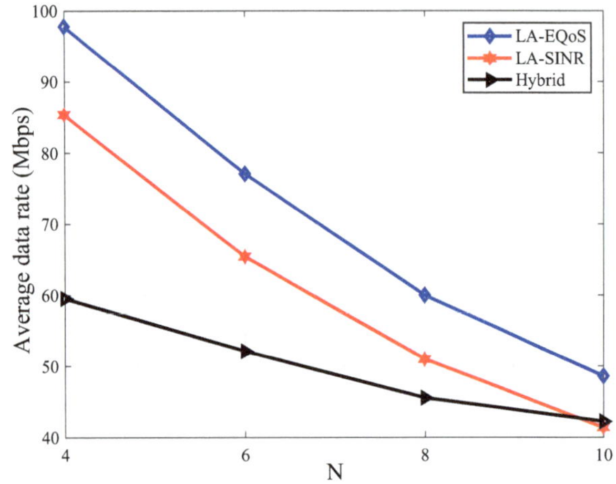

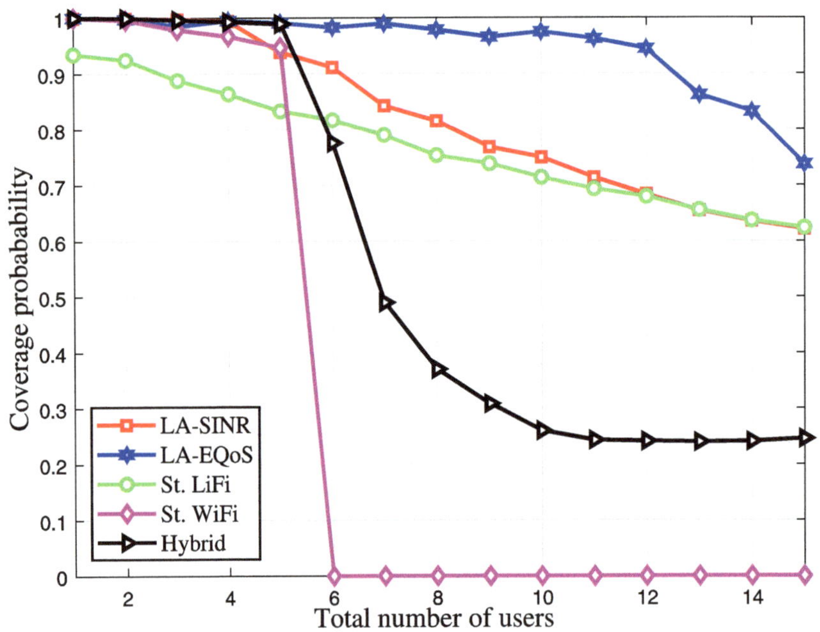

Fig. 8.17 Coverage for predefined QoS R_0—20 Mbps

2. Describe the different types of NOMA in the context of VLC systems. How do power-domain NOMA and code-domain NOMA (such as SCMA) differ in terms of their approach to assigning resources to users?
3. Discuss the impact of shot noise and thermal noise on VLC systems. How does the shot noise variance change with respect to the intensity of the received optical signal, and why is this particularly important in indoor VLC scenarios?
4. Explain how the factor graph representation for the 4×6 SCMA system, with six users (variable nodes) and four resource nodes, facilitates the joint multiuser detection process at the receiver. Discuss the significance of the mapping matrices and how they affect the codebook design and detection complexity.
5. In the 4×6 SCMA system described, the load factor is given as $\lambda = \frac{U}{K} = 1.5$. Explain the implications of this load factor on system performance. How would changing the number of users or resource elements affect the load factor, and what are the potential trade-offs in terms of spectral efficiency and receiver complexity?
6. Differentiate between Li-Fi and Wi-Fi according to the advantages and disadvantages?
7. How does Li-Fi coexist with Wi-Fi?
8. What is Link Aggregation?
9. How does LA-enabled heterogeneos network is different from hybrid Li-Fi/Wi-Fi network?
10. For a LA-enabled heterogeneous network, the overhead due to packet reordering is 0.2 i.e., 80% of packets are sccessfully aggregated. The bandwidth of Li-Fi and Wi-

Fi are 40 MHz and 20 MHz, respectively. Calculate data rate for a link aggregated network for given SNR of 40 dB and 20 dB for Li-Fi and Wi-Fi, respectively.

References

1. Karoti, N.M., Paramita, S., Ahmad, R., Bohara, V.A., Srivastava, A.: Improving the performance of heterogeneous LiFi-WiFi network using a novel link aggregation framework. In: Proceeding of the IEEE Wireless Communications and Networking Conference (WCNC), pp. 2322–2327 (2022)
2. Dimitrov, S., Sinanovic, S., Haas, H.: Clipping noise in ofdm-based optical wireless communication systems. IEEE Trans. Commun. **60**(4), 1072–1081 (2012)
3. Marshoud, H., Kapinas, V.M., Karagiannidis, G.K., Muhaidat, S.: Non-orthogonal multiple access for visible light communications. IEEE Photon. Technol. Lett. **28**(1), 51–54 (2016)
4. Lin, B., Tang, X., Zhou, Z., Lin, C., Ghassemlooy, Z.: Experimental demonstration of SCMA for visible light communications. Optics Commun. **419**, 36–40 (2018)
5. An, J., Chung, W.Y.: Single-led multichannel optical transmission with scma for long range health information monitoring. J. Lightwave Technol. **36**(23), 5470–5480 (2018)
6. Lin, B., Tang, X., Ghassemlooy, Z.: A power domain sparse code multiple access scheme for visible light communications. IEEE Wirel. Commun. Lett. **9**(1), 61–64 (2020)
7. Mitra, R., Sharma, S., Kaddoum, G., Bhatia, V.: Color-domain scma noma for visible light communication. IEEE Commun. Lett. **25**(1), 200–204 (2021)
8. Lou, S., Gong, C., Gao, Q., Xu, Z.: Scma with low complexity symmetric codebook design for visible light communication. In: 2018 IEEE International Conference on Communications (ICC), pp. 1–6 (2018)
9. Gao, Q., Hu, S., Gong, C., Serpedin, E., Qaraqe, K., Xu, Z.: Distance-range-oriented constellation design for vlc-scma downlink with signal-dependent noise. IEEE Commun. Lett. **23**(3), 434–437 (2019)
10. Hu, S., Gao, Q., Gong, C., Xu, Z., Boluda-Ruiz, R., Qaraqe, K.: Energy-efficient modulation for visible light scma system with signal-dependent noise. In: 2018 11th International Symposium on Communication Systems, Networks & Digital Signal Processing (CSNDSP), pp. 1–6 (2018)
11. SENIOR, J.M.: Optical Fiber Communications: Principles and Practice. 3rd edn. Pearson Education Limited, London (2009)
12. Gao, Q., Hu, S., Gong, C., Xu, Z.: Modulation designs for visible light communications with signal-dependent noise. J. Lightwave Technol. **34**(23), 5516–5525 (2016). https://opg.optica.org/jlt/abstract.cfm?URI=jlt-34-23-5516
13. Moser, S.M.: Capacity results of an optical intensity channel with input-dependent gaussian noise. IEEE Trans. Inf. Theory **58**(1), 207–223 (2012)
14. Wang, J.Y., Ge, H., Zhu, J.X., Wang, J.B., Dai, J., Lin, M.: Adaptive spatial modulation for visible light communications with an arbitrary number of transmitters. IEEE Access **6**, 37108–37123 (2018)
15. Ahn, K.I., Kwon, J.K.: Capacity analysis of m-pam inverse source coding in visible light communications. J. Lightwave Technol. **30**(10), 1399–1404 (2012)
16. Chaaban, A., Rezki, Z., Alouini, M.S.: Fundamental limits of parallel optical wireless channels: capacity results and outage formulation. IEEE Trans. Commun. **65**(1), 296–311 (2017)
17. Wang, J.B., Hu, Q.S., Wang, J., Chen, M., Wang, J.Y.: Tight bounds on channel capacity for dimmable visible light communications. J. Lightwave Technol. **31**(23), 3771–3779 (2013)
18. Anwar, D.N., Ahmad, R., Srivastava, A.: Energy-efficient coexistence of lifi users and light enabled iot devices. IEEE Trans. Green Commun. Netw. **6**(2), 930–950 (2022)
19. Grubor, J., Randel, S., Langer, K.D., Walewski, J.W.: Broadband information broadcasting using led-based interior lighting. J. Lightwave Technol. **26**(24), 3883–3892 (2008). https://opg.optica.org/jlt/abstract.cfm?URI=jlt-26-24-3883
20. Hanzo, L., Haas, H., Imre, S., O'Brien, D., Rupp, M., Gyongyosi, L.: Wireless myths, realities, and futures: from 3G/4G to optical and quantum wireless. Proc. IEEE **100**(Special Centennial Issue), 1853–1888 (2012)
21. Chen, H., Xu, Z.: A two-dimensional constellation design method for visible light communications with signal-dependent shot noise. IEEE Commun. Lett. **22**(9), 1786–1789 (2018)
22. Yu, L., Fan, P., Cai, D., Ma, Z.: Design and analysis of scma codebook based on star-qam signaling constellations. IEEE Trans. Veh. Tech. **67**(11), 10543–10553 (2018)
23. Vameghestahbanati, M., Marsland, I.D., Gohary, R.H., Yanikomeroglu, H.: Multidimensional constellations for uplink scma systems—a comparative study. IEEE Commun. Surveys Tutor. **21**(3), 2169–2194 (2019)
24. Luna-Rivera, J.M., Suarez-Rodriguez, C., Guerra, V., Perez-Jimenez, R., Rabadan-Borges, J., Rufo-Torres, J.: Low-complexity colour-shift keying-based visible light communications system. IET Optoelectron.: Special Issue on Opt.-Wirel. Commun. **9**, 191–198 (2015)
25. Chen, R.: Solution of minimax problems using equivalent differentiable functions. Comput. Math. Appl. **11**(12), 1165–1169 (1985). https://www.sciencedirect.com/science/article/pii/089812218590104X
26. Chaturvedi, S., Anwar, D.N., Bohara, V.A., Srivastava, A., Liu, Z.: Low-complexity codebook design for scma-based visible light communication. IEEE Open J. Commun. Soc. **3**, 106–118 (2022)
27. Monteiro, E., Hranilovic, S.: Constellation design for color-shift keying using interior point methods. In: 2012 IEEE Globecom Workshops, pp. 1224–1228 (2012)
28. Anwar, D.N., Srivastava, A.: Constellation design for single photodetector based csk with probabilistic shaping and white color balance. IEEE Access **8**, 159609–159621 (2020)
29. Paramita, S., Srivastava, A., Bohara, V.A., Mitra, A., M, H.R., Krishna Atluri, H., Paventhan, A.: A proof-of-concept demonstration of hybrid Li-Fi/Wi-Fi network on a prototype testbed. In: 2023 IEEE International Conference on Advanced Networks and Telecommunications Systems (ANTS), pp. 509–514 (2023)
30. Wu, X., Safari, M., Haas, H.: Access point selection for hybrid Li-Fi and Wi-Fi networks. IEEE Trans. Commun. **65**(12), 5375–5385 (2017)
31. Paramita, S., Bhattacharya, A., Srivastava, A., Bohara, V.A.: Hybrid CSMA/CA and HCCA uplink medium access control protocol for VLC based heterogeneous users. Comput. Commun. **225**, 54–64 (2024). https://www.sciencedirect.com/science/article/pii/S0140366424002275
32. Zeng, C., Licardie, S.: Systems and methods for performing layer one link aggregation over wireless links (2015). US Patent 9,125,084

Wireless Multiple Access Networks 9

9.1 Introduction

Wireless networks have enabled the development of groundbreaking innovations such as intelligent automobiles, intelligent homes, and smart cities with better security features. Furthermore, mobile data applications such as high-definition video streaming and machine-to-machine (M2M) communication require rapid and reliable internet connectivity [1, 2]. Successive generations of wireless networks have been developed to meet these growing demands. Simultaneously, various multiple access and modulation techniques have been proposed with each new generation to satisfy the continuously rising requirements of the mobile communication era. At present, these multiple access techniques are being refined to accommodate the heightened demands of next-generation wireless networks [3].

In telecommunications networks, multiple access strategies are essential for enabling optimal sharing of limited network resources among various users. These solutions distribute resources among multiple users while ensuring that the entire network's resources are not utilized by a single user. Since the inception of contemporary communications, academics have sought the optimum multiple access methods to promote resource sharing among users [4]. These schemes serve as a key technology that differentiate various wireless systems from the First generation (1G) to the 5G. The evolution of multiple access techniques is essential for the efficient separation of bandwidth, frequency, time, and code. Consequently, Frequency division multiple access (FDMA) was the foundational multiple access method in analog 1G cellular networks. TDMA became the predominant method in digital Second generation (2G) networks.Code division multiple access (CDMA) emerged as the leading technology in Third generation (3G) networks [5]. In Fourth generation (4G) mobile networks, downlink communication is facilitated by orthogonal frequency division multiple access (OFDMA), whereas single-carrier frequency division multiple access (SC-FDMA) is utilized for uplink transmission. Non-orthogonal multiple access (NOMA) and Rate splitting multiple access (RSMA) schemes are used for 5G networks [6, 7].

Multiple access schemes are classified into orthogonal multiple access (OMA) and NOMA. In OMA, wireless resources are allocated orthogonally to users through various combinations of time, frequency, and code domains. Nevertheless, the number of users that OMA can support is constrained by the available orthogonal resources. Furthermore, channel impairments frequently disrupt orthogonality, even when orthogonal resources are used in the time, frequency, or code domains [8]. Therefore, high-complexity "orthogonality restoration methods," like multi-user equalizers, are needed. This challenge impairs the ability of OMA methods to fulfill the extensive connectivity requirements and spectral efficiency of 5G networks.

NOMA has been introduced as a solution for 5G wireless systems to support a higher number of users. In contrast to traditional orthogonal transmission methods, NOMA employs non-orthogonal transmission techniques at the transmitter, intentionally introducing inter-cell and/or intra-cell interference. At the receiver end, multi-user signal separation is achieved using successive interference cancellation (SIC). NOMA techniques allow multiple users to simultaneously utilize non-orthogonal radio resources [9]. A comparison of various multiple access techniques across different generations, from 1G to 5G, is presented in Table 9.1. This evaluation offers a thorough analysis of various multiple access schemes across the spectrum from 1G to 5G.

High-quality services are essential for contemporary mobile communications. Implementing multiple access can potentially enhance the QoSs (QoS) provided to end users. Thus, a concurrent review of multiple access offers a valuable insights for various access techniques across all five generations. Additionally, this review discusses the advantages, disadvantages, research gaps, and proposed solutions associated with these technologies.

V. A. Bohara, Swaminathan R, *Optical and Wireless Access Networks*, Textbooks in Telecommunication Engineering, https://doi.org/10.1007/978-3-031-89696-5_9

Table 9.1 General comparison of cellular network

Generation	1G	2G	3G	4G	5G
Year	1980	1990	2000	2010	2020
Multiple Access	FDMA	TDMA	CDMA	OFDMA	OFDMA/RSMA
Bandwidth	30 kHz	200 kHz	5 MHz	20 MHz	60 GHz
Data rate	2.4 Kbps	64 Kbps	2 Mbps	200 Mbps	1 Gbps
Latency	N/A	629 ms	212 ms	60–98 ms	<1 ms
Switching	Circuit	Circuit	Circuit	Packet	Packet
Applications	AMPS	GSM	UMTS	LTE	New Radio (NR)

Thus, in this chapter we discuss the following:

- An analysis of multiple access standards across various generations of wireless networks from 1G to 5G.
- Review of conventional multiple access schemes (FDMA, TDMA, CDMA,).
- Review of current- and next-generation multiple access schemes (OFDMA,NOMA, and RSMA).
- Finally, a detailed survey of the benefits of multiple access, research challenges, and their summary for integrated optical wireless are addressed.

9.1.1 Advancement of Wireless Networks

In the coming years, a society is envisioned where all individuals will have universal access to data and the ability to share information, regardless of time or location. Achieving these ambitious goals requires investigating new technological advancements that can enhance current wireless technologies. Additionally, the rapid rise in mobile data traffic poses substantial challenges for mobile network operators in delivering higher data rates and reduced latency services to users. Consequently, advancing from 1G to 5G has become essential. The evolution in technology from 1G to 5G is explained in detail in Introduction chapter of the book.

9.1.2 Performance Requirement of Wireless Networks

The performance criteria for 5G cellular network standards, as determined by both academic and industrial researchers to ensure optimal functionality, are detailed as follows [10].

- **Data rate:** 5G wireless networks are designed to reach data rates upto 10 Gbps and 20 Gbps for uplink and downlink, respectively, marking a tenfold improvement over current LTE networks, which provide peak data rates of up to 150 Mbps.
- **Bandwidth**: To support numerous devices and deliver ultrahigh bandwidths over large duration, 5G networks demand a 1000-fold increase in bandwidth per unit area.
- **Latency**: 5G wireless systems demand latency below 1 ms, significantly improving upon the 10 ms round-trip time seen in current 4G networks.
- **Energy usage**: 5G networks are expected to be 100 times more energy-efficient than previous generations. Standardization bodies are already exploring various technologies to significantly reduce device energy consumption. Enhancing energy efficiency (EE) is crucial, as network power requirements increase with usage frequency. Time and frequency resource scheduling must align with power optimization techniques to maximize the efficiency of future systems.
- **Battery life**: 5G networks are engineered to significantly extend battery life, potentially up to 10 years, enabling support for a wider range of applications compared to 4G networks.
- **Network availability**: 5G standards envision a network that is always accessible, regardless of location, providing continuous coverage everywhere and at all times.
- **Massive connectivity**: 5G networks must be capable of connecting billions of devices, enabling the widespread adoption of device-to-device (D2D) and IoT technologies.
- **Quality of service**: QoS pertains to the individual user's perception of the performance and effectiveness of a service or application. Since QoS is specific to both the user and the application, 5G must deliver a superior wireless experience, ensuring high levels of user satisfaction and reliability.
- **Security**: Enhanced security measures are essential for 5G, particularly in areas like authentication, authorization, and accounting.

9.2 Conventional Multiple Access Technique

In telecommunications networks, resources are limited and are generally measured in terms of bandwidth. When many users simultaneously attempt to access this finite bandwidth, multiple access strategies are implemented to allocate the bandwidth among them, ensuring that all users can access network services without any single user dominating the resources. From the beginning of modern communication, researchers have concentrated on finding the most efficient multiple access scheme for resource allocation among multiple users [4]. As a result, multiple access in communication networks encompasses strategies that enable the effective distribution of limited network resources among multiple users. The foundational techniques for bandwidth sharing, such as time division and frequency division, have been established as the core of the multiple access framework. Mobile radio networks are characterized by their services, operational modes, applications, and multiple access techniques. The limited availability of radio spectrum greatly impacts the development of commercial radio communication technologies. Consequently, the goal is to maximize the number of users for each available radio frequency segment. This is achieved through different multiple access systems, which enable user multiplexing and resource allocation by time, frequency, or code, facilitating shared resource usage while reducing interference. Multiple access technologies are generally classified into two categories: OMA and NOMA [8]. Conventional OMA methods allow multiple users to utilize resources in an orthogonal manner, based on time, frequency, or code. The primary OMA techniques include FDMA, TDMA, and CDMA.

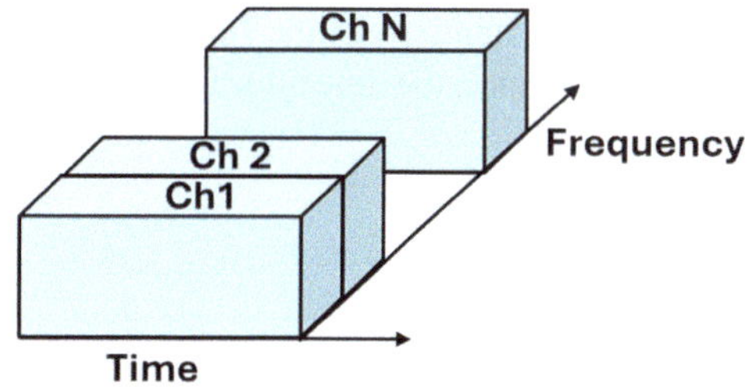

Fig. 9.1 Frequency division multiple access

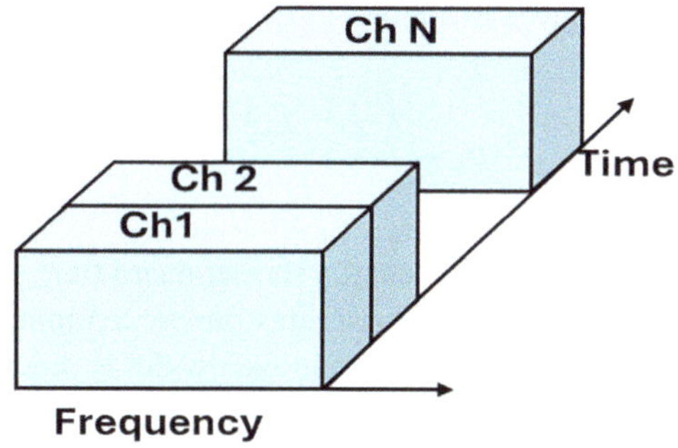

Fig. 9.2 Time division multiple access

9.2.1 Frequency Division Multiple Access

Since the advent of modern communication systems, researchers have been investigating multiple access techniques to enable resource sharing among numerous users as shown in Fig. 9.1. It was found that multiple access strategies are fundamental and effective for partitioning time, frequency, and bandwidth. In the earliest wireless networks employing multiple access, frequency modulation was utilized to divide the available frequency spectrum into distinct channels, with each channel allocated to an individual user, representing a segment of the total bandwidth [11]. This approach, known as FDMA, minimizes interference among users by assigning each user a separate frequency channel. In an FDMA system, the bandwidth assigned to each user is determined by the data rate and modulation scheme, while the total system bandwidth, excluding guard bands, is given by

$$B_s = KB \tag{9.1}$$

where K and $B = \alpha\left(\frac{R_b}{k}\right)$ represent the number of channels and the bandwidth per channel, respectively, α is a constant related to pulse shaping, filtering, and modulation order, R_b indicates the bit rate per channel, and k denotes the number of bits per symbol. When considering guard bands, the total bandwidth B_s is expressed as

$$B_s = K(B + B_g) \tag{9.2}$$

where B_g denotes the size of the guard band.

9.2.1.1 Limitations of FDMA

- Allocating multiple carriers within a single frequency channel poses significant challenges.
- Guard bands are necessary in the frequency domain to mitigate spectral emissions into adjacent frequency bands.
- Preference is given to narrowband channels, which are smaller than the coherence bandwidth of the wireless channel.
- The availability of orthogonal resources is inherently limited.

9.2.2 Time Division Multiple Access

In the early stages of digital communications, TDMA allocates distinct time slots along the time axis for the transmission of data to individual users. This approach allows a user to send large data files within the periodic frames defined by the TDMA system. As illustrated in Fig. 9.2, data from an individual user are collected and combined at the receiver to reconstruct the original transmitted packet. In TDMA, each user's data are consistently allocated the same time slot within each frame. During transmission, each transmitter utilizes the entire available bandwidth. The rate at which users access the channel and transmit data is known as the system bit rate, denoted by R_b^s, excluding the guard times.

The bit rate per channel is calculated as

$$R_b = \frac{R_b^s}{K}, \tag{9.3}$$

where K denotes the number of time slots per frame. Taking into account guard times, the time allocated for each channel is given by

$$T = \frac{T_f}{K} - T_g, \tag{9.4}$$

where T_f is the frame duration, and T_g is the guard time.

The data rate per channel is calculated by dividing the number of bits transmitted by each user by the duration of the frame, as given as follows [12]

$$R_b = \frac{\left(\frac{T_f}{K} - T_g\right) R_b^s}{T_f}. \tag{9.5}$$

To enhance system efficiency, a shorter guard time is preferable. However, practical constraints impose a minimum size for T_g. Consequently, the system bandwidth is directly proportional to the system data rate

$$B_s \propto \frac{R_b^s}{k}\alpha, \tag{9.6}$$

TDMA has been utilized as the multiple access scheme in several cellular standards, including 2G GSM and 2.5G GPRS.

9.2.2.1 Advantages of TDMA

- The bit capacity of a TDMA system remains unchanged regardless of the number of users accessing the system.
- TDMA systems gain an advantage from transponder power amplifiers operating in saturation mode, enhancing the capacity for multiple access.
- The high-speed logic processors and circuits used in TDMA systems provide enhanced flexibility and support high data rates.
- Compared to FDMA systems, TDMA systems are more cost-effective due to simpler multiplexing, independence from distance, and ease of interfacing with terrestrial services.
- Higher levels of interference noise are tolerated by TDMA systems.

9.2.3 Code Division Multiple Access

Both FDMA and TDMA techniques allocate a fixed number of channels or time slots per system, with each user being assigned a specific channel or time slot for the entire duration of communication. A dedicated channel or time slot ensures consistent service quality for real-time applications and constant-bit-rate voice services. However, as the range

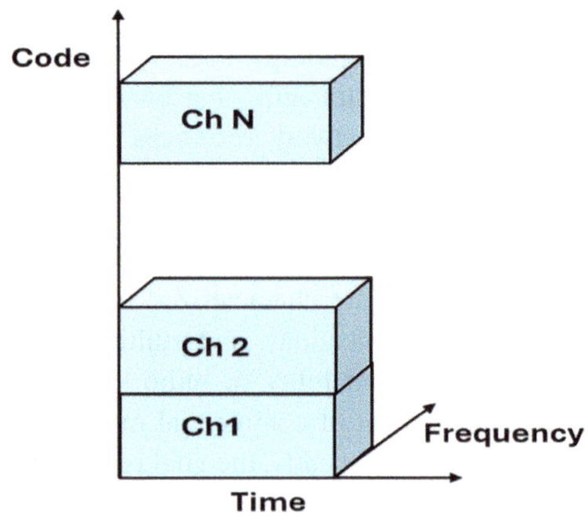

Fig. 9.3 Code division multiple access

of services expands from basic voice communication to intermittent data transmissions and the number of users increases, fixed-channel assignment becomes less efficient for spectrum utilization. This suggests that a more dynamic approach to channel assignment, where channels are allocated based on user demand, could enhance efficiency. In response to this challenge, 3G CDMA systems utilizing spread spectrum techniques were developed. CDMA extends relatively narrowband user information across a much broader spectrum by employing a high clock chip rate. By assigning unique, orthogonal codes to each user and ensuring the receiver is aware of the correct spreading code, multiple users can transmit simultaneously over the same frequency spectrum, as demonstrated in the Fig. 9.3. This method significantly facilitates signal detection at the receiver.

CDMA has evolved into a versatile multiple access scheme through the use of spread spectrum techniques, which do not impose rigid channel allocation constraints on individual users. Unlike TDMA and FDMA, where the number of users is fixed and predetermined, CDMA dynamically accommodates additional users. The total power of multi-user interference affects the maximum number of simultaneous users that can be supported by a CDMA system within the same frequency band. Due to its superior capacity compared to TDMA and FDMA, CDMA has become the preferred multiple access technique for 3G mobile cellular systems. In CDMA, multiple users share a single channel, often using unique orthogonal codes. During downlink transmissions, signals encoded with different codes reach the mobile station in an orthogonal manner, with a fixed base station transmitting and all codes being received synchronously. This orthogonality ensures minimal interference between transmissions on different codes, assuming no multipath propagation [13]. However, in the presence of multipath, the orthogonality of the codes can be disrupted, resulting in interference such as inter-user interference (IUI) or inter-symbol interference

(ISI). The challenge is even greater on the uplink, where varying propagation delays from different mobile users to the base station lead to a lack of guaranteed orthogonality and synchronization of received codes at the base station. Achieving orthogonality in a CDMA uplink necessitates sub-chip level synchronization, which can be as precise as a small fraction of a microsecond depending on the chip rate, posing significant practical challenges.

In a CDMA system, the asynchronous nature of the received signals can result in interference when multiple users transmit simultaneously. In 3G CDMA systems, although users are allocated with identical codes, these codes are scrambled with distinct pseudo-random sequences, resulting in non-orthogonal uplink transmissions.

The maximum uplink capacity of a CDMA system is given by

$$C_{\mathrm{CDMA}} = K \log_2 \left(1 + \frac{P}{(1+f)KP + (\alpha - 1)P + N_0} \right), \tag{9.7}$$

where P denotes the transmit power, N_0 denotes the power spectral density of noise, α is the fraction of the own-cell signal treated as interference, and f denotes the ratio of other-cell to own-cell signals.

9.3 Current- and Next-Generational Multiple Access Technique

Current- and next-generation wireless communication systems employ various multiple access techniques to enhance SE, connectivity, and user experience. OFDMA has been a cornerstone in 4G and 5G networks, providing efficient resource allocation by dividing the frequency spectrum into orthogonal subcarriers. NOMA introduces a paradigm shift by allowing multiple users to share the same frequency resources, differentiated by power levels or codes, thereby improving capacity and connectivity. RSMA further advances these techniques by enabling flexible user scheduling and interference management through the division of transmission rates, promising enhanced performance in highly congested networks. Together, these techniques are pivotal in meeting the demands of current and future wireless systems.

9.3.1 Orthogonal Frequency Division Multiple Access

In the LTE wireless downlink channel, OFDMA is utilized as the primary multiple access technique. Instead of transmitting data as a single continuous stream, OFDMA partitions it into multiple parallel sub-streams. Orthogonal frequency division multiplexing (OFDM), a specialized form of multicarrier modulation (MCM) and frequency division multiplexing (FDM), operates on the principle of dividing the signal's bandwidth into smaller bands or parallel sub-carriers, as illustrated in the Fig. 9.4. To address the limitations of CDMA, 4G systems employ two multiple access techniques: OFDMA for the downlink and single-carrier frequency division multiple access (SC-FDMA) for the uplink. OFDMA offers several advantages over CDMA, including enhanced flexibility, simpler equalization, better resilience to channel fading, and superior spectral efficiency. These characteristics render OFDMA especially effective for supporting high-speed data transmission.

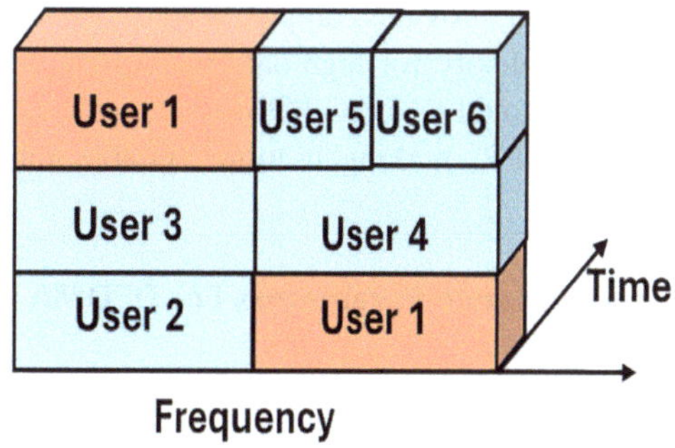

Fig. 9.4 Orthogonal frequency division multiple access

In OFDM, the entire frequency spectrum is segmented into orthogonal subcarriers, facilitating efficient utilization of the available bandwidth [14]. OFDM has been identified as the optimal choice for meeting the requirements of 4G wireless downlink transmissions. However, like any technology, it has its drawbacks. These include inter-carrier interference (ICI) among subcarriers, a high peak-to-average power ratio (PAPR), and increased susceptibility to frequency errors. The high PAPR results from the OFDM signal in the time domain being composed of parallel random signals. While high PAPR is not a significant issue for the downlink where power is less constrained, it becomes more problematic for the uplink. In the uplink scenario, where battery life is limited and efficient power amplification is crucial, the need for a cost-effective power amplifier exacerbates the challenge posed by high PAPR. To mitigate the high PAPR in OFDM systems, various techniques such as selective mapping, clipping, and filtering are employed. These methods aim to reduce PAPR but often involve a trade-off between system performance and computational complexity. In the uplink, where battery life and efficient power amplification are critical, this trade-off becomes particularly significant, as techniques to reduce PAPR may increase computational complexity or impact overall system performance.

To address the limitations associated with high PAPR, SC-FDMA was developed. SC-FDMA, which incorporates a discrete Fourier transform (DFT) pre-coding stage before

OFDM modulation, provides comparable throughput, performance, and complexity for high-data-rate uplink transmissions. Compared to OFDM, SC-FDMA offers the advantage of a reduced PAPR, making it more suitable for uplink transmissions. '

9.3.1.1 Performance Parameters For OFDMA Analysis

When evaluating the performance of OFDMA, the following parameters are considered:

1. **Bit error rate:** This is defined as the ratio of erroneous bits to the total number of bits transmitted over a specified time period. The bit error rate (BER) occurs when bits are corrupted due to factors such as interference, noise, distortion, or synchronization errors. It is determined by comparing the transmitted signal with the received signal and calculating the number of errors in relation to the total number of bits transmitted.
2. **Signal-to-noise ratio:** The signal-to-noise ratio (SNR) denotes the ratio of bit energy (E_b) to noise power spectral density (N_0), mathematically expressed as $\mathrm{SNR} = \frac{E_b}{N_0}$. It also represents the ratio of relevant information to extraneous data. An inverse relationship exists between BER and SNR ($\mathrm{SNR} \propto \frac{1}{\mathrm{BER}}$), where an increased BER can lead to packet loss, increased latency, and reduced throughput. Therefore, BER is commonly assessed in relation to SNR across different modulation techniques.
3. **Error probability:** The error probability quantifies the frequency of errors in the received signal. For example, in an additive white Gaussian noise (AWGN) channel, the error probability for $M-$ary quadrature amplitude modulation (QAM) can be approximated by

$$P_e \approx 2\left(1 - \frac{1}{\sqrt{M}}\right) \mathrm{erfc}\left(\sqrt{\frac{3E_{av}}{2(M-1)N_0}}\right), \quad (9.8)$$

   where N_0 represents the noise power spectral density in the AWGN channel, M denotes the modulation order, and E_{av} denotes the average energy of the transmitted symbols in M-ary QAM.
4. **Power limited capacity:** The capacity of a wireless channel with constant transmit power is given by

$$C = W \log_2\left(1 + \frac{\mathrm{SNR}}{W}\right) \text{ [Mb/s]}, \quad (9.9)$$

   where W indicates the bandwidth in MHz. With a constant transmit power, simply expanding the bandwidth does not result in a proportional rise in capacity. To achieve a linear enhancement in capacity, it is necessary to proportionally increase the transmit power with the bandwidth, necessitating the use of costly high-power broadband amplifiers. Although this approach is expensive, it facilitates multiple simultaneous uplink transmissions, thereby maximizing both the total transmitted power and channel capacity.
5. **Power spectral density:** Power spectral density (PSD) describes how a signal's power is distributed across various frequencies. It is crucial for making informed decisions about radio resource management at the base station in a wireless communication system, particularly when allocating transmission formats such as modulation schemes and bandwidth.
6. **Peak-to-average power ratio:** The PAPR represents the ratio of the maximum instantaneous power to the average power of the OFDM symbol. Elevated PAPR levels can lead to distortion and compromise the orthogonality among the OFDM subcarriers.

$$\mathrm{PAPR} = \frac{\text{Max power of } x(t)}{\text{Average power of } x(t)} \quad (9.10)$$

   where $P_{\max}$ and P_{avg} represent the maximum power and the average power in each OFDM frame, respectively.

   The PAPR for a continuous-time signal $x(t)$ can be expressed as

$$\mathrm{PAPR} = \frac{\max\{|x(t)|^2\}}{\mathbb{E}\{|x(t)|^2\}} \quad (9.11)$$

   where $0 \leq t \leq NT$, $\max\{|x(t)|^2\}$ denotes the peak power of $x(t)$, $\mathbb{E}\{\cdot\}$ is the expectation operator, and $\mathbb{E}\{|x(t)|^2\}$ represents the average power of $x(t)$.

In OFDMA, the system resources are distributed among several users by allocating only a segment of the total bandwidth to each individual user. This arrangement permits multiple users to transmit concurrently on orthogonal subcarriers. To preserve orthogonality between transmissions from different users, the relative delay of the received signals must be shorter than the length of the cyclic prefix (CP). The CP length, usually spanning several microseconds, accounts for multipath delay and ensures proper timing synchronization within the CP.

The uplink capacity for an OFDMA system is given by

$$C_{\mathrm{OFDMA}} = \sum_{i=1}^{K} \beta_i \log_2\left(1 + \frac{Pf_i}{P + \beta_i N_0}\right), \quad (9.12)$$

where the capacity is expressed in [B/s/Hz], β_i represents the fraction of bandwidth allocated to user i, K is the number of users transmitting simultaneously, P denotes the received power for a user, and f_i is the ratio between other-cell and own-cell signals.

In an OFDMA system, the use of orthogonal subcarriers and 1-tap OFDM subcarrier equalization eliminates intra-cell interference and ISI. However, the CP overhead, usually around 10%, must be considered. Consequently, the capacity of an OFDMA system is adjusted to account for this CP overhead and is given as

$$C_{\mathrm{OFDMA}} = \frac{T_S}{T_S + T_C} \log_2\left(1 + \frac{K P f_K}{P + N_0}\right), \tag{9.13}$$

where T_S and T_C denote the symbol duration and cyclic prefix duration of OFDM, respectively.

9.3.2 Non-Orthogonal Multiple Access

This section delves into the background and evolution of NOMA. Each new generation of communication technology has demanded innovative multiple access techniques to enable simultaneous spectrum reuse among users, ensuring high capacity by distributing available bandwidth (or channels) across multiple users without compromising their QoS. OMA schemes, while simplifying transceiver design and minimizing co-channel interference, have limitations including a restricted number of users that can be served simultaneously. Maintaining orthogonality requires precise scheduling of users and robust feedback mechanisms [15]. In contrast, NOMA provides a more efficient resource block (RB) allocation method by optimizing power, code signatures, and receiver design [16]. The non-orthogonal approach of NOMA reduces the need for exact multi-user multiplexing scheduling and simplifies the channel requirements (Fig. 9.5).

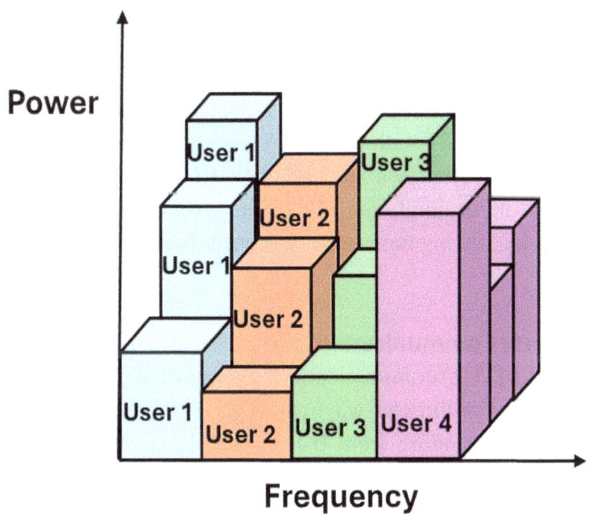

Fig. 9.5 NOMA

9.3.2.1 Basic Principle of NOMA

Downlink NOMA

In the downlink (DL) scenario, superposition coding (SC) is utilized at the base station (BS) to multiplex signals with different power allocation coefficients. Successive interference cancellation (SIC) is then employed at the receiver to separate these overlapping signals. Power allocation is based on the channel conditions of the users: users with poorer channel conditions receive higher power allocations, while users with better channel conditions receive lower power allocations [17], as illustrated in the Fig. 9.6.

Uplink NOMA

In the uplink scenario, each user equipment (UE) transmits signals to the BS. At the BS, the SIC technique is employed to differentiate between signals from various UEs, utilizing their designated power allocation coefficients [18], as depicted in Fig. 9.7. The signal received at the BS can be represented as

$$r_s = \sum_{p=1}^{N} h_p a_p P_m x_p + \eta, \tag{9.14}$$

where r_s denotes the received signal, x_p represents the transmitted signal, h_p signifies the channel gain, a_p is the power allocation coefficient, P_m is the transmitted power from the UE, and η denotes the AWGN.

9.3.2.2 Power-Domain NOMA (PD-NOMA)

This section outlines the different variations of PD-NOMA employed in 5G networks. PD-NOMA allows for simultaneous service to multiple users on the same subchannel by assigning distinct power levels to each. This approach is primarily utilized in 5G to enhance spectral efficiency, energy efficiency, and minimize latency. The following subsections provide detailed explanations of the applications of PD-NOMA in each wireless communication technology.

Visible Light Communication

Visible light communication (VLC) is a small cell technology utilized in 5G to provide extensive broadband connectivity in indoor environments. VLC operates in an unlicensed spectrum range of terahertz frequencies (430–790 THz), which remains untapped in wireless communication. According to the findings in [19], approximately 70% of data traffic originates from indoor locations. Therefore, integrating VLC with existing fiber networks allows its unlicensed spectrum to alleviate the load on data traffic. Typically, light-emitting

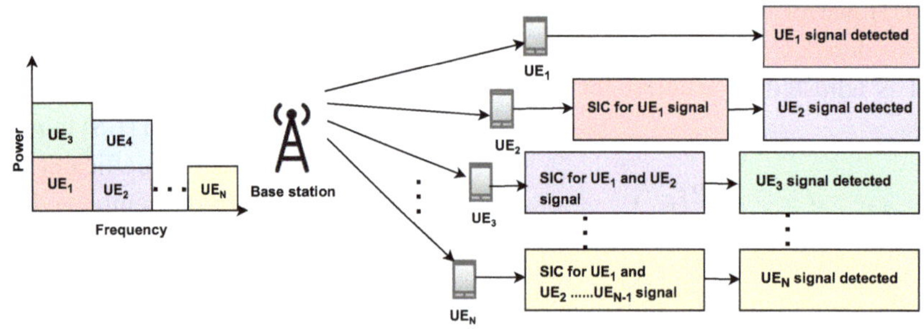

Fig. 9.6 Downlink NOMA

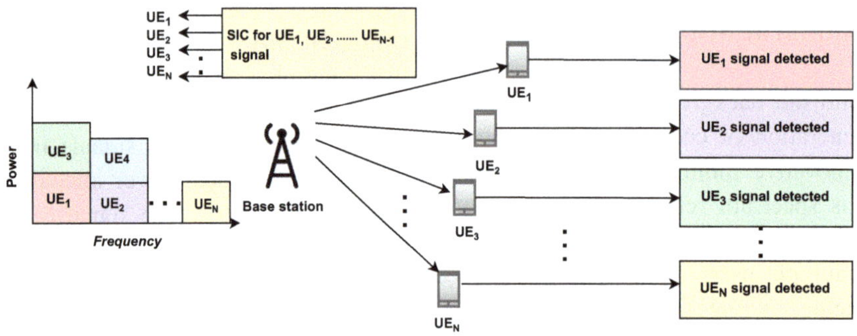

Fig. 9.7 Upwnlink NOMA

diodes (LEDs) are employed to facilitate VLC in both indoor and outdoor settings. The use of LEDs helps lower broadband connection costs and enhances data transmission rates over white light. VLC relies on two techniques: intensity modulation (IM) and direct detection (DD). In IM, the intensity of light emitted by LEDs varies depending on the users' positions, while in DD, photo-detectors on the receiving end detect changes in light intensity. Since light signals from LEDs cannot penetrate walls, inter-cell interference is eliminated. This feature enhances security and improves the QoS for cell-edge users. Despite these advantages, VLC has certain limitations that must be addressed to fully realize its potential. The primary constraints are a narrow modulation bandwidth that restricts the achievable data rate and severe inter-cell interference in multi-cell scenarios. To overcome these challenges, the integration of NOMA with VLC is essential. This integration leads to the development of systems such as OFDMA-NOMA and NOMA-SCFDM.

Simultaneous Wireless Information and Power Transfer

In addition to improving spectral efficiency, enhancing energy efficiency is another crucial aspect that needs to be tackled in 5G wireless networks. Most communication devices rely on batteries with limited lifespans. To extend the operational life of these energy-constrained devices, energy harvesting has been identified as an effective solution. Traditionally, energy harvesting techniques draw energy from renewable sources like solar or wind power; however, these methods are unreliable as they are subject to environmental conditions. In contrast, simultaneous wireless information and power transfer (SWIPT) has emerged as a popular energy harvesting approach. In SWIPT, nodes or terminals extract energy from the electromagnetic radiation of radio frequency signals during the data decoding process. Remarkably, SWIPT can even harvest energy from interfering signals [20]. However, cooperative NOMA (C-NOMA) systems revealed a challenge: when strong users act as relays for weak users, their batteries deplete rapidly, causing a decline in the signal quality for both. This issue has prompted researchers from academia and industry to explore the integration of SWIPT with C-NOMA to improve network energy efficiency. Techniques such as C-SWIPT-NOMA, C-SWIPT-multiple-input single-output (MISO)-NOMA, and NOMA-energy-harvesting have been proposed to address these challenges.

Terahertz Communication

Terahertz (THz) communications operate at higher frequency bands than millimeter-wave (mmWave) technologies, with frequencies ranging from 0.1 to 10 THz, situated between infrared and microwave regions. This band can provide communication bandwidths up to 10^3 to 10^4 times greater than those of current wireless communication bands commonly used in mobile devices [21]. The key advantages of THz communications are substantial capacity, precise directional capabilities, enhanced security, robust resistance to interference, and suitability for ultrahigh-speed, short-range indoor communication. Despite these benefits, THz communications face several challenges [22] such as significant transmission attenuation loss, high-speed broadband communication requirements, absorption attenuation by atmospheric molecules and water droplets, path loss peaks and drastic changes in communication distance, limited coverage area, substantial transmission burden, and high energy consumption. To address these challenges and achieve communication rates of 10 Gbps or more, researchers have integrated NOMA with THz technology. This integration offers several advantages like increased network capacity, improved resource utilization, support for ultra-massive machine-type communications, enhanced energy efficiency, and ultrahigh reliability and low latency. The different NOMA variants compatible with THz technology include THz-NOMA, THz-MIMO-NOMA, and Cooperative THz-MIMO-NOMA.

Intelligent Reflecting Surface

An intelligent reflecting surface (IRS) is an advanced technology designed to enhance network coverage. It comprises numerous passive elements that reflect incident signals by adjusting their reflection coefficients, thereby increasing the received signal power at the receiver. Unlike amplify-and-forward relays, IRS lacks signal processing capabilities [23]. Additionally, IRS differs from active, intelligent surface-based massive MIMO in terms of performance, hardware cost, and power consumption [24] When combined with NOMA, IRS improves channel conditions by adjusting the phase shifts and amplitudes of the IRS elements and creates additional pathways to strengthen the combined channel gain. Various IRS-NOMA configurations include IRS-NOMA and IRS-mmWave-NOMA.

Vehicle-to-Everything Communication

Vehicle-to-everything (V2X) communication represents a cutting-edge technology in 5G that enhances user travel experience while providing improved security, safety, and internet access services [25]. Despite these advantages, V2X networks face challenges related to energy efficiency, low latency, and rapid packet transmission and reception in ultra-dense environments. To address these challenges, researchers have integrated NOMA with V2X technology. This integration helps alleviate data congestion by enabling non-orthogonal channel access [26], offering benefits such as reduced transmission latency, support for extensive connectivity, and improved spectral efficiency and energy efficiency. However, issues related to link reliability, bandwidth efficiency, and collision reduction still need to be addressed. Proposed variants of NOMA for V2X communication include NOMA with spatial modulation and NOMA with mixed centralized/distributed (MCD) scheme.

Artificial Intelligence/Machine Learning

Machine learning (ML) is a branch of artificial intelligence (AI) that enables systems to learn and enhance their performance through experience without being explicitly programmed. ML emphasizes creating computer programs capable of accessing and learning from data independently. Compared to traditional optimization methods, ML algorithms offer significant advantages for resource optimization problems. Traditional approaches often face high costs and complexity as the number of parameters increases and are sensitive to parameter selection, requiring heuristics to be restarted from scratch with even minor changes in the system model. These limitations have driven researchers to investigate ML techniques for optimizing resources in communication systems [27]. In multi-objective optimization scenarios, the aim is to identify Pareto-optimal solutions, where improving one metric inevitably leads to a compromise in another. Given the extensive search space involved, ML presents a compelling approach for discovering these Pareto-optimal solutions. Variants of ML applied to NOMA include unsupervised learning-NOMA, federated learning-NOMA, deep learning (DL)-aided NOMA, and deep reinforcement learning-NOMA.

9.3.2.3 Code-Domain NOMA (CD-NOMA)

CD-NOMA is a technique where user-specific spreading sequences are employed to multiplex signals at the transmitter side. The key benefits of this approach include low density, minimal inter-correlation, and support for grant-free access. CD-NOMA, an advanced iteration of CDMA introduced in the 1990s, necessitates that receivers detect active codes and estimate the associated data. This technique relies on compressive sensing, Gaussian random codes, and sparse codes as its core components. CD-NOMA encompasses five variants such as scrambling-based NOMA, interleaving-based NOMA, spreading-based NOMA, coding-based NOMA, and lattice and beam-based NOMA. Each variant offers unique characteristics and applications within the framework of CD-NOMA.

Scrambling-Based NOMA

Scrambling-based NOMA includes several variants, such as low code rate and signature-based shared access and resource spread multiple access. Both employs distinct scrambling structures and low-rate channel coding for each user. Resource spread multiple access utilizes a long sequence of user signatures, which introduces high latency and decoding complexity at the receiver. In contrast, low code rate and signature-based shared access address these issues by reducing the length of the user signatures. To separate user signatures at the receiver, techniques such as the elementary signal estimator and minimum mean square error with successive interference cancellation are applied [8].

Interleaving-Based NOMA

The system encompasses four types of NOMA schemes: interleave division multiple access (IDMA), interleave-grid multiple access (IGMA), repetition division multiple access (RDMA), and low code rate spreading (LCRS). IGMA is designed by Samsung, IDMA by Nokia, RDMA by Mediatek, and LCRS by Intel. These schemes share common characteristics, including the use of different channel interleavers for overlapping multiple users and the application of low-rate channel coding for multiuser decoding [28,29].

Spreading-Based NOMA

Spreading-based NOMA is divided into seven distinct types: multi-user shared access, non-orthogonal coded multiple access, non-orthogonal coded access, group orthogonal coded access, welch-bound spreading multiple access, successive interference cancellation amenable access, and sequence spreading multiple access [30–32].

Coding-Based NOMA

The NOMA schemes include pattern division multiple access, sparse code multiple access, and low density spreading signature vector extension, developed by CATT, Huawei, and Fujitsu, respectively. These schemes are designed to utilize a sparse codebook that overlaps across multiple domains such as space, code, time, or frequency. At the receiver end, a message-passing algorithm is employed for multi-user detection. These schemes exhibit high complexity, which grows exponentially with the size of the codebook. To address this issue and reduce complexity from exponential to linear, a novel expectation propagation algorithm was proposed in [33]. This algorithm, based on an approximate interference method, has become increasingly relevant in the fields of ML and DL.

Lattice and Beam-Based NOMA

When users experience similar channel conditions, the performance of NOMA can deteriorate compared to OMA. To mitigate this issue, researchers have proposed a geometrical separation technique for superposition transmissions. This method employs beamforming to create clusters of users with varying channel conditions. However, as user density and traffic demand increase, this technique encounters limitations such as increased complexity and co-channel interference among users in adjacent clusters. To address these challenges, a new multiple access technique, low-power multiple access, has been introduced. Additionally, block-orthogonal multiple access utilizes distinct tiled building blocks to differentiate between users with poor channel state information (CSI) and those with good CSI. This approach enhances the BER performance for users with poor CSI [34].

9.3.2.4 Achievable Sum Rate Comparison of NOMA and OMA

In NOMA, the sum rate refers to the total data transmission rate achieved by all users sharing the same resources (such as frequency or time). The sum rate of uplink NOMA can be written as

$$\begin{aligned} R_{\text{sum}}^{\text{NOMA-u}} &= \sum_{n=1}^{N} \log_2(1+\text{SINR}_n) \\ &= \log_2\left(1+a_1\gamma|h_1|^2\right) \\ &\quad + \sum_{n=2}^{N} \log_2\left(1+\frac{a_n\gamma|h_n|^2}{\gamma\sum_{i=1}^{n-1}a_i|h_i|^2+1}\right) \\ &= \log_2\left(1+\gamma\sum_{n=1}^{N}a_l|h_l|^2\right) \end{aligned} \tag{9.15}$$

where SINR_n represents signal-to-interference-plus-noise ratio for the n-th user, γ denotes SNR, and h_n, and a_n represents the channel coefficient and power coefficient allocated for the n-th user, respectively.

When $\gamma \to \infty$, the sum rate of uplink NOMA becomes

$$R_{\text{sum}}^{\text{NOMA-u}} \approx \log_2\left(\gamma\sum_{n=1}^{N}|h_n|^2\right) \tag{9.16}$$

The achievable data rate for the N-th user in OMA for both uplink and downlink can be expressed as [35]

$$R_n^{\text{OMA}} = \alpha_n \log_2\left(1+\frac{\beta_n\gamma|h_n|^2}{\alpha_n}\right), \tag{9.17}$$

where β_n and α_n represent the power coefficient and the parameter associated with the specific resource of user U_n, respectively. The total sum rate for OMA is given by

$$R_{\text{sum}}^{\text{OMA}} = \sum_{n=1}^{N}\alpha_n \log_2\left(1+\frac{\beta_n\gamma|h_n|^2}{\alpha_n}\right). \tag{9.18}$$

In the case of OMA, such as FDMA, where the total bandwidth and power resources are evenly distributed among users, setting $\alpha_n = \beta_n = \frac{1}{N}$ results in

$$R_{\text{sum}}^{\text{OMA}} = \sum_{n=1}^{N}\frac{1}{N}\log_2\left(1+\gamma|h_n|^2\right). \tag{9.19}$$

As $\gamma \to \infty$, the sum rate for OMA approaches

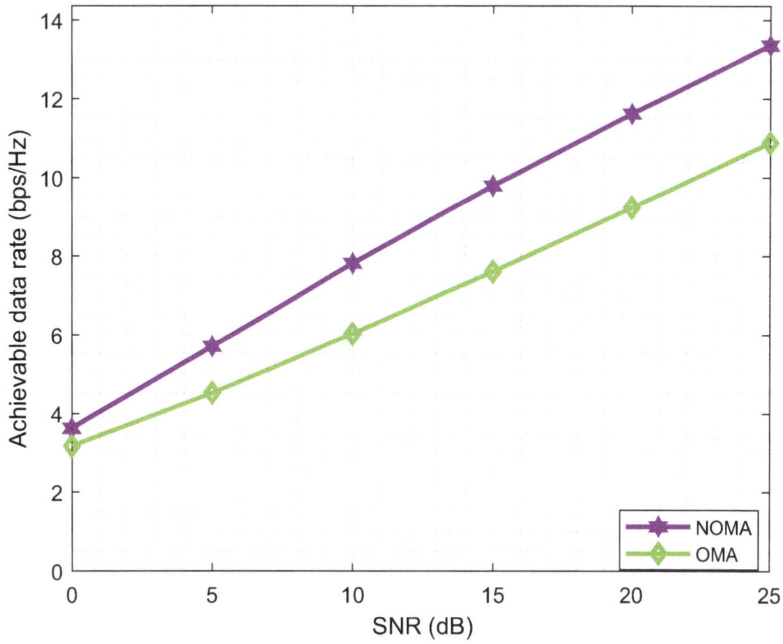

Fig. 9.8 Orthogonal multiple access vs. non-orthogonal multiple access

$$R_{\text{sum}}^{\text{OMA}} \approx \sum_{n=1}^{N} \frac{1}{N} \log_2\left(\gamma |h_n|^2\right). \tag{9.20}$$

Given that $|h_1|^2 \leq |h_2|^2 \leq \cdots \leq |h_N|^2$, we have

$$R_{\text{sum}}^{\text{OMA}} \approx \sum_{n=1}^{N} \frac{1}{N} \log_2\left(\gamma |h_n|^2\right) \leq \sum_{n=1}^{N} \frac{1}{N} \log_2\left(\gamma |h_n|^2\right)$$
$$= \log_2\left(\gamma |h_n|^2\right) \approx R_{\text{sum}}^{\text{NOMA-u}}. \tag{9.21}$$

Thus, it can be concluded that

$$R_{\text{sum}}^{\text{OMA}} \leq R_{\text{sum}}^{\text{NOMA-u}}. \tag{9.22}$$

To simplify, the sum rates of uplink NOMA and OMA are compared for two user scenario. Using Eqs. (9.16) and (9.20), the sum rates at high SNR are, respectively, given by

$$R_{\text{sum}}^{\text{NOMA-u}} \approx \log_2\left(\gamma |h_1|^2 + \gamma |h_2|^2\right), \tag{9.23}$$

$$R_{\text{sum}}^{\text{OMA}} \approx \frac{1}{2} \log_2\left(\gamma |h_1|^2\right) + \frac{1}{2} \log_2\left(\gamma |h_2|^2\right) \leq \log_2\left(\gamma |h_2|^2\right). \tag{9.24}$$

From (9.23) and (9.24), it is observed that

$$R_{\text{sum}}^{\text{OMA}} \leq R_{\text{sum}}^{\text{NOMA-u}}. \tag{9.25}$$

To justify the aforementioned observations, the sum rates of OMA and NOMA have been plotted, and Fig. 9.8 illustrates that NOMA surpasses OMA in terms of sum rate for uplink in a two-user network scenario.

9.3.2.5 Advantages of NOMA Over OMA

This section outlines the advantages of NOMA compared to OMA as follows:

1. **High spectral efficiency:** NOMA achieves higher spectral efficiency than OMA by allowing multiple users to share the same resource block. In contrast, OMA allocates one resource block per user, leading to potential bandwidth wastage. Furthermore, NOMA integrates seamlessly with advanced 5G technologies such as massive MIMO, millimeter wave, heterogeneous networks (HetNet), device-to-device communication, and cognitive radio, thereby enhancing network throughput.
2. **Massive connectivity:** NOMA supports billions of smart devices due to its non-orthogonal nature, making it ideal for the IoT and the tactile internet. It accommodates numerous devices within a single resource block through superposition coding, whereas OMA assigns a separate resource block to each device, resulting in resource block wastage.
3. **Fairness:** NOMA fosters fairness by allocating more power to users with weaker channel conditions and less power to users with stronger channels. This approach ensures that both strong and weak users meet their QoS requirements in terms of throughput.
4. **Ultra-low latency:** NOMA addresses stringent latency requirements in 5G networks, which are exacerbated by the HetNet architecture. OMA techniques, which depend on access-grant requests, introduce additional transmission latency and signaling overhead.

9.3.3 Rate-Splitting Multiple Access

The concept of rate splitting (RS) has its origins in early studies of the two-user interference channel, initially investigated by Carleial and later extended by Han and Kobayashi [36]. These seminal works introduced transmission strategies that employ RS to explore new communication rate regions. In the Han–Kobayashi framework, each transmitter divides its message into two components: a "private" segment and a "common" (or "public") segment [37]. These components are encoded together through superposition coding and transmitted simultaneously, optimizing communication rates under interference conditions.

The effectiveness of the Han–Kobayashi scheme lies in its dual approach to interference management. Each receiver not only decodes its intended message, which includes both private and common components, but also extracts part of the interfering signal specifically, the common portion from the other transmitter. This scheme's strength is its flexibility; it can shift between two extremes: treating interference purely as noise or fully decoding it [38]. Depending on the scenario, the Han–Kobayashi scheme can either employ one of these strategies exclusively or strike a balance between them, allowing it to adapt effectively to varying interference conditions.

Rooted in RS principles, RSMA has emerged as a promising physical layer transmission paradigm for 6G, focusing on non-orthogonal transmission, interference control, and management techniques. Unlike the more rigid interference management approaches of space division multiple access (SDMA) and NOMA, RSMA's main goal is to divide user messages into common and private segments. This allows for partial decoding of interference while treating the remaining interference as noise. RSMA's flexibility enables it to handle varying levels of interference by adjusting the power and content of the common and private streams. This adaptability allows RSMA to transition seamlessly between SDMA and NOMA based on interference conditions, effectively bridging the gap between these two techniques. RSMA is utilized in three primary contexts: uplink RSMA, downlink RSMA, and multi-cell RSMA.

In the downlink scenario, the transmission message is partitioned into sub-messages to optimize decoding and interference management [39]. For example, in 1-layer rate splitting, the message is divided into two sub-messages: a common part (CM) and a private part (PV). In 2-layer hierarchical rate splitting, the message is further divided into three sub-messages, incorporating an additional layer of hierarchical structure. The proposed approach represents a message as a bit string segmented into a CM and a PV. Correspondingly, the transmission stream is divided into a common stream (CS) and a private stream (PS). The K-user downlink transmitter is shown in Fig. 9.9. The CM is encoded into the CS using a shared codebook and is decoded by all users. Conversely, the PV is encoded into the PS using a distinct codebook and is decoded by individual users. At the receiver end, the common messages are decoded while treating the remaining private messages as interference. Similarly, a portion of the common message is decoded, with the remainder of the common message treated as noise as shown in Fig. 9.10. This methodology enhances efficient decoding and effective interference management in downlink RSMA systems.

In uplink RSMA, the technique is designed to access every point within the capacity region of the Gaussian multiple access channel without requiring time sharing among users [40]. The primary methods to achieve this include joint encoding and decoding, time sharing, and RS. Among these, RS is deemed the most efficient for reaching all points in the capacity region with reduced complexity. This efficiency is achieved through superposition coding and SIC. In uplink RSMA, the transmitter splits the message into sub-messages, encoding them into separate streams. The receiver then applies a specific decoding order to these streams, beginning with those assigned a lower decoding priority. This approach enables effective decoding of each user's message, even in the presence of interference, thereby optimizing channel capacity utilization.

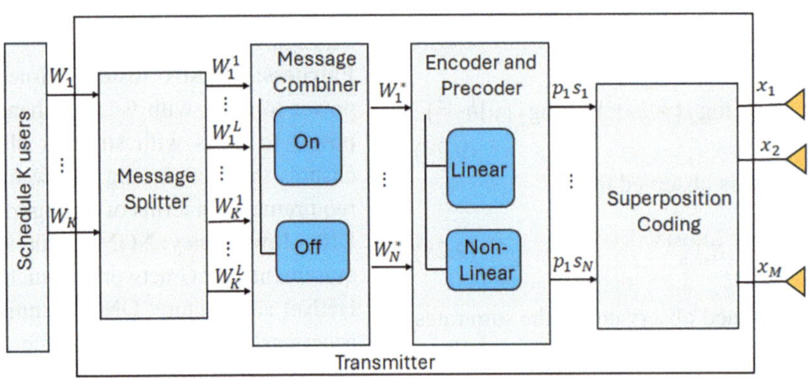

Fig. 9.9 K-user downlink RSMA transmitter

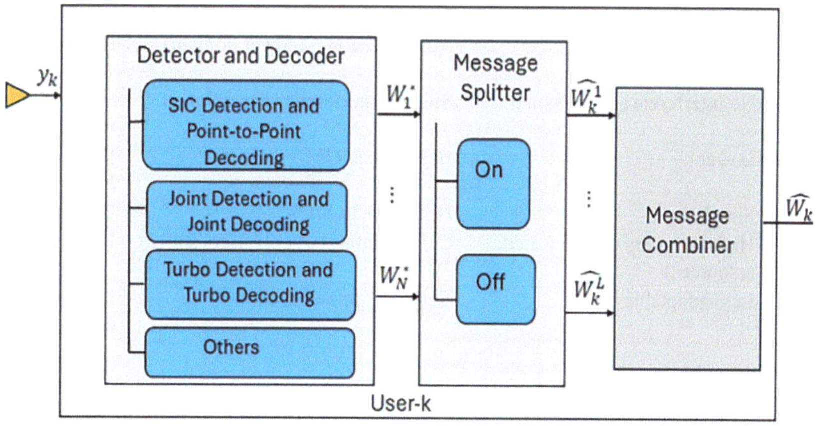

Fig. 9.10 Downlink RSMA receiver

Multi-cell RSMA encompasses both coordination and cooperative transmission techniques. Coordination transmission involves techniques such as the MISO network, where each transmitter, equipped with multiple antennas, serves a single user within its cell. This method requires minimal data exchange and CSI sharing between cells, focusing on coordinated transmission to manage interference. Cooperative transmission pertains to the MIMO network, where multiple transmitters collaborate to serve multiple users across cells. This approach demands extensive data and CSI sharing among cells to facilitate joint service and improve overall system performance. Applying the RS principle in multi-cell environments enables partial interference decoding, while treating the remaining interference as noise. Depending on the use of coordination or cooperation, these techniques can enhance interference management and processing efficiency across cells.

9.4 Comparisons Among RSMA, NOMA, and OMA Schemes

Table shows the comparisons among RSMA, NOMA, and OMA techniques. The key aspects of these techniques are summarized as follows:

- **OMA:** Best for simplicity and no interference, but limited by its lower spectral efficiency and inability to support many users efficiently.
- **NOMA:** Provides better spectral efficiency than OMA by allowing multiple users to share the same resources, but comes with more complexity and challenges in managing interference through SIC.
- **RSMA:** Strikes the best balance by splitting messages into common and private parts, offering better spectral efficiency and more flexible interference management than NOMA and OMA. It is the most adaptable solution for heterogeneous user conditions but is also more complex to implement.

RSMA is seen as a strong candidate for future 5G and beyond communication systems due to its flexibility, high efficiency, and better handling of interference.

9.5 Random Medium Access Control Protocol (CSMA/CA)

The primary Medium Access Control (MAC) mechanism within the IEEE 802.11 standard is the Distributed Coordination Function (DCF), which is designed as a random access scheme based on the Carrier Sense Multiple Access with Collision Avoidance (CSMA/CA) protocol. This protocol employs a binary slotted exponential back-off strategy to manage access to the communication medium, mitigating collisions caused by simultaneous transmissions of multiple user equipment (UE) [41].

DCF operates in two distinct modes for packet transmission. The first mode is the default two-way handshaking mechanism, commonly referred to as the *basic access mechanism*, which involves minimal overhead. The second, more sophisticated mode is the optional four-way handshaking mechanism, also known as *RTS/CTS* (Request to Send / Clear to Send). This method is specifically intended to address the issue of collisions, particularly in environments with higher network traffic density.

9.5.1 Random Back-Off and Collision Avoidance

In DCF, UE initiate transmission by first sensing the medium for a predefined time interval, known as the *Distributed Interframe Space (DIFS)*. If the medium is found to be idle, each

Feature	OMA	NOMA	RSMA
Resource allocation	Orthogonal (separate resources)	Non-orthogonal (power domain)	Hybrid (common and private parts)
Interference management	No interference (orthogonal)	SIC for interference cancellation	SIC + message splitting for flexibility
Spectral efficiency	Lower	Higher than OMA	Higher than both OMA and NOMA
Complexity	Simple	High (due to SIC)	Higher (due to rate-splitting)
Fairness	High (all users get dedicated resources)	Can be unfair (weaker users suffer)	Fairer resource allocation
Suitability for Heterogeneous networks	Less adaptable	Better for mixed users	Highly adaptable to different user conditions

UE generates a random backoff time, denoted by B_j, where $j = 1, 2, \ldots, N_{\text{UE}}$, and N_{UE} represents the number of UEs competing for access to the medium. The back-off value, B_j, is uniformly selected from the range $[0, w - 1]$, where w is the current contention window size.

The contention window w starts at a minimum value, $CW_{\min}$, and doubles after each unsuccessful transmission attempt, up to a maximum value, $CW_{\max}$. The back-off window for the l-th backoff stage is defined as $w = 2^l CW_{\min}$, where $l = 1, 2, \ldots, M_r$, and M_r represents the retry limit. In this context, M_r is given by $M_r = \log_2 \left(\frac{CW_{\max}}{CW_{\min}}\right)$. The vector of back-off values for the UE is represented as $\mathbf{B} = [B_j]_{1\times N_{\text{UE}}}$.

Once the back-off time $B_j \times t_{\text{slot}}$ has elapsed (where t_{slot} is the duration of a slot time), the UE with the smallest back-off value, $u_0 = \arg\min_j(\mathbf{B})$, is prioritized to access the medium. The chosen UE, denoted as u_0, proceeds by transmitting an RTS frame to the Access Point (AP). If the AP successfully receives the RTS frame, it responds after a Short Inter-frame Space (SIFS) with a CTS frame. The u_0 UE may then transmit its data frame following the SIFS period, contingent on receiving the CTS. Upon successful reception of the data, the AP transmits an acknowledgment (ACK) frame after another SIFS period, confirming the packet's successful delivery.

During this process, UE that detect the RTS transmission from u_0 will halt their countdown process, freezing their back-off timers until the channel becomes idle again. UE that fail to detect the transmission, particularly hidden nodes, may attempt transmission, leading to potential collisions.

9.5.2 Hidden Node and Collision Analysis

A collision can occur when a hidden UE, denoted u_1, initiates an RTS transmission without detecting the u_0-th UE's ongoing communication. This typically happens if $(B_{u_1} - B_{u_0}) \times t_{\text{slot}} < t_{\text{RTS}}$, where t_{RTS} is the time required to transmit an RTS frame, which depends on the frame length L_{RTS} and the uplink transmission rate R_u, i.e., $t_{\text{RTS}} = \frac{L_{\text{RTS}}}{R_u}$.

9.5.3 Network Allocation Vector (NAV) and Virtual Carrier Sensing

The RTS and CTS frames also include a *Network Allocation Vector (NAV)*, which serves as a virtual carrier sensing mechanism. The NAV field specifies the duration (in microseconds) for which the medium is reserved for the ongoing transmission. When a UE receives an RTS frame, it sets its NAV timer to account for the subsequent CTS, data, and ACK frames. Conversely, the CTS frame carries an updated NAV value to reflect the time already consumed, ensuring that neighboring UE defer their transmissions until the current data exchange is completed. This mechanism significantly reduces the likelihood of collisions.

However, in LiFi networks, a fundamental challenge arises due to the use of different transmission mediums for the uplink (infrared) and downlink (visible light). Unlike RF-based networks, UE in LiFi systems cannot directly sense each other's transmissions, exacerbating the hidden node problem. In this scenario, UE are effectively hidden from each other, as the downlink photodetector is tuned to receive visible light and cannot detect the infrared signals transmitted by other UE in the uplink. This limitation makes the NAV mechanism less effective in LiFi systems, as UE can only respond to the NAV field in the CTS frame, while the RTS frame remains undetected.

9.6 Conventional CSMA/CA

The main medium access control (MAC) mechanism of IEEE 802:11 is called distributed coordination function (DCF), which is a random access scheme based on the CSMA/CA protocol with a binary slotted exponential back off [41]. The default scheme of DCF for packet transmission is a two-way handshaking technique known as the basic access mechanism. The second scheme is an optional four-way handshaking technique known as the RTS/CTS mechanism. DCF adopts a slotted binary exponential back-off scheme to reduce collisions due to the simultaneous transmission of UE. This is known as the collision avoidance feature of the protocol.

Hence, prior to transmission, UE listen to the channel for a time interval called distributed inter-frame space (DIFS). Then, if the channel is found to be idle, the UE generate a random back off, B_j, for $j = 1, 2, \ldots, N_{UE}$, where N_{UE} is the number of competing UE served by the considered AP. The value of B_j is uniformly chosen in the range $[0, w - 1]$, where w is the current contention window size. Before the first transmission attempt, w is set to the minimum contention window, CW_{min}, and after each unsuccessful transmission, w is doubled up to the maximum contention window, CW_{max}. Thus, the size of the contention window at the lth back-off stage is equal to $w = 2^l CW_{min}$, for $l = 1, 2, \ldots, M_r$, where M_r is the retry limit and is equal to $M_r = \log_2\left(\frac{CW_{max}}{CW_{min}}\right)$. Let $\mathbf{B} = [B_j]_{1\times N_{UE}}$, be the backoff vector of the UEs. After sensing the channel for the time interval DIFS, UE should wait for $B_j \times t_{slot}$ seconds, where t_{slot} is the duration of each time slot. Obviously, the UE with the lowest back off is prioritized to transmit, i.e., the u_0th UE, where $u_0 = \arg\min_j(\mathbf{B})$. Therefore, the u_0th UE sends the RTS frame to the AP before $N_{UE} - 1$ other UE. If the RTS frame is received at the AP successfully, the AP replies after a short inter-frame space (SIFS) with the CTS frame. The u_0th UE only proceeds to transmit the data frame, after the time interval of SIFS, if it receives the CTS frame. Eventually, an acknowledgment (ACK) packet is transmitted after the period of SIFS by the serving AP to notify the successful packet reception. The UE that can hear the u_0th UE will freeze their back-off counter at the start of RTS transmission. The back-off counter will be reactivated when the channel is sensed to be idle again after the period of DIFS. If the u_1th UE cannot hear the u_0th UE, it will start to send the RTS frame after waiting for $B_{u_1} \times t_{slot}$ seconds. Here, the u_1th UE is called the hidden node, and a collision occurs if $(B_{u_1} - B_{u_0}) \times t_{slot} < t_{RTS}$, where t_{RTS} is the RTS transmission time, which is directly proportional to the length of RTS frame, L_{RTS}, and inversely proportional to the uplink rate, R_u, that is, $t_{RTS} = \frac{L_{RTS}}{R_u}$.

RTS and CTS frames carry a network allocation vector (NAV) field, which is a virtual carrier-sensing mechanism. The NAV indicates the number of microseconds that the channel is reserved by a UE. The NAV in an RTS frame includes the CTS frame, the data frame, and the subsequent ACK frame from the AP. Nodes that can hear the RTS frame will set their NAV timers corresponding to the RTS NAV. The CTS frame contains a new NAV updated to the time already elapsed. After the CTS frame is sent, all UE that can receive the CTS will update their NAV timer and defer transmission until their NAV timers reach zero. This keeps the channel free for the u0th UE to complete the process of transmitting data to the AP and alleviates the problem of hidden nodes. In contrast to RF networks, the UE cannot hear each other directly in LiFi networks since in the downlink and uplink visible light and IR are employed, respectively. Hence, the downlink filter is tuned in the visible light band and cannot sense the uplink transmission of other UE via IR. Hence, all UE are hidden from each other, and this problem is extreme in LiFi networks. Note that since UE cannot hear the RTS frame in LiFi networks, they can only set their NAV with the NAV of the CTS frame.

9.7 Medium Access Control in IEEE 802.11: Distributed Coordination Function (DCF)

The primary Medium Access Control (MAC) mechanism in IEEE 802.11 is the Distributed Coordination Function (DCF), a random access scheme based on Carrier Sense Multiple Access with Collision Avoidance (CSMA/CA) and binary slotted exponential back off [41]. DCF offers two main schemes for packet transmission: the *basic access mechanism*, a two-way handshaking process, and the optional *RTS/CTS* (Request to Send / Clear to Send) mechanism, which employs a four-way handshaking process. DCF employs a slotted binary exponential back-off strategy to reduce collisions due to simultaneous transmissions by User Equipment (UE), thus providing collision avoidance.

9.7.1 Random Back-Off Mechanism

Before initiating a transmission, UE monitor the channel for a Distributed Inter-frame Space (DIFS) interval. If the channel is idle, each UE generates a random back-off time, B_j, where $j = 1, 2, \ldots, N_{UE}$ and N_{UE} represent the number of UE competing for access to the Access Point (AP). The back-off value B_j is uniformly selected from the range $[0, w - 1]$, where w is the current contention window size. Initially, w is set to the minimum contention window, CW_{min}, and doubles after each unsuccessful transmission attempt, up to a maximum value CW_{max}. The size of the contention window at the lth back-off stage is defined as $w = 2^l CW_{min}$, for $l = 1, 2, \ldots, M_r$, where $M_r = \log_2\left(\frac{CW_{max}}{CW_{min}}\right)$. Let $\mathbf{B} = [B_j]_{1\times N_{UE}}$ represent the back-off vector of the UE.

After the DIFS interval, each UE waits for $B_j \times t_{slot}$ seconds, where t_{slot} is the slot time duration. The UE with the lowest back-off value, $u_0 = \arg\min_j(\mathbf{B})$, is given priority to transmit. The u_0 UE sends the RTS frame to the AP. Upon successful reception of the RTS, the AP responds after a Short Inter-frame Space (SIFS) interval with a CTS frame. If the u_0 UE receives the CTS frame, it proceeds to transmit its data after the SIFS. Subsequently, the AP sends an acknowledgment (ACK) frame after another SIFS to confirm successful data reception.

9.7.2 Hidden Node Problem and Collisions

The UE that can hear the u_0-th UE freeze their back-off counters upon detecting the RTS frame and resume when the channel is idle again after the DIFS period. However, if a hidden UE, u_1, cannot hear u_0's transmission, it may send an RTS frame after waiting for $B_{u_1} \times t_{\text{slot}}$. A collision occurs if $(B_{u_1} - B_{u_0}) \times t_{\text{slot}} < t_{\text{RTS}}$, where t_{RTS} is the RTS transmission time, calculated as $t_{\text{RTS}} = \frac{L_{\text{RTS}}}{R_u}$, where L_{RTS} is the length of the RTS frame, and R_u is the uplink rate.

9.7.3 Network Allocation Vector and LiFi Challenges

Both RTS and CTS frames contain a Network Allocation Vector (NAV), a virtual carrier-sensing mechanism. The NAV specifies the duration (in microseconds) for which the channel is reserved for a UE. The NAV field in an RTS frame includes the duration of the CTS frame, data transmission, and the subsequent ACK frame. UE that receive the RTS frame set their NAV timers accordingly, and the CTS frame provides an updated NAV value for the remaining duration of the transmission. This prevents other UE from transmitting, thereby alleviating collisions.

In LiFi networks, however, UE cannot directly detect each other's transmissions due to the nature of light-based communication, where visible light is used for downlink and infrared (IR) for uplink. This limitation exacerbates the hidden node problem, as UE cannot sense RTS frames transmitted by other UEs. Instead, they rely on the NAV in the CTS frame to defer their transmissions.

9.7.4 Modified CSMA/CA Protocol

In the modified CSMA/CA protocol, the access point transmits a CB tone simultaneously with receiving the RTS packet, signaling to other user equipment (UE) that the channel is occupied. UE that detect the CB tone will freeze their back-off counters, resuming only after the channel is sensed as idle following the distributed inter-frame space (DIFS) period. If the AP does not transmit the CB tone, other UE will continue counting down their back-off timers. When the back-off counter reaches zero, the UE will transmit their RTS packet, increasing the likelihood of a collision.

Figure 9.11 compares the conventional and modified CSMA/CA protocols with the RTS/CTS mechanism in a LiFi network. In the conventional CSMA/CA, the RTS from user 2 collides with the RTS from user 1, as user 2 is unable to detect user 1's transmission. The modified CSMA/CA, however, resolves this issue through the CB tone. In this case, user 2 hears the CB tone from the AP and freezes its back-off counter. Afterward, the AP transmits the CTS along with the network allocation vector (NAV). Upon receiving the CTS packet, user 2 defers its transmission for the NAV-specified duration. As a result, the modified CSMA/CA effectively addresses the hidden node problem in an LiFi network.

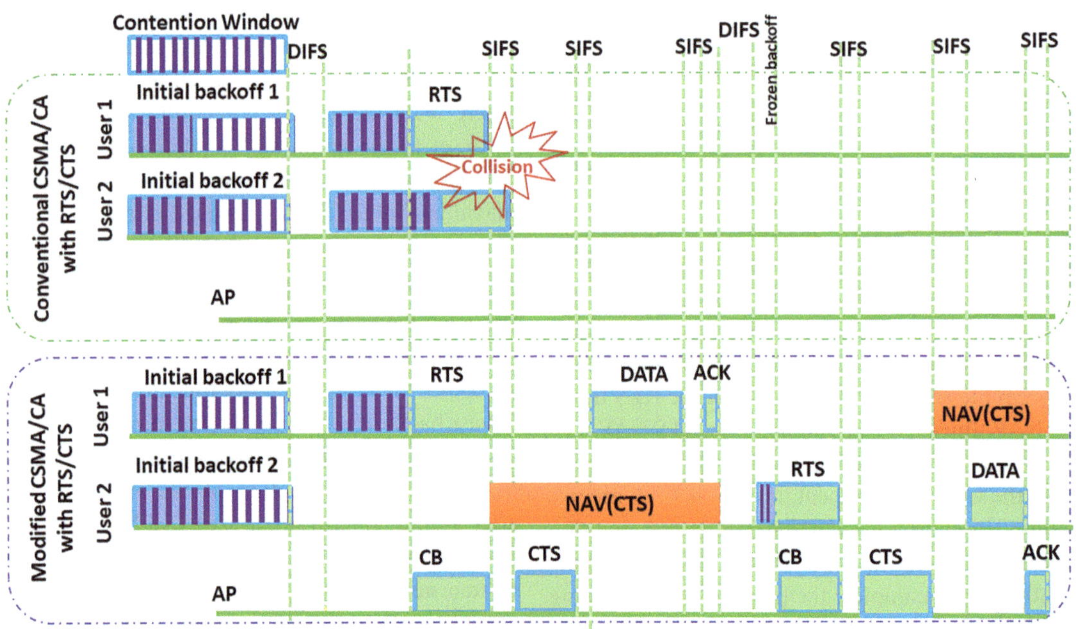

Fig. 9.11 Conventional vs. Modified CSMA/CA protocol for LiFi [41]

9.8 Integrated LiFi- WiFi

In an integrated LiFi-WiFi network, the MAC protocols must be adapted to accommodate the distinct communication mediums used by each technology. LiFi, which relies on light signals for bidirectional data transmission, operates over a vastly different medium compared to WiFi, which utilizes radio frequency signals. Due to these fundamental differences, standard MAC protocols like CSMA/CA, TDMA, and OFDMA need to function independently within the integrated network. This separation ensures that the MAC protocols are optimized for the specific characteristics of light-based and radio frequency–based communication, thereby maintaining efficient data flow and minimizing interference within the hybrid network.

9.9 Summary

As cellular technology continues to evolve at a rapid pace, ongoing research is crucial to sustain the advancement of mobile communication services. With users demanding enhanced service quality, faster data speeds, and improved connectivity, exploring various access methods has become imperative. This chapter aims to assess the performance of different multi-access techniques, including OMA, NOMA, and the recently developed RSMA. A comprehensive analysis of traditional access methods, such as FDMA, TDMA, CDMA, and OFDMA, is provided, detailing their functionality, benefits, and limitations. The study further delves into the intricacies of NOMA and RSMA, examining their potential impacts on future cellular networks. The chapter highlights that NOMA can outperform OMA in terms of spectral efficiency with accurate power distribution. Further, RSMA outperforms both OMA and NOMA in terms of spectral efficiency by allowing both common and private messages to coexist. Finally, random access protocol CSMA/CA is discussed in this chapter. Furthermore, an updated CSMA/CA was discussed in this chapter where the back-off period of CSMA/CA was modified to fulfill the user requirements.

Homework Problems

1. Discuss and compare the key features, technologies, and advancements of various generations of cellular networks, from 1G to 5G, highlighting their impact on communication capabilities, data transmission speeds, latency, and network capacity?
2. What are the key performance criteria for 5G cellular network standards, and how do they improve upon previous generations?
3. In the AMPS system, 416 channels are allocated to various operators, with a 10 KHz guard band between channels and a channel bandwidth of 30 KHz. What is the spectrum allocation per operator?
4. Given a frame duration T_f of 20 ms and a guard time T_g of 2 ms, what is the effective time allocated for each channel T in a TDMA system with 5 time slots per frame ($K = 5$)?
5. In wireless communication systems, which multiple access technique is employed by GPS to transmit signals effectively across different channels?
6. Explain the CDMA capacity?
7. Why is a cyclic prefix required in an OFDMA?
8. What are the key performance parameters to consider when analyzing OFDMA in wireless networks, and how do they impact overall system performance?
9. What are the primary challenges associated with decoding NOMA signals in wireless networks, and how do these challenges differ from those encountered in traditional OMA methods?
10. What are the differences between RSMA, NOMA, and OMA Schemes?
11. In CSMA/CA, the minimum and maximum contention window sizes are 11 and 1023, respectively. Calculate the number of back-off stages.
12. Explain hidden node problem and how to mitigate the problem.
13. How does conventional CSMA/CA differ from modified CSMA/CA in case of LiFi MAC protocol?

References

1. Tataria, H., Shafi, M., Molisch, A.F., Dohler, M.,H. Sjöland, Tufvesson, F.: 6G wireless systems: vision, requirements, challenges, insights, and opportunities. Proc. IEEE **109**(7), 1166–1199 (2021)
2. Wang, C.-X., You, X., Gao, X., Zhu, X., Li, Z., Zhang, C., Wang, H., Huang, Y., Chen, Y., Haas, H., Thompson, J.S., Larsson, E.G., Renzo, M.D., Tong, W., Zhu, P., Shen, X., Poor, H.V., Hanzo, L.: On the road to 6G: visions, requirements, key technologies, and testbeds. IEEE Commun. Surv. Tuts. **25**(2), 905–974 (2023)
3. Jamalipour, A., Wada, T., Yamazato, T.: A tutorial on multiple access technologies for beyond 3G mobile networks. IEEE Commun. Mag. **43**(2), 110–117 (2005)
4. Qasim, A., Karabulut, M.A., Ilhan, H., Islam, M.B.: Survey and performance evaluation of multiple access schemes for next-generation wireless communication systems. IEEE Access **9**, 113428–113442 (2021)
5. Sharma, S.K., Patwary, M., Chatzinotasand, S.: Multiple access techniques for next generation wireless: recent advances and future perspectives. EAI Endorsed Trans. Wirel. Spectr. **2**(7), 1–12 (2016)
6. Dai, L., Wang, B., Yuan, Y., Han, S., Chih-Lin, I., Wang, Z.: Nonorthogonal multiple access for 5G: solutions, challenges, opportunities, and future research trends. IEEE Commun. Mag. **53**(9), 74–81 (2015)

7. Mao, Y., Clerckx, B., Li, V.O.K.: Rate-splitting multiple access for downlink communication systems: bridging, generalizing, and outperforming SDMA and NOMA. EURASIP J. Wirel. Commun. Netw. **2018**, 133 (2018)
8. Ping, L., Liu, L., Wu, K.Y., Leung, W.K.: On interleave-division multiple access. In: Proceedings of IEEE International Conference on Communications, Paris (2004), pp. 2869–2873
9. Dai, L., Wang, B., Ding, Z., Wang, Z., Chen, S., Hanzo, L.: A survey of non-orthogonal multiple access for 5G. IEEE Commun. Surv. Tuts. **20**(3), 2294–2323 (2018)
10. Bojkovic, Z.S.: Research challenges for 5G cellular architecture. In: Proceedings of International Conference on Advanced Technologies, Systems and Services in Telecommunications, Nis, Serbia, pp. 215–222 (2015)
11. Kukushkin, A.: Introduction to Mobile Network Engineering: GSM, 3GWCDMA, LTE and the Road to 5G. Wiley, Hoboken (2018)
12. Buehrer, R.M.: CDMA. Morgan & Claypool, San Rafae (2006)
13. Gilhousen, K.S., Jacobs, I.M., Padovani, R., Viterbi, A.J., Weaver Jr., L.A., Wheatley III, C.E.: On the capacity of a cellular CDMA system. IEEE Trans. Veh. Technol. **40**(2), 303–312 (1991)
14. Li, J., Qualcomm, B., Wu, X.: OFDMA Mobile Broadband Communications A Systems Approach. Cambridge University Press, Cambridge (2013)
15. Al-Eryani, Y., Hossain, E.: The D-OMA method for massive multiple access in 6G: performance, security, and challenges. IEEE Veh. Technol. Mag. **14**(3), 92–99 (2019)
16. Yunzheng, T., Long, L., Shang, L., Zhi, Z.: A survey: several technologies of non-orthogonal transmission for 5G. China Commun. **12**(10), 1–15 (2015).
17. Islam, S.M.R., Zeng, M., Dobre, O.A., Kwak, K.-S.: Resource allocation for downlink NOMA systems: key techniques and open issues. IEEE Wirel. Commun. **25**(2), 40–47 (2018)
18. Mohammadkarimi, M., Raza, M.A., Dobre, O.A.: Signature-based nonorthogonal massive multiple access for future wireless networks: uplink massive connectivity for machine-type communications. IEEE Veh. Technol. Mag. **13**(4), 40–50 (2018)
19. Global mobile data traffic forecast update, 2018–2023, White Paper, Cisco, San Jose, CA (2019)
20. Kokare, M.B., Gautam, S.: On efficient resource allocation strategies for multi-RIS enhanced V2I SWIPT systems. In: Proceedings of IEEE Wireless Communications and Networking Conference, Dubai, UAE, pp. 1–6 (2024)
21. Sarieddeen, H., Alouini, M.-S., Al-Naffouri, T.Y.: Terahertz-band ultra-massive spatial modulation MIMO. IEEE J. Sel. Areas Commun. **37**(9), 2040–2052 (2019)
22. Ulgen, O., Erkucuk, S., Baykas, T.: Non-orthogonal multiple access for terahertz communication networks. In: Proceedings of IEEE Annual Ubiquitous Computing, Electronics & Mobile Communication Conference , New York, NY, pp. 0737–0742 (2020)
23. Wu, Q., Zhang, S., Zheng, B., You, C., Zhang, R.: Intelligent reflecting surface-aided wireless communications: a tutorial. IEEE Trans. Commun. **69**(5), 3313–3351 (2021)
24. Kokare, M.B., Swaminathan, R., Gautam, S.: Performance of multiple IRS-enabled V2V communication over double generalized gamma fading channel. In: Proceedings of IEEE International Conference on COMmunication Systems & NETworkS, Bangalore, pp. 913–919 (2024)
25. Singh, P.K., Nandi, S.K., Nandi, S.: A tutorial survey on vehicular communication state of the art, and future research directions. Veh. Commun. **18**, 1–39 (2019)
26. Qian, L.P., Wu, Y., Zhou, H., Shen, X.: Dynamic cell association for non-orthogonal multiple-access V2S networks. IEEE J. Sel. Areas Commun. **35**(10), 2342–2356 (2017)
27. Wang, J., Jiang, C., Zhang, H., Ren, Y., Chen, K.-C., Hanzo, L.: Thirty years of machine learning: the road to pareto-optimal wireless networks. IEEE Commun. Surv. Tuts. **22**(3), 1472–1514 (2020)
28. Kusume, K., Bauch, G., Utschick, W.: IDMA vs. CDMA: analysis and comparison of two multiple access schemes. IEEE Trans. Wirel. Commun. **11**(1), 78–87 (2012)
29. Zhang, R., Hanzo, L.: Three design aspects of multicarrier interleave division multiple access. IEEE Trans. Veh. Technol. **57**(6), 3607–3617 (2008)
30. Eid, E.M., Fouda, M.M., Eldien, A.S.T., Tantawy, M.M.: Performance analysis of MUSA with different spreading codes using ordered SIC methods. In: Proceedings of International Conference on Computer Engineering and Systems, Cairo, pp. 101–106 (2017)
31. Dai, X., Chen, S., Sun, S., Kang, S., Wang, Y., Shen, Z., Xu, J.: Successive interference cancelation amenable multiple access (SAMA) for future wireless communications. In: Proceedings of IEEE International Conference on Communication Systems, Macau, pp. 222–226 (2014)
32. Naim, M.A., Fonseka, J.P., Dowling, E.M.: A building block approach for designing multilevel coding schemes. IEEE Commun. Lett. **19**(1), 2–5 (2015)
33. Meng, X., Wu, Y., Chen, Y., Cheng, M.: Low complexity receiver for uplink SCMA system via expectation propagation. In: Proceedings of IEEE Wireless Communications and Networking Conference, San Francisco, pp. 1–5 (2017)
34. Fang, D., Huang, Y.-C., Ding, Z., Geraci, G., Shieh, S.-L., Claussen, H.: Lattice partition multiple access: a new method of downlink nonorthogonal multiuser transmissions. In: Proceedings of IEEE Global Communications Conference , Washington, DC, pp. 1–6 (2016)
35. Kizilirmak, R.: Non-orthogonal multiple access (NOMA) for 5G networks. In: Towards 5G Wireless Networks - A Physical Layer Perspective, pp. 83–98. InTech (2016)
36. Han, T., Kobayashi, K.: A new achievable rate region for the interference channel. IEEE Trans. Inf. Theory. **27**(1), 49–60 (1981)
37. Clerckx, B., Joudeh, H., Hao, C., Dai, M., Rassouli, B.: Rate splitting for MIMO wireless networks: a promising PHY-layer strategy for LTE evolution. IEEE Commun. Mag. **54**(5), 98–105 (2016)
38. Mao, Y., Clerckx, B., Li, V.O.: Energy efficiency of rate-splitting multiple access, and performance benefits over SDMA and NOMA. In: Proceedings of International Symposium on Wireless Communication Systems, Lisbon, pp. 1–5 (2018)
39. Dizdar, O., Mao, Y., Han, W., Clerckx, B.: Rate-splitting multiple access for Downlink multi-antenna communications: physical layer design and link-level simulations. In: Proceedings of International Symposium on Personal, Indoor and Mobile Radio Communications, Toronto, ON, pp. 1–6 (2020)
40. Mao, Y., Dizdar, O., Clerckx, B., Schober, R., Popovski, P., Poor, H.V.: Rate-splitting multiple access: fundamentals, survey, and future research trends. IEEE Commun. Surv. Tutor. **24**(4), 2073–2126 (2022)
41. Soltani, M.D., Safari, M., Haas, H.: Analysis of random orientation and user mobility in LiFi networks. Ph.D. Dissertation (2019)

Multiuser Diversity

10

10.1 Introduction

Diversity combining signal paths that are independently fading is one of the most effective methods for mitigating the effects of fading. Diversity combining capitalizes on the low likelihood of concurrent deep declines in the independent signal paths. Consequently, this chapter extends the concept of multiple access networks to multiuser diversity. Multiuser diversity scheme enables the transmission of identical data over distinct fading channels. The fading of the resultant signal is mitigated by the combination of these independent paths. For instance, consider a system that is subject to independent fading and has two antennas at either the transmitter or receiver. It is improbable that both antennas will experience deep fades simultaneously if they are adequately spaced apart. Selection combining, a method that involves selecting the antenna with the greatest signal, results in a significantly superior signal compared to the use of a single antenna. This chapter concentrates on the conventional techniques employed by the transmitter and receiver to attain diversity in optical wireless systems.

10.2 Types of Diversity

This section describes the various various ways to achieve independent fading paths in the network. The types are diversity are also shown in Fig. 10.1.

10.2.1 Space Diversity

One approach is to employ an antenna array, which is a collection of multiple transmit or receive antennas that are spaced apart. Space diversity is the term used to describe this form of diversity. It is important to note that receiver space diversity enables the realization of independent fading paths without requiring an increase in transmit signal intensity or bandwidth. Furthermore, the signal-to-noise power ratio at the receiver is enhanced by the coherent combining of the diversity signals, surpassing the SNR that would be achieved with a single receive antenna. This SNR increase, referred to as array gain, can also be achieved through transmitter space diversity (sometimes referred as maximal ration transmission or transmit beam forming) by appropriately weighting the antenna transmit powers in relation to the channel gains. Space diversity, in addition to array gain, also offers diversity gain, which is defined as the change in slope of the error probability that originates from the diversity combining. In the following sections, we will provide a detailed explanation of the array gain and diversity gain for specific diversity-combining procedures. Typically, the maximum diversity gain for either transmitter or receiver space diversity necessitates that the separation between antennas be such that the fading amplitudes corresponding to each antenna are approximately independent.

10.2.2 Polarization

A second approach to attaining diversity involves the use of two transmit antennas or two receive antennas with varying polarization, such as vertically and horizontally polarized waves. The two waves that are transmitted adhere to the same trajectory. Nevertheless, the average received power corresponding to either polarized antenna is approximately the same, as the power is distributed nearly equally relative to both polarizations by the multiple random reflections. Given the random nature of the scattering angle in relation to each polarization, it is exceedingly unlikely that the signals received on the two diversely polarized antennas would be in a deep fade at the same time. Polarization diversity has two drawbacks. Initially, the maximum number of diversity branches is two, which corresponds to the two forms of polarization. The second drawback is that polarization diversity

V. A. Bohara, Swaminathan R, *Optical and Wireless Access Networks*, Textbooks in Telecommunication Engineering, https://doi.org/10.1007/978-3-031-89696-5_10

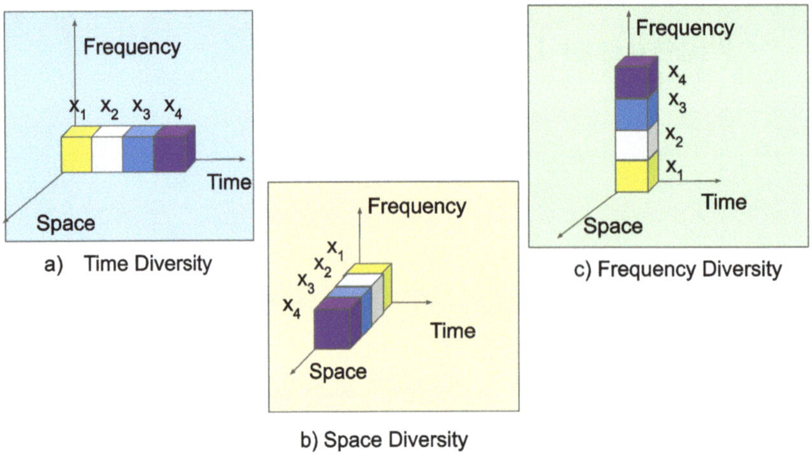

Fig. 10.1 Different types of diversity techniques

results in a loss of approximately 50% of the power (3 dB) due to the division of the transmit or receive power between the two antennas with varying polarization. By constraining the beamwidth of the receive antenna to a specific angle, directional antennas offer angle (or directional) diversity. In the most extreme case, if the angle is extremely small, only one of the multipath rays will fall within the receive beamwidth. Consequently, there is no multipath fading from multiple rays. Nevertheless, this diversity technique necessitates either a sufficient number of directional antennas to cover all potential directions of arrival or a single antenna whose reactivity can be directed to the arrival angle of one of the multipath components (preferably the strongest). It is also important to consider that the SNR may decrease as a result of the loss of multipath components that are outside the receive antenna beamwidth when using this technique. This is unless the directional gain of the antenna is sufficiently high to compensate for the lost power. Smart antennas are directional antennas that can be steered to the incoming angle of the strongest multipath component. These antennas are composed of antenna arrays with adjustable phases at each antenna element [1].

10.2.3 Frequency Diversity

Frequency diversity is attained by transmitting the same narrowband signal at different carrier frequencies, where the carriers are separated by the channel's coherence bandwidth. In order to transmit the signal across multiple frequency bands, this method necessitates an increase in transmit power. Spread-spectrum techniques are occasionally described as offering frequency diversity, which is achieved by varying the channel gain across the bandwidth of the transmitted signal [1,2]. Nevertheless, this is not the same as transmitting the same information signal over paths that are diminishing independently.

10.2.4 Time Diversity

Time diversity is achieved by transmitting the same signal at various times where the time difference exceeds the channel coherence time (the inverse of the channel Doppler spread). While time diversity does not necessitate an increase in transmit power, it does result in a decrease in data rate. This is due to the fact that data are repeated in diverse time periods rather than new data being transmitted during that period. Coding and interleaving are also viable methods for achieving time diversity [1,3]. Clearly, stationary applications cannot utilize time diversity, as the channel coherence time is infinite, and fading is strongly correlated over time. In this chapter, we emphasize space diversity as a reference when describing diversity systems and the various combining techniques, despite the fact that the techniques can be applied to any form of diversity. Consequently, the combining procedures will be characterized as operations on an antenna array. The system models and combining methodologies of receiver and transmitter diversity are treated separately due to their significant differences.

10.3 Space Diversity in Indoor Optical Wireless System

A combination of an array of receivers and a line arrangement of transmitters has also been employed in indoor environments to contemplate spatial multiplexing [4]. The transmitters were regarded as a linear array of lasers, while the receiver was equipped with an array of photo-detecting elements. A coupled IM/DD channel was established between each pair of transmitters and receivers. Each photodiode receives a linear combination of all transmit intensities. Consequently, this Optical wireless (OW) Multiple input multiple output (MIMO) system was accurately modeled as a channel with spatial inter-channel interference. A linear equalizer was developed to mitigate channel interference and separate them. It is important to acknowledge that this method is computationally intensive, particularly as the number of elements increases, and necessitates that the receiver be aware of the channel characteristics. Subsequently, a comparable methodology was implemented, which involved the a priori measurement of the MIMO channel and its subsequent inversion at the receiver [5, 6]. There has been a surge in interest in point-to-point OW MIMO on VLC channels in recent years. Despite the fact that these channels are significantly similar to infrared channels, the primary distinction is the positioning of illumination elements and the device bandwidth constraint rather than multipath distortion. A comprehensive introduction to point-to-point MIMO for VLC channels is provided in [7]. The authors utilize both imaging and non-imaging receivers and treat a collection of lighting fixtures on the ceiling as an MIMO emitter. Again, the receiver assumes complete channel knowledge, which merely inverts the channel matrix.

Optoelectronic components are employed in optical wireless systems to convert signals from baseband to transmission wavelengths. Typically, an electrical current signal is converted to a time-varying optical intensity using lasers and LEDs. An optical intensity, which is the normalized power in Watts per steradian, is the transmitted signal of concern. A current proportional to the impinging optical power is outputted by the receiver, which admits emissions within its FOV. The vast majority of OW systems contain these channels, which are referred to as Intensity modulated/direct-detection (IM/DD).

The instantaneous intensity, $I(t)$, is the quantity of interest at the transmitter. Data are applied to the instantaneous emitted optical power $x(t)$, which is the integration of $I(t)$ over the solid angle of the transmitter. A current $y(t)$ is output at the receiver in proportion to the voltage on the detector surface. The irradiance, measured in Watts per square meter, is integrated over the detector surface to determine the received power. Two primary constraints are imposed on all emitted signals by the IM/DD channel. Given that $x(t)$ is a power, it is necessary that it be nonnegative, i.e.,

$$x(t) \geq 0. \tag{10.1}$$

Safety limits are established by international organizations to guarantee the safety of the epidermis and eyes by regulating the permissible peak and average optical power. In the majority of instances, the average optical power constraint is more significant than the peak constraint for modulated laser sources that operate at a frequency exceeding 55 kHz [8]. A thorough examination of the eye safety of numerous commercial OW devices can be found in [9]. Consequently, it is necessary to establish a safety limit on the average optical power that is emitted. More specifically,

$$\lim_{u\to\infty} \frac{1}{2u} \int_{-u}^{u} x(t)dt \leq \mathcal{P}, \tag{10.2}$$

where $\mathcal{P}$ is the average optical power's eye-safety limit. All signaling designs for IM/DD OW systems must take these channel constraints into account.

Ambient illumination from fluorescent and incandescent sources is the primary source of noise in indoor OW systems. These devices emit emissions into the infrared spectrum that are detrimental to OW communications [10]. Fluorescent sources exhibit substantial powers in the near-infrared band and are modulated at tens to hundreds of kilohertz [11]. Moreover, rooms that have windows may permit sunlight to penetrate, which can exacerbate the corruption of OW signaling. It is a widely held belief that the ambient light is of high intensity. This results in a Poisson shot noise with a large mean, which is well modeled by a Gaussian distribution [12]. Consequently, the noise is typically modeled as white, signal-independent, and Gaussian distributed in the majority of indoor OW communication studies [11].

10.3.1 MIMO for diffuse OW Channels

In order to establish a link, diffuse OW links in indoor environments rely on the reflections of transmitted signals from objects in the room. Consequently, there is no stringent requirement for a line of sight between the transmitter and receiver [11]. The overwhelming majority of optical signal reflections from surfaces in the room are diffuse, despite the fact that they may be specular. This is because the scale of surface roughness of typical surfaces in a space (e.g., walls and ceiling) is roughly equivalent to the wavelength of light. This results in diffuse reflections, in which the angle of reflection is nearly independent of the incident angle [13].

The Lambertian radiation can be employed to accurately approximate the intensity distribution that results from a diffuse reflection.

$$I_{\text{Lambertian}}(\theta, \phi) = \mathcal{P}_{\text{elem}} \frac{n+1}{2\pi} \cos^n(\theta), \tag{10.3}$$

The optical intensity reflected at a polar angle $\theta \in \left[0, \frac{\pi}{2}\right]$ and azimuthal angle $\phi \in \left[0, \frac{\pi}{2}\right)$ is denoted by $I_{\text{Lambertian}}(\theta, \phi)$, as illustrated in Fig. 10.2. A surface element's total reflected power is denoted as $\mathcal{P}_{\text{elem}}$ [W], and the mode number of the Lambertian pattern is denoted as n [14–16]. The reflected

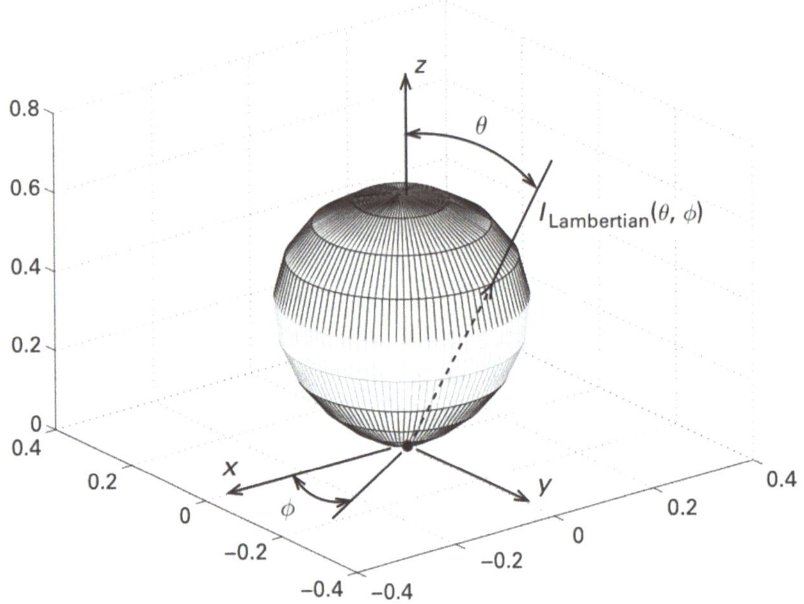

Fig. 10.2 A Lambertian reflection pattern with $n = 3$ and normalized to unit power. The reflecting element is in the $x - y$ plane [17]

Lambertian pattern is not affected by ϕ as a result of the symmetry in the azimuthal coordinate, as illustrated in Fig. 10.2.

Diffuse OW communication links do not direct their emissions; rather, they permit them to reflect diffusely from surfaces within the room. The emitter's unstructured emissions are dispersed throughout the room as they ricochet off the floor, ceiling, walls, and room objects. Consequently, it has been demonstrated that these connections are insensitive to both shadowing and pointing. The most significant aspect of this property is the extensive degree of user mobility.

Indoor diffuse OW connections are not impeded by multipath fading, as is the case with their radio counterparts. This is primarily due to the fact that the receiver photodiode is significantly larger than the light's narrow wavelength. The detector diameter is typically measured in thousands of wavelengths received. Consequently, the received intensity of a detector can exhibit numerous variations that are proportional to the wavelength of light. These local fades are effectively averaged out over the surface of the photodiode, providing an inherent degree of spatial diversity against multipath fading as a consequence of the large size of the detector relative to the wavelength [11].

Multipath-induced dispersion is a significant impairment in diffuse indoor connections, despite the fact that multipath fading is not a significant impairment. As illustrated in Fig. 10.3, the reception of multiple delayed and attenuated versions of the transmitted signal can be effectively modeled as a low-pass channel. In addition, receivers typically have a wide FOV in order to capture an adequate amount of power. In addition to capturing the transmitted signal, these receivers accumulate a substantial quantity of ambient light, which, in conjunction with the substantial path loss, diminishes the SNR of these connections.

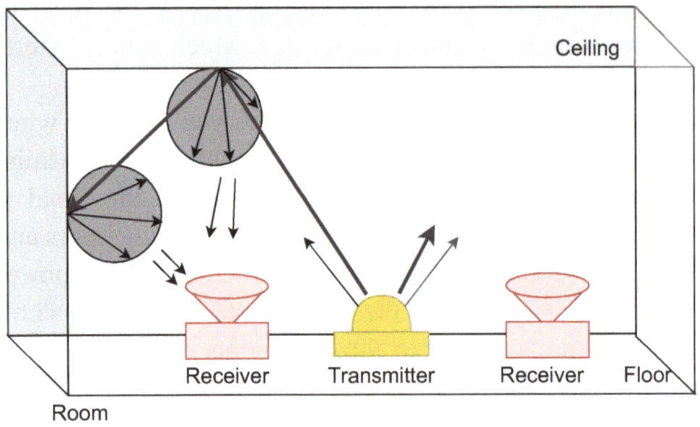

Fig. 10.3 An example of an indoor diffuse OW interconnection. The receiver accumulates multipath components, while the transmitter emits radiation over a wide angle

An accurate model for indoor diffuse OW channels is the linear baseband system

$$y(t) = Rx(t) \circledast h(t) + z(t) \tag{10.4}$$

where the impulse response is $h(t)$, the responsivity of the photodiode is R [A/W], and the continuous-time convolution operator is denoted by $\circledast$. The channel noise, $z(t)$, is a high-intensity shot noise that is generated by ambient light. It is modeled as being signal-independent and approximately Gaussian distributed [11, 16, 18].

A DC gain for the channel impulse response, which is low-pass is defined as follows:

$$H_0 = \int_{-\infty}^{\infty} h(t)dt, \tag{10.5}$$

which ranges from 10^{-7} to 10^{-5} [19]. The root-mean-square (RMS) delay spread of $h(t)$ is frequently reported to characterize the degree of multipath dispersion. This term is defined as,

$$\mathcal{D} = \sqrt{\frac{\int_{-\infty}^{\infty}(t-Q)^2 h^2(t)dt}{\int_{-\infty}^{\infty} h^2(t)dt}} \tag{10.6}$$

where the mean delay is represented as [19]

$$Q = \frac{\int_{-\infty}^{\infty} th^2(t)dt}{\int_{-\infty}^{\infty} h^2(t)dt}. \tag{10.7}$$

In measurements of typical room configurations, this parameter fluctuates within the 2–10 nanosecond range [19].

There are two prevalent models for $h(t)$: the exponential-decay model and the ceiling-bounce model. In the exponential-decay paradigm,

$$h(t) = \frac{H_0}{2\mathcal{D}} \exp\left(\frac{-t}{2\mathcal{D}}\right) U(t) \leftrightarrow H(f) = \frac{H_0}{1 + j4\pi\mathcal{D}f}, \tag{10.8}$$

where $\mathcal{D}$ is the root-mean-square (RMS) delay spread of $h(t)$, $U(t)$ is the unit step and $H(f)$ is the continuous-time Fourier transform of $h(t)$. The exponential-decay model is analogous to the ceiling bounce model, with the tail of $h(t)$ diminishing as t^{-7}. In the case of functional models, the delay spread can be readily correlated with the 3-dB cutoff of the channel. For instance, the exponential model (10.8)

$$f_{3dB} = \frac{1}{4\pi\mathcal{D}}. \tag{10.9}$$

For this model, a $\mathcal{D}$ in the range of 2–10 ns [11, 19] yields a channel bandwidth of approximately 10–40 MHz.

In any case, the impulse response is contingent upon the orientation and positions of the transmitters and receivers in the room. Nevertheless, $h(t)$ is regarded as constant in time for a specific dimension of room and transmitter and receiver configuration [11].

10.3.2 Spot-diffusing OW MIMO Systems

One of the earliest instances of OW MIMO systems for indoor networking applications is the spot-diffusing configuration. It integrates the benefits of mobility from diffuse links with the reduced path loss and multipath distortion of LoS links [20].

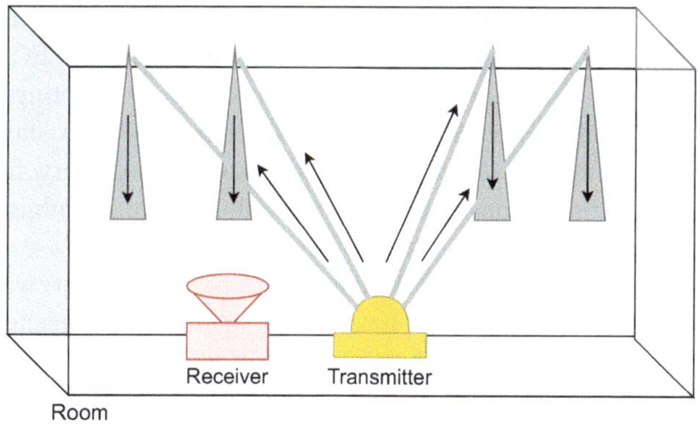

Fig. 10.4 The spot-diffusing OW MIMO architecture

An indoor OW spot-diffusing system is illustrated in Fig. 10.4. A series of low-divergence beams are employed by the transmitter to illumine the ceiling with an array of spots. Each point on the ceiling can be regarded as a Lambertian source that is diffusely reflecting. The data are detected by directing the receiver toward the ceiling and imaging the designated location(s). With the appropriate system design, the receiver can be configured to image a minimal number of locations, ideally one, during any reception. The arrangement and number of spots are selected to ensure that the receiver images at least one spot in its field of view for each position in the room, and identical data are transmitted to all locations in the room. Spot-diffusing MIMO OW links were initially introduced in [21], utilizing a "fly-eye" receiver. They are frequently referred to as multi-spot diffusing or quasi-diffuse links. Therefore, spot-diffusing links can be regarded as MIMO systems that employ spatial repetition of data to enhance channel characteristics and facilitate user mobility.

It is important to note that diffuse links and spot-diffusing links rely on diffuse reflections from room surfaces to establish a connection from the transmitter to the receiver. Nevertheless, the emitted optical signals in spot-diffusing links exhibit a well-defined structure that is leveraged to generate substantial link gain. Initially, spot-diffusing generates a sequence of spots by directing narrow divergence beams toward the ceiling, whereas diffuse systems emanate optical power in an unstructured manner. In this scenario, spot-diffusing links experience significantly less path loss than diffuse links, as the emitter loses minimal power to the ceiling. Furthermore, diffuse emitters must possess a broad divergence in order to encompass all regions of the room, which inevitably results in a greater path loss.

The receiver is the location of the second significant distinction between diffuse and spot-diffusing connections. Spot-diffusing systems can utilize narrow FOV receivers as long as a single spot is within the FOV as a result of the structured character of the optical emissions on the ceiling. These receivers with a confined FOV reject a greater proportion of ambient light interference, thereby reducing the noise at the receiver's input. Furthermore, the receiver's narrower FOV improves the bandwidth of such connections by rejecting a significant portion of multipath distortion. This is in stark contrast to the wide FOV receivers that are necessary for diffuse connections, which are constrained by high ambient light noise and multipath distortion.

In previous computer simulation studies [23], it was shown that spot-diffusing links are not significantly impacted by multipath and have bandwidths exceeding 2 GHz, while the bandwidth of diffuse links is restricted to 50 MHz. The simulated impulse response for a spot diffusing channel in a prototypical room is depicted in Fig. 10.5. It should be noted that the line-of-sight component dominates the impulse response, and the bandwidth exceeds 2 GHz. The efficacy of spot-diffusing links can be significantly enhanced by the use of an imaging receiver, which enables the spatial rejection of ambient light and multipath components [24].

Spot-diffusing connections are resistant to blockage and allow for a significant degree of mobility. Users can navigate a room while maintaining a high-speed connection as long as the FOV is sufficient to capture at least one spot in each position, and there are an adequate number of spots on the ceiling. It has been demonstrated that spot-diffusing systems are resistant to shadowing and that the probability of obstruction is less than 2% [25].

10.3.3 Dynamic Spot Diffusing Architecture

The general spot-diffusing model is a special case of the dynamic spot-diffusing (DSD) architecture [26]. A block diagram of the DSD channel is illustrated in Fig. 10.6. The transmitter in this architecture is composed of a single or a limited number of points that are translated across the ceiling in a closed path. The detector is a multi-element imaging receiver that implements select-best combining across all elements in the array. When the point is within the receiver's FOV, data are received; otherwise, no data are received. The primary distinction between DSD and conventional spot-diffusing and diffuse channels is that the latter is time-varying, meaning that the spot motion introduces fading into the link. In reality, this supplementary fading simplifies the implementation of the transmitter, enables adaptive transmission, and enables the use of higher transmit powers.

An array of spots is necessary to cover the ceiling in traditional spot-diffusing techniques. The multipath distortion of the DSD system is significantly reduced due to the

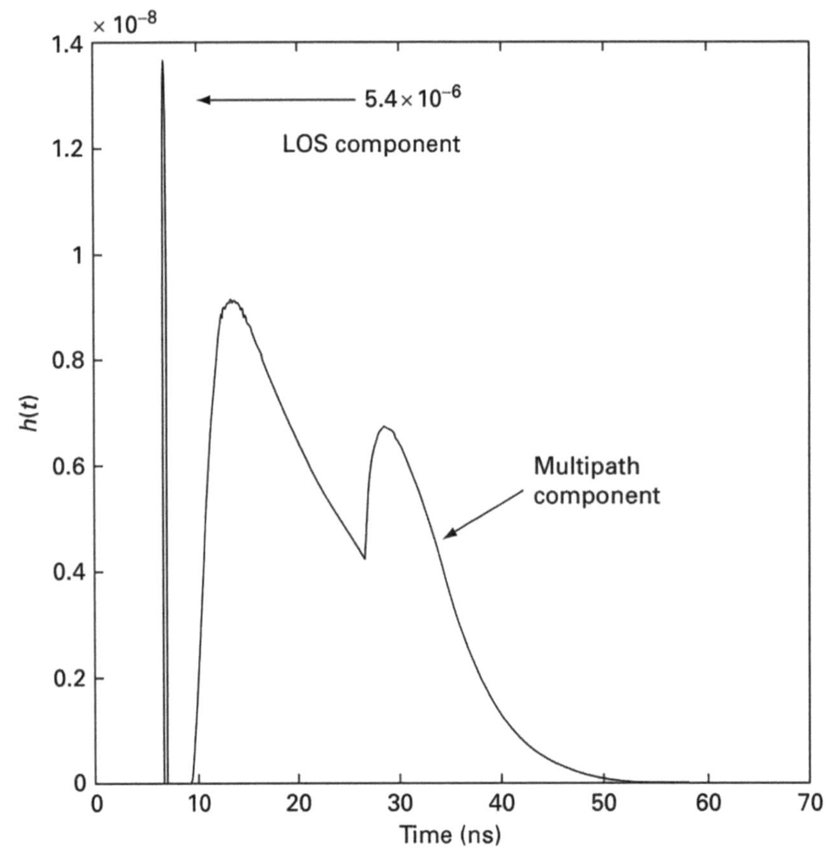

Fig. 10.5 Simulated impulse response, $h(t)$, for a spot-diffusing system with 10×10 spots, in a $6 \times 6 \times 3$ m room. The receiver is single element with $FOV = 11.2°$ and $A_R = 1\ \text{cm}^2$ [22]

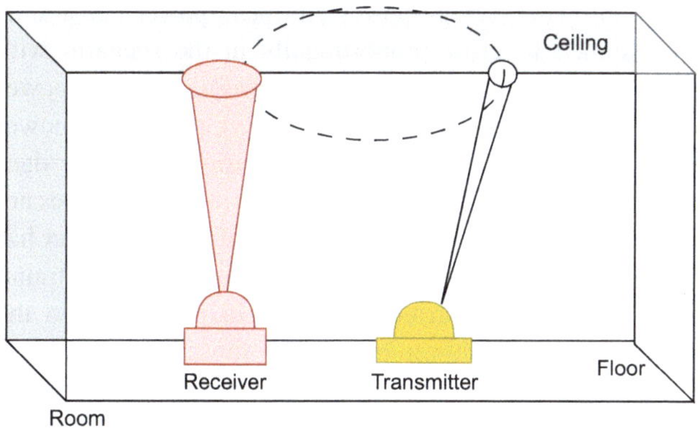

Fig. 10.6 A dynamic spot-diffusing link with one rotating spot on the ceiling and an imaging receiver

fact that it contains significantly fewer places. Therefore, high-speed modulators can be employed to transmit data immediately. Furthermore, the spot's movement alleviates eye-safety restrictions and enables the immediate emission of greater optical powers. The distribution of optical power over the ceiling can be defined with greater flexibility than previous spot-diffusing techniques, and the spot path can be adjusted in response to user feedback.

The DSD channel transitions between two states: a high data rate, a high SNR channel when the spot is in the receiver's field of view, and a channel with near-zero capacity. In order to facilitate dependable communication, the DSD channel is equipped with rateless codes. Luby transform (LT) codes and Raptor codes are two examples of these codes, which are highly resilient in the face of a wide range of channel conditions. This channel topology allows for user mobility, and the rateless codes adapt to the channel's capacity at the specified location in a seamless manner.

10.4 Capacity in Optical Wireless Multiuser Wireless Channels

In general, the capacity of a channel is the maximum rate at which it is possible to communicate reliably over the channel. In the case of rates below channel capacity, it is feasible to achieve an arbitrarily low probability of error by employing channel codes with a block length that is suitably large. However, this is not feasible for rates that exceed channel capacity. Consequently, channel capacity is a critical metric for practical systems and serves to characterize the integrity of the channel. The capacity of discrete-time channels is typically measured in nats per channel use. The channel model and the input constraints are used to determine the capacity. The capacity with allowed peak power $\mathcal{A}$ and allowed average power ϵ is denoted as $C(\mathcal{A}, \epsilon)$. In accordance with [27]:

$$C(\mathcal{A}, \epsilon) = \underset{\sup}{Q} I(Q, W), \qquad (10.10)$$

where the supremum is over all the probability distributions Q that satisfy the given input constraints, and $I(Q, W) = I(X; Y)$ denotes the mutual information [28] between the input X and the output Y of a channel of law $W(|)$ when the input X is distributed according to Q. This section provides an overview of three distinct discrete-time, pulse-intensity modulated optical channel models: the discrete-time Poisson channel, the free-space optical intensity channel, and the optical intensity channel with input-dependent Gaussian noise.

10.4.1 Discrete-time Poisson Channel

The channel output is the sum of two independent stochastic processes in the continuous-time Poisson channel. The initial process is a conditional Poisson process, which is also referred to as a doubly stochastic Poisson process, and its rate function is equivalent to the input. The second process is a homogeneous Poisson process that takes into consideration spurious counts that are the result of background radiation and receiver noise. The number of counts registered in the intervals $\{[kT, (k+1)T)\}$ is a sufficient statistic for predicting the input waveform based on the channel output if we use pulse intensity modulation, where the input remains constant over length-T time intervals of the form $[kT, (k+1)T)$ for integers k. This results in the discrete-time Poisson channel.

The discrete-time Poisson channel is modeled as a nonnegative sequence $\{x_k\}$, where x_k is the integral of the continuous-time input over the interval $[kT, (k+1)T)$. Consequently, x_k is the product of T and the instantaneous power of the transmitted signal during the interval $[kT, (k+1)T)$, as we are considering pulse intensity modulation. The output is a sequence $\{Y_k\}$, where Yk represents the number of counts registered within the interval $[kT, (k+1)T)$. We presume that Y_k has a Poisson distribution of mean $x_k + \lambda$, where λ is the anticipated number of spurious counts during the interval $[kT, (k+1)T)$, conditional on the transmitted waveform. At times, we refer to λ as "dark current." Thus, the discrete-time Poisson channel is a memoryless discrete-time channel whose input X is a set of nonnegative real numbers $\mathbb{R}_0^+$ and whose output Y is a value in the set of nonnegative integers $\mathbb{Z}_0^+$. The output is Poisson with a mean of $x + \lambda$, provided that the input x is nonnegative is given by:

$$W(y|x) = e^{-(x+\lambda)} \frac{(x+\lambda)^y}{y!}, \quad y \in \mathbb{Z}_0^+, \; x \in \mathbb{R}_0^+. \qquad (10.11)$$

The optical average power and the maximal instantaneous power must be restricted for practical implementation and safety reasons. Consequently, we implement the limitations as follows:

$$\mathbb{E}[X] \le \epsilon, \quad (10.12)$$

$$\Pr[X > \mathcal{A}] = 0, \quad (10.13)$$

for certain fixed parameters ϵ, $\mathcal{A} > 0$, and denote ϵ as the allowed average power and $\mathcal{A}$ as the allowed peak power[1] In order to quantify this in energy per second, we must multiply by the energy of each photon ω, where ω is the frequency of the transmitter in radians per second. If the detector's quantum efficiency is not 1, we can also divide by it. It is important to reiterate that the average-power constraint is based on the expectation of the channel input, rather than its square.

We denote the ratio of the allowed average power to the allowed peak power by α,

$$\alpha \triangleq \frac{\epsilon}{\mathcal{A}}, \quad (10.14)$$

where $0 < \alpha \le 1$. Please be advised that the average-power constraint is inactive at $\alpha = 1$, meaning that it has no effect on the system and is automatically satisfied when the peak-power constraint is satisfied. Consequently, the case with only a peak-power constraint is represented by $\alpha = 1$. In the same vein, a stringent average-power constraint and only a mild peak-power constraint are associated with $\alpha << 1$.

We will focus on the low-power asymptotic regime in our discussion of capacity. The low-power regime, in which $\epsilon \downarrow 0$ or $\mathcal{A} \downarrow 0$, can be achieved in two distinct ways: either from the continuous-time scenario with pulses of extremely short duration $T << 1^2$ or from a continuous-time scenario with a fixed pulse duration and extremely low transmit power. This is due to the fact that the input power of the discrete-time Poisson channel is equivalent to the product of the light intensity and the pulse duration in the continuous-time model. The dark current λ is also extremely small in the former case, as the number of illusory counts in each (short) pulse interval is also very small. Conversely, in the latter scenario, the number of spurious counts does not decrease with the power. Similarly, the high-power regime, which is also of interest to us, can result from two distinct scenarios, contingent upon whether the pulse duration is maintained at a constant value.

In the subsequent subsections, we present the asymptotic high- and low-power capacity of the discrete-time Poisson channel. We present results for the high-power regime in three scenarios: the scenario with both an average- and a peak-power constraint, the scenario with only a peak-power constraint, and the scenario with only an average-power constraint. The average-power and peak-power constraints are assumed to grow proportionally in the scenario with both constraints, i.e., with a constant average-to-peak power ratio α (refer to (10.14). The precise asymptotic high-power capacity is known for all three scenarios when the dark current is maintained at a constant value (i.e., when it does not tend to infinity with the power). The only scenario that has been investigated is the absence of a peak-power constraint, in which the dark current increases in proportion to the power. Upper and lower bounds on the asymptotic high-power capacity have been proposed for this configuration; however, they do not align.

The asymptotic low-power capacity is examined in two scenarios: one in which the dark current λ is fixed and positive, and another in which it tends to zero proportionally to the average-power constraint ϵ, i.e., with a constant ratio:

$$c \triangleq \frac{\lambda}{\epsilon}. \quad (10.15)$$

Both scenarios are examined with a fixed peak-power constraint (i.e., the peak-power constraint does not tend to 0) or without peak-power constraint.

10.4.1.1 High-power Results

The parameter ν^* is defined as the unique solution to the equation for a given $\alpha \in (0, 1/3)$ prior to the presentation of the theorem:

$$\alpha = \frac{1}{2\nu} - \frac{e-\nu}{\sqrt{\nu}\sqrt{\pi}\mathrm{erf}(\sqrt{\nu})}. \quad (10.16)$$

Here, erf(.) denotes the error function defined as [1]:

$$\mathrm{erf}(\xi) \triangleq \frac{2}{\pi}\int_0^{\xi} e^{-t^2}dt, \quad \forall \xi \in \mathbb{R}. \quad (10.17)$$

Note that the function $\nu \Longrightarrow \frac{1}{2\nu} - \frac{e-\nu}{\sqrt{\nu}\sqrt{\pi}\mathrm{erf}(\sqrt{\nu})}$ is monotonically decreasing in $[0, \infty)$ and tends to $\frac{1}{3}$ for $\nu \downarrow 0$ and to 0 for $\nu \uparrow \infty$. Hence, for $\alpha \in (0, \frac{1}{3})$, a solution always exists and is unique.

Theorem 10.1 (Constant Dark Current, [29]) *If the channel is peak-power constrained to $\mathcal{A}$ and average-power constrained to $\epsilon = \alpha\mathcal{A}$ for some $\alpha \in (0, \frac{1}{3})$, then*

$$C_{\mathrm{Poisson}}(\lambda, \mathcal{A}, \alpha\mathcal{A}) = \frac{1}{2}\log\left(\frac{2\mathcal{A}}{\pi e}\right) - (1-\alpha)\nu^* - \log(1-2\alpha\nu^*) + o_{\mathcal{A}}(1), \quad (10.18)$$

where $o_{\mathcal{A}}(1)$ is a term that tends to 0 as the power $\mathcal{A}$ tends to infinity and where ν^* is the unique solution to (10.17).

[1] In the corresponding continuous-time model, a peak-power constraint $\mathcal{A}$ and an average-power constraint ϵ in the discrete-time model correspond to a peak-power constraint $\mathcal{A}/T$ and an average-power constraint ϵ/T. Additionally, the term "power in the continuous-time model" denotes the "average number of photons emitted per second."

If the channel is only peak-power constrained to $\mathcal{A}$ (which corresponds to $\alpha = 1$), or if the channel is peak-power constrained to $\mathcal{A}$ and average-power constrained to $\epsilon = \alpha\mathcal{A}$ for some $\alpha \in [0, \frac{1}{3}]$, then:

$$C_{\text{Poisson}}(\lambda, \mathcal{A}, \alpha\mathcal{A}) = \frac{1}{2}\log\left(\frac{2\mathcal{A}}{\pi e}\right) + o_{\mathcal{A}}(1) \quad (10.19)$$

If the channel is only average-power constrained to ϵ, then

$$C_{\text{Poisson}}(\lambda, \epsilon) = \frac{1}{2}\log\epsilon + o_{\epsilon}(1) \quad (10.20)$$

The dark current λ does not affect the high-power asymptotes (10.18), (10.19), and (10.20). Additionally, the asymptote (10.19) is not influenced by the average-to-peak power ratio α when $\alpha \geq \frac{1}{3}$. Therefore, the average-power constraint is not active in the high-power asymptote if $\epsilon \geq \frac{\mathcal{A}}{3}$.

Note that the high-power-capacity asymptotes of the classical additive white Gaussian noise (AWGN) channel are congruent with the asymptotes (10.19) and (10.20) when the noise variance $\sigma^2 = 1$ and the channel input is either peak-power constrained to $\mathcal{A}$, i.e., $0 \leq X_2 \leq \mathcal{A}$, or average-power constrained to ϵ, i.e., $\epsilon[X_2] \leq \epsilon$. We can observe that the asymptotic high-power capacity is reduced by $(1-\alpha)\upsilon^* + \log(1 - 2\alpha\upsilon^*)$ when an average-power constraint $\epsilon = \alpha\mathcal{A}$ is imposed in addition to a peak-power constraint $\mathcal{A}$ for $\alpha \in (0, \frac{1}{3})$. Similarly, when comparing (10.19) and (10.20), it is evident that the high-power capacity is increased by $\frac{1}{2}\log\left(\frac{\pi e}{2}\right)$ when a (weaker) average-power constraint $\epsilon = \mathcal{A}$ is imposed in place of a (stronger) peak-power constraint $\mathcal{A}$.

Theorem 10.2 (Dark Current Proportional to ϵ, [30]) *The following two bounds are valid if the average power is restricted to ϵ and the dark current λ grows proportionally with ϵ, i.e., $\lambda = c\epsilon$ for some $c > 0$: Given that $\varepsilon > 0$, there is a ϵ such that the capacity is bounded by $\epsilon > \epsilon_\varepsilon$ in all cases.*

$$C_{\text{Poisson}}(\lambda = c\epsilon, \epsilon) \geq \frac{1}{2}\log\left(\frac{\epsilon}{2pi}\right) - \frac{1}{2}\log(1+c) - \varepsilon, \quad (10.21)$$

$$\begin{aligned} &C_{\text{Poisson}}(\lambda = c\epsilon, \epsilon) \\ &\leq \frac{1}{2}\log\left(\frac{\epsilon}{2pi}\right) + \log\left(\frac{1}{\sqrt{c}}\left(1 + \frac{1}{\epsilon_\varepsilon}\right) + \sqrt{c}\right) + 1 \\ &+ \log\left(\frac{3}{2}\right) + \varepsilon. \end{aligned} \quad (10.22)$$

10.4.1.2 Low-power Results

The initial low-power result addresses the scenario in which the dark current λ is both positive and fixed.

Theorem 10.3 (Constant Nonzero Dark Current, [31]) *If a peak-power constraint $\mathcal{A}$ and an average-power constraint ϵ are imposed, then for a fixed dark current $\lambda > 0$:*

$$\lim_{\epsilon\downarrow 0}\frac{C_{\text{Poisson}}(\lambda, \epsilon, \mathcal{A})}{\epsilon} = \left(1 + \frac{\lambda}{\mathcal{A}}\right)\log\left(1 + \frac{\mathcal{A}}{\lambda}\right) - 1. \quad (10.23)$$

Under the assumption that only an average-power constraint ϵ is enforced, the following holds for a fixed dark current $\lambda > 0$:

$$\underline{\lim}_{\epsilon\downarrow 0}\frac{C_{\text{Poisson}}(\lambda, \epsilon)}{\epsilon\log\log\left(\frac{1}{\epsilon}\right)} \geq \frac{1}{2}, \quad (10.24)$$

$$\overline{\lim}_{\epsilon\downarrow 0}\frac{C_{\text{Poisson}}(\lambda, \epsilon)}{\epsilon\log\log\left(\frac{1}{\epsilon}\right)} \leq 2. \quad (10.25)$$

Our upper and lower bounds on the asymptotic low-power capacity are not tight when only an average-power constraint ϵ is imposed; rather, they exhibit similar behavior.

On–off signaling can be employed to accomplish both (10.23) and (10.24). The "on" signal is equivalent to $\mathcal{A}$ in the case of (10.23), whereas it tends to infinity as a constant times $\log\left(\frac{1}{\epsilon}\right)$ in the case of (10.24).

The asymptotic low-power capacity is greatly influenced by the peak-power constraint $\mathcal{A}$ for a constant (nonzero) dark current λ. That is, it is linear in the average-power constraint ϵ when a peak-power constraint is present, and it is proportional to $\epsilon\log\log\left(\frac{1}{\epsilon}\right)$ when it is absent.

The second low-power result addresses the scenario in which the dark current is proportional to the average-power constraint ϵ and, as $\epsilon \downarrow 0$, it tends to zero.

Theorem 10.4 (Dark Current Proportional to ϵ, [31]) *Given a peak-power constraint $\mathcal{A}$ and an average-power constraint ϵ, a dark-current λ that is proportional to ϵ, i.e., $\lambda = c\epsilon$ for some real number $c \geq 0$, is as follows:*

$$\lim_{\epsilon\downarrow 0}\frac{C_{\text{Poisson}}(\lambda = c\epsilon, \epsilon, \mathcal{A})}{\epsilon\log\left(\frac{1}{\epsilon}\right)} = 1. \quad (10.26)$$

It is important to note that (10.26) is valid regardless of the permitted (positive) peak power $\mathcal{A}$. In reality, (10.26) can be accomplished through on-off signaling, in which the "on" signal is a positive constant that is arbitrarily small (i.e., it does not tend to 0 as $\epsilon \downarrow 0$).

The proportionality constant $c \geq 0$ does not affect the asymptote (10.26). In particular, it is not contingent upon whether the dark current λ is zero or if it only tends to zero as $\epsilon \downarrow 0$. This is due to the fact that our constant "on" signal dominates the dark-current floor $\lambda = c\epsilon$ for a sufficiently small ϵ, irrespective of the constant c.

The implications of Theorem 10.4 for the continuous-time Poisson channel are the subject of our final discussion in this section. We concentrate on the scenario in which the maximal power is not restricted. As previously stated, the discrete-time model examined in Theorem 10.4 can be employed to characterize a pulse intensity modulated system over a continuous-time Poisson channel. In this scenario, the pulse duration T approaches zero, and the average power $\frac{\epsilon}{T}$ (i.e., the average intensity) remains constant. Theorem 10.4 thereby implies that the data rate $\frac{C_{\text{Poisson}}}{T}$ tends to infinity in a continuous-time system as the pulse duration T approaches zero, similar to a constant time $\log\left(\frac{1}{T}\right)$. This observation is in accordance with the fact that the capacity of a continuous-time Poisson channel is infinite in the absence of a peak-power constraint.

10.4.1.3 Nonasymptotic results

The following lower bounds are applicable to arbitrary peak- and/or average-power constraints and arbitrary dark current.

Theorem 10.5 (Arbitrary Dark Current, [29]) *If the channel is peak-power constrained to $\mathcal{A}$ and average-power constrained to $\epsilon = \alpha\mathcal{A}$ for some $\alpha \in \left(0, \frac{1}{3}\right)$, then*

$$\begin{aligned} C_{\text{Poisson}}(\lambda, \mathcal{A}, \alpha\mathcal{A}) \geq \frac{1}{2}\log\mathcal{A} - (1-\alpha)\nu^* \\ - \log\left(\frac{1}{2} - \alpha\nu^*\right) \\ - e^{\nu^*}\left(\frac{1}{2} - \alpha\nu^*\right)\left[\log\left(1 + \frac{\lambda + 1/12}{\mathcal{A}}\right)\right. \\ \left. + 2\sqrt{\frac{\lambda + 1/12}{\mathcal{A}}}\arctan\left(\sqrt{\frac{\mathcal{A}}{\lambda + 1/12}}\right)\right] \\ (\epsilon + 1)\log\left(1 + \frac{1}{\epsilon}\right) - 1 - \frac{1}{2}\log(2\pi e), \end{aligned} \tag{10.27}$$

where ν^* is the unique solution to (10.16).

If the channel is peak-power constrained to $\mathcal{A}$ and average-power constrained to $\epsilon = \alpha\mathcal{A}$ for some $\alpha \in \left[\frac{1}{3}\right]$, or if the channel is only peak-power constrained to $\mathcal{A}$ (which corresponds to $\alpha = 1$), then

$$\begin{aligned} C_{\text{Poisson}}(\lambda, \mathcal{A}, \alpha\mathcal{A}) \geq \frac{1}{\mathcal{A}} + \left(\frac{\mathcal{A}}{3} + 1\right)\log\left(1 + \frac{3}{\mathcal{A}}\right) - 1 \\ - \sqrt{\frac{\lambda + 1/12}{\mathcal{A}}}\left(\frac{\pi}{4} + \frac{1}{2}\log 2\right) - \frac{1}{2}\log\left(\frac{\pi e}{2}\right). \end{aligned} \tag{10.28}$$

If the channel is only average-power constrained to ϵ, then

$$\begin{aligned} &C_{\text{Poisson}}(\lambda, \epsilon) \\ &\geq \frac{1}{2}\log\epsilon - \frac{\pi\,(\lambda + 1/12)}{2\epsilon} + (1+\epsilon)\log\left(1 + \frac{1}{\epsilon}\right) - 1. \end{aligned} \tag{10.29}$$

For the special case without dark current and under an average-power constraint only, [32] established an upper bound and a tighter lower limit on the capacity of the discrete-time Poisson channel.

Theorem 10.6 (Zero Dark Current, [32]) *If only the average power is constrained to ϵ and the dark current $\lambda = 0$, then*

$$C_{\text{Poisson}}(\lambda = 0, \epsilon) \geq \frac{1}{2}\log(1+\epsilon), \tag{10.30}$$

$$\begin{aligned} &C_{\text{Poisson}}(\lambda = 0, \epsilon) \\ &\leq \left(\epsilon + \frac{1}{2}\right)\log\left(\epsilon + \frac{1}{2}\right) - \epsilon\log\epsilon - \frac{1}{2} \\ &\quad + \log\left(1 + \frac{\sqrt{2e} - 1}{\sqrt{1 + 2\epsilon}}\right). \end{aligned} \tag{10.31}$$

10.4.2 Optical Intensity Channel With Input-Dependent Gaussian Noise

In our third model, we continue to assume that the signal at the receiver is impeded by additive Gaussian noise. However, we now assume that the variance of the noise is contingent upon the channel input. In this way, we endeavor to more accurately represent the implicit relationship between the signal and the random distortion in an optical channel. Thus, this model integrates certain characteristics of the preceding two models: the discrete-time Poisson channel's double-stochastic (signal-dependent) nature and the free-space optical intensity channel's Gaussian noise assumption.

The optical intensity channel with input-dependent Gaussian noise is the name of the model, and its conditional channel law is as follows:

$$W(y|x) = \frac{1}{\sqrt{2\pi(\sigma^2 + x)}}e^{-\frac{(y-x)^2}{2(\sigma^2+x)}}, \quad y \in \mathbb{R},\ x \geq 0. \tag{10.32}$$

Thus, the channel is described by

$$Y = x + \sqrt{x}Z_1 + Z_0, \tag{10.33}$$

where Y represents the channel output, $x \geq 0$ represents the channel input, and Z_0 is a zero-mean, variance-σ^2 random variable. The Gaussian random variable used to describe the input-independent noise is Z_1, which is a zero-mean, unit-variance random variable. The input-dependent noise is described by a Gaussian random variable. Here, Z_0 and Z_1 are independent. It is important to note that the assumption that Z_1 has a unit variance does not result in a loss of generality. In reality, the input-dependent noise component of each channel of the form (10.33) can be converted into a channel with unit variance by scaling the channel input and output appropriately.

We continue to evaluate both an average-power constraint (10.12) and a peak-power constraint (10.13), and we refer to the average-to-peak power ratio as α (10.14).

We present asymptotic capacities for the optical intensity channel with input-dependent Gaussian noise, as well as firm (nonasymptotic) lower constraints on the capacity for arbitrary power in both the high-power and low-power regimes. The three scenarios discussed are as follows: with peak- and average-power constraints, with peak-power constraints only, and with average-power constraints only.

10.4.2.1 High-Power Analysis

Theorem 10.7 ([33]) *If a peak-power constraint $\mathcal{A}$ and an average-power constraint $\epsilon = \alpha\mathcal{A}$ is imposed for some $\alpha \in \left(0, \frac{1}{3}\right)$, then*

$$C_{\text{DepNoise}}(\mathcal{A}, \alpha\mathcal{A}) = \frac{1}{2}\log\left(\frac{2\mathcal{A}}{\pi e}\right) - (1-\alpha)\nu^* - \log(1-2\alpha\nu^*) + o_{\mathcal{A}}(1), \tag{10.34}$$

where $o_{\mathcal{A}}(1)$ is a term that tends to 0 as the power A tends to infinity and ν^* is the unique solution to (10.26).

If only a peak-power constraint $\mathcal{A}$ is imposed (which corresponds to $\alpha = 1$), or if a peak-power constraint $\epsilon = \alpha\mathcal{A}$ is imposed for some $\alpha \in \left[\frac{1}{3}, 1\right]$, then

$$C_{\text{DepNoise}}(\mathcal{A}, \alpha\mathcal{A}) = \frac{1}{2}\log\left(\frac{2\mathcal{A}}{\pi e}\right)) + o_{\mathcal{A}}(1). \tag{10.35}$$

If only an average-power constraint ϵ is imposed, then

$$C_{\text{DepNoise}}(\epsilon) = \frac{1}{2}\log\epsilon + o_{\epsilon}(1). \tag{10.36}$$

We observe that the asymptotic high-power capacity is not influenced by α for $\alpha \in \left[\frac{1}{3}, 1\right]$. Therefore, the average-power constraint is dormant in the high-power asymptote when $\alpha \in \left[\frac{1}{3}, 1\right]$. The results of our asymptotic high-power capacity estimates (10.34)–(10.36) for the optical intensity channel with input-dependent Gaussian noise are consistent with the asymptotic high-power capacity estimates (10.18)–(10.20) for the discrete-time Poisson channel. This correspondence can be intuitively understood by recognizing that the cumulative distribution function of a Poisson random variable with mean x approximates the cumulative distribution function of a Gaussian random variable with mean x and variance x for large values of x [33].

10.4.2.2 Low-Power Results

Theorem 10.8 ([33]) *If a peak-power constraint $\mathcal{A}$ and an average-power constraint $\epsilon = \alpha\mathcal{A}$ are imposed for some $\alpha \in \left(0, \frac{1}{2}\right)$, then*

$$\lim_{\mathcal{A}\downarrow 0} \frac{C_{\text{DepNoise}(\mathcal{A},\alpha\mathcal{A})}}{\mathcal{A}^2} = \alpha(1-\alpha)\frac{1+2\sigma^2}{4\sigma^4}. \tag{10.37}$$

If both a peak-power constraint $\mathcal{A}$ and an average-power constraint $\epsilon = \alpha\mathcal{A}$ are imposed for some $\alpha \in \left[\frac{1}{2}, 1\right]$, or if only a peak-power constraint $\mathcal{A}$ is imposed (which corresponds to $\alpha = 1$), then

$$\lim_{\mathcal{A}\downarrow 0} \frac{C_{\text{DepNoise}(\mathcal{A},\alpha\mathcal{A})}}{\mathcal{A}^2} = \frac{1+2\sigma^2}{16\sigma^4}. \tag{10.38}$$

We observe that the asymptotic low-power capacities (10.37) and (10.38) are highly similar to the asymptotic capacities (10.48) and (10.49), particularly for large values of σ^2. The threshold between the case with both a peak and an average-power constraint and the case where the average-power constraint is inactive is $\alpha = \frac{1}{2}$, similar to the free-space optical intensity channel. However, unlike the high-power regime, it is not $\alpha = \frac{1}{3}$.

Therefore, we can deduce that the capacity of the optical intensity channel with input-dependent Gaussian noise is similar to the capacity of the discrete-time Poisson channel at high power, but it is similar to the capacity of the free-space optical intensity channel at low power.

10.4.2.3 Nonasymptotic Results

Theorem 10.9 ([33]) *If a peak-power constraint $\mathcal{A}$ and an average-power constraint $\epsilon = \alpha\mathcal{A}$ are imposed for some $\alpha \in \left(0, \frac{1}{2}\right)$, then*

$$\begin{aligned} C_{\text{DepNoise}}(\mathcal{A}, \epsilon) \geq{}& \frac{1}{2}\log\mathcal{A} - \frac{1}{2}\log(2\pi e) - (1-\alpha)\nu^* \\ &- \log\left(\frac{1}{2} - \alpha\nu^*\right) \\ &- e^{\nu^*}\left(\frac{1}{2} - \alpha\nu^*\right)\left(2\sqrt{\frac{\sigma^2}{\mathcal{A}}}\arctan\left(\sqrt{\frac{\mathcal{A}}{\sigma^2}}\right)\right. \\ &\left. + \log\left(1 + \frac{\sigma^2}{\mathcal{A}}\right)\right) \\ &+ \frac{1}{2}\log\left(1 + \frac{2}{\epsilon}\right) + \sqrt{\epsilon(2+\epsilon)} - \epsilon - 1 \end{aligned} \tag{10.39}$$

where v^* is defined to be the solution to (9.11).

If there is only a peak-power constraint (which corresponds to $\alpha = 1$) or if both a peak-power constraint $\mathcal{A}$ and an average-power constraint $\epsilon = \alpha\mathcal{A}$ are imposed with $\alpha \in \left[\frac{1}{3}, 1\right]$, then

$$C_{\text{DepNoise}}(\mathcal{A}, \epsilon) \geq \frac{1}{2}\log\mathcal{A} - \frac{1}{2}\log(\frac{\pi e}{2}) - \frac{1}{2}\log\left(1 + \frac{\sigma^2}{\mathcal{A}}\right) + \frac{1}{2}\log\left(1 + \frac{6}{\mathcal{A}}\right) - \sqrt{\frac{\sigma^2}{\mathcal{A}}}\arctan\left(\sqrt{\frac{\mathcal{A}}{\sigma^2}}\right) + \sqrt{\frac{\mathcal{A}}{3}\left(2 + \frac{\mathcal{A}}{3}\right)} - \frac{A}{3} - 1 \tag{10.40}$$

If only an average-power constraint ϵ is imposed, then

$$C_{\text{DepNoise}}(\epsilon) \geq \frac{1}{2}\log\epsilon - \sqrt{\frac{\pi\sigma^2}{2\epsilon}} + \frac{1}{2}\log\left(1 + \frac{2}{\epsilon}\right) + \sqrt{\epsilon(2+\epsilon)} - \epsilon - 1 \tag{10.41}$$

10.4.3 Free-Space Optical Intensity Channel

In our third model, we consider three sources of noise: thermal noise of the post-detection circuitry following the photodetection, which is well-modeled by a Gaussian distribution; relative-intensity noise, which models random intensity fluctuations inherent to low-cost laser sources and can also be assumed to be Gaussian [13]; and shot noise caused by the ambient light. The shot noise's impact is only noticeable at high intensities, where its distribution is Gaussian and, in a first approximation, can be presumed to be independent of the signal. The thermal noise is the limiting factor at moderate intensity. Independent and memoryless additive Gaussian noise can effectively simulate the sum of these three noise sources:

$$Y = x + Z, \tag{10.42}$$

where Y represents the channel output, and x represents the channel input, which is proportional to the optical intensity and, as a result, cannot be negative,

$$x \in \mathbb{R}_0^+; \tag{10.43}$$

and where the additive noise is zero-mean Gaussian with a variance of σ^2. Unlike the input, the output Y may be negative due to the negative noise introduced at the receiver. This is a critical point to consider. α represents the average-to-peak power ratio (10.14), and the input is once again restricted in both its average power (10.12) and its peak power (10.13).

Sometimes, this model is employed to characterize optical communication through the air, and it is referred to as the free-space optical intensity channel. This channel is also known as the optical direct-detection channel with Gaussian post-detection noise [34, 35].

For a channel model that also considers the random components of the ambient light and the effects of an optical preamplifier at the receiver [36, 37].

We provide firm (nonasymptotic) constraints on capacity for arbitrary power, as well as results on the asymptotic capacity in the high and low-power regime, for the free-space optical intensity channel. The average-to-peak power ratio $\alpha = \frac{\epsilon}{\mathcal{A}}$ is maintained at a constant value in the high- and low-power regimes, resulting in a proportional increase or decrease in both powers.

By utilizing the symmetry of the channel model, it is possible to demonstrate that the average-power constraint is inactive when it exceeds half of the peak-power constraint. Consequently, the capacity $C_{\text{FreeSpace}}(\mathcal{A}, \epsilon)$ with a peak-power constraint $\mathcal{A}$ and an average-power constraint E is equivalent to the capacity $C_{\text{FreeSpace}}(\mathcal{A}, \mathcal{A})$ with only a peak-power constraint when the average-to-peak power ratio $\alpha \geq \frac{1}{2}$.

The capacity of a continuous-time version of the free-space optical intensity channel was investigated by [38], who also calculated upper and lower bounds on this capacity under the constraints of an average-power constraint and the use of a specific set of pulse shapes for the pulse intensity modulation scheme.

10.4.3.1 High-power Results

For every $\alpha \in \left(0, \frac{1}{2}\right)$ define μ^* as the unique solution to

$$\alpha = \frac{1}{\mu} - \frac{e^{-\mu}}{1 - e^{-\mu}}. \tag{10.44}$$

(It is easily shown that the solution to (10.44) always exists and is unique.)

Theorem 10.10 ([39]) *If a peak-power constraint $\mathcal{A}$ and an average-power constraint $\epsilon = \alpha\mathcal{A}$ are imposed for some $\alpha \in \left(0, \frac{1}{2}\right)$, then*

$$C_{\text{FreeSpace}}(\mathcal{A}, \alpha\mathcal{A}) = \frac{1}{2}\log\left(\frac{\mathcal{A}^2}{2\pi e\sigma^2}\right) - (1-\alpha)\mu^* \log(1 - \alpha\mu^*) + o_{\mathcal{A}}(1), \tag{10.45}$$

where μ^* is the unique solution to (10.44).

If only a peak-power constraint $\mathcal{A}$ is imposed (which corresponds to $\alpha = 1$), or if an average-power constraint $\epsilon = \alpha\mathcal{A}$ is imposed for some $\alpha \in \left[\frac{1}{2}, 1\right]$, then

$$C_{\text{FreeSpace}}(\mathcal{A}, \alpha\mathcal{A}) = \frac{1}{2}\log\left(\frac{\mathcal{A}^2}{2\pi e\sigma^2}\right) + o_{\mathcal{A}}(1). \quad (10.46)$$

If only an average-power constraint ϵ is implemented, then

$$C_{\text{FreeSpace}}(\epsilon) = \frac{1}{2}\log\left(\frac{\epsilon^2}{2\pi e\sigma^2}\right) + o_{\epsilon}(1). \quad (10.47)$$

From the well-known asymptotic behavior of the capacity of the classical AWGN channel under a peak-power constraint [40], the asymptotic capacity (10.46) can be directly determined. By comparing (10.45) and (10.46), it is evident that the asymptotic high-power capacity is reduced by $(1-\alpha)\mu^* + \log(1-\alpha\mu^*)$ when an average-power constraint $\epsilon = \alpha\mathcal{A} < \frac{\mathcal{A}}{2}$ is imposed in addition to a peak-power constraint $\mathcal{A}$. In the same vein, the asymptotic high-power capacity is increased by 1 nat when an average-power constraint is imposed in place of a peak-power constraint, as evidenced by the comparison between (10.45) and (10.46).

10.4.3.2 Low-power Results

Theorem 10.11 ([39]) *If a peak-power constraint $\mathcal{A}$ and an average-power constraint $\epsilon = \alpha\mathcal{A}$ are imposed for a value of $\alpha \in \left(0, \frac{1}{2}\right)$, then*

$$\lim_{\mathcal{A}\downarrow 0} \frac{C_{\text{FreeSpace}}(\mathcal{A}, \alpha\mathcal{A})}{\mathcal{A}^2} = \frac{\alpha(1-\alpha)}{2\sigma^2} \quad (10.48)$$

If both a peak-power constraint $\mathcal{A}$ and an average-power constraint $\epsilon = \alpha\mathcal{A}$ are imposed for some $\alpha \in \left[\frac{1}{2}, 1\right]$, or if only a peak-power constraint $\mathcal{A}$ is imposed (which corresponds to $\alpha = 1$), then

$$\lim_{\mathcal{A}\downarrow 0} \frac{C_{\text{FreeSpace}}(\mathcal{A}, \alpha\mathcal{A})}{\mathcal{A}^2} = \frac{1}{8\sigma^2} \quad (10.49)$$

If only an average-power constraint ϵ is imposed, then

$$\underline{\lim}_{\epsilon\downarrow 0} \frac{C_{\text{FreeSpace}}(\epsilon)}{\frac{\epsilon}{\sigma}\sqrt{\log\left(\frac{\sigma}{\epsilon}\right)}} \geq \frac{1}{\sqrt{2}}, \quad (10.50)$$

$$\overline{\lim}_{\epsilon\downarrow 0} \frac{C_{\text{FreeSpace}}(\epsilon)}{\frac{\epsilon}{\sigma}\sqrt{\log\left(\frac{\sigma}{\epsilon}\right)}} \leq 2. \quad (10.51)$$

The asymptotic upper and lower bounds at low power do not coincide in the sense that their ratio is $2\sqrt{2}$ and not 1, when only an average-power constraint ϵ is imposed.

The well-known asymptotic capacity of the classical AWGN channel with peak-power constraint [40] provides an easy way to derive the expression (10.49).

Binary signaling can be employed to attain asymptotes (10.49), (10.50), and (10.51). The "on" signal is selected as $\mathcal{A}$ in the presence of a peak-power constraint $\mathcal{A}$; in its absence, it is selected as a constant times $\sqrt{\log\left(\frac{\sigma}{\epsilon}\right)}$, which tends to infinity as $\epsilon \downarrow 0$.

The results of (10.49), (10.50), and (10.51) demonstrate that the asymptotic low-power capacity is significantly influenced by the peak-power constraint in the presence of an average-power constraint. In reality, the low-power capacity increases quadratically in ϵ due to a peak-power constraint that trends to 0 proportionally to ϵ. Conversely, the low-power capacity expands sublinearly in the absence of a peak-power constraint, i.e., at a significantly higher rate for small values of ϵ.

Lastly, it is evident from the comparison of (10.48) and (10.49) that the asymptotic low-power capacity is diminished by a factor of $4\alpha(1-\alpha)$ when an average-power constraint $\epsilon = \alpha\mathcal{A} < \frac{\mathcal{A}}{2}$ is imposed in addition to a peak-power constraint $\mathcal{A}$.

10.4.3.3 Nonasymptotic Results

In [41,42], a numerical method was proposed for the efficient computation of channel capacity under a peak-power and/or average-power constraint. The analytic bounds of [39] are presented in this section.

Theorem 10.12 ([39]) *The capacity $C_{\text{FreeSpace}}(\mathcal{A}, \alpha\mathcal{A})$ is lower-bounded by the sum of the average-power constraint $\epsilon = \alpha\mathcal{A}$ and the peak-power constraint $\mathcal{A}$ for some $\alpha \in \left(0, \frac{1}{2}\right)$.*

$$C_{\text{FreeSpace}}(\mathcal{A}, \alpha\mathcal{A}) \geq \frac{1}{2}\log\left(1 + \mathcal{A}^2 \frac{e^{2\alpha\mu^*}}{2\pi e\sigma^2}\left(\frac{1-e^{-\mu^*}}{\mu^*}\right)^2\right), \quad (10.52)$$

and upper-bounded by each of the two bounds

$$C_{\text{FreeSpace}}(\mathcal{A}, \alpha\mathcal{A}) \leq \frac{1}{2}\log\left(1 + \alpha(1-\alpha)\frac{\mathcal{A}^2}{\sigma^2}\right), \quad (10.53)$$

$$\begin{aligned} C_{\text{FreeSpace}}(\mathcal{A}, \alpha\mathcal{A}) \leq & \left(1 - Q\left(\frac{\delta + \alpha\mathcal{A}}{\sigma}\right)\right. \\ & \left. - Q\left(\frac{\delta + (1-\alpha)\mathcal{A}}{\sigma}\right)\right) \\ & \times \log\left(\frac{\mathcal{A}\left(e^{\frac{\mu\delta}{\mathcal{A}}} - e^{-\mu\left(1+\frac{\delta}{\mathcal{A}}\right)}\right)}{\sqrt{2\pi}\mu\sigma\left(1 - 2Q\left(\frac{\delta}{\sigma}\right)\right)}\right) \\ & + \frac{\delta e^{-\frac{\delta^2}{2\sigma^2}}}{\sqrt{2\pi}\sigma} + \frac{\sigma\mu}{\sqrt{2\pi}\mathcal{A}}\left(e^{\frac{\delta^2}{2\sigma^2}} - e^{-\frac{(\mathcal{A}+\delta)^2}{2\sigma^2}}\right) \end{aligned}$$

$$-\frac{1}{2}+Q\left(\frac{\delta}{\sigma}\right)+\mu\alpha\left(1-2Q\left(\frac{\delta+\frac{\mathcal{A}}{2}}{\sigma}\right)\right). \quad (10.54)$$

Here, $\mu > 0$ and $\delta > 0$ are free parameters, μ^* is the unique solution to (10.44), and $Q(\cdot)$ represents the Gaussian Q-function.

$$Q(\xi) \triangleq \int_{\xi}^{\infty} \frac{1}{2\pi} e^{\frac{t^2}{2}} dt - \frac{1}{2}\left(1-erf\left(\frac{\xi}{\sqrt{2}}\right)\right), \quad \forall \xi \in \mathbb{R} \quad (10.55)$$

For the free parameters δ and μ in (10.54), [43] provides a suboptimal but valuable choice. This decision is sufficient to demonstrate that the right-hand side of (10.45) is the upper bound of the asymptotic high-power capacity. The bounds of Theorem 10.12 for $\alpha = 0.1$ are illustrated in Fig. 10.7

Theorem 10.13 ([39]) *If a peak-power constraint $\mathcal{A}$ and an average-power constraint $\epsilon = \alpha\mathcal{A}$ are imposed for some $\alpha \in \left[\frac{1}{2}, 1\right]$, then $C_{\text{FreeSpace}}(\mathcal{A}, \alpha\mathcal{A})$ is lower-bounded by*

$$C_{\text{FreeSpace}}(\mathcal{A}, \alpha\mathcal{A}) \geq \frac{1}{2}\log\left(1+\frac{\mathcal{A}^2}{2\pi e\sigma^2}\right), \quad (10.56)$$

and upper-bounded by each of the two bounds

$$C_{\text{FreeSpace}}(\mathcal{A}, \alpha\mathcal{A}) \leq \frac{1}{2}\log\left(1+\frac{\mathcal{A}^2}{4\sigma^2}\right), \quad (10.57)$$

$$C_{\text{FreeSpace}}(\mathcal{A}, \alpha\mathcal{A}) \leq \left(1-2Q\left(\frac{\delta+\frac{\mathcal{A}}{2}}{\sigma}\right)\right) \log\left(\frac{\mathcal{A}+2\delta}{\sqrt{2\pi}\sigma(1-2Q(\frac{\delta}{\sigma}))}\right)-\frac{1}{2} + Q\left(\frac{\delta}{\sigma}\right)+\frac{\delta e^{-\frac{\delta^2}{2\sigma^2}}}{\sqrt{2\pi}\sigma}, \quad (10.58)$$

where $\delta > 0$ is a free parameter.

A suboptimal but valuable choice for the free parameter $\delta > 0$ in (10.58) is provided in [39]. This decision is sufficient to demonstrate that the right-hand side of (10.27) is the upper bound of the asymptotic high-power capacity.

The bounds of Theorem 10.13 are illustrated in Fig. 10.8, where the upper bound (10.58) is numerically minimized over $\delta > 0$. The maximum disparity between the upper and lower bounds is numerically identified as ≈0.50 nats (for $\frac{\mathcal{A}}{\sigma} \approx 6.4$ dB).

Theorem 10.14 ([39]) *The capacity $C_{\text{FreeSpace}}(\epsilon)$ is lower-bounded by the average-power constraint ϵ:*

$$C_{\text{FreeSpace}}(\epsilon) \geq \frac{1}{2}\log\left(1+\frac{\epsilon^2 e}{2\pi\sigma^2}\right), \quad (10.59)$$

and upper-bounded by each of the bounds

$$C_{\text{FreeSpace}}(\epsilon) \leq \log\left(\beta E^{-\frac{\delta^2}{2\sigma^2}}+\sqrt{2\pi}\sigma Q\left(\frac{\delta}{\sigma}\right)\right) - \log(\sqrt{2\pi}\sigma)-\frac{\delta\epsilon}{2\sigma^2} \frac{\delta^2}{2\sigma^2}\left(1-Q\left(\frac{\delta}{\sigma}\right)-\frac{\epsilon}{\sigma}Q\left(\frac{\delta}{\sigma}\right)\right) +\frac{1}{\beta}\left(\epsilon+\frac{\sigma}{sqrt2\pi}\right), \quad (10.60)$$

$$C_{\text{FreeSpace}}(\epsilon) \leq \log\left(\beta E^{-\frac{\delta^2}{2\sigma^2}}+\sqrt{2\pi}\sigma Q\left(\frac{\delta}{\sigma}\right)\right) +\frac{1}{2}Q\left(\frac{\delta}{\sigma}\right)+\frac{\delta}{2\sqrt{2\pi}\sigma}e^{-\frac{\delta^2}{2\sigma^2}} \frac{\delta^2}{2\sigma^2}\left(1-Q\left(\frac{\delta+\epsilon}{\sigma}\right)\right)+\frac{1}{\beta}\left(\delta+\epsilon+\frac{\sigma}{\sqrt{2\pi}}e^{-\frac{\delta^2}{2\sigma^2}}\right) -\frac{1}{2}\log(2\pi e\sigma^2), \quad (10.61)$$

where $\beta > 0$ and δ are free parameters. The bound (10.61) is only valid for $\delta \leq -\sigma e^{-\frac{1}{2}}$, whereas the bound (10.61) is only valid for $\delta \geq 0$.

Lapidoth and Moser [39] provides a suboptimal but useful choice for the free parameters β and δ in (10.61), and [39] provides a suboptimal choice for the free parameters β and δ in (10.34. The first option is sufficient to demonstrate that the asymptotic low-power capacity is upper-bounded, as indicated in (10.51), while the second option is sufficient to demonstrate that the asymptotic high-power capacity is upper-bounded by the right-hand side of (10.47).

The bounds of Theorem 10.14 are illustrated in Fig. 10.9 when the upper bounds (10.61) and (10.34) are numerically minimized over the permitted values of β and δ. The maximum difference between the upper and lower bounds is numerically identified as ≈0.57 nats (for $\frac{\epsilon}{\sigma} \approx 2.8$ dB).

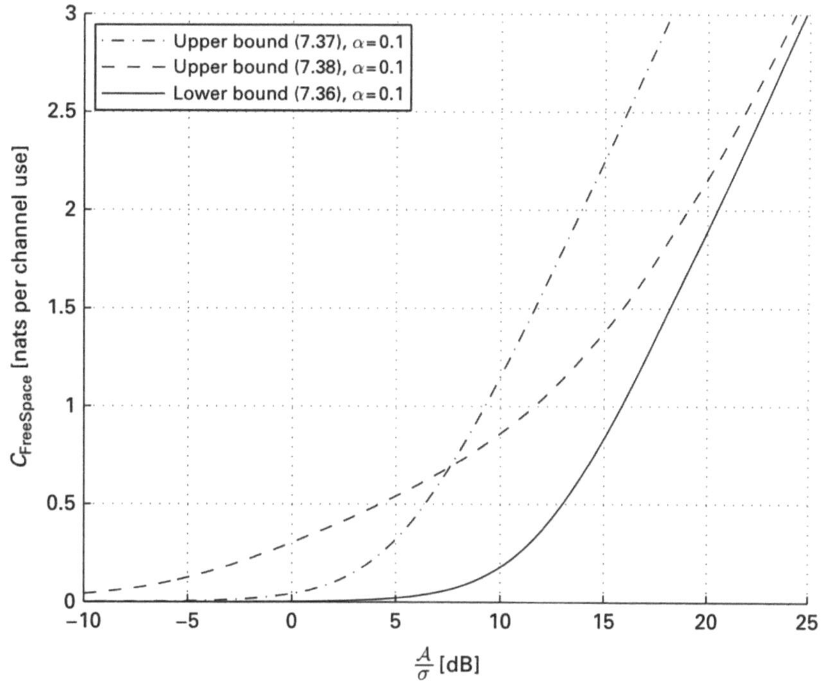

Fig. 10.7 Illustration of Theorem 10.12 for $\alpha = 0.1$ [22]

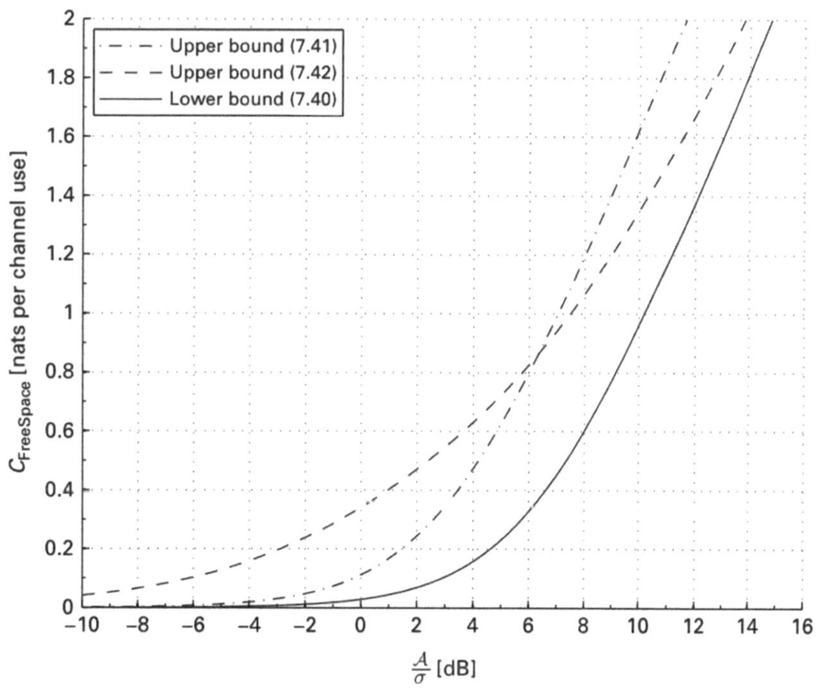

Fig. 10.8 Illustration for bounds of Theorem 10.13 for $\alpha \in [0.5, 1]$

10.5 Summary

This chapter studies the multiuser diversity utilized in optical wireless networks. We first discuss the various diversity schemes utilized in wireless systems, such as space diversity, frequency diversity, time diversity, and polarization. Further, the schemes are extended to space diversity schemes for indoor optical wireless systems. Moreover, the capacity in optical wireless multiuser systems like discrete-time Poisson channels, optical intensity channels with input-dependent Gaussian noise, and free-space optical intensity channels are also discussed in this chapter.

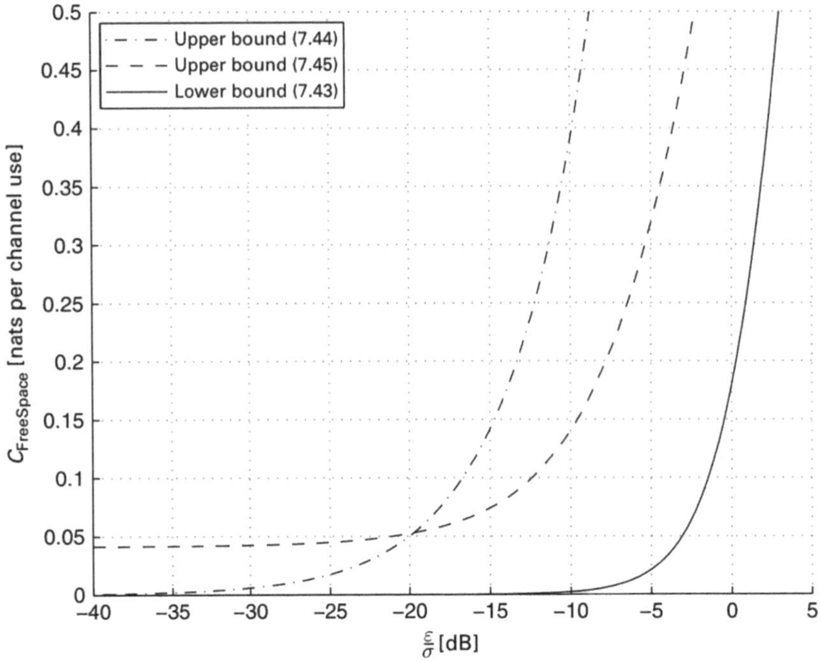

Fig. 10.9 Illustration for bounds of Theorem 10.14

Homework Problems

1. What is diversity? What are the advantages of diversity? Discuss the different types of diversity schemes.
2. Consider the potential model for MIMO for diffuse OW system with the root-mean-square (RMS) delay spread of channel $h(t)$ be 5 ns. Find the 3-dB cutoff frequency of the channel using the exponential model.
3. Explain the spot-diffusing configuration for OW MIMO systems.
4. Derive the expression for the capacity of the discrete-time Poisson channel.
5. Mention the constraints to find the upper limit for discrete-time Poisson channel considering the dark current $\lambda = 0$.
6. Derive the expression high-power results on the capacity of free-space optical intensity channel.
7. Derive the expression upper-bound and lower-bound capacity of free-space optical intensity channel under average-power constraint.

References

1. Goldsmith, A.: Wireless Communications. Cambridge University Press, Cambridge (2005)
2. Lee, W.C.Y.: Mobile Communications Engineering. McGraw-Hill, New York (1982)
3. Winters, J.: Signal acquisition and tracking with adaptive arrays in the digital mobile radio system is-54 with flat fading. IEEE Trans. Veh. Tech. **42**(4), 377–384 (1993)
4. Takase, D., Ohtsuki, T.: Spatial multiplexing in optical wireless mimo communications over indoor environment. IEICE Trans. Commun. **89**(4), 1364–1371. (2006)
5. O'Brien, D.: Multi-input multi-output (mimo) indoor optical wireless communications. In: 2009 Conference Record of the Forty-Third Asilomar Conference on Signals, Systems and Computers, pp. 1636–1639 (2009)
6. O'Brien, D.: Optical multi-input multi-output systems for short-range free-space data transmission. In: 2010 7th International Symposium on Communication Systems, Networks & Digital Signal Processing (CSNDSP 2010), pp. 517–521 (2010)
7. Zeng, L., O'Brien, D.C., Minh, H.L., Faulkner, G.E., Lee, K., Jung, D., Oh, Y., Won, E.T.: High data rate multiple input multiple output (mimo) optical wireless communications using white led lighting. IEEE J. Sel. Areas Commun. **27**(9), 1654–1662 (2009)
8. IEC 60825-1, safety of laser products—part 1: equipment classification, requirements (2001)
9. Boucouvalas, A.: IEC 825-1 eye safety classification of some consumer electronic products. In: IEE Colloquium on Optical Free Space Communication Links, pp. 13/1–13/6 (1996)
10. Chu, T.S., Gans, M.: High speed infrared local wireless communication. IEEE Commun. Mag. **25**(8), 4–10 (1987)
11. Kahn, J., Barry, J.: Wireless infrared communications. Proc. IEEE **85**(2), 265–298 (1997)
12. Hubbard, W.: The approximation of a poisson distribution by a gaussian distribution. Proc. IEEE **58**(9), 1374–1375 (1970)
13. A photon modeling method for the characterization of indoor optical wireless communication. Progr. Electromagnet. Res. **92**, 121–136 (2009)
14. Gfeller, F., Bapst, U.: Wireless in-house data communication via diffuse infrared radiation. Proc. IEEE **67**(11), 1474–1486 (1979)
15. Barry, J., Kahn, J., Krause, W., Lee, E., Messerschmitt, D.: Simulation of multipath impulse response for indoor wireless optical channels. IEEE J. Sel. Areas Commun. **11**(3), 367–379 (1993)
16. Hranilovic, S.: Wireless Optical Communication Systems. Springer, Berlin (2004)

17. Mohamed, M.D.A.: Modulation and detection in wireless optical channels using temporal and spatial degrees of freedom (2010). Ph.D. Dissertation, Department of Electrical and Computer Engineering
18. Kahn, J., Barry, J., Audeh, M., Carruthers, J., Krause, W., Marsh, G.: Non-directed infrared links for high-capacity wireless lans. IEEE Pers. Commun. **1**(2), 12– (1994)
19. Kahn, J., Krause, W., Carruthers, J.: Experimental characterization of non-directed indoor infrared channels. IEEE Trans. Commun. **43**(2/3/4), 1613–1623 (1995)
20. Alqudah, Y., Kavehrad, M.: MIMO characterization of indoor wireless optical link using a diffuse-transmission configuration. IEEE Trans. Commun. **51**(9), 1554–1560 (2003)
21. Yun, G., Kavehrad, M.: Spot-diffusing and fly-eye receivers for indoor infrared wireless communications. In: 1992 IEEE International Conference on Selected Topics in Wireless Communications, pp. 262–265 (1992)
22. Arnon, S.: Advanced Optical Wireless Communication Systems. Cambridge University Press, Cambridge (2012)
23. Jivkova, S., Kavehrad, M.: Receiver designs and channel characterization for multi-spot high-bit-rate wireless infrared communications. IEEE Trans. Commun. **49**(12), 2145–2153 (2001)
24. Jovkova, S., Kavehard, M.: Multispot diffusing configuration for wireless infrared access. IEEE Trans. Commun. **48**(6), 970–978 (2000)
25. Jivkova, S., Kavehrad, M.: Shadowing and blockage in indoor optical wireless communications. In: GLOBECOM '03. IEEE Global Telecommunications Conference (IEEE Cat. No.03CH37489), vol. 6, pp. 3269–3273 (2003)
26. Khozeimeh, F., Hranilovic, S.: Dynamic spot diffusing configuration for indoor optical wireless access. IEEE Trans. Commun. **57**(6), 1765–1775 (2009)
27. Carruthers, J., Kahn, J.: Modeling of nondirected wireless infrared channels. IEEE Trans. Commun. **45**(10), 1260–1268 (1997)
28. Cover, T.M., Thomas, J.A.: Elements of Information Theory, 2nd edn. John Wiley & Sons, London (2006)
29. Lapidoth, A., Moser, S.M.: On the capacity of the discrete-time poisson channel. IEEE Trans. Inform. Theory **55**(1), 303–322 (2009)
30. Brady, D., Verdú, S.: The asymptotic capacity of the direct detection photon channel with a bandwidth constraint.In: Proceedings of the 28th Allerton Conference on Communication, Control and Computing, Monticello, IL, USA (1990)
31. Lapidoth, A., Shapiro, J.H., Venkatesan, V., Wang, L.: The discrete-time poisson channel at low input powers. IEEE Trans. Inform. Theory **57**(6), 3260–3272 (2011)
32. Martinez, A.: Spectral efficiency of optical direct detection. J. Opt. Soc. Am. B **24**(4), 739–749 (2007). https://opg.optica.org/josab/abstract.cfm?URI=josab-24-4-739
33. Moser, S.M.: Capacity results of an optical intensity channel with input-dependent gaussian noise. IEEE Trans. Inform. Theory **58**(1), 207–223 (2012)
34. Ho, K.P.: Phase-Modulated Optical Communication Systems. Springer, Berlin (2005)
35. Kahn, J., Barry, J.: Wireless infrared communications. Proc. IEEE **85**(2), 265–298 (1997)
36. Katz, M., Shamai, S.: On the capacity-achieving distribution of the discrete-time non-coherent and partially-coherent AWGN channels. In: The 22nd Convention on Electrical and Electronics Engineers in Israel, 2002, pp. 148–150 (2002)
37. Lapidoth, A.: On phase noise channels at high SNR. In: Proceedings of the IEEE Information Theory Workshop, pp. 1–4 (2002)
38. Hranilovic, S., Kschischang, F.: Capacity bounds for power- and band-limited optical intensity channels corrupted by gaussian noise. IEEE Trans. Inform. Theory **50**(5), 784–795 (2004)
39. Lapidoth, A., Moser, S.M.: On the capacity of the discrete-time poisson channel. IEEE Trans. Inform. Theory **55**(1), 303–322 (2009)
40. Shannon, C.E.: A mathematical theory of communication. Bell Syst. Tech. J. **27**(3), 379–423 (1948)
41. Farid, A.A., Hranilovic, S.: Capacity of optical intensity channels with peak and average power constraints. In: 2009 IEEE International Conference on Communications, pp. 1–5 (2009)
42. Farid, A.A., Hranilovic, S.: Channel capacity and non-uniform signalling for free-space optical intensity channels. IEEE J. Sel. Areas Commun. **27**(9), 1553–1563 (2009)
43. Lapidoth, A., Moser, S.M., Wigger, M.A.: On the capacity of free-space optical intensity channels. IEEE Trans. Inform. Theory **55**(10), 4449–4461 (2009). https://doi.org/10.1109/TIT.2009.2027522

IEEE 802 and ITU-T Optical Wireless Standards 11

11.1 Introduction

In the preceding chapters, we examined various multiple-access technologies and categorized multiuser diversity techniques. However, we have not yet discussed the established standards that govern the operation of communication networks. In this chapter, we will explore the development and key features of various wireless and wired communication standards. In communication systems, it is essential to standardize the interface that governs information exchange between the networked nodes, especially when dealing with products from different manufacturers. The primary organizations responsible for such standardization include the European Telecommunication Standardization Institute (ETSI), the IEEE, and the ITU. The ITU is divided into ITU-T, which focuses on wired communication, and ITU-R, which specializes in radio communication systems. ETSI has been instrumental in developing standards such as Global system for mobiles (GSM) for wireless telephony and Universal mobile telecommunications system (UMTS) for mobile broadband. In the field of local area networks, ETSI introduced Hiperlan (High-Performance local area network (LAN)), offering data rates up to 2 Mbit/s, and later Hiperlan/2, which increased data rates to 54 Mbit/s. The IEEE, on the other hand, introduced the 802 series of standards, which define various communication networks. This series is maintained by multiple working groups that address different types of networks, with IEEE 802 primarily encompassing standards for the physical and data link layers of technologies such as Ethernet and wireless communication.

The IEEE 802 committee's work is currently organized into the following working groups:

- **802.1 Higher Layer LAN Protocols Working Group:** Tasked with developing standards and recommended practices in the following areas: 802 LAN/Metropolitan Area Network (MAN) architecture, internetworking between 802 LANs, MANs, and other wide area networks, overall network management for 802, and protocol layers above the medium access control (MAC) and Logical Link Control (LLC) layers.
- **802.2 LLC Working Group:** Currently inactive.
- **802.3 Ethernet Working Group:** Responsible for developing standards for LANs based on carrier sense multiple access with collision detection (CSMA/CD) (Ethernet) technology.
- **802.4 Token Bus Working Group:** Responsible for developing standards for token bus LANs. Currently inactive.
- **802.5 Token Ring Working Group:** Responsible for developing standards for token ring LANs.
- **802.6 Metropolitan Area Network Working Group:** Responsible for developing standards for MANs. Currently inactive.
- **802.7 Broadband TAG:** This technical advisory group developed the IEEE Std 802.7-1989, titled "IEEE Recommended Practices for Broadband Local Area Networks" (reaffirmed in 1997). The group is currently inactive and has no ongoing projects. Maintenance of IEEE Std 802.7-1989 is now supported by 802.14.
- **802.8 Fiber Optic TAG:** Responsible for developing recommended practices for fiber optics.
- **802.9 Isochronous LAN Working Group:** Responsible for developing standards for isochronous LANs. Currently inactive.
- **802.10 Security Working Group:** Responsible for developing standards for interoperable LAN/MAN security.
- **802.11 Wireless LAN Working Group:** Responsible for developing standards for wireless LANs.
- **802.12 Demand Priority Working Group:** Responsible for developing standards for demand priority LANs (called 100VG-AnyLAN; this type of LAN is no longer used).

V. A. Bohara, Swaminathan R, *Optical and Wireless Access Networks*, Textbooks in Telecommunication Engineering, https://doi.org/10.1007/978-3-031-89696-5_11

- **802.13:** Currently not used. Reserved for Fast Ethernet development.
- **802.14 Cable Modem Working Group:** Tasked with creating standards for data transport over traditional cable TV networks, this group focuses on a hybrid fiber/coax architecture with an 80 km radius from the head end. The primary goal is to design a network protocol that supports IEEE 802.2 LLC traffic, such as Ethernet. While it no longer exists as an active group, it played a crucial role in the early development of cable modem technology.
- **802.15 Wireless Personal Area Networks Working Group:** Responsible for developing personal area network standards for short-distance wireless networks.
- **802.16 Broadband Wireless Access Working Group:** Responsible for developing standards for broadband wireless access.
- **802.17 Resilient Packet Ring Working Group:** The working group is responsible for standardizing the Resilient Packet Ring (RPR) architecture for use in metropolitan and wide area networks. RPR is a network topology based on fiber optic rings, designed to efficiently and resiliently transmit data packets at high speeds.
- **802.18 Radio Regulatory:** The IEEE 802.18 Radio Regulatory Technical Advisory Group (RR-TAG) supports the IEEE 802 LAN/MAN Standards Committee (LMSC) and the IEEE 802 wireless Working Groups—IEEE 802.11 (WLAN), IEEE 802.15 (WSN), IEEE 802.16 (WMAN), IEEE 802.20 (Wireless Mobility), IEEE 802.21 (Handoff/Interoperability Between Networks), and IEEE 802.22 (WRAN). The RR-TAG actively monitors and engages in radio regulatory issues worldwide, advocating on behalf of IEEE 802.
- **802.19 Wireless Coexistence:** IEEE 802.19 is the Wireless Coexistence Working Group within the IEEE 802 LAN/MAN Standards Committee. IEEE 802.19 develops standards to ensure coexistence between wireless standards for unlicensed devices. The group reviews coexistence assurance (CA) documents created by working groups that develop new wireless standards for unlicensed devices. This standard outlines radio technology-independent methods for network-based coexistence among dissimilar or independently operated networks of unlicensed devices. It is applicable to geolocation-capable devices operating under the general authorization, such as TV band white spaces, the 5 GHz license-exempt bands, and the general authorized access in the 3.5 GHz band.
- **802.20 Mobile Broadband Wireless:** It was proposed to develop mobile broadband networks. It is currently disbanded.

11.2 Services and Protocols

A protocol is a convention or standard that governs the connection, communication, and data transfer between communication devices. Essentially, a protocol consists of rules that define the syntax, semantics, and synchronization of communication. Protocols can be implemented through hardware, software, or a combination of both. At its most basic level, a protocol specifies how hardware connections should behave.

Although protocols can differ widely in purpose and complexity, most define one or more of the following properties:

- Detection of the underlying physical connection (wired or wireless), or the existence of the other endpoint or node
- Handshaking (dynamically setting parameters of a communications channel)
- Negotiation of various connection characteristics
- How to start and end a message
- How to format a message
- What to do with corrupted or improperly formatted messages (error correction)
- How to detect unexpected loss of the connection, and what to do next
- Termination of the session and/or connection.

11.2.1 Switching Level Convergence Protocol

There are three major forms of switching techniques used in communication and networking:

- Circuit Switching
- Packet Switching
- Cell Switching

Circuit switching is connection-oriented. Before information transfer begins, a dedicated circuit with two channels for two-way communication is established between the communicating entities. This circuit is created using dedicated physical resources, which may include copper wires, optical fibers, radio or satellite links, or a combination of these media. The primary advantage of circuit switching is that once the circuit is set up, there is a direct, transparent connection between the entities, ensuring smooth and orderly information flow. The information is delivered in the correct sequence, with no risk of out-of-sequence delivery [1].

In packet switching, messages are divided into packets at the source and sent through the network to their destination. Packet-switched networks use two main approaches for transporting packets: the Datagram approach and the Virtual Circuit approach. The Virtual Circuit approach aims to combine the in-sequence delivery benefits of circuit switching with more efficient use of physical resources. Instead of establishing a dedicated circuit, it establishes a fixed route from the source to the destination. Since packets follow this predetermined route, they are delivered in order, eliminating the need for re-sequencing [2].

Cell switching utilizes the Virtual Circuit principle, ensuring that cells are delivered in sequence. This method significantly reduces switching overhead and accelerates switching processes, earning it the nickname "fast packet switching." Networks that use cell switching are known as Asynchronous Transfer Mode (ATM) networks. Unlike fixed-time switching, ATM networks handle cells as they arrive, which can be from different messages or services. This asynchronous transfer allows for efficient resource utilization, as network resources are shared among multiple services rather than being dedicated to a single one [3].

11.2.2 Transmission Control Protocol/Internet Protocol (TCP/IP)

The Internet Protocol Suite, commonly known as TCP/IP, is the set of communication protocols used for the Internet and similar networks. It can be viewed as a layered architecture, with each layer addressing specific aspects of data transmission and providing well-defined services to the layers above, while relying on services from the layers below. The suite consists of four layers, listed from lowest to highest: the Link Layer, the Internet Layer, the Transport Layer, and the Application Layer.

Efforts to map the Internet Protocol Suite to the seven-layer OSI Model generally result in the following correspondences: the TCP/IP Link Layer aligns with the OSI Data Link and Physical layers, the Internet Layer corresponds to the OSI Network Layer, and the Transport Layer maps directly to OSI Layer 4. The OSI Application, Presentation, and Session Layers are combined into the TCP/IP Application Layer.

TCP, a connection-oriented service, operates at a higher level than IP and ensures reliable and error-free communication. It handles duplicate datagrams by accepting only the first error-free copy and uses acknowledgment and timer mechanisms to detect lost datagrams. Each datagram received is acknowledged by the destination.

IP handles the signaling and routing functions necessary for transmitting and receiving data across networks. Signaling protocols include Multiprotocol Label Switching (MPLS) and Generalized Multiprotocol Label Switching (GMPLS), while routing protocols consist of Open Shortest Path First (OSPF), Intermediate System-to-Intermediate System (IS-IS), and Border Gateway Protocol (BGP).

11.2.3 User Datagram Protocol (UDP)

User Datagram Protocol (UDP) provides a connectionless service at the user level, utilizing IP for its operations. In this context, UDP operates at a higher level compared to the Internet Protocol (IP). A user sends their entire message to UDP with a request for delivery to the specified destination. The user message serves as the payload for UDP. UDP encapsulates this payload with its own header and forwards it to IP as a payload. User datagrams are distinct from IP datagrams and do not adhere to the IP standard; they are simply chunks of information of any size. UDP can encapsulate a user datagram with its own header to create a UDP datagram, and it may split a user datagram into multiple UDP datagrams to comply with IP standards. As a connectionless service, UDP operates on a best-effort basis, meaning it does not provide delivery acknowledgments or guarantees of delivery.

11.2.4 Address Resolution Protocol (ARP)

The Address Resolution Protocol (ARP) is a communication protocol essential for resolving Internet layer addresses into link layer addresses within the Internet protocol suite. ARP also refers to the program used for managing these addresses in most operating systems.

ARP maps a network address, such as an IPv4 address, to a physical address like an Ethernet address (also known as a MAC address). It has been implemented across various network and data link layer technologies, including IPv4, Chaosnet, DECnet, and Xerox PARC Universal Packet (PUP) using IEEE 802 standards, FDDI, X.25, Frame Relay, and ATM, with IPv4 over IEEE 802.3 and IEEE 802.11 being the most common use cases. In IPv6 networks, the ARP functionality is replaced by the Neighbor Discovery Protocol (NDP), which performs similar address resolution tasks. Generally, ARP is used to translate IP addresses to MAC addresses.

11.2.5 FTP (File Transfer Protocol)

FTP (File Transfer Protocol) is a standard Internet protocol used to transfer files between computers over TCP/IP connections. It operates on a client-server model, using two communication channels: a command channel to control the session and a data channel to transfer file content. Clients initiate sessions by requesting file downloads from the servers.

FTP supports two modes: active and passive. In active mode, after the client initiates a session via the command channel, the server establishes a data connection back to the client to transfer data. In passive mode, the server provides the client with the necessary information to open a data channel through the command channel. FTP is compatible with firewalls and Network Address Translation (NAT) gateways.

11.2.6 Internet Control Message Protocol (ICMP)

The Internet Control Message Protocol (ICMP) is used to provide feedback for diagnostics and to report logical errors. One of the most common uses of ICMP is the ping command. ICMP was designed to be a useful and informative protocol, but it is often exploited by hackers. They may use ICMP to execute attacks such as sending ping of death packets, creating Smurf DoS attacks, querying system timestamps or netmasks, or even sending ICMP type 5 packets to redirect traffic.

11.2.7 Dynamic Host Configuration Protocol (DHCP)

This protocol helps users configure multiple network devices from a single source. It is designed to help manage and configure multiple networks efficiently. This facilitates the management and maintenance of large networks, ensuring that devices can communicate effectively without network setting conflicts. DHCP is essential in modern networks, as it streamlines the process of connecting devices and efficiently managing network resources.

11.2.8 Hypertext Transfer Protocol Secure (HTTPS)

HTTPS is the secure version of HTTP, widely used to access website content. It enhances HTTP by incorporating SSL or TLS encryption, which provides a secure connection. When HTTPS is in use, the URL begins with "https," and a padlock icon appears in the browser's address bar. HTTPS is the global standard for secure online transactions, including payment processing and other data-sensitive activities.

11.2.9 Real-time Transport Protocol (RTP)—Voice over Internet Protocol (VoIP)

RTP (Real-time Transport Protocol) is a network protocol designed for delivering audio and video over IP networks. It is used in communication and entertainment systems that involve streaming media, including telephony, video conferencing applications like WebRTC, television services, and web-based push-to-talk features.

11.3 OSI Networking Reference model

The Open Systems Interconnection (OSI) model provides a standardized framework for designing networks [4]. Jointly specified by the International Organization for Standardization (ISO) and ITU-T, the OSI model defines seven layers, each with specific functions as shown in Fig. 11.1. At the top

	Layers	Function	Data Unit	
7	Application	Network process to application	Data	Host Layers
8	Presentation	Data representation and encryption		
5	Session	Interhost communication		
4	Transport	End-to-end connection and reliability	Segments	
3	Network	Path determination and logical addressing	Packets	Media Layers
2	Data Link	Physical Addressing	Frames	
1	Physical	Media, signal, and binary transmission	Bits	

Fig. 11.1 Description of the layers in the OSI reference model [5]

of this hierarchy is the Application Layer (Layer 7), which allows users to access and interact with network services, such as database management and network management, through applications. Below it, the Presentation Layer (Layer 6) translates data to offer a standardized interface for the Application Layer. The Session Layer (Layer 5) manages the dialogues (sessions) between devices, establishing, controlling, and terminating connections between local and remote applications. It also supports duplex or half-duplex operation and handles checkpointing, suspension, termination, and restart procedures. The top three layers—5, 6, and 7—are considered data-driven levels, focusing on data interaction and management.

The fourth layer, known as the Transport Layer, ensures the transparent transfer of data between the end users, freeing the upper layers from concerns related to data transport while providing reliable data transfer. This layer manages the reliability of communication through flow control, segmentation and reassembly, and error correction. Some protocols at this layer are stateful and connection-oriented, meaning they can track packets and retransmit those that fail. A well-known example of a Layer 4 protocol is the (TCP), which is connection-oriented and offers basic data transfer services between nodes, including interconnection, flow control, multiplexing, and reliability with priority and security [6, 43].

The third layer, the Network Layer, handles the transfer of variable-length data sequences from a source to a destination across one or more networks, ensuring the QoS requested by the Transport Layer. This layer is responsible for network routing and may also manage segmentation and reassembly, as well as report delivery errors. Routers operate at this layer to facilitate data transmission across extended networks, enabling Internet connectivity.

The remaining two layers are the Data Link Layer and the Physical Layer. The Data Link Layer provides the mechanisms for transferring data between network entities and detecting and correcting errors that occur in the Physical Layer. It is where network bridges and switches operate, forming Layer 2 domains for unicast or broadcast forwarding among locally attached network nodes. Protocols at this layer include CSMA/CD used in Ethernet, as well as High-level Data Link Control (HDLC) and Advanced Data Communications Control Procedure (ADCCP) for point-to-point or packet-switched networks [7].

The Physical Layer is the lowest layer in the OSI model hierarchy. It defines all electrical, optical, and physical media specifications for devices, such as fiber types, cable specifications, hubs, repeaters, and network adapters. This layer establishes and terminates connections to the communications medium and ensures efficient sharing of communication resources among multiple users. It also handles contention resolution and flow control and converts data between the equipment and the signal requirements of the communications channel.

11.4 Wireless LAN (IEEE 802.11)

The IEEE 802.11 standard, first introduced in 1999, has become the most widely deployed wireless communication standard and has undergone numerous extensions, with many more in development [8, 9]. Initially designed for home and office wireless local area connectivity, the original standards offered a maximum data rate of 2 Mbps per access point (AP). This rate increased to 11 Mbps with the IEEE 802.11b extension. Subsequent standards, such as IEEE 802.11g and IEEE 802.11a, further boosted the maximum data rate to 54 Mbps per AP using various techniques.

WLAN devices based on IEEE 802.11g typically offer data rates of 100–125 Mbps, while the newer IEEE 802.11n standard achieves a maximum data rate of around 540 Mbps. Subsequent standards addressed many QoS and security issues from earlier versions. For instance, IEEE 802.11e and IEEE 802.11i introduced mechanisms for improved QoS support and security. IEEE 802.11n also included MAC enhancements to address limitations in previous standards, while IEEE 802.11s added support for mesh network topologies. IEEE 802.11u improved connectivity with external non-802.11 networks, and IEEE 802.11w enhanced management security.

The IEEE 802.11ad standard introduced "fast session transfer," allowing seamless transitions between the 2.4 GHz and 5 GHz bands and the 60 GHz frequency band. Meanwhile, IEEE 802.11ac increased multi-station WLAN throughput to at least 1 Gbps and single-link throughput to at least 500 Mbps.

Despite these advancements, the original 2.4 GHz frequency band faced high interference from other devices operating on the same frequency, which limited the distance between mobile users and APs and constrained wireless network topology with a maximum link bandwidth of 11 Mbps.

11.4.1 IEEE 802.11a

Launched shortly after IEEE 802.11b, this standard employs orthogonal frequency division multiplexing (OFDM) and operates at a 5 GHz frequency. This frequency offers several benefits, including reducing the interference seen in earlier standards [10]. The new technology enhances spectral efficiency without the need for complex equalization, thanks to adaptive filtering, and also minimizes multipath effects and narrowband interference [11].

11.4.2 IEEE 802.11b

The IEEE 802.11b standard was designed to support wireless communications between mobile nodes and access points (APs). It provides specifications for the PHY and MAC layers, utilizing carrier-sense multiple access with collision avoidance (CSMA/CA).

11.4.3 IEEE 802.11c

The IEEE 802.11c standard addresses bridge operations, which are simpler and more efficient than IP routers. Completed in 2003 by the 802.11c task group, this standard was later incorporated into the IEEE 802.1d standard for LAN bridges. IEEE 802.11c served as a supplement to IEEE 802.1d, adding specific requirements for bridging 802.11 wireless client devices.

11.4.4 IEEE 802.11d

Introduced in 2001, IEEE 802.11d is an amendment to the IEEE 802.11 specification that extends support for additional regulatory domains [12]. It addresses regulatory differences across various countries and defines operational parameters for the 5 MHz frequency band, specifically for North America, Europe, and Japan. The amendment also includes the addition of a country information element to beacons, probe requests, and probe responses.

11.4.5 IEEE 802.11e

Deployed in 2005, IEEE 802.11e is a significant amendment to the IEEE 802.11 standard that introduced QoS enhancements for wireless LAN applications by modifying the Media Access Control (MAC) layer [13]. IEEE 802.11e supports time-scheduled and polled communication during idle periods when no other data are being transmitted. It also enhances polling efficiency and channel robustness. These improvements are designed to ensure the quality needed for services like IP telephony and video streaming.

11.4.6 IEEE 802.11f

The IEEE 802.11f, or Inter-Access Point Protocol, was an optional extension to the IEEE 802.11 standard designed to enable communication among wireless access points from different vendors. It also facilitated seamless roaming of a device from one access point to another, ensuring uninterrupted transmission.

11.4.6.1 IEEE 802.11g

Designed to provide enhanced simultaneous response capabilities, it employs OFDM technology[20, 23]. Although it operated exclusively in the 2.4 GHz band, raising concerns about interference, it increased bandwidth to 54 Mbps and maintained support for legacy standards [24].

11.4.7 IEEE 802.11h

The European Union Military utilizes part of the 5 GHz frequency band for satellite and radar communications, alongside IEEE 802.11a signals. To manage these situations, IEEE 802.11h introduced two mechanisms: Dynamic Frequency Selection (DFS) and Transmit Power Control (TPC). DFS allows the AP to detect other networks operating in the same frequency band and adjust the WLAN's operating frequency to avoid interference. Meanwhile, TPC helps ensure that the signal level remains below a preset threshold to prevent interference with nearby satellite signals [14].

11.4.8 IEEE 802.11i

The IEEE 802.11i standard was designed to address security issues by implementing the Advanced Encryption Standard (AES) block cipher. This contrasts with Wired Equivalent Privacy (WEP) and Wi-Fi Protected Access (WPA), which use the RC4 stream cipher [15].

11.4.9 IEEE 802.11j

The IEEE 802.11j standard enables Wireless LAN operation in the 4.9–5 GHz band, ensuring compliance with Japanese regulations for radio operation across indoor, outdoor, and mobile applications.

11.4.10 IEEE 802.11k

IEEE 802.11k enhances radio resource management by providing mechanisms accessible to protocol layers above the physical layer. Additionally, it extends the transmit-power control procedures defined in the IEEE 802.11h standard to other regulatory domains and frequency bands, aiming to reduce power consumption, minimize interference, and manage range control [11].

11.4.11 IEEE 802.11m

IEEE 802.11m is an ongoing initiative aimed at providing a unified perspective on the IEEE 802.11 base standards through continuous monitoring, management, and maintenance.

11.4.12 IEEE 802.11n

IEEE 802.11n introduced new MAC layer mechanisms to enhance throughput. The standard employs MIMO technology and various modulation and coding techniques to boost data rates [16]. It operates with a fixed channel bandwidth of 20 MHz, ensuring backward compatibility with older standards, and also offers an optional 40 MHz channel. To achieve data rates up to 600 Mbps, the standard requires a maximum of four spatial streams and one 40 MHz-wide channel [17].

11.4.13 IEEE 802.11p

IEEE 802.11p is an approved amendment to the IEEE 802.11 standard that introduces enhancements needed to support Intelligent Transportation Systems (ITS) applications. It facilitates data exchange between high-speed passenger vehicles, ambulances, and roadside infrastructure within the licensed ITS band of 5.9 GHz (5.85–5.925 GHz) [18].

11.4.14 IEEE 802.11r

IEEE 802.11r facilitates secure and rapid handoffs between stations (STAs) for wireless connectivity. This standard supports fast roaming, even for moving vehicles. It also improves the mobile client transition process between access points (APs) by redefining the security key negotiation protocol, allowing for more efficient negotiations and requests for wireless resources [19].

11.4.15 IEEE 802.11s

IEEE 802.11s is an IEEE 802.11 standard for mesh networking. It outlines how wireless devices can interconnect to form a WLAN mesh network, suitable for both static topologies and ad-hoc networks. A network device based on IEEE 802.11s is referred to as a mesh STA. Mesh STAs establish mesh links with each other, enabling the creation of mesh paths using a routing protocol. IEEE 802.11s specifies the Hybrid Wireless Mesh Protocol (HWMP) as the default mandatory routing protocol [20]. However, it also permits the use of alternative protocols by vendors. HWMP is derived from a combination of AODV and tree-based routing approaches. IEEE 802.11s also provides mechanisms for deterministic network access, a framework for congestion control and power save [21].

11.4.16 IEEE 802.11u

The IEEE 802.11u standard was developed to enhance internetworking with external non-802.11 networks. It achieves this through network discovery and selection, as well as the distribution of QoS maps. The QoS mechanism maps the IP differentiated services code point (DSCP) to Layer 2 priority on a per-device basis, enabling end-to-end QoS over the air [22].

11.4.17 IEEE 802.11v

Introduced in 2011, the IEEE 802.11v standard enables the configuration of clients while they are connected to the network.

11.4.18 IEEE 802.11w

The IEEE 802.11w standard introduces protected management frames through mechanisms that ensure data origin authenticity, data integrity, and protection against replay attacks.

11.4.19 IEEE 802.11y

The IEEE 802.11y standard adapts existing layers to operate in a new frequency band and introduces three new concepts: Contention-Based Protocol (CBP), Extended Channel Switch Announcement (ECSA), and Dependent Station Enablement (DSE) [23].

11.4.20 IEEE 802.11z

The IEEE 802.11z standard extends Direct Link Setup (DLS) to enable two stations to communicate directly with each other without needing an access point (AP). It enhances the IEEE 802.11 MAC and PHY to support direct-link setups independent of the AP and introduces power-saving features.

11.4.21 IEEE 802.11aa

The IEEE 802.11aa specifies enhancements to the IEEE 802.11 medium access control (MAC) for robust audio video (AV) streaming at the same time maintaining coexistence with other types of traffic.

11.4.22 IEEE 802.11ad

The IEEE 802.11ad specification introduces a "fast session transfer" feature, allowing wireless devices to seamlessly switch between the legacy 2.4 GHz and 5 GHz bands and the 60 GHz frequency band [24]. This capability ensures optimal performance and range, enabling devices to maintain the best possible connection by moving between bands as needed. It allows numerous users in dense environments to achieve top-speed performance without interference or bandwidth-sharing issues commonly seen in legacy frequency bands [25].

11.4.23 IEEE 802.11ac

IEEE 802.11ac is designed to deliver a multi-station WLAN throughput of at least 1 Gbps and a single-link throughput of at least 500 Mbps [26]. This is accomplished by enhancing the air interface concepts from 802.11n, including wider RF bandwidth (up to 160 MHz), additional MIMO spatial streams (up to 8), multiuser MIMO, and high-density modulation [26–28].

11.4.24 IEEE 802.11ae

The IEEE 802.11ae standard introduces a mechanism for prioritizing management frames and specifies a protocol for communicating management frame prioritization policies.

11.4.25 IEEE 802.11af

The IEEE 802.11af standard is designed to operate in TV White Spaces, which are spectrum bands allocated to TV broadcasters but unused at specific locations and times. It will employ cognitive radio technology to identify usable white spaces, based on an authorized geolocation database. This database will provide details on which frequencies can be used, at what times, and under what conditions [29].

11.4.26 IEEE 802.11ah

The IEEE 802.11ah standard focuses on creating a global WLAN network that enables users to access subcarrier frequencies below 1 GHz within the ISM band. Additionally, it will enhance coverage range, enabling new applications such as wide-area sensor networks, sensor backhaul systems, and potential Wi-Fi offloading functions to emerge [30].

11.4.27 IEEE 802.11ai

IEEE 802.11ai includes a Fast Initial Link Setup (FILS) feature that enables a station to establish a secure link in less than 100 milliseconds [24].

11.4.28 IEEE 802.11aj

The IEEE 802.11aj standard will introduce updates to the IEEE 802.11ad Physical (PHY) layer and Medium Access Control (MAC) layer to support operation in the Chinese Millimeter Wave (CMMW) frequency bands, including the 59–64 GHz range. The amendment is designed to ensure backward compatibility with 802.11ad when operating in the 59–64 GHz frequency band. It will also outline changes to the PHY and MAC layers to support operation in the Chinese 45 GHz frequency band.

11.4.29 IEEE 802.11aq

The IEEE 802.11aq standard will introduce mechanisms to support pre-association service discovery by providing methods to advertise the availability of services and deliver information about them. This will allow stations on IEEE 802.11 wireless networks to access details about services before establishing a connection.

11.5 LiFi

The Light Communications IEEE 802.11bb Task Group was established in 2018, chaired by pureLiFi, and supported by Fraunhofer HHI—two leading organizations in LiFi development. Their goal is to promote accelerated adoption and interoperability, not only among LiFi vendors but also with WiFi technologies, through these standardization efforts [31].

Although the original IEEE 802.11 standard included a physical layer operating in the infrared part of the light spectrum, it was discontinued in 2016 due to limited commercial success. In 2011, the IEEE 802.15 Working Group introduced the IEEE 802.15.7 standard, which operates in the visible light spectrum, but initially garnered little industry interest. Advances in visible-light communications (VLC) technology later led to a revision of this standard. In December 2015, the IEEE 802.15.7r1 revision project was initiated and subsequently split into two groups: 802.15.7m, which focuses on optical camera communications with lower data rates, and 802.15.13, which targets high-speed applications using photodetectors, particularly for specialized industrial environments. The 802.15.7m standard was published in 2018, and the 802.15.13 standard was released in 2023 [32, 33].

The IEEE 802.11bb standard outlines the physical layer specifications and system architectures for wireless communication using light waves. This new standard establishes a foundation for the broad adoption of LiFi technology and facilitates interoperability between LiFi systems and the established WiFi standard. The standard was ratified in June 2023 [32, 33].

11.5.1 Usage Models

The IEEE 802.11.bb Task Group identified four key usage models where radio frequencies are either undesirable or inefficient: industrial wireless, wireless access in medical environments, enterprise networks, and home networks.

11.5.1.1 Industrial Wireless

In this usage scenario, devices in factories may encounter unstable radio frequency (RF) connections due to electromagnetic interference (EMI). To overcome this challenge, LC can be implemented to deliver reliable wireless connectivity for industrial networks. It can support applications such as ultrahigh-definition video streaming for production monitoring, video collaboration for team, customer, and supplier meetings, and real-time communication with factory sensors and robots [32].

11.5.1.2 Wireless Access in Medical Environments

In medical environments, RF interference can disrupt the operation of medical equipment. The IEC 60601-1-2 standard, in fact, advises maintaining a minimum separation distance of 30 cm between medical electrical devices and RF wireless communications equipment. To address this concern, LC can be utilized for wireless data exchange in medical settings, enabling the transmission of medical multimedia and diagnostic information for telemedicine services. Additionally, various medical devices can be wirelessly controlled via LC, providing Intranet/Internet access and facilitating audio or video calls for doctors, nurses, and patients [32].

11.5.1.3 Enterprise Network

In enterprise environments, high user density can result in slow speeds, dropped connections, or even lost connections. To resolve this, LC can be deployed to provide reliable wireless connectivity, supporting applications such as video conferencing, virtual desktops, Intranet/Internet access, and file exchanges between two or more users within the same room [32].

11.5.1.4 Secure Home Network

In a home environment, RF signals can be intercepted by nearby devices, whether belonging to legitimate users or hackers. This potential risk raises concerns among wireless users who are particularly mindful of data security and privacy. To mitigate this issue, LC can be implemented to offer wireless connectivity within a confined area, allowing users to engage in various activities, such as Internet access and online video chats, without the worry of wireless signals leaking into unintended locations [32].

11.5.2 Channel Modeling

The Lambertian model is commonly used in literature, though it falls short in accurately modeling real-world luminaires. In contrast, some researchers utilize site-specific channel models developed in software like Zemax® for specific scenarios. To facilitate consistent performance comparisons across different proposed algorithms and systems, the IEEE standardization committee intends to define usage scenarios and provide site-specific channel models that researchers can universally apply [32, 33].

11.5.3 Physical Layer

IEEE 802.11bb Light-Communication Task Group (TGbb) presents two options: the first involves adapting existing 802.11 PHYs to operate at LC frequencies, while the second proposes using an LC-optimized (LCO) PHY, as standardized in ITU-T G.9991 (G.vlc). The first option requires minimal effort to implement, whereas the second enhances throughput and robustness through adaptive bit loading. This technique allows modulation to vary independently across subcarriers, assigning higher-order modulations to subcarriers with a high SNR and lower-order modulations to those with a low SNR. As a result, adaptive bit loading increases the resilience of LC channels, even when LoS paths are obstructed, and boosts throughput in mobile environments.

11.5.4 Medium Access Control Layer

In addition to addressing the physical layer (PHY), 802.11bb also updates MAC protocols. IEEE 802.11.bb Task Group focuses on key challenges such as hidden stations and LoS blockages.

11.5.4.1 Hidden Stations Problem

IEEE 802.11.bb Task Group has introduced a solution for the hidden station problem called Repetition Clear Channel Assessment (RCCA). With RCCA, the AP retransmits any detected uplink signal using an amplify-and-forward method, thereby notifying all associated stations that the channel is occupied.

11.5.4.2 LoS Blockages Problem

One potential solution is to deploy multiple spatially distributed transmitters and receivers, reducing the likelihood that all connections will be blocked simultaneously. Another approach involves alternating between optical and radio links: if the LoS path is obstructed, the devices can switch to radio communication and vice versa. Combining 802.11bb with other technologies from the 802.11 family will enhance user mobility, expand network coverage, and accelerate link setup.

11.6 ITU-T PON Recommendation

11.6.1 Asynchronous Transfer Mode-Passive Optical Network (A-PON)/Broadband-Passive Optical Network (B-PON)

In 1995, the FSAN were established to establish a unified framework for access networks. In 1998, they proposed a standard to the ITU-T that was based on the ATM protocol for PON. ATM-PON or A-PON was the name of this standard. Afterward, the standard was renamed to Broadband PON or B-PON. The incorporation of additional services, such as Ethernet or broadcast video, over the PON was the reason for this name change. In 2005, the standard was revised, and certain recommendations were combined. B-PON is capable of supporting downlink rates of 155.52, 622.08, and 1244.16 Mb/s, as well as transmit rates of 155.52 and 622.08 Mb/s. B-PON employs TDM, and its time slots are dedicated to ATM or Physical layer operations, administration and maintenance (PLOAM) cells. TDMA is employed to multiplex the various ONU in the uplink, and a DBA mechanism is implemented. On the basis of ATM virtual circuits, communication between OLT and ONUs can be implemented at varying levels of QoS. B-PON has a greatest physical reach of 20 km. The downlink wavelength should be between 1480 and 1500 nm, while the uplink wavelength should be between 1260 and 1360 nm. The wavelength of analog video broadcasting has been established within the range of 1539–1565 nm. This standard is based on the following ITU recommendation [34]:

1. G.983.1—Broadband optical access systems based on PON. This recommendation was published in 1998 and reviewed in 2005.
2. G.983.2—ONT management and control interface specification for B-PON. This recommendation was published in 2002 and reviewed in 2005, merging it with the old recommendations G.983.6 (2002), G.983.7 (2001), G.983.8 (2003), G.983.9 (2004), and G.983.10 (2004)).
3. G.983.3—A broadband optical access system with increased service capability by wavelength allocation. This recommendation was published in 2001.
4. G.983.4—A broadband optical access system with increased service capability using DBA. This recommendation was published in 2001.
5. G.983.5—A broadband optical access system with enhanced survivability. This recommendation was published in 2002.

11.6.2 Gigabit-Capable Passive Optical Network (G-PON)

In order to address the requirements of PON, FSAN established a new standard that employed a variable length of packages, increased rates, and a greater variety of services. This standard was designed to optimize bandwidth utilization. In 2003, the ITU standardized G-PON standard, this is also illustrated in Fig. 11.3. This new standard was referred to as Gigabit-capable PON or G-PON. Compatibility with preceding standards is one of the objectives of this standard. This standard supports rates of 1.24 and 2.8 Gb/s at the downlink and 155.52 Mb/s, 622.08 Mb/s, 1.24 Gb/s, and 2.8 Gb/s at the uplink [35]. It is capable of operating in either symmetrical or asymmetrical mode. This standard employs the G-PON transmission convergence (GTC) layer framing, which offers a variety of functions, including transport multiplexing between the OLT and the ONUs, PLOAM functions, a DBA interface, ONU ranging and registration or FEC (Reed-Solomon (255,223)), and downstream data encryption as an optional function. The G-PON standard supports the operational modes of Asynchronous transfer mode (ATM) and GEM. GEM is the most prevalent mode and is comparable to the GFP. The segmentation capability of the GEM encapsulation enables the use of TDM circuits. The uplink is implemented using TDMA. This standard supports a maximum physical reach of 20 km with a split ratio of 1:64, with the option of a maximum reach of 60 km or a split ratio of 1:128 for future requirements. The wavelength range for the downlink is 1480–1500 nm. The uplink wavelength range is as follows: 1260–1360 nm for the regular band option (utilizing Fabry–Perot lasers), 1290–1330 nm for the reduced band option (utilizing an ordinary Distributed-feedback (DFB) laser), and 1300–1320 nm for the narrow band option (utilizing wavelength-selected lasers). The wavelength range for video broadcasting should be 1550–1560 nm. The necessary switching laser timing is 13 ns. For a B+ class budget, the loss budget for G-PON is 28 dB, while for a C+ class budget, it is 32 dB. This standard is based on the following ITU recommendation [34, 35]:

1. G.984.1—Gigabit-capable passive optical networks (GPON): General characteristics. This recommendation was published in 2003 and reviewed in 2008.
2. G.984.2—Gigabit-capable Passive Optical Networks (GPON): Physical Media Dependent (PMD) layer specification. This recommendation was published in 2004 and reviewed in 2014, including several amendments.
3. G.984.4—Gigabit-capable passive optical networks (GPON): ONT management and control interface specification. This recommendation was published in 2004 and reviewed in 2014.
4. G.984.5—Gigabit-capable passive optical networks (GPON): Enhancement band. This recommendation was published on 2007 and reviewed on 2014.
5. G.984.6—Gigabit-capable passive optical networks (GPON): Reach extension. This recommendation was published on 2008.
6. G.984.7—Gigabit-capable passive optical networks (GPON): Long reach. This recommendation was published on 2010.

11.6.3 10-Gigabit-capable Passive Optical Network (XG-PON) or Next-Generation Passive Optical Network 1 (NG-PON1)

ITU-T standardized a new standard in 2010 that addressed 10 Gb/s communications at the instance of FSAN, known as the XG-PON, as illustrated in Fig. 11.3. This standard can be categorized into two options: XG-PON1, which supports a 10 Gb/s downlink and 2.5G b/s uplink, and a future development, XG-PON2, which supports a symmetrical 10 Gb/s PON. This final option (XG-PON2) has ultimately been established as an independent standard (XGS-PON). The previous standard, GPON, must coexist with XG-PON. XG-PON facilitates the coexistence of both standards by utilizing G-PON frames and protocols, in addition to supporting Ethernet frames. The encapsulation method is referred to as XGEM, and the framing sublayer is referred to as XG-PON transmission convergence (XGTC). XGEM permits data privacy, fragmentation, and individual traffic flows. The PON and power savings options on the apparatus must be safeguarded by a robust mutual authentication required by this standard. This standard employs an RS (255,223) Forward error correction (FEC). The XG-PON upstream is burst-oriented and utilizes TDMA to facilitate access to the shared medium. The split ratio must be 1:64 or 1:128, and the utmost physical reach of this standard is between 20 km and 60 km. In the same scenario, the utmost requirements must not be simultaneously met. According to Fig. 11.2, the uplink wavelength range is 1260–1280 nm, while the downlink wavelength range is 1575–1580 nm. For PIN and APD receiver cases, the loss budget for XG-PON is 29 dB for an N1 class budget, 31 dB for an N2 class budget, 33 dB for an E1 class budget, and 35 dB for an E2 class budget. This standard is based on the following ITU recommendation [34, 36]:

1. G.987—10-Gigabit-capable passive optical network (XG-PON) systems: Definitions, abbreviations, and acronyms. This recommendation was published in 2010 and reviewed in 2012.
2. G.987.1—10-Gigabit-capable passive optical networks (XG-PON): General requirements. This recommendation was published in 2010 and reviewed in 2016.
3. G.987.2—10-Gigabit-capable passive optical networks (XG-PON): Physical Media Dependent (PMD) layer specification. This recommendation was published in 2010 and reviewed in 2016.
4. G.987.3—10-Gigabit-capable passive optical networks (XG-PON): Transmission Convergence (TC) layer specification. This recommendation was published in 2010 and reviewed in 2014.
5. G.987.4—10-Gigabit-capable passive optical networks (XG-PON): Reach extension. This recommendation was published in 2012.
6. G.988—ONU Management and Control Interface (OMCI) specification. This recommendation was published in 2010 and reviewed in 2012.

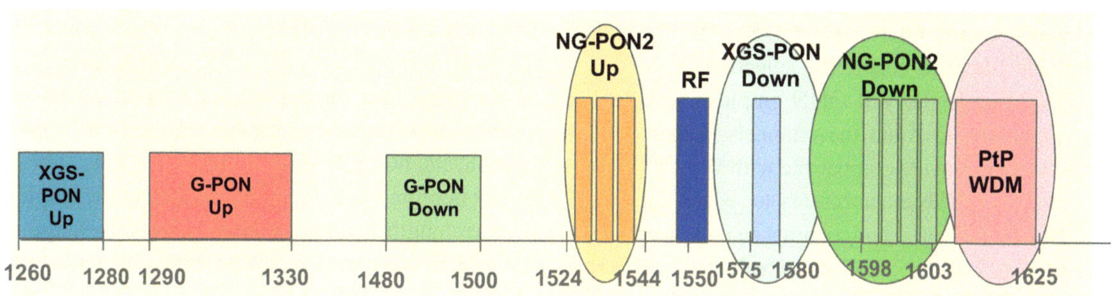

Fig. 11.2 Types of PON architectures

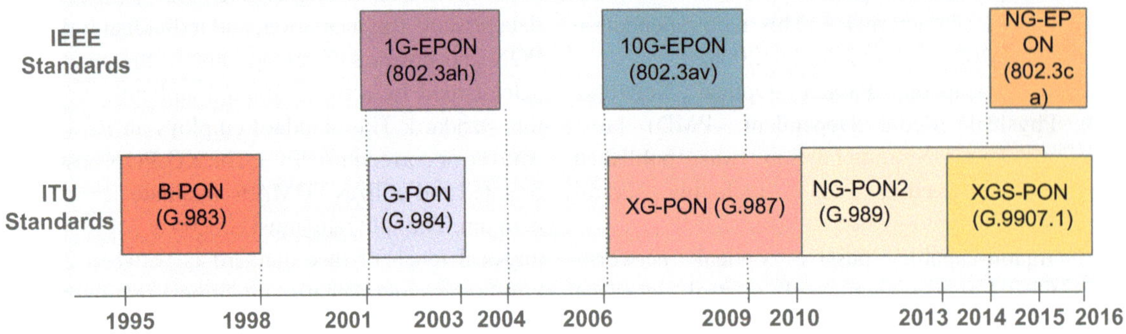

Fig. 11.3 Evolution of standards

11.6.4 Next-Generation Passive Optical Network 2 (NG-PON2)

ITU standardized a new standard proposed by FSAN in 2015, as illustrated in Fig. 11.3. This standard is referred to as the 40 Gigabit-capable Passive Optical Network (NG-PON2). NG-PON2 is capable of aggregating downlinks for mobile back-hauling, residential and commercial applications, and other services. This standard has been developed to be compatible and coexist with all the previous standards. NG-PON2 can be viewed as an orderly augmented version of XG-PON1, with a significant quantity of component reuse, as some companies have described. The DBA is also present in the previous ITU standards, and the Transmission convergence (TC) layer is the protocol layer of NG-PON2. This standard suggests the use of two categories of links: Time and wavelength multiplexing (TWDM) links and P2P WDM links. The up-link and downlink of TWDM links comprise four or eight TWDM channels, with rates of 2.5 or 10 Gb/s. The NG-PON2 must be capable of supporting a minimum of 40 Gb/s of aggregated traffic at the downstream and 10 Gb/s at the upstream. Depending on the P2P WDM client, P2P WDM channels can accommodate rates of 1.25, 2.5, or 10 Gb/s. The physical reach distance of this standard should be a minimum of 40 km, with the potential to extend to 60 km, and a differential fiber distance of 40 km. Additionally, it must be capable of operating with a split ratio of 1:256. Two varieties of optical distribution networks (ODN) can be supported by the P2P WDM PON: Wavelength-Selected ODN (WS-ODN) and Wavelength-Routed ODN (WR-ODN). The WS-ODN employs tunable filters to enable the selection of the channel at the ONU, while the WR-ODN employs wavelength splitters at the ODN to route the channels. Legacy ODN necessitates systems that are compatible with WS-ODN. Furthermore, colorless ONUs are necessary to reduce the cost of operations. NG-PON2's loss budget must be compatible with that of XG-PON (29 dB for an N1 class budget, 31 dB for an N2 class budget, 33 dB for an E1 class budget, and 35 dB for an E2 class budget). There are two varieties of upstream links in NG-PON2: Type A, which is devoid of a preamplifier, and Type B, which is equipped with a preamplifier at the OLT. Type A connections necessitate more potent ONUs, while Type B links necessitate more sensitive OLT. Furthermore, NG-PON2 establishes three wavelength-tuning time classes: Class 1, which necessitates less than 10 μs of tuning time, Class 2, which necessitates between 10 μs and 25 μs of tuning time, and Class 3, which necessitates between 25 μs and 1 s of tuning time. Thermal effects are typically the basis for Class 3 devices. Class 2 devices will permit protection with a sub-50 μs timeframe, while Class 1 devices will facilitate the allocation of future dynamic bandwidth and wavelengths. Crosstalk requirements are introduced by NG-PON2 due to its utilization of wavelength multiplexing. The Upstream Out-Of-Channel (OOC) interference penalty must be less than 1 dB, while the Downstream OOC interference penalty must be less than 0.1 dB. The Out-of-Band (OOB) interference penalty must be less than 0.1 dB in both directions. The Maximum Spectral Excursion (MSE) has been employed to define the wavelength stability requirements in NG-PON2 in order to prevent power leakage between channels and to guarantee that the connections operate appropriately within their respective channels. The TWDM downstream must have an MSE of ±20GHz and a channel spacing of 100 GHz. The MSE of a channel spacing of 50GHz is ±12.5GHz, that of a channel spacing of 100GHz is ±20GHz, and that of a channel spacing of 200GHz is ±25GHz. The channel spacing of the TWDM upstream can be anywhere from 50 to 200 GHz. A short-term spectral excursion may result from the laser's on/off operation, which must also satisfy the MSE requirements. The TX tunable laser of the ONUs is classified into three categories based on its wavelength calibration accuracy: sufficient calibration, imprecise calibration, and no calibration. If the laser is not calibrated, it may be outside the MSE limit, necessitating the implementation of wavelength-locking systems. The wavelength range for TWDM downstream is 1596–1603 nm, while for TWDM upstream, it is 1524–1544 nm for the wideband option, 1528–1540 nm for the reduced option, and 1535–1540 nm for the narrow option, as illustrated in Fig. 11.2. The wavelength range for the expanded spectrum option is 1524–1625 nm, while the shared spectrum option is 1603–1625 nm in the case of P2P WDM up/downstream. This standard is capable of accommodating a variety of wavelengths and is sufficiently adaptable to the future requirements of the 100G and beyond era. This standard is predicated on the subsequent ITU recommendation [34, 37]:

1. G.989—40-Gigabit-capable passive optical networks (NG-PON2): Definitions, abbreviations and acronyms. This recommendation was published in 2015.
2. G.989.1—40-Gigabit-capable passive optical networks (NG-PON2): General requirements. This recommendation was published in 2013 and reviewed in 2016.
3. G.989.2—40-Gigabit-capable passive optical networks 2 (NG-PON2): Physical Media Dependent (PMD) layer specification. This recommendation was published in 2014.
4. G.989.3—40-Gigabit-capable passive optical networks (NG-PON2): Transmission convergence layer specification. This recommendation was published in 2015.

11.6.5 10-Gigabit-Capable Symmetric Passive Optical Network (XGS-PON)

Since 2013, FSAN has been engaged in the development of the new standard XGS-PON, which was initially referred to as XG-PON2 in the XG-PON standard. The XGS-PON is a new standard that was released in 2016. Figure 11.3 illustrates this. This standard is intended to facilitate mobile backhauling and residential and business access services. This standard employs the XGS-PON TC framing sublayer, which enables the implementation of DBA. XGEM framing is implemented in this standard. In order to guarantee compatibility with previous specifications (XG-PON), the OLT must accommodate dual line-rates (2.5 Gb/s and 10 Gb/s). The downlink employs TDM to multiplex the various ONUs, while the upstream employs TDMA to facilitate access to the common medium by the various ONUs. Additionally, the Reed–Solomon FEC (255,223) is implemented. For the purpose of ensuring coexistence with G-PON, the minimal split ratios for XGS-PON should be 1:64. However, operators have expressed interest in split ratios of 1:128 and 1:256. Furthermore, in order to facilitate coexistence with XG-PON, the TC layer must support a minimum of 20 km of maximum fiber distance, while the physical reach distance of XGS-PON standard should be a minimum of 40 km, with the potential to extend to 60 km, and a differential fiber distance of 40 km.

Two wavelength ranges are available for this standard: fundamental and optional. The fundamental wavelength range is identical to that of XG-PON (downstream=1575–1580 nm, upstream=1260–1280 nm), as illustrated in Fig. 11.2. Dual-upstream rate TDMA and TDM at the downstream are employed to support XGS-PON and legacy XG-PON ONUs in the fundamental wavelength. The coexistence of legacy G-PON ONUs is direct when the fundamental range is employed. The optional wavelength range is identical to that of the restricted upstream range of G-PON (downstream = 1480–1500 nm, upstream = 1300–1320 nm). Legacy G-PON ONUs are not supported in the optional range, but legacy XG-PON is supported with wavelength multiplexing. This two-wavelength range can be employed concurrently. The wavelength range selected determines the loss budget of XGS-PON. The loss budget is equivalent to the XG-PON when the fundamental wavelength range is employed (29 dB for an N1 class budget, 31 dB for an N2 class budget, 33 dB for an E1 class budget, and 35 dB for an E2 class budget). The loss budget is the G-PON loss budget when the optional wavelength range is employed (28 dB for a B+ class budget and 32 dB for a C+ class budget). This standard is predicated on the subsequent ITU recommendation [34, 38]:

1. G.9807.1—10-Gigabit-capable symmetric passive optical network (XGS-PON). This recommendation was published in 2016.

11.7 IEEE PON Standards

11.7.1 Gigabit Ethernet Passive Optical Network (1G-EPON)

Ethernet in the first mile, also known as the 802.3ah task force group, was established by IEEE in 2001. In 2004, this group established the 1G-EPON in accordance with the IEEE 802.3ah-2004 standard, which specifies the physical layers, management parameters, and media access control parameters for subscriber access networks. The 1G-EPON is a PON that features a symmetrical 1 Gb/s transmission rate. The reach distances for 1G-EPON can be 20 km (with 1000 BASE-PX20-D PMD and 1000 BASE-PX20-U PMD) or 10 km (1000 BASEPX10-D PMD and 1000 BASE-PX10-U PMD) with a minimum split ratio of 1:16 [39]. The downlink and uplink band options have wavelength allocation ranges of 1480–1500 nm and 1260–1360 nm, respectively (Fig. 11.2). The loss budget for 1G-PON is 29 dB for a PX30 class budget and 33 dB for a PX40 class budget. The laser must be turned on and off at a rate of 512 ns. 1G-EPON is a Multipoint control protocol (MPCP) that coordinates communication between the OLT and ONUs in a shared PON medium. It is analogous to the P2P Ethernet system. The 1G-EPON upstream operates in a burst mode, which is equivalent to

TDMA. The ONU utilizes a loop timing for the upstream burst mode transmission, obtaining the clock from the downstream received signal, and the 1G-EPON downstream has a continuous signal stream and clock synchronization. The DC balanced output and simple clock recovery of the 1G-EPON are achieved through the use of an 8B/10B code line, which encodes 8 data bits as 10 line bits. However, in order to transmit 1 Gb/s of data, the symbol rate must be increased to 1.25 Gb/s. The FEC is an optional feature that employs an RS (255,239). It is not specified in the standard, but 1G-EPON can implement dynamic bandwidth allocation (DBA), Operation and management (OAM) sublayer, encryption, and protection mechanisms. Circuit emulation is essential for the implementation of fixed bandwidth TDM circuits.

11.7.2 10 Gigabit Ethernet Passive Optical Network (10G-EPON)

The 10G-EPON study group was established in 2006 and standardized the IEEE Std 802.3av-2009 physical layer specifications and management parameters for 10G-EPON in 2009 [40]. The 10G-EPON standard takes into account a symmetrical 10G/10G transmission rate PON and an asymmetrical 10G/1G transmission rate PON, with a downstream bandwidth of 10 Gb/s and an upstream bandwidth of 1 Gb/s. The loss budget for 10G-PON is 29 dB for a PR30 and PRX30 class budget and 33 dB for a PR40 and PRX40 class budget. At this standard, the minimum reach distances are 10 km or 20 km, and the division ratios are 1:16 and 1:32. The necessary laser switching timing is 512 ns. In order to ensure coexistence with 1G-EPON, the wavelength ranges for the uplink are 1260–1280 nm, and for the downlink, are 1575–1580 nm. The coexistence of 10G-EPON and 1G-EPON downstream is resolved through the use of WDM multiplexing, as their wavelength bands do not intersect. In the upstream, the dual-rate burst receiver and dual-rate DBA must be employed due to the overlap of both standard bands, which requires the relocation of all coexistence issues to the electrical domain. The communication between OLT and ONU is orchestrated by the 10G-EPON protocol, which is a modified MPCP. The 64B/66B code line of the 10G-EPON generates less latency than the 8B/10B code line of the 1G-EPON. Consequently, the symbol rate must be 10.3125 Gb/s to transmit a 10 Gb/s of data. The FEC is mandatory in 10G-EPON and employs an RS (255,223).

11.7.3 Next-Generation Ethernet Passive Optical Network (NG-EPON)

IEEE Std 802.3ca next-generation Ethernet passive optical network (NG-EPON) is the next standard for EPON networks that the NG-EPON considered in 2014. This standard takes into account a multi-wavelength EPON with an aggregate capacity of at least 40 Gb/s in both directions (40G-EPON) and the potential to increase the aggregated capacity to 100 Gb/s (100G-EPON). A single-wavelength EPON is also taken into account, with symmetric rates of at least 25 Gb/s (25G-EPON) or a downstream rate of 25 Gb/s and an upstream rate of 10 Gb/s (25/10GEPON) [42].

The division ratios that NG-EPON supports must be comparable to the previous EPON standards (1G-EPON and 10G-EPON), and the reach distance must be a minimum of 20 km, with a differential reach of at least 10 km. The division ratio must be 1:16, and the reach distance must be a minimum of 10 km if the loss budget is low class ($\leq$20 dB). The division ratio must be 1:32, and the reach distance must be a minimum of 10 km if the loss budget is categorized as a medium class ($\leq$24 dB). A division ratio of 1:32 and a minimum reach distance of 20 km are required when the loss budget is high class ($\leq$29 dB). The division ratio must be 1:64, and the reach distance must be a minimum of 20 km if the loss budget is for extended class ($\leq$33 dB). In the event that the Optical distribution node (ODN) contains legacy 1G-EPON and 10G-EPON ONUs, a triple-rate burst receiver at the OLT will be necessary to ensure compatibility with the previous standards. This standard contemplates the use of NRZ as a modulation, as well as multi-level modulation schemes such as Orthogonal frequency division multiplexing (OFDM), 4-PAM, or duobinary. Additionally, the utilization of Wavelength switching ODN (WS-ODN) and Wavelength routing ODN (WR-ODN) as ODN is currently being assessed in the context of NG-EPON.

11.8 Summary

This chapter provides a brief overview of the standards under IEEE 802 and ITU-T PON, along with description of the protocols associated with these standards. The functioning of each layer of the OSI reference model for designing a network is explained in this chapter. Additionally, the classification and features of IEEE 802.11 standards are explored. The chapter also covers the LiFi standard for light-based communication.

Homework Problems

1. Describe the architecture of Open Systems Interconnection (OSI) and explain its main components.
2. Compare the operational differences between GPON and XGPON, two types of passive optical networks.
3. Classify the types of Passive optical network (PON) and describe the PON features that provide the preferred network strategy for the implementation of FTTC / FTTH.

4. Explain the two modes through which a wireless terminal can detect the presence of an access point in the IEEE 802.11 standard.
5. Compare the different IEEE 802.11 standards.
6. Based on the IEEE 802.11 standard, what are the two methods for creating a wireless network, and what are their characteristics?
7. Ethernet is the most widely used LAN protocol. At which OSI layer(s) does it operate?
8. Summarize the key features of the Optical Transport Network (OTN) and explain its hierarchy as defined by the ITU-T.
9. What services does IEEE 802.11 WLAN provide?
10. Provide a detailed explanation of the layer in which IEEE 802.11 operates within the OSI model.
11. List the key features of IEEE 802.11b and IEEE 802.11g.

References

1. Joel Jr, A.: Circuit-switching fundamentals. In: Fundamentals of Digital Switching, pp. 1–39. Springer, Berlin (1983)
2. Kleinrock, L., Tobagi, F.: Packet switching in radio channels: Part I—carrier sense multiple-access modes and their throughput-delay characteristics. IEEE Trans. Commun. **23**(12), 1400–1416 (1975)
3. Rainer, H., et al.: Atm Networks: Concepts, Protocols, Applications, 3/E. Pearson Education India (1999)
4. ISO, I.: IEC 7498-1 information technology-open systems interconnection-basic reference model: the basic model. Int. Organ. Standard. **2** (1994)
5. SENIOR, J.M.: Optical Fiber Communications: Principles and Practice, 3rd edn. Pearson Education Limited (2009)
6. Callegati, F., Casoni, M., Raffaelli, C.: Packet optical networks for high-speed TCP-IP backbones. IEEE Commun. Mag. **37**(1), 124–129 (1999)
7. Rodriguez-Moral, A., Bonenfant, P., Baroni, S., Wu, R.: Optical data networking: protocols, technologies, and architectures for next generation optical transport networks and optical internetworks. J. Lightwave Technol. **18**(12), 1855 (2000)
8. Association, I.S., et al.: IEEE 802.11, IEEE standard for local and metropolitan area networks specific requirements—part 11: wireless LAN medium access control (MAC) and physical layer (PHY) specifications (1999). http://standards.ieee.org/
9. IEEE Standards Association et al.: IEEE standard for local and metropolitan area networks-specific requirements: Part 11 (IEEE 802.11-2007). http://standards.ieee.org/ (2007)
10. Fluhrer, S., Mantin, I., Shamir, A.: Weaknesses in the key scheduling algorithm of RC4. In: Selected Areas in Cryptography: 8th Annual International Workshop, SAC 2001 Toronto, Ontario, Canada, August 16–17, 2001 Revised Papers 8, pp. 1–24. Springer, Berlin (2001)
11. McNair, J., Tugcu, T., Wang, W., Xie, J.L.: A survey of cross-layer performance enhancements for mobile IP networks. Comput. Netw. **49**(2), 119–146 (2005)
12. Vassis, D., Kormentzas, G., Rouskas, A., Maglogiannis, I.: The IEEE 802.11 g standard for high data rate WLANs. IEEE Netw. **19**(3), 21–26 (2005)
13. IEEE Standards Association et al.: IEEE standard for local and metropolitan area networks specific requirements—part 11: wireless LAN medium access control (MAC) and physical layer (PHY) amendment 3: specification for operation in additional regulatory domains (2001). http://standards.ieee.org/
14. Ni, Q., Romdhani, L., Turletti, T.: A survey of QOS enhancements for IEEE 802.11 wireless LAN. Wirel. Commun. Mobile Comput. **4**(5), 547–566 (2004)
15. Konsgen, A.J., Hossain, Z., Gorg, C.: Transmit power control algorithms in IEEE 802.11 h based networks. In: 2005 IEEE 16th International Symposium on Personal, Indoor and Mobile Radio Communications, vol. 3, pp. 1441–1445. IEEE, Piscataway (2005)
16. Xiao, Y.: IEEE 802.11 n: enhancements for higher throughput in wireless LANs. IEEE Wirel. Commun. **12**(6), 82–91 (2005)
17. Rabaey, J., Wolisz, A., Ercan, A., Araujo, A., Burghardt, F., Mustafa, S., Parsa, A., Pollin, S., Wang, I.H., Malagon, P.: Connectivity brokerage-enabling seamless cooperation in wireless networks. White paper (2010)
18. Wellens, M., Westphal, B., Mahonen, P.: Performance evaluation of IEEE 802.11-based WLANs in vehicular scenarios. In: 2007 IEEE 65th Vehicular Technology Conference-VTC2007-Spring, pp. 1167–1171. IEEE, Piscataway (2007)
19. Hartenstein, H., Laberteaux, L.: A tutorial survey on vehicular ad hoc networks. IEEE Commun. Mag. **46**(6), 164–171 (2008)
20. Molta, D.: Primer: mesh configurations applications make a mesh of your WLAN. Netw. Comput. **11**(17), 67–71 (2005)
21. Djukic, P., Valaee, S.: 802.16 mcf for 802.11 a based mesh networks: a case for standards re-use. In: 23rd Biennial Symposium on Communications, 2006, pp. 186–189. IEEE, Piscataway (2006)
22. Maniezzo, D., Villa, G., Gerla, M.: A smart MAC-routing protocol for WLAN mesh networks. UCLA-CSD Technical Report **40032** (2004)
23. Stankiewicz, R., Cholda, P., Jajszczyk, A.: Qox: what is it really? IEEE Commun. Mag. **49**(4), 148–158 (2011)
24. Kim, E., Seok, Y.: Method for operating a station in a white space, and apparatus for same (2017). US Patent 9,681,424
25. Cordeiro, C., Akhmetov, D., Park, M.: IEEE 802.11 ad: Introduction and performance evaluation of the first multi-GBPs WIFI technology. In: Proceedings of the 2010 ACM international workshop on mmWave communications: from circuits to networks, pp. 3–8 (2010)
26. Zhao, F., Médard, M., Hundebøll, M., Ledet-Pedersen, J., Rein, S.A., Fitzek, F.H.: Comparison of analytical and measured performance results on network coding in IEEE 802.11 ad-hoc networks. In: 2012 International Symposium on Network Coding (NetCod), pp. 43–48. IEEE, Piscataway (2012)
27. Ong, E.H., Kneckt, J., Alanen, O., Chang, Z., Huovinen, T., Nihtilä, T.: IEEE 802.11 ac: Enhancements for very high throughput WLANs. In: 2011 IEEE 22nd International Symposium on Personal, Indoor and Mobile Radio Communications, pp. 849–853. IEEE, Piscataway (2011)
28. Park, M.: IEEE 802.11 ac: dynamic bandwidth channel access. In: 2011 IEEE International Conference on Communications (ICC), pp. 1–5. IEEE, Piscataway (2011)
29. Leith, D.J., Clifford, P.: Using the 802.11 e EDCF to achieve TCP upload fairness over WLAN links. In: Third International Symposium on Modeling and Optimization in Mobile, Ad Hoc, and Wireless Networks (WiOpt'05), pp. 109–118. IEEE, Piscataway (2005)
30. Van Nee, R.: Breaking the gigabit-per-second barrier with 802.11 ac. IEEE Wirel. Commun. **18**(2), 4 (2011)
31. Purwita, A.A., Haas, H.: Studies of flatness of LiFi channel for IEEE 802.11bb. In: 2020 IEEE Wireless Communications and Networking Conference (WCNC), pp. 1–6 (2020)
32. Miramirkhani, F., Baykas, T., Elamassie, M., Uysal, M.: IEEE 802.11bb reference channel models for light communications. IEEE Commun. Standards Mag. **7**(4), 84–89 (2023)
33. Khorov, E., Levitsky, I.: Current status and challenges of Li-Fi: IEEE 802.11 bb. IEEE Commun. Standards Mag. **6**(2), 35–41 (2022)

34. Altabas, J.: Passive optical networks, introduction. https://upcommons.upc.edu/bitstream/2117/121280/5/Consolidated_Version_Passive%20Optical%20Networks%20Introduction_JAL_SSH_final.pdf
35. IEEE Standard for Information Technology—local and metropolitan area networks—specific requirements—part 3: CSMA/CD access method and physical layer specifications amendment 1: physical layer specifications and management parameters for 10 Gb/s passive optical networks. IEEE Std 802.3av-2009 (Amendment to IEEE Std 802.3-2008) pp. 1–227 (2009)
36. G.987.3: 10-gigabit-capable passive optical networks (XG-PON): transmission convergence (TC) layer specification (2020). https://www.itu.int/rec/T-REC-G.987.3
37. G.989 NG-PON2 passive optical networks. ITU-T Study Group 15 (2020). https://www.itu.int/en/ITU-T/studygroups/2017-2020/15/Documents/flyers/Flyer_ITU-T_G.989.pdf
38. Series G: Transmission systems and media, digital systems and networks, 10-Gigabit-capable symmetric passive optical network (XGS-PON), G.9807.1, ITU-T (2016)
39. IEEE draft standard for information technology–telecommunications and information exchange between systems–local and metropolitan area networks–specific requirements part 3: Carrier sense multiple access with collision detection (CSMA/CD) access method and physical layer specifications (revision of IEEE std 802.3-2005 including all approved amendments) section 1. IEEE Unapproved Draft Std P802.3/D2.1, Feb 2008 (2008)
40. 10G-EPON (2010). https://www.zeitgeistlab.ca/doc/10G-EPON.html#:~:text=The%2010G%2DEPON%20standard%20provides,1%20Gbps%20upstr\eam%20data%20rates
41. Patibandla, A., Guptha, M., M.Nagma: Optical Fiber Communication: Overview, Advantages, and Signal Distortion, Study Guides, Projects, Research of Communication
42. IEEE/ISO/IEC international standard-telecommunications and exchange between information technology systems—requirements for local and metropolitan area networks—part 3: Standard for ethernet amendment 9: Physical layer specifications and management parameters for 25 gb/s and 50 gb/s passive optical networks. ISO/IEC/IEEE 8802-3:2021/Amd.9:2021(E), pp. 1–270 (2021)
43. Wu, Z.Z., Chen, H.C.: Design and implementation of tcp/ip offload engine system over gigabit ethernet. In: Proceedings of 15th International Conference on Computer Communications and Networks, pp. 245–250. IEEE, Piscataway (2006)

12 Next-Generation Integrated Optical and Wireless Access Networks

12.1 Introduction

The previous chapter introduces the various ITU-T and IEEE standards developed for enabling optical and wireless communication. Moreover, the functioning of each layer of the OSI reference model for designing a network is also discussed in the previous chapter. However, along with the various IEEE and ITU-T standards, the future optical and wireless integrated networks also need to be studied. Some of the most important OWC technologies are RoF, Integrated LiFi/WiFi networks, Integrated fiber-wireless networks, broadband access networks, hybrid wirlesss-optical broadband access networks, some of which will be covered in this chapter.

12.2 Radio-over-fibre (RoF) Networks

The concept of RoF originated in the 1980s, shortly after the introduction of radio frequency/microwave subcarrier multiplexing, which quickly gained popularity for distributing cable TV channels [1–95]. The ever-increasing demand for wireless bandwidth has put significant pressure on the lower microwave spectral region, prompting a shift toward the mmWave region to take advantage of its inherent bandwidth. In the early days, the primary motivation for RoF was to tackle the problem of congested spectrum in the lower microwave band. RoF offered a more effective method of distributing mmWave signals via optical fiber compared to wireless transmission, which is hindered by high propagation losses. RoF combines wireless and optical systems to provide high data rates, high capacity, and enhanced mobility [80].

Figure 12.1 shows a typical RoF system. It consists of a central station (CS) that includes data resources, optical Transmitter (Tx) equipped with lasers, and Receiver (Rx) with PDs. The CS is interconnected with the Trunk network, the Internet, and possesses switching capabilities. When communicating with the base stations (BSs) in the downlink direction, the CS up-converts the electrical signal to the optical domain and transmits it through ODN (optical distribution network) to the BSs. In some configurations, resource nodes (RNs) is employed; in active networks, these RNs amplify the optical signal, while in passive networks, they split or demultiplex the signal to the appropriate BS. The BS then converts the optical signal back into the electrical domain and transmits it to the mobile terminal (MT) end user using mmWave frequencies. In the uplink direction, the MT transmits the mmWave signal to the BS. Depending on the BS configuration, the mmWave signal may be down-converted before being modulated onto an electro-optical device, which then transmits the uplink information through the ODN back to the CS.

RoF communications systems are superior to wireless or traditional coaxial cable systems in a number of ways:

- The use of optical fibers results in low attenuation.
- Resources are centralized at the CS, enabling shared use and reducing the complexity and cost of remote BSs, which mainly consist of optical-to-electrical converters, RF amplifiers, and antennas.
- Since they are essentially modulation format independent, they have low-cost expandability.
- RoF systems can transport higher frequency signals, allowing for greater data rates and accommodating future service demands.
- The same RoF network can be used to distribute traffic from multiple operators and services, allowing for independent infrastructure providers and multi-service operations.
- Dynamic resource allocation is possible since functions like modulation, switching, and other key processes are performed centrally at the CS, the system can dynamically allocate capacity based on demand.

V. A. Bohara, Swaminathan R, *Optical and Wireless Access Networks*, Textbooks in Telecommunication Engineering, https://doi.org/10.1007/978-3-031-89696-5_12

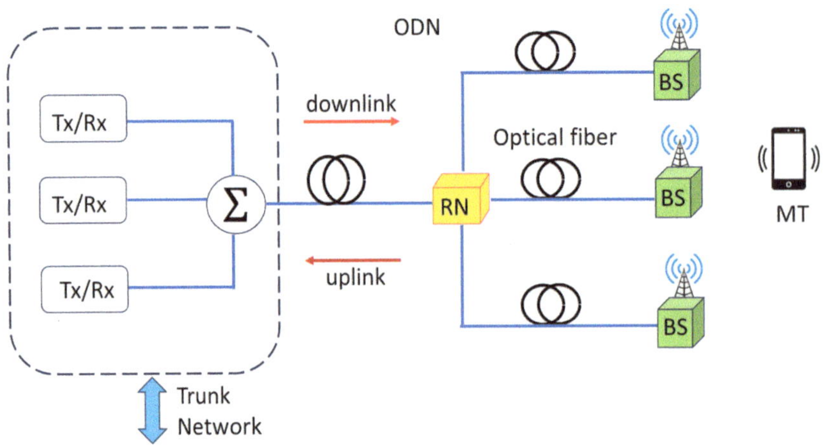

Fig. 12.1 Architecture of a general RoF system

The main disadvantages of an RoF system are linked to the increasing complexity and requirements as RF frequencies increase. Higher frequencies necessitate the use of high-speed optical components, advanced mmWave generation techniques, and a large number of BSs equipped with large bandwidth PDs to effectively provide coverage to a service area. Additionally, chromatic dispersion effects across the ODN become problematic, even over relatively short fiber spans, especially when using conventional double-sideband modulation. To mitigate these issues, advanced modulation and transmission schemes are required, leading to the development of optical components that operate in the mmWave region. However, multiple system architectures have been proposed to centralize functions like signal routing, processing, handover, and frequency allocation at the CS, thereby simplifying the BSs.

12.2.1 Generation and Distribution of RoF Signals

The mmWave RoF is based on a centralized controlled architecture. This architecture allows for centralized mmWave signal generation and control in the CO. However, all the connected users have to share the infrastructure cost. In the early stages of RoF research, significant attention was given to the optical generation of mmWave signals, leading to the development of various schemes. These schemes include frequency doubling based on carrier suppression, frequency quadrupling, optical single sideband modulation, dual-mode sources, mode-locked lasers, and pulsed lasers.

Maintaining transparency for wireless users while transporting the optically generated mmWave signals seamlessly is another challenge. The effects of fiber chromatic dispersion, increased noise, and nonlinear distortion associated with analog optical link transmission all affect the optically generated mmWave signals. Numerous strategies have been proposed to address these issues. The severity of impairments and the complexity of BS hardware for mmWave RoF systems could be determined by the technique used for the transport of mmWave signals.

12.2.1.1 Millimeter-wave Opical Signal Generation

In [69], for the first time, the mmWave RoF link was demonstrated with a wireless frequency of up to 26 GHz. The difficulty in putting such a mmWave RoF link into practice is finding an appropriate optical source that can produce mmWave signals. Subsequently, several techniques for producing optical mmWave have been presented. These include generation based on nonlinear effects through Stimulated Brillouin scattering, four-wave mixing, direct modulation of dual-mode or mode-locked lasers, and external modulation techniques.

12.2.1.2 RoF Signal Transport Schemes

It is challenging to transport optically generated mmWave signals over fiber. Traditionally, techniques such as RF-over-fiber, intermediate frequency-over-fiber, and baseband-over-fiber have been used to transport mmWave signals across optical links. Figure 12.2 provides a graphic representation of these transport schemes.

In the RF-over-fiber scheme, mmWave signals are transmitted directly over the optical fiber at the desired wireless transmission frequency. This method simplifies the architecture design by eliminating the need for frequency up/down conversion at remote antenna sites. However, high-speed PDs and optical transmitters are required for mmWave RoF systems, which can increase the deployment costs. Additionally, the effect of fiber chromatic dispersion worsens the link performance, particularly as signal frequency increases, leading to performance degradation [43, 81].

Both IF-over-fiber and baseband-over-fiber schemes necessitate frequency up/down conversion at the antenna sites,

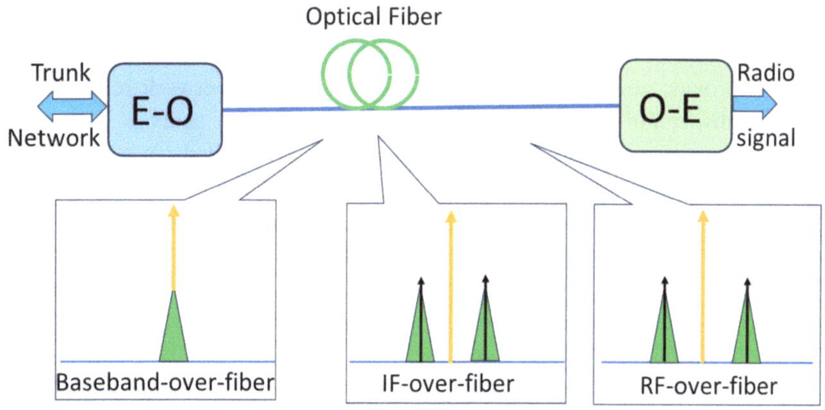

Fig. 12.2 RoF signal transport schemes

but these schemes utilize lower-speed optoelectronic devices and mitigate the effects of fiber chromatic dispersion. Despite these benefits, the complexity at the antenna sites increases due to the need for mmWave local oscillators (LOs) and high-speed mixers for frequency translation, which limits the scalability and upgradability of the RoF link. This limitation affects the ability to add more mmWave wireless channels or adjust carrier frequencies in the future. To address this, the LO required at the BS for the up-and-down conversion to/from mmWave RF is delivered remotely to provide greater flexibility in managing these transport schemes. For many years, the IF-over-fiber transport scheme has been preferred for uplink communication due to the scarcity of low-cost optical mmWave signal generators.

The concept of Radio over fiber (RoF) originated in the 1980s, shortly after the introduction of radio frequency/microwave subcarrier multiplexing, which quickly gained popularity for distributing cable TV channels [4, 95]. The ever-increasing demand for wireless bandwidth has put significant pressure on the lower microwave spectral region, prompting a shift toward the millimeter wave (mmWave) region to take advantage of its inherent bandwidth. In the early days, the primary motivation for RoF was to tackle the problem of congested spectrum in the lower microwave band. RoF offered a more effective method of distributing mmWave signals via optical fiber compared to wireless transmission, which is hindered by high propagation losses. RoF combines wireless and optical systems to provide high data rates, high capacity, and enhanced mobility [80].

Figure 12.1 shows a typical RoF system. It consists of a central station (CS) that includes data resources, optical transmitters (Tx) equipped with lasers, and receivers (Rx) with photodetectors (PDs). The CS is interconnected with the Trunk network, the Internet, and possesses switching capabilities. When communicating with the base stations (BSs) in the downlink direction, the CS up-converts the electrical signal to the optical domain and transmits it through ODN (optical distribution network) to the BSs. In some configurations, remote nodes (RNs) are employed; in active networks, these RNs amplify the optical signal, while in passive networks, they split or demultiplex the signal to the appropriate BS. The BS then converts the optical signal back into the electrical domain and transmits it to the Mobile Terminal (MT) end user using mmWave frequencies. In the uplink direction, the MT transmits the mmWave signal to the BS. Depending on the BS configuration, the mmWave signal may be down-converted before being modulated onto an electro-optical device, which then transmits the uplink information through the ODN back to the CS.

RoF communications systems are superior to wireless or traditional coaxial cable systems in a number of ways:

- The use of optical fibers results in low attenuation.
- Resources are centralized at the CS, enabling shared use and reducing the complexity and cost of remote BSs, which mainly consist of optical-to-electrical converters, RF amplifiers, and antennas.
- Since they are essentially modulation format independent, they have low-cost expandability.
- RoF systems can transport higher frequency signals, allowing for greater data rates and accommodating future service demands.
- The same RoF network can be used to distribute traffic from multiple operators and services, allowing for independent infrastructure providers and multi-service operations.
- Dynamic resource allocation is possible since functions like modulation, switching, and other key processes are performed centrally at the CS, the system can dynamically allocate capacity based on demand.

The main disadvantages of an RoF system are linked to the increasing complexity and requirements as RF frequencies increase. Higher frequencies necessitate the use of high-speed optical components, advanced mmWave generation techniques, and a large number of BSs equipped with large bandwidth PDs to effectively provide coverage to a service area. Additionally, chromatic dispersion effects across the ODN become problematic, even over relatively short fiber spans, especially when using conventional double-sideband modulation. To mitigate these issues, advanced modulation and transmission schemes are required, leading to the development of optical components that operate in the mmWave region. However, multiple system architectures have been proposed to centralize functions like signal routing, processing, handover, and frequency allocation at the CS, thereby simplifying the BSs.

12.2.2 Optical Distribution Network Architecture

The connection between Base stations (BSs) and BSs controllers in cellular systems is mainly based on three types of links: coaxial, RF, and optical fiber. However, due to the increasing number of customers and data-hungry applications, the distribution network requires more capacity. To address this, optical fiber links are deployed in densely populated urban and suburban areas. The optical fiber deployments help in the planning and design of the future mmWave RoF ODN. By virtue of their ability to cohabit with current hybrid optical-wireless networks, RoF technologies exhibit superior performance in terms of overall capacity, bandwidth allocation, and user count [54].

12.2.2.1 Optical Fiber

Over the past few years, a number of optical fiber designs have been explored for RoF systems, with the most significant among those are: single mode fiber (SMF), multi mode fiber (MMF), and Polymer Optical Fibers. The network characteristics and resource availability decide which optical fiber is the best suited for the RoF systems. SMF has been previously proven to carry mmWave signals reliably for both indoor and outdoor (long-distance) applications. However, the installation cost of SMF can become a significant concern when it is deployed. As a result, several research groups have come up with novel solutions for using low-cost fibers in designing RoF systems.

In most offices and buildings, there is a network of optical fiber already installed; this fiber is typically MMF, which is carrier for Ethernet data. In several commercial building, MMF is being used for transmitting broadband wireless IEEE 802.11b/g signals for indoor applications. POFs have also been suggested as an affordable substitute for in-building networks with short reach [12]. The POFs offer the benefits of optical fibers with the comparable ease of installation as that of copper wires.

12.2.2.2 Network Topology

RoF systems should have a network topology that offers high flexibility, network scalability, and reliability. In [29], a comprehensive analysis of network topologies for RoF systems is provided. A comparison of various topologies including a) star, (b) ring, (c) multilevel rings, (d) multilevel stars, (e) hybrid multilevel star-rings, (f) and ring-stars is given in Fig. 12.3. It has been demonstrated that for large RoF networks with thousands of BSs connected to the CS, multilevel ring topology yields the best performance.

12.2.2.3 Optical Amplification

Optical amplifiers play a crucial role in long-distance data transmission by amplifying weak signals. This helps in maintaining signal quality and and ensuring signal quality over vast distances. Further, in optical access networks, signals often need to be split to serve multiple BSs or destinations. This splitting process, typically done using optical splitters, introduces branching losses—a reduction in signal power because the optical signal is divided among multiple paths. Optical amplifiers can mitigate these losses by boosting the signal strength after splitting. Optical amplifiers are crucial for maintaining signal quality, extending network reach, and improving reliability. In [63], the effect of Distributed Raman Amplifiers (DRA) and EDFAs amplification schemes on WDM RoF networks are investigated. Results revealed that DRA has better performance in uplink due to the suppression of gain saturation and has better noise performance in the downlink, in comparison to EDFA.

12.3 Integrated LiFi/WiFi networks

As per the latest visual networking index by Cisco Systems, by 2022, mobile data traffic is expected to account for 71% of all Internet protocol traffic, with over 80% of that traffic occurring indoors [11]. This highlights the importance of short-range wireless technologies, such as WiFi, as a crucial element in the 5G and beyond. However, the scarcity of RF spectrum creates a bottleneck for dense deployment of WiFi hotspots. Hence, the RF system is under pressure to meet the rapidly rising demands for mobile data traffic.

To address the impending spectrum shortage in RF, the researchers are focusing on wireless communication technologies that utilize extremely high frequencies. Light fidelity, or LiFi, is one such technology [9]. LiFi takes advantage of the large optical spectrum of around 300 THz and uses light as a carrier of information. The primary advantage is its capacity to incorporate LiFi APs into the current lighting infrastructure, allowing APs to serve the dual purpose of illumination and communication. It has been shown in recent studies that LiFi can reach peak data speeds of 10 Gigabit per second

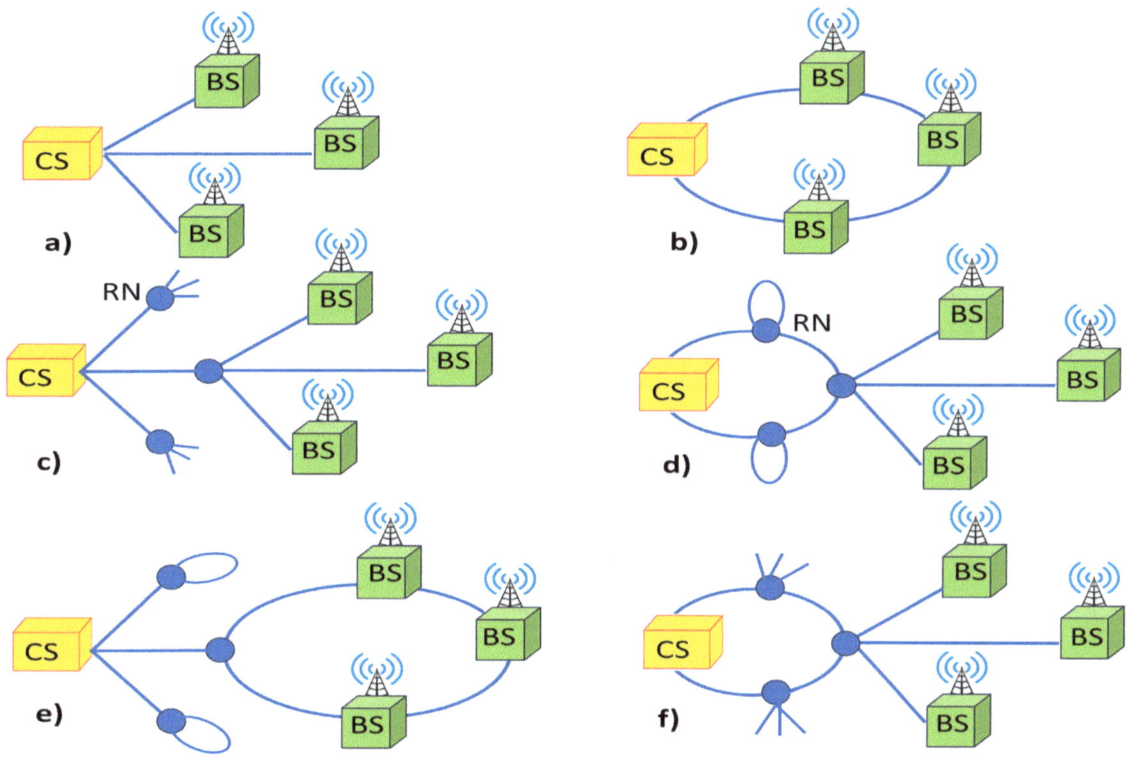

Fig. 12.3 ODN Topologies: (**a**) Star, (**b**) ring, (**c**) multilevel stars, (**d**) multilevel rings, (**e**) hybrid multilevel star-rings, and (**f**) hybrid multilevel ring-stars

(Gbps) with just one LED [49]. Furthermore, as compared to its RF counterpart, LiFi offers various advantages, such as: i) license-free spectrum, ii) secure wireless communications (as light does not pass through opaque objects), and iii) used in RF-restricted zones (hospitals, underwater). Despite the advantages, LiFi has some limitations. LiFi typically has a range of a few meters with a single AP. Light's physical attribute, which offers the security benefit, turns into its biggest disadvantage due to signal loss caused by barriers.

The Hybrid LiFi WiFi Networks (HLWNets) was introduced in 2011 by Rahim et al. [73]. Subsequently, in [90], the combination of LiFi and femtocells is considered. The research demonstrated that such hybrid networks outperformed stand-alone RF or LiFi systems [25]. Figure 12.4 shows the concept of combining LiFi with wireless networks in 5G and beyond communication scenarios. For outdoor communication, macro/micro cells or LiFi-enabled street lamps and satellites can be used to serve mobile users. As they move indoor, they can be served by an HLWNet in order to improve QoS.

12.3.1 Framework of HLWNet

This section introduces a framework for HLWNets across five key components: network architecture, cell deployment, multiple access and modulation schemes, illumination requirements, and backhaul.

12.3.1.1 Network Architectures

LiFi can be integrated into current WiFi infrastructure in two ways: either autonomous and centralized. The first approach is to add LiFi to the existing WiFi autonomous network topology. The AP can use any available channel, and the user is free to select an AP from the two technologies. Although this approach simplifies network management, it affects the network performance in terms of connectivity and user handover. As an alternative, a central control unit can be used to control both WiFi and LiFi APs that are owned by the same person. At the network level, resource allocation and routing can be optimized. The architecture is based on Software-defined networking (SDN), which separates the control plane from the data plane of forwarding devices. In this architecture, SDN-enabled switch connects WiFi APs and LiFi using SDN agents. The SDN controller then utilizes this information to make routing decisions, as shown in Fig. 12.5. This decoupling allows for more flexible and centralized management of the network, improving its efficiency and adaptability.

12.3.1.2 Cell Deployments

LiFi APs have a limited coverage area, often a few meters in diameter, whereas WiFi APs, in indoor scenarios, can extend up to 50 meters. To achieve high-quality network performance, the LiFi APs must be positioned correctly. The cell deployment in practice depends on environmental factors

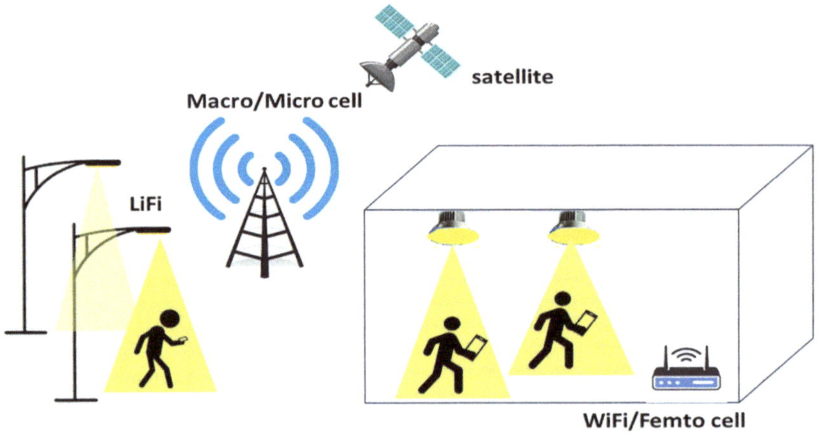

Fig. 12.4 Hybrid LiFi/WiFi network in 5G and beyond communication

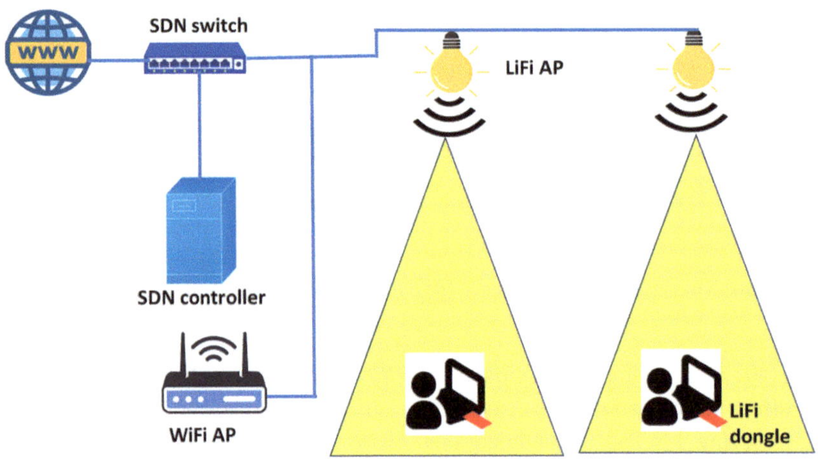

Fig. 12.5 Hybrid LiFi/WiFi indoor network

such as the layout of rooms. The current literature typically considers four cell deployment models: hexagon, matrix and Poisson point process (PPP), and hardcore point process (HCPP).

- The hexagon deployment model is used for cellular networks and offers the maximum signal-to-interference-plus-noise ratio (SINR) coverage probability in LiFi [30]. Even though the hexagon structure of lamps is not common in everyday life, the hexagon deployment serves as a benchmark for upper bound performance analysis.
- The ease of installation makes matrix deployment of lamps a popular choice. Most VLC and HLWNets research works take this deployment into account. It can achieve an SINR coverage probability that is nearly as high as that of the hexagon deployment [30].
- The PPP deployment is used for the simulation of randomly positioned APs. However, achieving uniform illumination is challenging in the case of PPP deployment. In addition, the PPP deployment provides the lowest SINR coverage probability compared to the hexagon and matrix deployments [30].
- The HCPP model is considered a more realistic, albeit pessimistic, approximation for cell deployment in scenarios where two light sources might be placed adjacent to each other—something that rarely happens. To address the limitations of the homogeneous PPP model, the HCPP provides a real-world approximation and has, therefore, been utilized in the analysis of cellular systems [47].

12.3.1.3 Multiple Access and Modulation Schemes

In broad terms, the modulation schemes for LiFi standards falls into two categories: single-carrier and multi-carrier.

Single-carrier modulation includes on-off keying, unipolar Pulse amplitude modulation (PAM), Pulse position modulation (PPM), Pulse width modulation (PWM), and PAM-Discrete multi-tone (DMT) modulation. Single-carrier modulation techniques have low complexity and are better suited for low-rate applications. These modulation techniques are more robust against carrier frequency offset and phase noise than multi-carrier modulation schemes. OOK modulation uses LED diode to represent data. When LED is "ON," it indicates logic "1," and when it is "OFF," it indicates logic "0." OOK modulation is the most commonly used modulation scheme as its implementation is much easier as compared to other schemes. However, single-carrier modulation schemes suffer from Inter symbol interference (ISI) [9]. Therefore, with the increase in the amount of data, there is a need for advanced equalizers in order to implement these modulation schemes and to prevent the ISI over the dispersive channel.

On the other hand, multi-carrier modulation schemes offer high data rates and high spectral efficiency. OFDM is the most commonly used technique for the multi-carrier modulation. In LiFi systems, light signals can reflect off walls, furniture, and other objects, leading to multi-path distortion. OFDM's resilience to multi-path effects is beneficial in LiFi, as it reduces ISI and maintains signal integrity, even in complex indoor environments.

In hybrid LiFi/WiFi systems, multiple access schemes are crucial for managing how multiple devices or users can share the communication medium. There are several multiple access schemes that can be applied in LiFi and WiFi networks. Only CSMA/CA and time division multiple access are being used in LiFi standard. However, using conventional CSMA/CA causes collisions. This issue arises from different transmission environments, as LiFi uses IR in the uplink and visible or IR in the downlink. To address this, a modified CSMA/CA can be used where a channel busy can be broadcast to reduce collision [39]. A more effective approach is to use a different multiple-access scheme altogether. TDMA offers improved bandwidth utilization and power consumption. Other multiple access schemes like OFDMA and NOMA have also been explored for use in LiFi. OFDMA allocated different subcarriers to different users simultaneously, allowing multiple users to access the network concurrently, while NOMA allows multiple users to share the same time and frequency resources by superimposing their signals at different power levels. NOMA demonstrates better performance only when the difference in channel conditions is significant.

12.3.1.4 Illumination Requirements

One of the best ways to increase the energy efficiency in LiFi or hybrid LiFi/WiFi systems that use visible light from LEDs as a downlink is to strategically place or arrange the LEDs within the building. By strategically placing the LEDs and managing their output power levels, energy consumption can be minimized while achieving a desired lighting pattern. When building an LiFi network that uses LEDs as a downlink carrier, it is essential to take room lighting requirements into account. For an indoor setting, the illumination should range between 300 and 1500 Lux [60]. In most rooms, the light is highest in the center and lowest in the corners. Thus, meticulous planning is necessary to ensure uniform brightness.

12.3.1.5 Backhaul Design

Backhaul refers to the links connecting APs to the core network. The main challenges in the implementation of backhaul for HLWNets are: i) network's heterogeneous structure, ii) comparatively large number of APs, and iii) large network capacity, as a single AP can provide link data rates in Gbps. Several technologies, such as power-line communications (PLC), plastic optical fiber (POF), power-over-ethernet (PoE), mmWave, IR, and VLC, have been presented as backhaul solutions for indoor wireless networks.

12.3.2 Performance of HLWNets

The performance of wireless networks is evaluated using various metrics, such as SINR coverage probability, spectral efficiency, area spectral efficiency, energy efficiency, network capacity, QoS, and user fairness. This section reviews these metrics in the context of HLWNet.

12.3.2.1 SINR Coverage

The SINR coverage probability is the probability that the SINR is higher than a certain threshold. RF systems use omnidirectional antennas, and the received signal strength depends on the shadowing and link distance. However, for LiFi and HLWNets, the orientation of user also plays an important role. Specifically, when light is incident normally, it results in peak received signal power. Conversely, no signal is received if the incident direction falls outside the FOV of the PDs. Consequently, variation in the user's orientation can substantially alter the coverage areas of LiFi APs [87].

12.3.2.2 Spectral Efficiency

Spectral efficiency is a key metric for assessing the effectiveness of a communication link. It quantifies the efficiency with which a given amount of frequency spectrum is utilized. A hybrid RF/VLC network offers higher spectral gain as compared to stand-alone RF or VLC networks. In some studies, it is demonstrated through Monte Carlo simulations that the hybrid RF/VLC system can offer superior spectral performance and data speed as compared to purely RF networks.

12.3.2.3 Area Spectral Efficiency

Unlike WiFi, LiFi can achieve high spectral reuse in spatial dimensions, since a single LiFi AP can cover only a small region of 2-3 meters in diameter. Area spectral efficiency (ASE) provides a fair comparison of spectral efficiency between LiFi and RF technologies. ASE measures the sum of the maximum average data rates per unit bandwidth per unit area. According to Stefan et al. [89], LiFi can produce an ASE that is atleast 10 times higher than that of an RF femtocell system. In comparison to a stand-alone RF system, the hybrid LiFi and RF femtocell system described in [90] can raise the ASE by atleast two times. This is due to the inherent interference mitigation in LiFi caused by solid walls in a typical indoor environment, which results in a substantial increase in ASE as compared to an RF system. In other words, the slight obstructions may help the ASE, since they might block more optical interference than the intended optical signals. Wang et al. [94] examined the effects of light path obstructions when analyzing the ASE of HLWNets.

12.3.2.4 Energy Efficiency

Although ASE can be improved by increasing the number of AP deployments, but it also leads to a significant rise in energy consumption. This creates a trade-off between energy efficiency and spectral efficiency, as achieving higher spectral efficiency demands more energy per bit. HLWNets have the potential to enhance energy efficiency by allowing switching of users between different networks. This optimization can be approached by formulating a problem that focuses on allocating bandwidth and power in a way that maximizes the energy efficiency of the network. The authors in [57] used this approach to show that a hybrid RF/VLC network outperforms an RF-only system, achieving up to a 75% improvement in energy efficiency.

12.3.2.5 Network Capacity

Network capacity in wireless networks refers to the maximum data rate that the network can support under specific constraints, such as maintaining a required BER. Network capacity is crucial for ensuring that the network performs well and can handle the data demands placed upon it. HLWNets have the potential to boost the network capacity, particularly in scenarios with dense deployment of WiFi APs. Moreover, the presence of WiFi can help mitigate capacity degradation in LiFi networks due to light-path blockages, as shown in [94].

12.3.2.6 Quality of Service

Another metric that can be used to measure the performance of VLC and hybrid RF networks is QoS. QoS gauges how well users perform on a network. Different metrics such as bit rate, operational efficiency, transmission delay, and link availability can be utilized to assess and quantify the network QoS. Furthermore, additional QoS assessments in the data link layer will need to be taken into account due to the increasing demand for latency-sensitive applications, such as games and video streaming.

12.3.2.7 User Fairness

Fairness measures how evenly resources are allocated among users. Users may require varying latency, throughput, user experience, etc. Therefore, it is imperative to guarantee that every user has an equitable portion of the system's resources. Typical measures to evaluate fairness in resource allocation include Jain's fairness index, max-min fairness, and quality of experience (QoE) fairness. Jain's fairness index is the commonly used metric to evaluate the throughput fairness among users. Maximizing network capacity alone often leads to resource allocation that favors users with better channel quality, which can create significant unfairness, especially in HLWNets where many users are competing for limited WiFi resources. To address this issue and enhance user fairness, proportional-fairness schemes are typically employed for resource allocation in HLWNets.

12.3.3 Effect of User Behvaviour on the Performance of HLWNets

Efficiency and QoS in LiFi cellular networks and, consequently, hybrid WiFi/LiFi networks are influenced by several factors, including user mobility, random orientation of the user's device, and light-path blockage. Numerous mobility models have been discussed in various studies [10], with the Random Way Point (RWP) model being one of the simplest and most widely recognized models used to simulate the movement of users in both indoor and outdoor scenarios. Additionally, the performance of users in LiFi network is decided by two crucial factors: device orientation and link blocking.

12.3.3.1 User Mobility

A lot of research has been done on mobility models to analyze the characteristics of wireless ad hoc networks. These models can be categorized into three groups based on whether the movement has memory or restrictions: random models, models with dependency, and models with geographic restrictions. In indoor settings, the behavior of users is very random and unpredictable as compared to outdoor mobility. In [86], an advanced mobility model was presented based on a rule that simulates realistic indoor movements inside a multi-room building, with users following predetermined paths from one room to another. However, it is challenging to assess a wireless network's overall performance because these models are tailored for certain conditions. As an alternative, random models have been extensively employed in the literature to assess the HLWNets' network performance.

12.3.3.2 Device Orientation

The orientation of user device is a significant factor affecting LiFi network performance, unlike current WiFi systems,

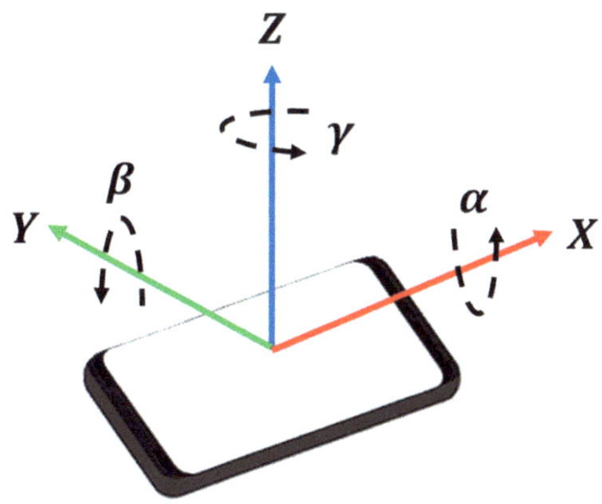

Fig. 12.6 Axial rotation of a user equipment

where device orientation has minimal impact, LiFi systems are highly sensitive to device orientation due to the short wavelength of light. Despite this, there is a lack of suitable models for device orientation, leading many studies related to LiFi to overlook the effects of random device orientation. A few studies have evaluated the performance of Hybrid LiFi-WiFi networks (HLWNets) considering randomly oriented devices, such as those discussed in [38], which is based on Euler's rotation theorem. According to this theorem, any rotation in 3D space can be uniquely described by a combination of three axial rotations in a 3D Cartesian coordinate system as shown in Fig. 12.6.

12.3.3.3 Light-Path Blockage Orientation

Similar to mmWave and THz communications, LiFi is vulnerable to channel obstructions brought on by opaque objects like furniture, walls, and human bodies. Researchers have shown particular interest in the impact of human bodies on LiFi performance, since mobile users are primarily responsible for link blocking in indoor networks. In such cases, the light path to a device can be obstructed either by the person using the device or by others nearby. To model these scenarios, the human body is typically represented as a cylindrical or rectangular object [53]. Figure 12.7 illustrates a rectangular and cylindrical blockage model for human body. This blockage model when integrated with the orientation-aware RWP model can be used to form comprehensive joint model that effectively analyzed the impact of user behavior on LiFi network performance. Additionally, a statistical model of blockage is available, which characterized blockage using two key metrics: occurrence rate, indicating how frequently the blockage happens, and occupation rate, indicating the duration of the blockage. Several strategies can mitigate performance degradation due to blockages. For instance, using an LED with a wider half-intensity angle can help to expand the coverage area and reduce the impact of blockages [48]. However, this approach causes more interference. As an alternative, an omnidirectional receiver with PDs on each side of the handset can be used to make the device more resilient to blockages [31]. With regard to HLWNets, if there is a significant light-path obstruction, the user can be moved to the WiFi system and then back once LiFi access is restored.

12.3.4 Applications of HLWNets

HLWNets offer a versatile platform that can support a variety of applications and services, and they can be categorized into the following groups: communication-based, functional, and advanced applications.

1. **Communication-based applications:** HLWNets are ideal for high-speed wireless communications, such as 4K/8K video streaming, VR, and holographic display. For instance, an integrated system combining mmWave and LiFi has been reported to efficiently transmit data for holographic 3D displays [65]. Additionally, HLWNets can alleviate resource constraints in high-density environments like stadiums, conference venues, and airports.
2. **Functional applications:** These include indoor positioning system (IPS) and physical layer security (PLS). IPS and PLS form the foundation of advanced applications in the IoT. IPS in IoT is crucial for applications requiring precise location tracking within indoor environments. This includes smart homes, healthcare facilities, warehouses, and industrial IoT systems. With the proliferation of IoT devices, ensuring secure communication is essential, especially in environments with limited computational power and energy resources. PLS, in the context of IoT, aims to protect communications against various security threats such as eavesdropping, jamming, and unauthorized access, often through the innovative uses of signal processing, channel properties, and lightweight cryptographic methods.
3. **Advanced applications:** HLWNets facilitate advanced IoT applications, such as consumer IOT and industrial IoT. For example, a hybrid VLC/RF indoor IoT system that harvests solar energy has been demonstrated, highlighting the potential of HLWNets to power a diverse array of future technologies [70].

12.4 Integration of PON and WiFi

The convergence of the optical and wireless technologies can be used to realize Fiber wireless (FiWi) architecture systems. This architecture has gained a lot of attention for supporting a large number of potential subscribers with universal access to information and advanced services in a cost-efficient manner [78]. By utilizing next-generation WLANs, such as IEEE

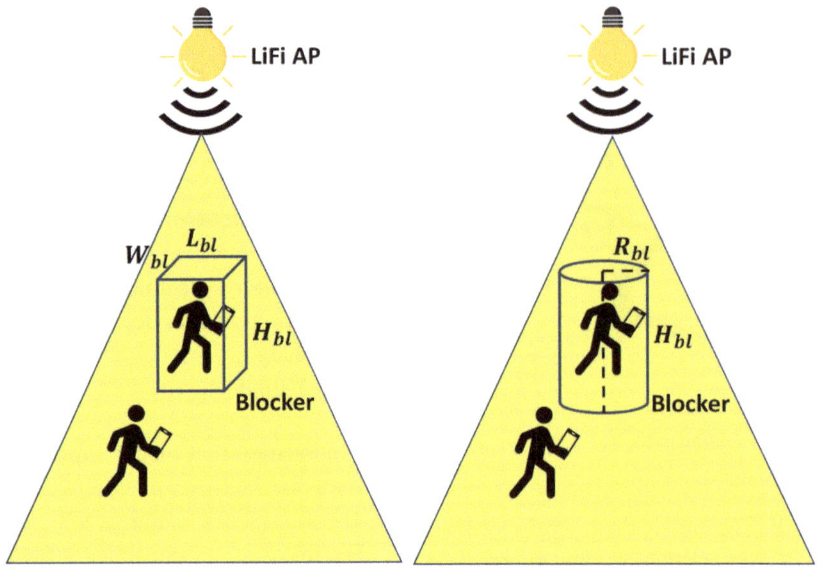

Fig. 12.7 Blockage modelling in LiFi network

802.11n and 802.11ac and XG-PON, it is aimed to achieve high-quality broadband connectivity to provide critical information related to education, health services, agriculture, e-government, etc., to the citizens in rural areas at low cost. WiFi-based WLANs operate on unlicensed bands, possess broadband capabilities, and are easy to deploy, which makes them a much more cost-effective choice compared to the broadband cellular networks. In order to provide QoS differentiation, IEEE 802.11e was introduced that defines two mechanisms, Enhanced distributed channel access (EDCA) and hybrid coordination function (HCF) controlled channel access (HCCA), both backward compatible with the legacy DCF access mechanism to provide QoS support at Media access control (MAC) layer for different kinds of applications in WLANs. IEEE 802.11e EDCA standard specifies four access categories (ACs): voice, video, BE, and background as shown in Fig. 12.8. To impart service differentiation among wireless stations, each AC employs a specific parameter set that includes: (a) Minimum contention window ($CW_{\min}$), (b) Maximum contention window ($CW_{\max}$), (c) Arbitrary inter-frame space number (AIFSN), and (d) Transmission opportunity ($TXOP$). Based on the values assigned to these parameters, the ACs' performance can be altered.

In this section, the performance of integrated XG-PON and EDCA-based WiFi network with deficit DBA in XG-PON considering Poisson Pareto Burst Process (PPBP) that realistically models the real time traffic burstiness of the upstream global IP traffic is analyzed. This work is an extension of [59] in which the authors dealt with converged XG-PON and WiFi architecture considering Pareto on–off traffic model for all incoming services (voice, video, BE) with different frame sizes and data rates and implemented deficit DBA [23] in XG-PON to reduce the average delay for BE service while maintaining the average delays for real-time voice and video services within tolerable limit. The extensions include (a) incorporating realistic traffic models to model voice, video, and BE traffic each with their particular specifications as per the incoming global IP traffic and (b) accordingly tune the values of parameters belonging to EDCA and XG-PON so that e-services can be served to a large number of potential subscribers within tolerable delay and PLR limit. This is quite a challenging task since there is a need to ensure that considerable capacity of XG-PON is effectively utilized by WiFi applications and simultaneously, the resource allocation mechanism is implemented in XG-PON according to the incoming global IP traffic pattern in order to cater to the bandwidth requirement of WiFi clients [23].

12.4.1 System Description

As shown in Fig. 12.9, the system is created by integrating XG-PON and EDCA based WiFi modules in NS-3. XG-PON consists of active elements like OLT in CO and ONUs at/near customer sites that are connected via an ODN using shared optical fiber and passive optical splitter. OLT broadcasts downstream traffic to all ONUs by employing the concept of TDM, whereas TDMA principle is adopted to transmit frames in upstream direction, i.e., from ONUs to OLT.

In upstream, traffic multiplexing functionality is distributed, whereas in downstream direction, it is centralized [24]. Each ONU is connected to multiple WiFi APs (WAPs) via P2P ethernet links. Number of wireless clients connected to a WAP are varied, and each WiFi client generates three kinds of traffic, i.e., voice, video, and BE. An ethernet link is used to connect OLT to the gateway that is linked to internet

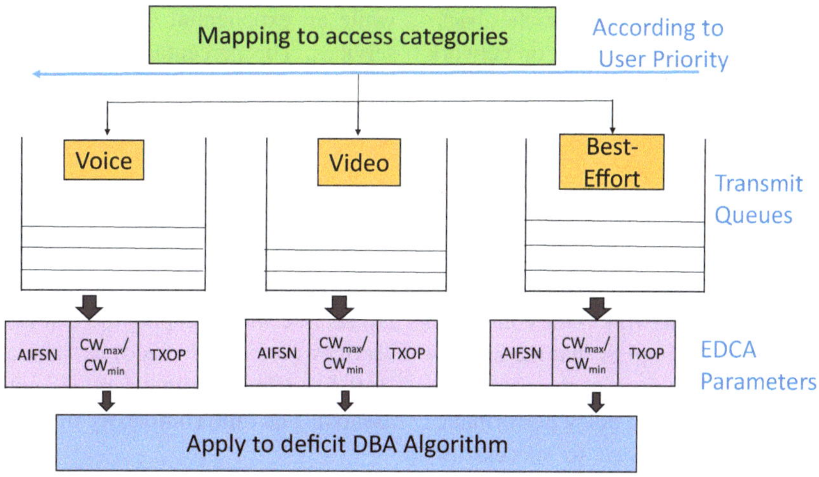

Fig. 12.8 EDCA schematic architecture

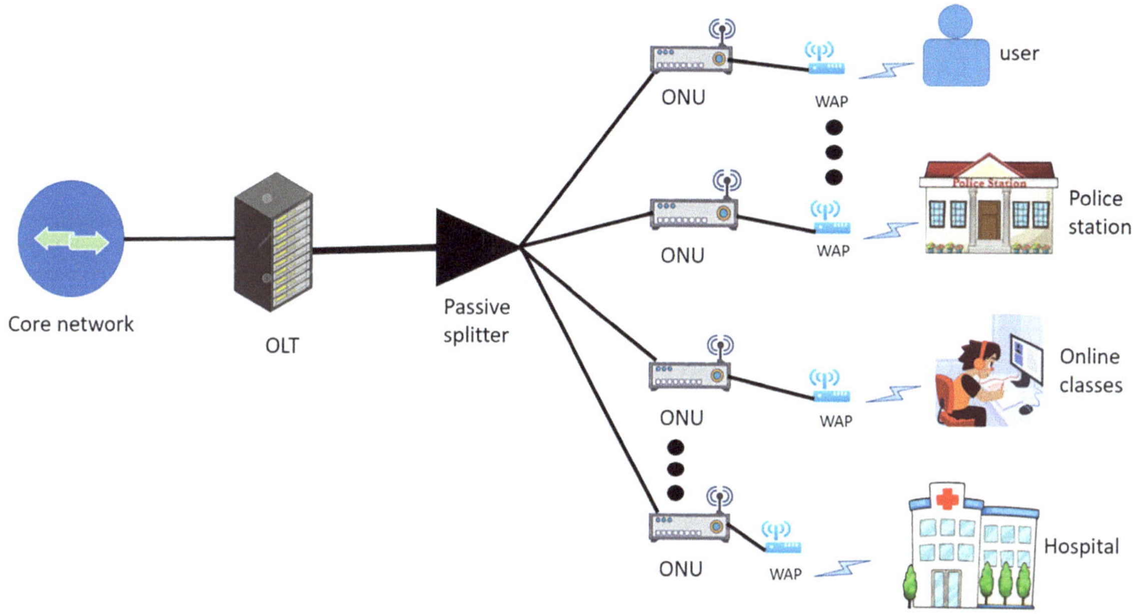

Fig. 12.9 System architecture

servers. Voice, video, and BE upstream internet traffic is generated from WiFi clients. WiFi packets generated from each WiFi client are linked with user priority (UP) value having specific marking from 0 to 7. Based upon this UP value, they are marked with unique differentiated services code point (DSCP) value to classify packets on differentiated services (Diff-Serv) network. They are forwarded to ONU via an ethernet link and placed in appropriate T-CONT queue according to their DSCP value. UP to DSCP mapping is followed as given in [74]. Depending on the UP value and number of clients, packets are assigned $TXOP$, AIFSN, and CW value within EDCA network. $TXOP$ limit is the time period during which a particular wireless station, post the contention process, has an uninterrupted access to the channel. In order to precisely implement $TXOP$ limits and burst sequences, duration of each frame that is en-queued must be calculated and fixed. If frame duration exceeds $TXOP$ limit, queuing is rejected. As soon as a $TXOP$ is granted with $TXOP$ limit > 0, the transmission duration of impending frame and the next frame also needs to be verified. If transmission of both frames with a short inter-frame spacing (SIFS) interval is possible within the $TXOP$, then this frame of burst sequence is ready to be transmitted. There is a remarkable role played by duration of $TXOP$ allocation in modifying the performance of the network. Greater the $TXOP$ allocations for high priority traffic classes, more will be the freezing time for lower priority traffic, which may lead to avalanche effect. This is because during the freezing time of lower priority traffic, a large burst of high-priority traffic might arrive and is expected to be processed before lower priority traffic class finishes back-off countdown. which can have a negative effect on network fairness [66]. Therefore, a large $TXOP$ value cannot be straightforwardly assigned to the higher-priority traffic (such as video and voice) as compared to lower priority BE traffic as it is the majority traffic, which also has to be given enough opportunity to be transmitted while maintaining the delay and PLR limits.

The next parameter that has a notable impact on the performance of the EDCA-based WiFi network is the contention window (CW). CW differentiation implies that the average time needed to successfully attain access to the channel is changed. In order to improve the overall delay performance of the network, the average time needed to successfully deliver a packet is curtailed. This can be achieved by reducing the value of CW [28]. Tuning of $CW_{\min}$ and $CW_{\max}$ depends on the number of competing stations in the network, the priority that needs to be given to the services and other back-off parameters, that is, $TXOP$ and AIFSN. As more stations reduce the CW value, will also increase collision probability as well, overall reducing the effectiveness of the random access mechanism. There needs to be a trade-off among these back-off parameter values according to the given traffic scenario in order to get a well-performing wireless network that can provide the three services, i.e., voice, video, and BE effective to potential subscribers in terms of delay and PLR according to ITU-T QoS standard.

Regarding XG-PON QoS parameters, XG-PON provides differentiated QoS to four service classes, which are specified as T-CONT types. T-CONT 1 corresponds to fixed bandwidth, T-CONT 2 is attributed with assured bandwidth, T-CONT 3 is allotted assured as well as non-assured bandwidth, and T-CONT 4 is assigned to BE traffic. Based on the proportion of traffic entering into T-CONT queue that is identified by an $Alloc-ID$, tuning of parameters, i.e., service interval (SI) that conforms to the number of grant size allotment to each $Alloc-ID$, λ that conforms to the ratio of assured to non-assured bandwidth portion of T-CONT 3 and $ratio$ parameter, which is the proportion of maximum data rate assigned to each T-CONT is done. Bandwidth is dynamically assigned to the T-CONTs according to the deficit algorithm as discussed in [23].

The procedure for tuning the essential performance parameters of integrated architecture according to incoming upstream traffic pattern to get the desired performance in terms of latency and PLR as per ITU-T QoS standard is discussed in Algorithms 4, 5 and 6.

Each wireless station has three different ACs corresponding to three different services, namely voice, video and BE. For service differentiation, each AC has its own set of back-off parameters ($CW_{\min}$, $CW_{\max}$, AIFSN, and $TXOP$). AC1 and AC3 correspond to highest- and lowest-priority traffic, respectively. The PPBP model for video and BE services while an exponential ON-OFF model was used for voice services. The percentage of BE is kept highest by tuning parameter r. Algorithm 4 deals with tuning for BE traffic being dominant traffic and keeping the rest of parameters for other services as per standard 802.11n as given in Table 12.2. $TXOP$ parameter for BE service is varied in discrete multiples of 32 µs [1] with increasing number of wireless stations in order to get the idea how it is impacting the delay and PLR of the network. It ensures the optimum value of $TXOP$ for BE service while keeping the rest of the parameters of other services (voice and video) as standard such that delay and PLR for delay-sensitive services within tolerable limits are assured. The time complexity of Algorithm 4 is $O(n)$.

Algorithm 4 Tuning of $TXOP$ parameter

1: **for** each WiFi client p upto n at WAP **do**
2: **if** WiFi client generates BE traffic **then**
3: Perform step 6.
4: **else**
5: Go to step 7.
6: **end if**
7: Modify $TXOP$ parameter at discrete values 0, 1504, 3008, and 6016µs and perform step 8 after each modified $TXOP$ parameter.
8: Keep the $TXOP$ parameter as standard and perform step 8.
9: Obtain average delay and average PLR for each WiFi client to observe the impact of modification of $TXOP$ parameter for BE traffic on each EDCA service.
10: **end for**

Algorithm 5 Tuning of CW parameter

Require: Upstream traffic with tuned $TXOP$ for each EDCA service
Ensure: Tuned value of CW for each EDCA traffic type.
1: **for** each WiFi client p upto n at WAP **do**
2: **if** WiFi client generates BE traffic **then**
3: Perform step 6.
4: **else**
5: Go to step 7.
6: **end if**
7: Keep $CW_{\min}$ = 15 and decrease $CW_{\max}$ in binary exponential pattern ($2^t - 1$) where t decreases from 10 till 5, and perform step 8 after each decrement.
8: Swap CW ranges for voice and video traffic and perform step 8.
9: Obtain average delay and average PLR for each WiFi client to observe the impact of modification of CW on each EDCA service.
10: **end for**

Once the tuning of $TXOP$ parameter is done, the incoming traffic with tuned $TXOP$ value for all EDCA services is given as an input to Algorithm 5 that deals with fixing of CW parameter for all services. CW parameter plays quite an important role in providing priority-based differentiation to incoming traffic. Initially, the default values of $CW_{\min}$ and $CW_{\max}$ are considered keeping all other parameters of

Algorithm 6 Dynamic Bandwidth Allocation

1: Classify EDCA traffic and associate with appropriate T-CONT based on UP to DSCP mapping policy [74]
2: Implement Deficit algorithm [23,59] given below to assign bandwidth to each T-CONT according to proportion of incoming EDCA traffic.

1. **T-CONT 1**: Fixed periodic bandwidth assignment = 128 Kbps.
2. **T-CONT 2**: $BW_{assign} = \min [AB_{min}, BW_{req}, FB]$. AB_{min} depends on SI and r parameters decided as per incoming traffic proportion and is given as $AB_{min} = (SI \times 125\,\mu s \times r \times C_{XG-PON})/N_k$.
3. **T-CONT 3**: $BW_{assign} = \min [AB_{max}, BW_{req}]$
   - First round; i.e., assured round
     - $AB_{max} = (T2\_BW_{unused} + \lambda \times r \times C_{XG-PON})/N_k$ corresponding to each $Alloc-ID$ of T-CONT 3.
     - Go to step 4).
   - Second round; i.e. non-assured round
     - $AB_{max} = (T2\_BW_{unused} + (1\text{-}\lambda) \times r \times C_{XG-PON})/N_k$ corresponding to each $Alloc-ID$ of T-CONT 3.
     - Go to step 4).
4. **T-CONT 4:**
   - BW_{remain} after provisioning T-CONTs 2 and 3 $Alloc-IDs$ is calculated once per upstream frame.
   - BW_{thres} is calculated based on $BW_{remain}/T4\_N_{unserved}$.
   - BW_{def} is computed based on difference between BW_{thres} and BW_{req}.
   - $tot\_BW_{def} += BW_{def}$

3: **if** $BW_{def} > 0$ **then**
4: $BW_{thres} = [BW_{remain} + tot\_BW_{def}]/T4\_N_{unserved}$
$Grant\_sz = \min [BW_{thres}, BW_{req}]$
5: **end if**

each EDCA service unchanged. Then, the value of CW_{max} is decremented in an exponential pattern as given in step 2 of Algorithm 5 in order to reduce the delay and PLR of BE service while maintaining the delay and PLR values for real-time voice and video services within tolerable limit. The time complexity of Algorithm 5 is $O(n)$.

Tuned EDCA traffic is given as an input to the Algorithm 6 defined as Dynamic Bandwidth Allocation, and appropriate bandwidth is assigned to each T-CONT by the concept of deficit DBA [23]. T-CONT 1 provides a fixed amount of bandwidth statically in a periodic manner irrespective of whether there is demand or not. T-CONT 2 corresponds to assured bandwidth, so accordingly, voice traffic that has the highest priority in EDCA standard is mapped to it. Guaranteed bandwidth is provided to it. Video traffic is managed by T-CONT 3 that provides guaranteed as well as surplus bandwidth decided as per the demand. Real-time video streaming is served by guaranteed part of bandwidth. Non assured part of bandwidth handles the non-real-time video services such as downloadable video-on-demand (VoD) services. T-CONT 4 handles best effort services such as VoIP and OTT messaging applications. After serving the T-CONTs 2 and 3 services, the remaining bandwidth is utilized effectively by calculating the threshold bandwidth taking into consideration the number of unserved T-CONT 4 $Alloc-IDs$. The BW_{def} dynamically adjusts the threshold in order to cater to burstiness of traffic fairly among $Alloc-IDs$. Amount of bandwidth that is granted to a particular $Alloc-ID$ is based on minimum of threshold and requested bandwidth. The time complexity of Algorithm 6 is $O(n^2)$.

Table 12.1 Incoming traffic parameters

AC	Traffic model	Encoding bit rate	Data packet size
BE	PPBP, $H = 0.5$	2 Mbps	1472 bytes
Voice	ON-OFF model where ON and OFF durations are exponential with a mean of 0.35s and 0.65s respectively	64 Kbps	160 bytes
Video	PPBP, $H = 0.9$	500 Kbps	795 bytes

12.4.2 Simulation Environment

The standards-compliant converged system using preexisting XG-PON and EDCA-based WiFi network is implemented in NS-3. XG-PON network is composed of an OLT connected with 16 ONUs via splitter and shared optical cable and ONUs are connected via ethernet links to some APs having varied number of WiFi clients on the WLAN side. Each WLAN has bandwidth of 180 Mbps. Each T-CONT's queue size is taken as 1 MB [2]. There is a propagation delay between OLT and ONU of 0.4 ms that represents 60 km distance. As per the upstream global traffic scenario, specified in [58], majority BE traffic is generated followed by video and then voice considering the parameters mentioned in Table 12.1. In order to properly design and analyze the performance of converged XG-PON and WiFi networks, it is needed to clearly understand the nature of network traffic that is being generated from WiFi clients.

The PPBP traffic model used for generating BE and video traffic characteristics satisfies pivotal statistical properties of real-life IP networks. It enables ease of use and is computationally efficient to allow scalable scenarios [7]. The incoming traffic has characteristic burst length that is present on a wide range of timescales and is effectively captured in PPBP traffic model. It can be statistically interpreted by principle of self-similarity, and it exhibits long range dependence [35]. The auto-correlation function of such self-similar process follows power law delay, which is much slower than exponential delay. Degree of time series of self similar model is expressed by hurst parameter (H). It expresses speed of decay of series auto-correlation function and for long-range dependence (LRD), its value varies from $1/2 < H < 1$. As it approaches 1, degree of both self-similarity and LRD increases. Regarding PPBP modelling, bursts arrive according to poisson process with rate λ_P, and the burst length follows Pareto distribution given by H parameter. Each burst flow has a constant bit rate (CBR) r_b [8]. Number of wireless stations generating voice, video, and BE traffic are taken in such a way that approximate 5–8%, 25–30%, and 65–70% are voice, video, and BE traffic, respectively. Each wireless station would operate only one application at a given instant of time. Each wireless station is connected to a single WAP, which in turn links to a specific ONU.

12.4.3 Performance Evaluation

The performance of the integrated architecture for each class using the simulation environment as provided in Sect. 12.4.3 is evaluated in terms of PLR, mean queuing-delay, end-to-end delay, per class throughput, aggregate throughput, and fairness index. Mean queuing-delay of a given $Alloc - ID$ is the average queuing delay of all the packets entered into and forwarded from the T-CONT queue in the ONU. Throughput is the average throughput per station for each AC. Jain's fairness index is used to evaluate the extent upto which the DBA allocates bandwidth to all WiFi clients. It indicates the fairness of Bandwidth Allocation to all the users. As its value of fairness index becomes more closer to 1, it implies that demand and allocation almost match one another, thus providing greater QoS guarantee. Each experiment is executed for 60 seconds until upstream channel gets overloaded. The experiment is repeated 100 times, and for each experiment, it is considered 10 distinct seed values so that every pattern of packet generation in applications is mutually exclusive with one another. Confidence interval of 95% is considered for obtaining the simulation results. Margin of error for the obtained results is $< 7\%$.

The default EDCA parameters are given in Table 12.2. The three parameters of EDCA, $TXOP$, AIFS, and CW have a complicated relationship with each other. Similarly, there are some important parameters like SI, λ and r parameters that play a crucial role in deciding the bandwidth assignment to different incoming services according to their priority. Since one parameter affects the performance of another parameter, hence joint optimization of the integrated architecture is needed. This involves a feedback mechanism to $TXOP$ and CW values in order to get the system performance like the average end-to-end delay, mean queuing-delay, and average PLR maintained according to ITU-T standard.

Table 12.2 Standard EDCA parameters for 802.11n

AC	CW_{min}	CW_{max}	AIFSN	Max. $TXOP$ (μs)
BE	15	1023	3	0
Voice	3	7	2	1504
Video	7	15	2	3008

Taking into consideration the modeling parameters of BE traffic, which also is the dominant of all, the performance of Algorithm 4 and observed that by increasing the $TXOP$ within tolerable range of BE traffic from 0 to 3008 μs with a step size of 1504 μs, due to bursty arrivals, $TXOP$ allocations at low as well as high loads are larger due to which the queuing delay of the service is lower. But as $TXOP$ increases beyond $TXOP = 3008$ μs, it was observed that there is no noticeable improvement in the delay with increase in $TXOP$, as $TXOP$ allocation is incremental, and maximum allocation is only done at higher loads depending upon the burstiness of traffic. As the WiFi clients increase, it may not be a good idea to have high $TXOP$ as the chances of occurrence of avalanche effect increase, which may strongly impact the PLR of low priority traffic due to high back-off time for respective AC. So, there is a limit to which $TXOP$ can be increased and gets its positive impact on the performance of the network, and since BE traffic is of the lowest priority, it cannot improve its performance at the cost of other services. Therefore, the TXOP value is selected in such a way that the values of $TXOP$ remain in permissible range according to the ITU-T standard [1] and still can achieve a performance improvement.

Besides $TXOP$ setting, the initial back-off window size tuning is also very crucial according to network size [41] to minimize delay in the network. CW_{min} and CW_{max} values have to be jointly tuned to minimize the mean-queuing delay, end-to-end delay, and average PLR as per the number of wireless stations in the integrated network. As discussed in Algorithm 5, it is needed to first tune the BE traffic's CW range, being the dominant traffic. Since voice has the highest priority according to standard EDCA, it has a minimum CW range, as can be observed from Table 12.2, followed by video traffic. Depending on the proportion of incoming voice and video traffic and their respective traffic models, the CW range keeps in view the ITU-T QoS standard delay and PLR values that need to be maintained for both services. It was observed that the average delay for voice as well as video traffic was reduced, whereas the delay for BE service increased. Although there is a difference in initial CW sizes, the two ACs, voice and video, have high probability of successful transmission of their packets since it is decided based upon the activities done by each node in the network [41] and also on the proportion of kind of traffic being generated in the network. Due to decrease in initial CW size for video traffic, the stations generating video traffic aggressively contend for the channel along with voice, due to which the waiting time and hence the back-off time of BE traffic further increase. Also, voice traffic incurs increased delay but is still within the tolerable limit. With fixed CW size, as the number of wireless stations increases, collisions to increase due to the rate at which packets are lost in the network, and PLR increases. In order to reduce PLR and delay of BE service, being the

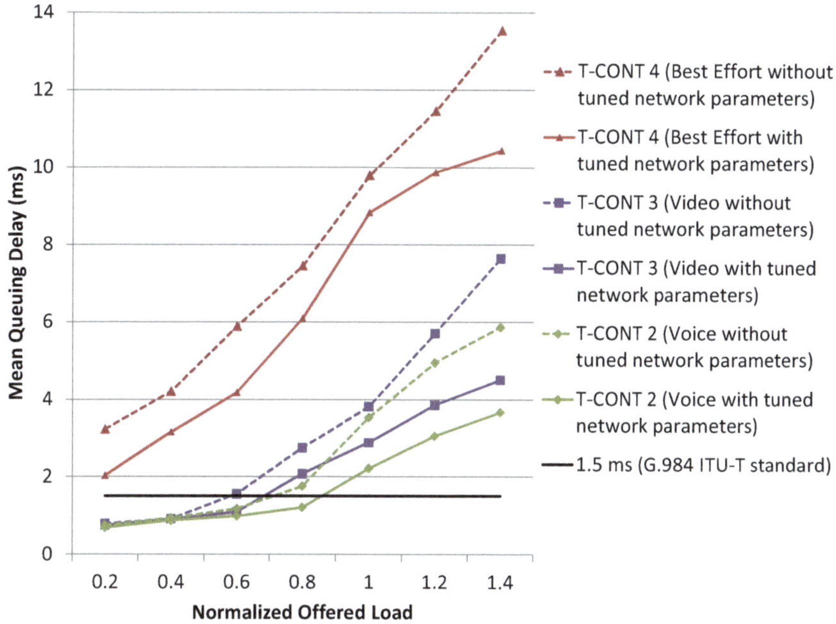

Fig. 12.10 Mean queuing-delay vs normalized offered load

dominant traffic, it is needed to adaptively tune the initial CW size and linearly increase it with an increase in a number of contending stations [41].

After tuning the EDCA network parameters, a tuning of the XG-PON parameters as per the incoming traffic needs to be done. Voice traffic flows via the T-CONT 2 queue, and according to the deficit DBA, the T-CONT 2 traffic is served in every allocation cycle. Video traffic corresponds to T-CONT 3 and BE to T-CONT 4. r parameter also has to be tuned based on the proportion and priority of incoming traffic type. After exhaustive simulations, it was found that to maintain the priority trend for T-CONT 2, T-CONT 3, T-CONT 4 while maintaining the proportion of incoming traffic, r = T-CONT 2:T-CONT 3:T-CONT 4 = 60:50:70 needs to be set. By setting ratio T-CONT 2 $>$ T-CONT 3, a priority of T-CONT 2 over T-CONT 3 for all loads is to be maintained. T-CONT 4 is also given a good proportion of bandwidth because it is the dominant one, and this ratio also provides a moderate overprovisioning factor. The overprovisioning factor should also not be too high; otherwise, it will induce DBA burstiness in the XG-PON backhaul for WiFi [23]. In alternate allocation cycles, the maximum allocation bytes are assigned depending upon the value of λ, which is equal to 0.4. The unused bandwidths of T-CONT 2 and T-CONT 3 are calculated once in every upstream allocation cycle, and accordingly, the threshold is also calculated, which also considers a number of unused T-CONT 4 $Alloc - IDs$.

Figure 12.10 shows the mean queuing-delay performance of deficit DBA in the designed integrated architecture for all loads. It can be observed that fine-tuning of XG-PON and EDCA parameters results in an improvement in the system performance for all the three types of e-services maintaining the priority among $Alloc - IDs$ of all T-CONT types. It can also be observed that for real-time services like voice and video, the mean queuing-delay performance is maintained within the tolerable limits of ITU-T standard. The maximum margin of error for mean queuing-delay is 3%.

Figure 12.11 shows the end-to-end delay performance of deficit DBA. According to [18], the tolerable limit for delay for voice, video, and BE services is 150 msec, 400 msec, and 5 sec, respectively. It can be observed from Fig. 12.11 that using deficit DBA, the end-to-end delay performance is maintained within these limits for each offered load. The tuning of network parameters further enhances the performance of the system. The margin of error for end-to-end delay analysis is less than 3%.

The performance of the integrated architecture in terms of PLR and throughput as shown in Figs. 12.12 and 12.13, respectively. It is observed that after thorough tuning of XG-PON and EDCA parameters, the achieved PLR of $<1\%$ [24] for video upto 60% total offered load for video traffic without compromising on the standard PLR (without tuning) performance. For voice, the PLR of $<3\%$ [24] and for best effort, PLR of $<6\%$ is achieved upto full load. The maximum margin of error for each service is 5%. It can be observed that due to fine-tuning of integrated architecture parameters, an improvement in throughput performance for all three types of service. The throughput evaluation for average throughput for each AC, voice, video, and best effort in Fig. 12.13 implies that during light load conditions, all three ACs are able to achieve approximately same throughput with margin of error upto 7%. But the offered load is

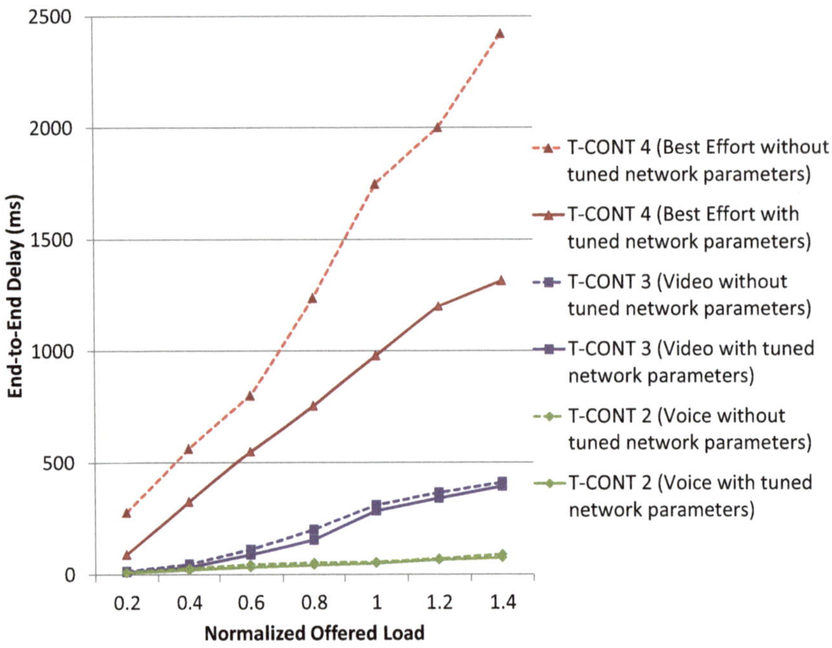

Fig. 12.11 End-to-end delay vs normalized offered load

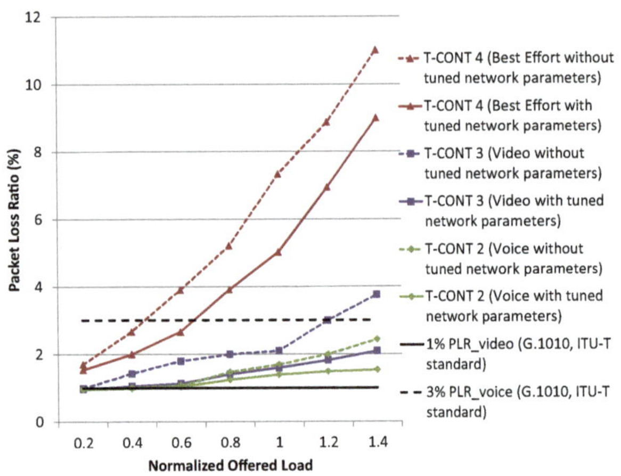

Fig. 12.12 Packet loss rate (%) vs normalized offered load

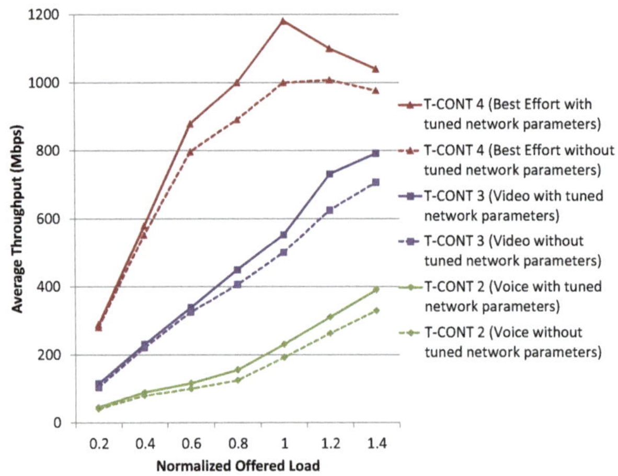

Fig. 12.13 Average throughput vs normalized offered load

increased, the average throughput of voice and video traffic increases almost linearly, which happens at the cost of best-effort traffic. This is due to the fact that when the network is overloaded, the priority of voice and video traffic will affect the performance of the best-effort traffic.

Figure 12.14 demonstrates the performance of integrated system with deficit DBA in terms of fairness index. Fairness is calculated by using Jain's fairness index as:

$$f(\gamma) = \frac{\sum_{i=1}^{n} \gamma_i}{n * \sum_{i=1}^{n} \gamma_i^2}, \gamma_i > 0$$

where, $\gamma_i = a_i/d_i$ for $a_i < d_i$ and $\gamma_i = 1$ otherwise, n is the number of contending users, a_i is the resource allocated to the user, and d_i is the corresponding demand. By using deficit DBA and proper tuning of integrated network parameters, the fairness performance of BE traffic is not at the cost of real-time voice and video services; instead, the fairness of voice and video traffic also improves. This indicates that bandwidth is allocated based on the incoming traffic distribution and its priority. The maximum margin of error for the fairness index is 4%.

Figure 12.15 demonstrates the aggregate throughput over the entire range of offered load considering all three service classes. The figure shows the effective utilization of upstream bandwidth capacity upto maximum of 2.20 Gbps for fully as well as overloaded conditions.

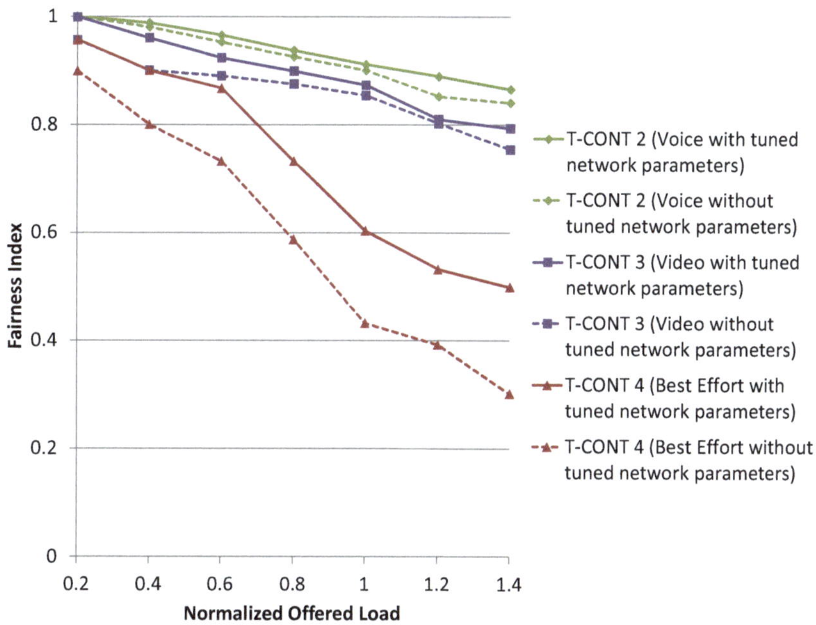

Fig. 12.14 Fairness index vs normalized offered load

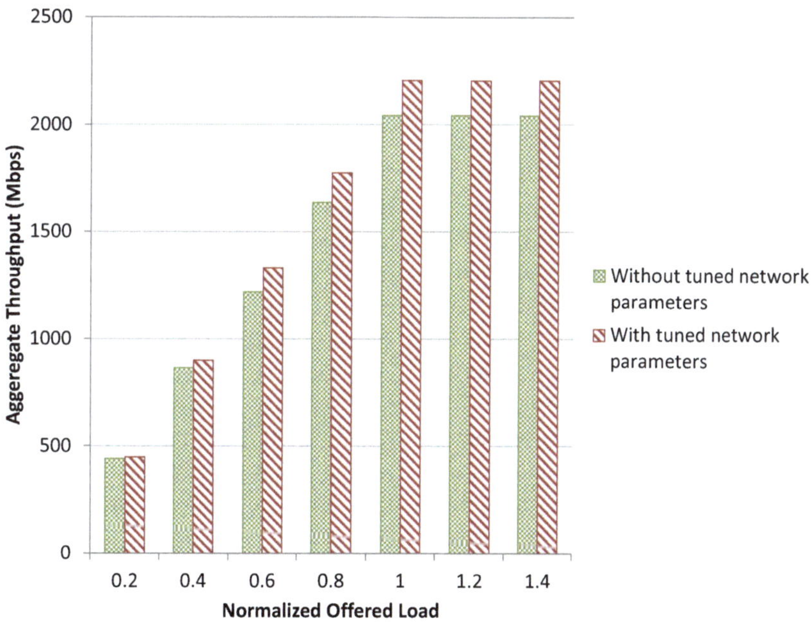

Fig. 12.15 Aggregate throughput vs normalized offered load

12.5 Broadband Access Networks

With the exponential increase in network traffic, network service providers are facing numerous challenges, prominent among them are an increase in network capacity, low latency, high data rates, and higher deployment and maintenance costs. FiWi architecture is a promising solution to mitigate some of the above challenges [45]. In the conventional FiWi architecture, next-generation access network technologies such as Next generation-PON2 (NG-PON), XG-PON are integrated with WiFi based front-end networks such as IEEE 802.11n/ac/ax [58]. XG-PON is a promising technology that provides data rates upto 10 Gbps and 2.48 Gbps in the downstream and upstream directions, respectively [19]. In FiWi architecture, an OLT is placed at the central office and is connected to a passive splitter using a feeder fiber. Further, this optical splitter is connected to the ONUs using distribution fibers [85]. Each ONU is colocated with a wireless AP or a cellular BS, which provides the required

QoS to the users via wireless links [44]. Specifically, when a wireless AP is utilized for the front end wireless links, the architecture is known as FiWi architecture. It is worth noting that large-scale deployment of optical fibers in the conventional FiWi architecture may not be appropriate for unconventional deployment scenarios such as difficult terrains, remote locations, or urban city centers where additional fiber deployment may be prohibited due to limited flexibility and high installation cost. One possible solution to alleviate the above is to introduce the Free Space Optics (FSO) link in AON. Unlike PON, AON is based on Ethernet, in which active splitter is equipped with Ethernet switch. An active splitter is also connected to multiple ONUs and the central office. Each ONU has a dedicated fiber link to the active splitter. The Ethernet switch at the active splitter aggregates the traffic from the ONUs and connects via fiber link to another Ethernet switch at the central office [91, 93].

FSO is an attractive solution as it offers greater flexibility and easier deployment. FSO has numerous benefits such as unlicensed spectrum, high bandwidth, immune to electromagnetic interference, high-speed communication, and quick installation [40]. FSO uses LD as transmitter and PD as receiver. Due to the coherent nature of laser light, FSO provides long distance point-to-point communication with higher data rates. Therefore, the integration of FSO-based back-end and WiFi-based front-end in XG-PON architecture can exploit the advantages of both FSO and WiFi to provide flexibility along with higher data rates [84].

In VLC, transmitter utilizes LEDs luminaries or LDs to generate an optical signal. The optical signal received at the PD is transformed into a corresponding electrical signal for further processing. VLC possesses numerous benefits such as unlicensed bandwidth, no electromagnetic interference, and ease of availability [14]. Light fidelity (LiFi), as discussed previously in Chap. 6, is a networking variant of VLC, which provides point-to-multipoint communication. LiFi builds over the physical layer of VLC to avail reliable, bidirectional, and fully networked indoor wireless communications [14, 33]. Therefore, FSO-based back-end can complement LiFi-based front-end in XG-PON architecture to provide a high transmission rate along with improvement in overall system performance.

12.5.1 Description of Different Architectures

As mentioned before, in addition to the conventional FiWi architecture, five architectures such as FoLi, FoWi, FiLi, FoLA, and FiLA are considered. Table 12.3 shows the backhaul and fronthaul technology used in aforementioned architectures. These architectures are described in detail in the subsequent subsections.

Table 12.3 Different architectures

Different architectures	Backhaul technology	Fronthaul technology	Shown in figure
FiWi	Fiber	WiFi	Fig. 12.16a
FoLi	FSO	LiFi	Fig. 12.16b
FoWi	FSO	WiFi	Fig. 12.17a
FiLi	Fiber	LiFi	Fig. 12.17b
FoLA	FSO	LA	Fig. 12.17c
FiLA	Fiber	LA	Fig. 12.17d

12.5.1.1 FiWi Architecture

Figure 12.16a shows the conventional FiWi network composed of optical fiber-based XG-PON back-end network and WiFi-based front-end network. In FiWi architecture, the OLT is situated at the central location, which is further connected to a passive optical splitter using feeder fiber. A passive optical splitter is connected to the number of ONUs through distribution fiber. These ONUs are colocated and connected with WiFi APs. XG-PON utilizes TDMA and TDM for upstream and downstream traffic flow, respectively. For indoor communication, a room of 10 m $\times$ 10 m $\times$ 3 m size is considered in which one WiFi AP is supposed to provide connectivity to the users inside the room. Further, N mobile users are distributed uniformly within the coverage area of WiFi AP.

12.5.1.2 FoLi Architecture

In this architecture, FSO-based XG-PON back-end network is connected to LiFi based front-end access network as shown in Fig. 12.16. In FoLi architecture, it is assumed that OLT consists of LD for generating an optical signal. The optical signal is transmitted to the ONUs through ODN. In ODN, each ONU is connected to the active splitter through FSO link, whereas active splitter is connected to OLT using another FSO link. At the ONUs, PDs are employed to receive optical signal. Further, the ONU is connected/colocated with LiFi APs. As shown in Fig. 12.16b, four LiFi APs are installed inside the room to provide high data rate connectivity to the users present inside the room.

12.5.1.3 FoWi Architecture

As shown in Fig. 12.17a, in FoWi architecture, OLT is connected to an active splitter through FSO link, and further, active splitter is connected to ONUs through another FSO link in ODN. In FoWi, OLT utilizes LD to generate an optical signal. This optical signal reaches to ONUs through FSO-based ODN. At the ONUs, an optical signal is received by PD and processed, further ONUs are connected/colocated with WiFi AP. For indoor communication, one WiFi AP is assumed to provide seamless wireless connectivity to the users present inside the room.

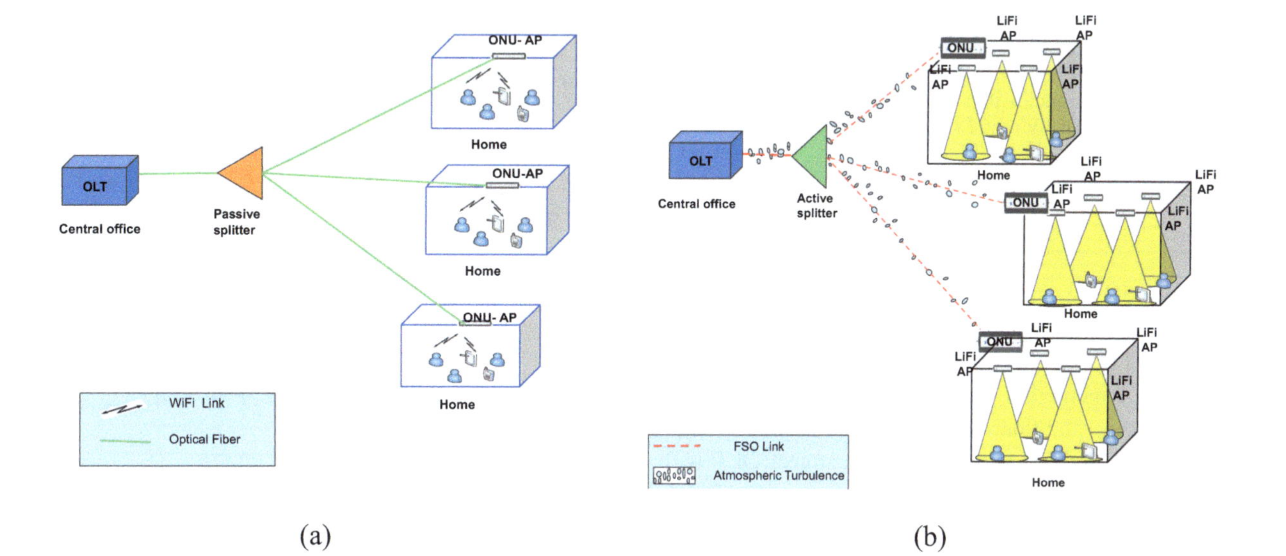

Fig. 12.16 (**a**) Conventional FiWi architecture (**b**) FoLi architecture

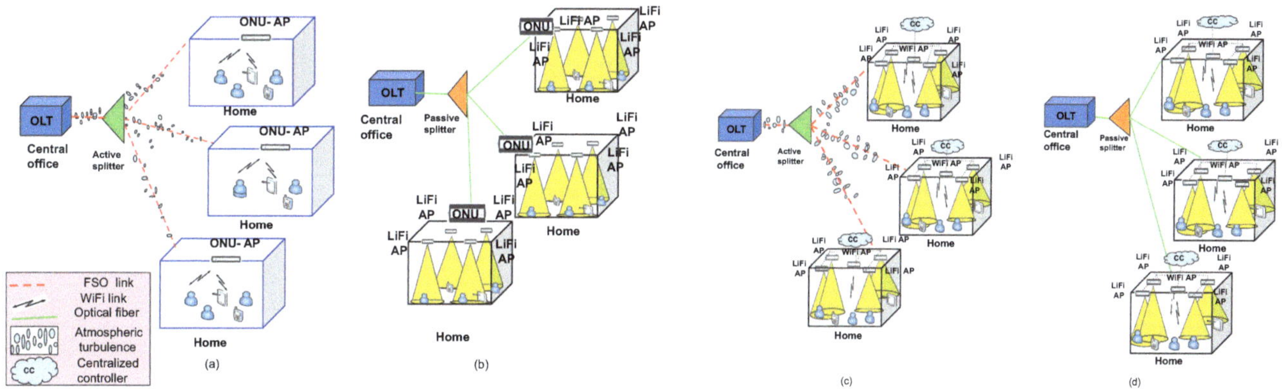

Fig. 12.17 System model for (**a**) FoWi, (**b**) FiLi, (**c**) FoLA, (**d**) FiLA

12.5.1.4 FiLi Architecture

This architecture consists of an optical fiber-based back-end and LiFi-based front-end access network. As shown in Fig. 12.17b, OLT is connected to a passive optical splitter using feeder fiber. In ODN, a passive splitter is connected to ONUs via distribution fibers. In the front-end network, like FoLi architecture, it is considered that ONU is colocated and connected with four LiFi APs in the indoor space.

12.5.1.5 FoLA Architecture

As LiFi and WiFi operate in the non-overlapping spectrum, it is possible that WiFi and LiFi can coexist together and complement each other. Therefore, in LA scheme, both WiFi AP and LiFi APs are connected to a central controller (CC) through an error-free link. A user is allowed to receive data concurrently from WiFi AP and LiFi APs using a LA scheme [15, 56]. In the FoLA system, LA scheme is implemented to improve the system performance in terms of average throughput and fairness. In Fig. 12.17c, OLT is connected to the splitter using FSO link, and the splitter is connected to ONUs via another FSO link. Further, ONU is connected with a CC to enforce LA scheme over WiFi and LiFi APs in the room. Like FoLi and FiLi architectures, there are four LiFi APs installed inside the room. Moreover, along with LiFi APs, a single WiFi AP is also placed inside the room. It may be noted that LA-enabled users can access data from both WiFi AP and LiFi APs concurrently with the help of CC.

12.5.1.6 Algorithm for Link Aggregation

The implementation of LA scheme in FiLA and FoLA architectures is considered as given in [56]. The LA-enabled user aggregates data from both WiFi AP (D_u^{WiFi}) and LiFi AP(D_u^{LiFi}) concurrently. LiFi and WiFi receiver in LA-enabled user will work independently. To include effect of practical realization, average LA overhead (β_{oh}) is included. As per literature [16], the value of β_{oh} is considered to be 0.8.

Algorithm LA algorithm

Input: Given total number of the users, N.

1: **for** $u = 1, ...,$ maximum N **do**

2: Calculate data rate for each LA enabled user

$$D_u^{LA} = \beta_{oh} \times (D_u^{\text{WiFi}} + D_u^{\text{LiFi}}),$$

3: **end for**

12.5.1.7 FiLA Architecture

As shown in Fig. 12.17d, like FoLA architecture, the LA scheme is implemented at the front-end access network in FiLA architecture, and fiber-based back-end network is considered. In ODN, feeder fiber is used to connect OLT with passive splitter, further passive spiltter is connected to ONUs using distribution fibers.

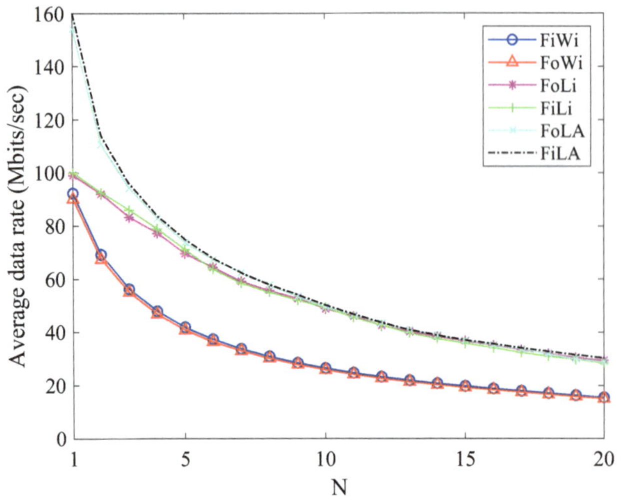

Fig. 12.18 Average data rate vs the number of users

12.5.2 Channel Modeling

The channel modeling will depend on the hybrid architecture considered for deployment. This section describes the channel model used for FSO, WiFi, and LiFi links, along with different factors that affect them. FSO link is affected by atmospheric turbulence mainly caused by random refractive index fluctuation of the atmosphere. The effect of atmospheric turbulence on FSO link is characterized by the Gamma–Gamma distribution fading [82]. Apart from the atmospheric turbulence effect, pointing error also impacts the performance of FSO link. The pointing error is induced by FSO transmitter and FSO receiver's misalignment, resulting from the strong wind, earthquakes, and thermal expansion. To mitigate the effect of misalignment, an acquisition, tracking, and pointing system can be installed to align the optical beam [17]. However, the misalignment problem can also be alleviated by the installation of multiple FSO apertures at FSO transmitter [27]. In addition, LiFi and WiFi channels for indoor communication can be modeled as [46] and [20], respectively.

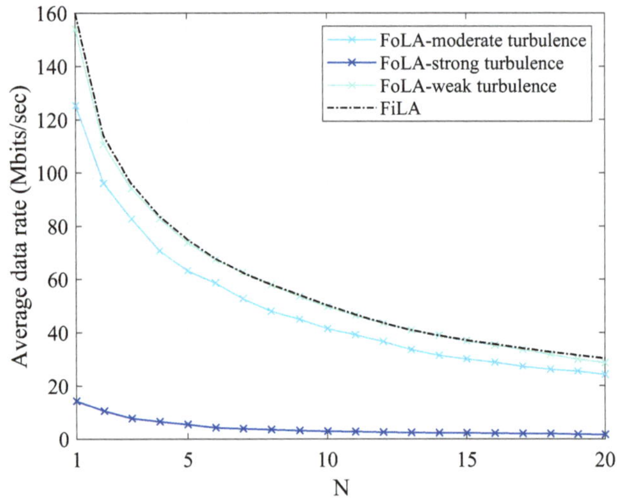

Fig. 12.19 Average data rate vs the number of users

12.5.3 Performance Evaluation

In this subsection, the simulation results of the FoLi, FoWi, FiLi, FoLA, and FiLA are compared with the conventional FiWi architecture in terms of average data rate and fairness, and the CAPEX of the aforementioned hybrid architectures is compared.

Figure 12.18 illustrates the average data rate of the users with respect to the number of users (N) for the aforementioned architectures and the conventional FiWi architecture. As shown in Fig. 12.18, the average data rate per user in downlink decreases with rise in N. It can be observed that FiLA architecture outperforms other hybrid architectures under weak turbulence with less pointing error ($\xi_2 = 43.96$). It may also be noted that FiLA and FoLA architectures outperform FiLi and FoLi architectures with respect to average data rates per user for $N < 10$. Further, it is important to note that for $N > 10$, the performance of FiLi, FoLi, FiLA, and FoLA architectures saturates. Architectures with LiFi and LA front-end, i.e., FiLi, FoLi, FiLA, and FoLA, outperform FiWi and FoWi architectures in terms of average data rate, specifically in the presence of weak turbulence with lower pointing error. However, the performance of the architecture deteriorates with rise in atmospheric turbulence from weak to moderate/strong turbulence. This moderate/strong turbulence effect can be alleviated by implementing a proper optical amplifier [3, 13]. Figure 12.19 shows a comparison between FiLA and FoLA architectures under the impact of weak, moderate, and strong turbulence. It is clearly observed

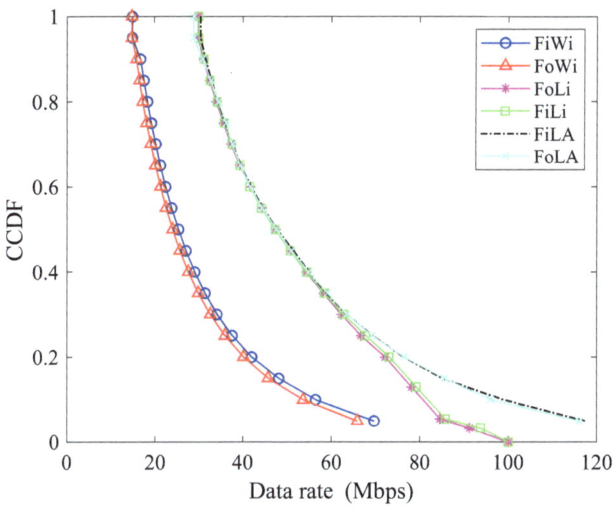

Fig. 12.20 CCDF of user's data rate (Mbps) for different architectures and the conventional FiWi architectures

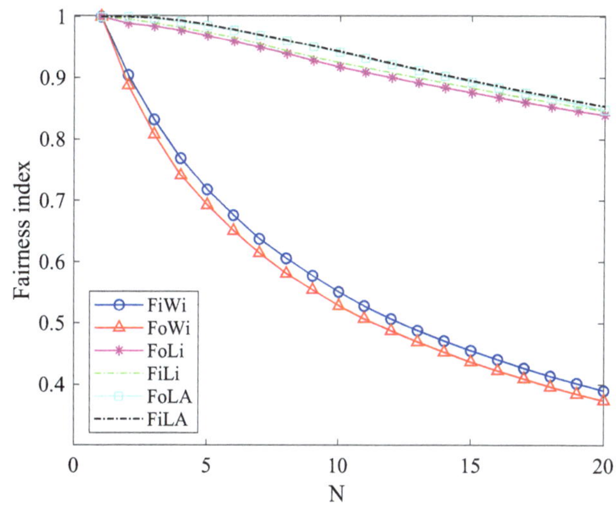

Fig. 12.21 Fairness index vs the number of users

that as the strength of turbulence increases, the average data rate per user in FoLA architecture deteriorates as compared to FiLA architecture. It may also be noted that the performance of FoLA architecture in weak turbulence is very close to the FiLA architecture. Another practical solution for mitigating the effect of moderate/strong atmospheric turbulence is to implement spatial diversity techniques, e.g., repetition coding [72] or orthogonal space-time block codes [88]. Figure 12.20 depicts the Complementary cumulative distribution function (CCDF) of the user's data rate for different architectures. It can be noted that 50% of the users achieve more than 24 Mbps and 26 Mbps data rates in FiWi and FoWi architecture, respectively. While in the case of FoLi, FiLi, FoLA, and FiLA architectures, 50% of the users can achieve more than 46 Mbps. Therefore, from the CCDF plot, it can be seen that aforementioned architectures perform better than FiWi and FoWi architectures. It is also interesting to observe that for the less number of users, FoLA and FiLA architectures perform much better than FoLi and FiLi architectures.

12.5.3.1 Fairness Analysis

In order to compare the performance of the architectures with the conventional FiWi architectures, other architectures in terms of fairness are evaluated. For the fairness analysis, the Jain's fairness index is adopted for measuring fairness among the users, and it is given as [51]

$$J = \frac{\left(\sum_{u=1}^{N} D_u\right)^2}{N \sum_{u=1}^{N} D_u^2}, \qquad (12.1)$$

where D_u denotes the individual data rate achieved by the uth user, N indicates the number of users. It can be observed from Fig. 12.21 that the FiLi, FoLi, FiLA, and FoLA architectures can achieve better fairness among the users than FiWi and FoWi architectures. It is observed that LA-based front-end architectures such as FoLA and FiLA architectures achieve full fairness when $N = 3$. Furthermore, as N increases, fairness among the users is compromised. However, for $N = 20$, as compared to FiWi and FoWi, LiFi and LA-based front-end architectures such as FoLi, FiLi, FoLA, and FiLA can achieve a fairness index of more than 0.8. Furthermore, it is also observed that the FiLA and FoLA outperform LiFi-based front-end architecture such as FiLi and FoLi in terms of fairness. For $N = 10$, fairness index of 0.94 is achieved in case of FiLA and FoLA. While in the case of FiLi and FoLi, fairness index of 0.92 is achieved. The cost analysis of the hybrid FoLi, FiLi, FoLA, FiLA, and the conventional FiWi architecture having the same split ratios is performed. 1 : 16 as a split ratio for the cost analysis is assumed. The CAPEX of FiWi architectures C_{FiWi} is given as [55]:

$$C_{\text{FiWi}} = C_{\text{OLT}} + C_{\text{splitter}} + L_{ff}C_{ff} + \sum_{m \epsilon N_{\text{ONU-AP}}} L_{df}^{m} C_{df} + N_{\text{ONU-AP}} C_{\text{ONU-AP}}, \qquad (12.2)$$

where C_{OLT} is cost of OLT, C_{splitter} is cost of splitter, L_{ff} is feeder fiber length, C_{ff} is cost of feeder fiber per km, L_{df}^{m} is distribution fiber length between splitter and mth ONU, C_{df} is cost of distribution fiber per km, $N_{\text{ONU-AP}}$ is the number of ONU-APs, and $C_{\text{ONU-AP}}$ is cost of ONU-AP. While for the architectures such as FoWi, FiLi, FoLi, FiLA, and FoLA, (12.2) can be modified to calculate CAPEX, C_{PA} as

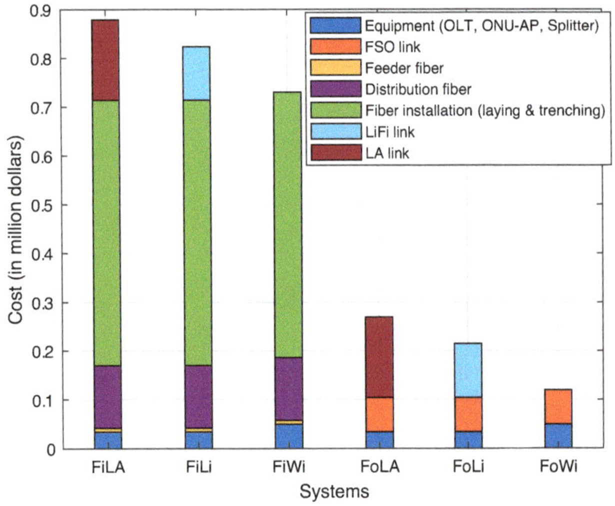

Fig. 12.22 Total deployment cost of the architectures and FiWi architecture

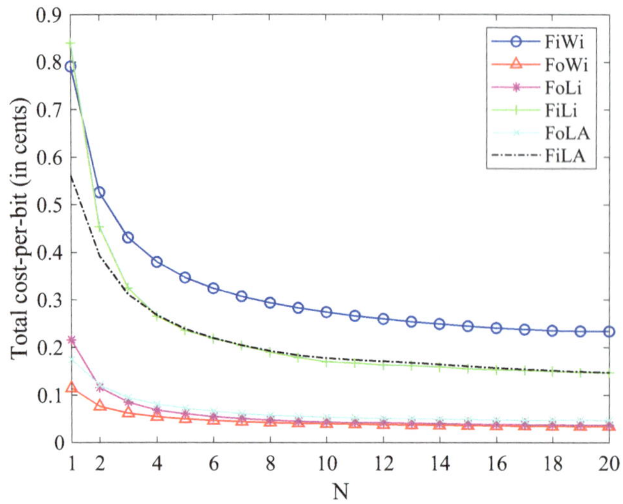

Fig. 12.23 Total cost-per-bit vs the number of users

Table 12.4 Cost values of network components [9,21,50]

Network Component	Unit cost ($)
OLT	27560
Splitter (1:16)	750
ONU	350
ONU-AP	1350
Fiber (/km)	4000
Fiber laying and trenching (/km)	16000
FSO link	4124
Hybrid LiFi and WiFi link (with LA capability)	10337
LiFi link (4 LiFi APs, 4 luminaries, 4 dongles)	6885

$$C_{PA} = C_{\text{OLT}} + C_{\text{splitter}} + C_{FSO1} + \sum_{m \in N_{\text{ONU-Eq}}} C^{m}_{FSO2} + N_{\text{ONU-Eq}} C_{\text{ONU-Eq}}, \quad (12.3)$$

where C_{FSO1}, C^{m}_{FSO2} indicates the cost of FSO link between OLT and splitter, and splitter and mth ONU, respectively. $N_{\text{ONU-Eq}}$ is the number of equipment such as ONU-AP, LiFi-AP, and LA-AP. $C_{\text{ONU-Eq}}$ denotes the cost of equipment such as ONU-AP, LiFi-AP, LA-AP. Figure 12.22 compares the total deployment cost of the FoWi, FiLi, FoLi, FiLA, and FoLA with the conventional FiWi architecture. For the CAPEX analysis, Table 12.4 lists the cost values of network components. It can be observed that the primary contribution in the total deployment cost of the conventional FiWi architecture is attributed to the cost of fiber infrastructure, which includes digging, trenching etc. It is observed that optical fiber purchasing and installing costs are high in fiber-based back-end architectures such as FiLA, FiLi, and FiWi architecture. While in FoLA, FoLi, and FoWi architecture, the cost associated with fiber and its installation cost can be minimized by using FSO links in the back-end network. Therefore, it is worth noting that by deploying FSO-based back-end hybrid architecture, nearly 60.9% of CAPEX can be saved compared to the fiber-based back-end architecture. Figure 12.23 shows the cost efficiency of the architectures such as FoLi, FiLi, FoLA, FiLA, and the conventional FiWi architecture in terms of cost-per-bit. The cost-per-bit is defined as the ratio of the CAPEX to the capacity of the architecture. It can be noted that the installation of FSO link in the back-end of the architecture makes the system viable by achieving a reduction in cost-per-bit. It can be clearly seen from Fig. 12.23 that FoWi provides a minimum cost-per-bit of 0.123 cents. However, FoWi achieves the minimum average data rate among other architectures. Therefore, there is a trade-off between cost-per-bit and average data rate for FoWi architecture. As compared to FoWi architecture, it is observed that FoLA and FoLi perform better than fiber-based back-end architecture (FiWi, FiLi, FiLA) in terms of cost-per-bit.

In this work, the reliability of the FoLi architecture is investigated, and its performance is compared with that of conventional FiWi architecture. In FiWi architecture, the passive splitter is more reliable component since it is a passive component [75]. However, the reliability of the conventional FiWi may suffer from link failures due to optical fiber cut, failure of components such as AP and ONU. Similarly, in the case of the FoLi architecture, FSO link, ONU, AP, and Active optical splitter (AOS) failures affect the reliability of the proposed FoLi architecture. Assume that P_{cut} denotes fiber cut probability between the splitter and ONU-AP, P_F is the probability of failure of components such as ONU, AP, AOS, and FSO link. To evaluate reliability of FiWi and FoLi architecture, ONU, AOS, and AP's failure probability is assumed as 10^{-6} [26]. With the advancement in FSO, the

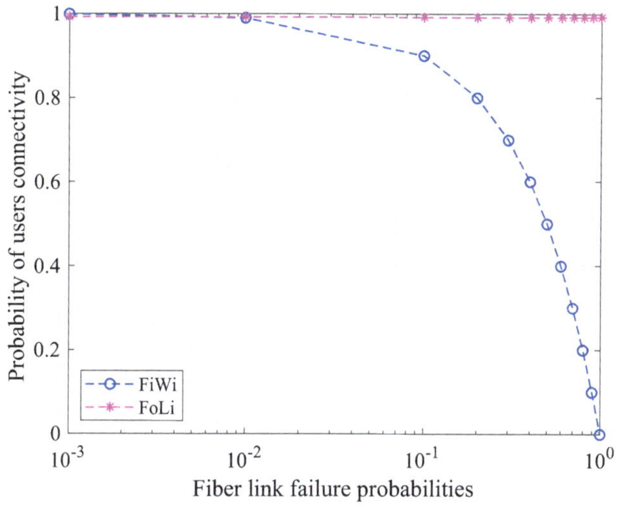

Fig. 12.24 Probability of the users connectivity vs Fiber link failure probabilities

availability of FSO link is considered as 99.999% [21, 92]. Accordingly, FSO link's failure probability is assumed as 10^{-5}. In the case of the conventional FiWi architecture, the probability of the users connectivity to the OLT is given as

$$P_{\text{con}}^{U} = (1 - P_{\text{cut}}) \times (1 - P_F^O) \times (1 - P_F^{AP}), \tag{12.4}$$

where P_F^O and P_F^{AP} represent the failure probability of ONU and AP, respectively. When fiber is replaced by FSO link and WiFi is replaced by LiFi in the FoLi architecture, the probability of the users connectivity in the case of FSO link, ONU, AOS, and AP failures is depicted in Fig. 12.24 and computed as follows

$$P_{\text{con}}^{U} = (1 - P_F^{FSO}) \times (1 - P_F^{AOS}) \times (1 - P_F^{AP}) \times (1 - P_F^{O}), \tag{12.5}$$

where P_F^{FSO} is the probability of failure of FSO link. Figure 12.24 depicts that the probability of user connectivity to the OLT with an increase in probability of fiber cut in FiWi and FoLi architecture. It is clearly seen that as compared to FiWi, FoLi achieves 0.99 of the probability of the users connectivity when probability of fiber cut is 0.1. Therefore, FoLi architecture has a better performance than FiWi in terms of reliability.

12.6 Hybrid Wireless-Optical Broadband Access Network

P2MP optical networks, sometimes referred to as PONs, are the leading broadband-access network emerging out of current research and development efforts. Figure 12.25 depicts

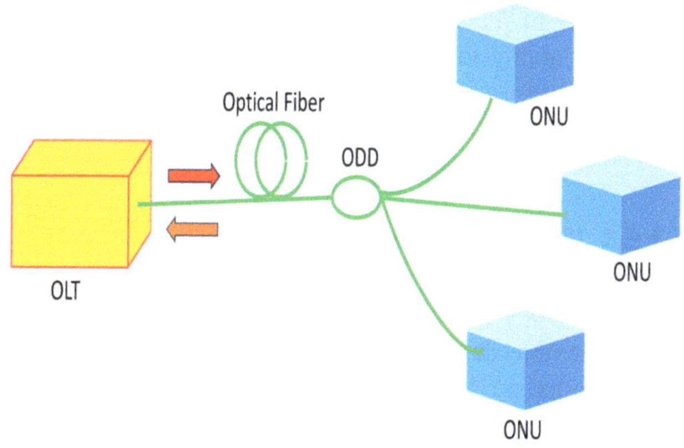

Fig. 12.25 General architecture of PON

the architecture of PON. A PON consists of an OLT at the CO, an optical fiber, an optical distribution node (ODD), and multiple ONUs located near the users' premises. The fundamental design of a PON uses two wavelength channels to connect the telecom CO to commercial and residential users. One wavelength channel is used to go downstream, from the CO's OLT to the ONUs, and another wavelength channel is used to go upstream, from ONUs to the OLT. The ODD, also known as the RN, multiplexes the upstream traffic from the ONUs to the optical fiber and demultiplexes the downstream optical signal received from the optical fiber to multiple ONUs [62]. A PON is robust because there isn't an active component involved in the signal's transmission from source to destination. Passive couplers, combiners, and splitters are the only interior components utilized in such a network.

Compared to existing systems like DSL and cable modem (CM), a PON offers substantially higher bandwidth for data applications along with wider fiber coverage. A PON can span a distance of up to 20 km between the OLT and the ONU. While solutions like FTTH, FTTB, or fiber-to-the-PC are designed to bring fiber directly to user's location, fiber-to-the-curb might be a cost-effective option for deployment. [22, 61].

The time-division-multiplexed PON, or TDM-PON, is another name for the standard single-wavelength PON. TDM provides high capacity with low cost as it uses a combination of optical fiber and passive infrastructure. Similar to most other distribution networks, such as those for power, video, and audio, the PON architecture has a tree-like structure since all the users share the OC via a passive splitter. As a result, the number of ONUs is constrained by the splitting loss and the bit rate of the transceivers in both OLT and the ONUs. Since all end users share the bandwidth, the per-user cost of such a network can be low. However, as user demand for high bandwidth increases, it becomes necessary to upgrade the PON architectures—such as EPON, BPON, GPON, GFP

PON, etc.—to WDM PON. A WDM PON is highly scalable due to its ability to support multiple wavelengths over the same fiber infrastructure, inherent transparency to the channel bit rate, and potential lack of power-splitting losses.

12.6.1 WOBAN Architecture

Future access solutions must have substantial backbone capacity, flexible deployment, upgradeability, and scalability to accommodate increasing user numbers and demands. The Wireless-optical broadband-access network (WOBAN) architecture is achieved by combining an optical network at the back-end and a wireless network at the front-end. The WOBAN architecture leverages the advantages of both systems: it combines the robustness, reliability, and high capacity of wireline optical communication with the flexibility and cost savings of a wireless network, offering "anytime-anywhere" approach [79]. For local and urban areas, this hybrid architecture provides a potential access solution.

Figure 12.26 illustrates the architecture of a WOBAN. The wireless front-end and CO are connected by a tree network that serves as the optical backhaul. An OLT at the CO, a Single mode fiber (SMF), a RN, and several APs make up the optical backhaul link. The wireless front-end features numerous APs strategically placed to reach a wide range of wireless end users (WEUs). The main advantages of a WOBAN over wireless networks and wired optical networks can be summed up as follows:

1. When it comes to cost, a WOBAN might be far more economical than a conventional wireline network. A WOBAN architecture (see Fig. 12.26) does not require costly "FTTH" connectivity because it could be expensive to deploy and maintain fiber all the way to every user. A user connecting wirelessly to a nearby ONU under WOBAN may do so over a number of hops via different wireless routers. The optical-fiber infrastructure at the ONU will be used to process and send the data from the wireless user to the OLT.
2. The wireless component of WOBAN enables smooth connections between users. As a result, a WOBAN is more adaptable than the optical-access network. The "anytime-anywhere" approach also applies to WOBAN. Therefore, WiFi is a suitable technology for the front end of the WOBAN, leveraging its flexibility and multi-hopping capability. WiMax is used as an alternative to WiFi since it offers flexibility and a higher bit rate than that of WiFi.
3. In comparison to a conventional wired network, a WOBAN is ought to be more resilient. An ONU in a conventional PON will go down if a fiber that connects to the splitter breaks. Worse still, if a trunk from the OLT to the splitter fails, all ONUs (and their connected users) will be affected. However, in a WOBAN, since users can form a multi-hop mesh topology, they may adjust using wireless connectivity and can locate an active neighboring ONU. In this case, users can connect to the operational ONU, which will then relay their communication to another OLT at the CO.
4. The WOBAN has high-capacity optical trunk due to which it offers significantly greater capacity compared to the relatively lower capacity of the wireless network.
5. A WOBAN is more reliable than a wireless network, which helps reduce congestion and information loss compared to current wireless networks. Additionally, a user in WOBAN can connect with other nearby ONU, if their primary ONU fails or becomes overloaded, hence providing superior load-balancing capabilities.

Two primary schemes have been proposed in order to transmit wireless signal over the optical fiber [64, 67]: (1) Baseband-over-fiber (BBOF) transmission scheme and (2) RoF transmission scheme. In BBOF transmission scheme, the bit stream that is received at the ONU/AP is modulated by an RF transmitter before being transmitted to the wireless end user. The wireless signal in the RoF transmission technique is upconverted over fiber using an optical carrier with a predetermined wavelength. The optical data received by the AP are downconverted to the appropriate wireless signal for transmission. After amplification, the AP directly transmits this wireless signal to the end user through wireless channel.

RF wireless signals are carried via fiber as a baseband signal using the BBOF transmission technique. In this system, the baseband signal detected from the optical backhaul is modulated to the appropriate RF at the gateway ONU-AP only. The WOBAN architecture comprises an optical backhaul that supports a Wireless mesh network (WMN) in the wireless front-end.

For a WOBAN utilizing the RoF transmission scheme for the transmission of wireless signals, at the OLT, the wireless signal is directly upconverted to an optical signal before being sent over fiber as an RoF signal. In this method, the AP simply recognizes the received RoF signal, amplifies it, and then transmits it via a wireless channel to the end user.

12.6.2 Optical Access Networks

Optical access networks have developed over time in order to attain high capacity and scalability with respect to both connection and user count [42]. PON enhances the bandwidth of access networks by providing favorable optical access networks. PONs can be classified into four primary schemes based on the techniques they employ: TDM, WDM, CDMA, and OFDM. The method employed and the kind of optical signal distribution at the ODD are the primary distinctions

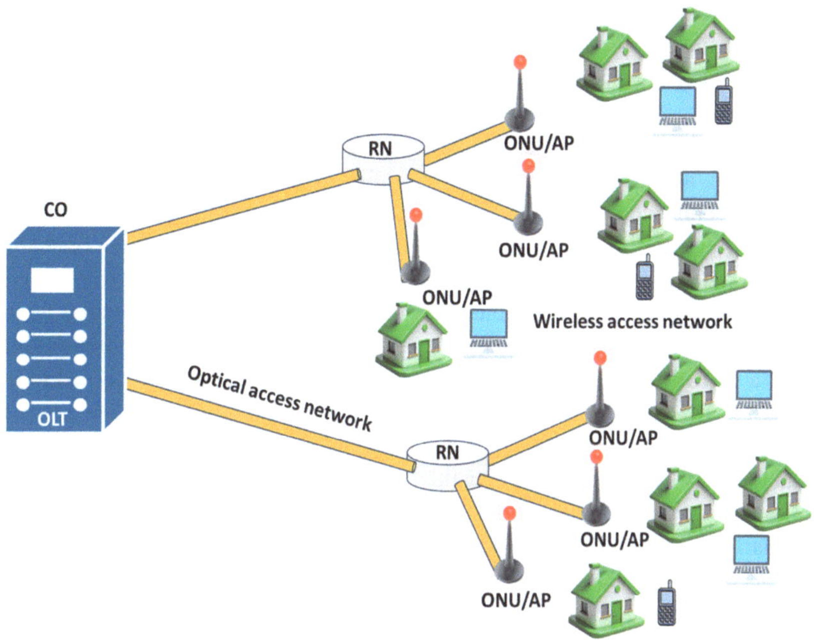

Fig. 12.26 WOBAN architecture

among these four approaches. Of all these techniques, TDM PON, WDM PON, and OFDM PON are mostly considered for the wide deployment in optical access networks.

12.6.2.1 TDM PON

TDM PON is an example of simple and cost-effective PON [79]. The TDM PON is limited to 32, 64, or 128 ONUs since it utilizes the passive optical power at the RN. In a TDM PON, downstream data are transported from OLT to ONUs via single downlink wavelength, while upstream data are transported from ONUs to OLT via separate uplink wavelength. The OLT allocates time slots to N users (ONU_1, ONU_2, ..., ONU_N) that are connected to the TDM PON. The RN uses a $1 \times N$ passive optical power splitter/combiner (PS/C) to distribute the optical signal to and from multiple ONUs. The upstream and downstream signal at the CO and ONUs are divided using optical circulator in a bidirectional TDM. All of the linked ONUs receive a broadcast of the downstream data. Every ONU chooses the designated stream slot and discards the slots that are meant for other ONUs. It typically employs burst-mode transmission for upstream communication, where the OLT allocates specific time slots for each ONU's upstream data transmission.

The three types of standardized TDM PONs are EPON, BPON, and GPON. In these networks, the optical power is divided among the subscribers, who access their data within designated time slots. The three variants of TDM PON differ from each other in terms of available data rate. The BPON supports a low data rate of 622/155 Mbps for downstream and upstream. The EPON supports a data rate of 1.0 Gbps for symmetrical operation, while GPON supports a data rate of 2.5/1.5 Gbps for asymmetric downstream/upstream operation [62].

Based on Ethernet technologies, EPON allows for seamless integration of Ethernet and IP networks. IEEE ratified EPON as the IEEE 802.3ah standard in June 2004 [77]. EPON uses a multipoint-to-multipoint architecture, where a single OLT at the CO can serve multiple ONUs at the customer premises. EPON's reliance on Ethernet technology makes it a cost-effective solution for network operators. It eliminates the need for complex protocols like ATM, which are used in other PON technologies. EPON supports a range of services, including high-speed internet access, VoIP, IPTV, and enterprise connectivity. It is commonly used in FTTH and FTTB applications.

BPON, as outlined in the ITU-T G.983 series, is an enhancement of the APON system. Designed with the aim of facilitating the timely and cost-effective implementation of broadband optical access systems, BPON provides a wide range of broadband services such as video distribution ATM, and Ethernet access. BPON utilizes WDM for transmitting data downstream, capable of accommodating up to 16 wavelengths with 200 GHz spacing or 32 wavelengths with a 100 GHz spacing between channels. Furthermore, BPON offers improved security by implementing a churning technique, which involves the encryption key being replaced at least once per second between the OLT and ONT situated at the customer premises.

GPON is one of the most widely deployed PON technologies globally. It is standardized under the ITU-T-G.984 series and is designed to provide high-capacity broadband access [5]. GPON employs a more efficient protocol for transmitting data compared to older PON technologies. It uses GEM for encapsulating user data, reducing overhead and improving efficiency. It uses WDM technology to separate downstream and upstream traffic, allowing for higher bandwidth utilization. GPON offers enhanced security features, including encryption of data between the OLT and ONUs to prevent eavesdropping and data interception. GPON can carry multiple types of services simultaneously, including ATM, Ethernet, and TDM traffic. This flexibility makes it suitable for a wide range of applications.

12.6.2.2 WDM PON

In a TDM PON, the available bandwidth is shared among the ONUs at the end users, which can result in lower average bandwidth per user, particularly when there are large number of ONUs. To deal with this issue, the WDM PON was introduced. It is highly scalable as it can accommodate several wavelengths over the same fiber infrastructure. Additionally, by employing P2P technique, it may avoid power-splitting losses [6]. It also offers upgradability, strong protocol transparency [52], and high data speeds upto 10 Gbps [37].

The WDM PON consists of N distinct wavelength channels from the CO to the ONUs in the downstream direction, referred to as downlink wavelength. The ONUs send the uplink wavelengths upstream to the OLT. In a WDM, each ONU can utilize the entire bit rate of its assigned wavelength channel (line rate), allowing for high data rates. Additionally, WDM PONs can enable long reach to the ONUs and are less susceptible to optical power budget constraints. An arrayed-waveguide-grating (AWG) router is used at both the OLT and the RN as a wavelength division Multiplexer (MUX)/Demultiplexer (DEMUX). In the RN, the WDM MUX/DEMUX combines the upstream optical signals to transport them via the optical fiber to the OLT and distributes the downstream optical signals based on the downlink wavelengths to the designated ONU (ONU_1, ONU_2, ..., ONU_N).

12.6.2.3 OFDM PON

Several wired and wireless access technologies use OFDM modulation. The OFDM offers high tolerance to the fiber chromatic dispersion and has high spectral efficiency. It is highly flexible in terms of dynamic bandwidth allocation and multiple services provisioning. OFDM effectively addresses ISI brought on by dispersive channels. However, its key shortcomings are high peak-to-power ratio and its sensitivity to frequency offset and phase noise [36]. OFDM PONs can provide extremely high data rates of 53 Gbps [34] and 100 Gbps [32].

The setup of OFDM PON is similar to that of conventional PON with one wavelength being utilized for uplink and another for downlink. To understand the working of OFDM PON, we assume that there are N ONUs at the end users. Using the inverse fast Fourier transform (IFFT), the OLT produces multiple orthogonal subcarriers. This process is known as OFDM modulation. As the entire OFDM bandwidth is split into N sub-bands, the OLT allots a certain number of orthogonal subcarriers to each ONU. These orthogonal frequencies propagate via an optical fiber after being upconverted over the optical carrier wavelength. Using the PS/C, the downlink signal is divided at the RN into N signals. After receiving the downstream signal, each ONU selects the subcarriers that are assigned to it. At the ONU receiver, the original transmitted data are generated using the fast Fourier transform (FFT). This process is known as OFDM demodulation. ONU creates a set of orthogonal subcarriers based on the assigned bandwidth, which is managed and controlled by the OLT. The orthogonal subcarriers generated are then upconverted by optical modulators utilizing the same uplink wavelength. At the RN, the upstream signals from the ONUs are combined and sent to the OLT via optical fiber.

12.6.3 Technologies for Wireless Access

Wireless access provides ubiquitous connectivity to end users, offering effective solutions to tackle mobility problems in access networks. WiFi and Worldwide Interoperability for Microwave Access (WiMaX) are the two main techniques used for wireless-access networks. Each of these technologies has its own set of merits and demerits.

12.6.3.1 WiFi

WiFi is one of the most widely used wireless technology, based on standards IEEE 802.11a/b/g. These standards were briefly discussed in Chapter 11. It is primarily utilized for wireless local-area networks. WiFi has a limited range of 100–200 meters and a maximum bit rate of 54/11/54/600 Mbps for 802.11a, b, g, and n, respectively [71]. While 802.11b/g operates at the unlicensed 2.4 GHz frequency, 802.11a operates at the 5 GHz band. Despite having higher throughput and more channels available for frequency than 802.11b/g, 802.11a is more susceptible to multi-path propagation. The direct sequence spread spectrum (DSSS) technique with 25MHz bandwidth is used to implement 802.11b, while the OFDM modulation with 20 MHz bandwidth is used to implement 802.11a/g.

The 802.11n standard aims to enhance medium access control (MAC) throughput to atleast 100 Mbps, with data rates going up to 600 Mbps. It employs two key concepts to achieve higher physical layer (PHY) data rates: multiple input multiple output (MIMO) technology and 40 MHz bandwidth channels. According to Perhia and Stacey (2008),

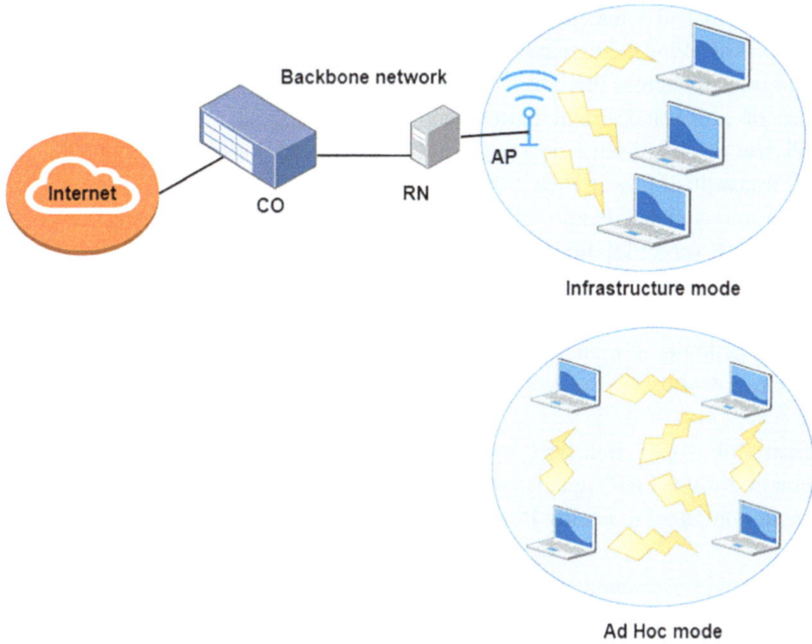

Fig. 12.27 WiFi structure modes

there is a four times increase in data rate when moving from one spatial stream and one transmit antenna to four spatial stream and four antennas [71]. By adopting spatial division multiplexing (SDM) technique, the transmitter multiplexes these streams, which are then demultiplexed at the receiver [68]. The 802.11n uses a channel bandwidth of 20 or 40 MHz and operates in the unlicensed 5 GHz and 2.4 GHz frequency bands.

WiFi is capable of functioning in both "Ad Hoc" and "Infrastructure" modes [76] as shown in Fig. 12.27. An access point or central authority is needed to manage the network in infrastructure mode. However, in ad hoc mode, there is no notion of an administrator, and the users are self-managed.

12.6.3.2 WiMAX

WiMax is based on IEEE 802.16 standard, and it is particularly suitable for wireless metropolitan area networks (WMANs) [83]. A WiMax BS can provide data rates up to 75 Mbps within a range of 3–5 km. However, bit rates reduce significantly over longer distance transmission. This is because WiMax signal attenuates heavily for longer distances significantly degrading its bit error rates. WiMax is highly efficient for single-hop communication, but it faces challenges such as higher latency and lower throughput when used for multi-hop communication. It operates in a less crowded spectrum and offers high bandwidth.

IEEE 802.16 networks can operate in two modes based on the applications and network investment: P2MP or mesh mode. In P2MP mode, a BS serves multiple subscriber stations within its coverage area. Subscriber stations can interact with one another in a multi-hop fashion in the mesh mode without the direct involvement of BSs. To optimize error performance and transmission rate, WiMax uses adaptive modulation. In adaptive modulation, a transmitter can automatically switch to a more reliable, although less efficient modulation technique when conditions are unfavorable.

12.7 Summary

This chapter studies the most important OWC technologies, which are Radio-over-fiber (RoF), Integrated LiFi/WiFi networks, Integrated fiber-wireless networks, broadband access networks, and hybrid wireless-optical broadband access networks. Moreover, the performance evaluation of different architecture for the upcoming technologies is also studied in this chapter. It can be observed that integrated wireless and optical networks have been utilized in various indoor and outdoor scenarios.

Homework Problems

1. Find the Rytov variance for spherical wave, where FSO link distance is 2 km, λ=1550 nm, and $C_n^2 = 4 \times 10^{-15} m^{-2/3}$.
2. Calculate the diameter of the spot of the optical beam at an optical link distance of 2 km when the wavelength is 1550 nm and the diameter of the transmitter is 3 cm.
3. Assume a 10 km single mode fiber link at 1310nm with 2 connector pairs and 2 splices. With a fiber loss 0.35

dB/km, splice loss 0.1 dB/km, and connector loss of 0.75 dB with safety margin of 3 dB. Find the required transmit power (link budget) to transmit across this link.
4. For the given values of noise and signal parameters responsivity R = 0.65 A/w, diode current $I_d = 1$ nA, load resistance $R_L = 1$K, transmitted power $P = 500$ MW, wavelength $\lambda = 850$ nm, and bandwidth of $\Delta f = 100$ MHz, calculate the SNR for a PIN diode.
5. Draw and explain the architecture of an RoF system.
6. What are the disadvantages of an RoF system?
7. What are the various topologies for an RoF system? Which one of them has the best and the worst performance and why?
8. Describe the different RoF signal transport schemes. Why the transmission of digitized RF signals over fiber is a better alternative as compared to analog RoF signal transport?
9. What are the advantages and disadvantages of LiFi?
10. State the two ways in which LiFi can be integrated into WiFi infrastructure.
11. In hybrid LiFi/WiFi systems, describe the importance of multiple access schemes.
12. What are the main challenges in the implementation of backhaul for HLWNets?
13. What are the advantages of a WOBAN over the wireline optical and wireless networks?
14. Explain the working of OFDM PON.
15. Explain the two modes in which a WiFi can work.
16. Why does the bit rate in WiMax reduces significantly over longer distance transmission?

References

1. IEEE 802.11e-2005 (Revision of IEEE Std 802.11-1999) IEEE Standard for Information technology—telecommunications and information exchange between systems-local and metropolitan area networks—specific requirements—Part 11: wireless LAN medium access control (MAC) and physical layer (PHY) specifications amendment 8: medium access control (MAC) quality of service enhancements (2005)
2. Arokkiam, J.A., Brown, K.N., Sreenan, C.J.: Refining the GIANT dynamic bandwidth allocation mechanism for XG-PON. In: 2015 IEEE International Conference on Communications (ICC), pp. 1006–1011 (2015)
3. Alnajjar, S.H., Noori, A.A., Moosa, A.A.: Enhancement of FSO communications links under complex environment. Photon. Sensors **7**(2), 113–122 (2017)
4. Baack, C., Elze, G., Grobkopf, G., Kraus, F., Krick, W., Küller, L.: Analogue optical transmission of 26 t.v. channels. Electron. Lett. **15**(10), 300–301 (1979)
5. IEEE Standard for Information Technology—local and metropolitan area networks—specific requirements—part 3: CSMA/CD access method and physical layer specifications amendment 1: Physical layer specifications and management parameters for 10 Gb/s passive optical networks. IEEE Std 802.3av-2009 (Amendment to IEEE Std 802.3-2008), pp. 1–227 (2009)
6. Banerjee, A., Park, Y., Clarke, F., Song, H., Yang, S., Kramer, G., Kim, K., Mukherjee, B.: Wavelength-division-multiplexed passive optical network (WDM-PON) technologies for broadband access: a review. J. Opt. Netw. **4**(11), 737–758 (2005). https://opg.optica.org/jon/abstract.cfm?URI=jon-4-11-737
7. Yuan, Z., Muntean, G.: A prioritized adaptive scheme for multimedia services over IEEE 802.11 WLANs. IEEE Trans. Netw. Service Manag. **10**(4), 340–355 (2013)
8. Ammar, D., Begin, T., Guerin-Lassous, I.: A new tool for generating realistic internet traffic in NS-3. In: Proceedings of the 4th International ICST Conference on Simulation Tools and Techniques, SIMUTools '11, pp. 81–83. ICST (Institute for Computer Sciences, Social-Informatics and Telecommunications Engineering), ICST, Brussels, Belgium (2011). http://dl.acm.org/citation.cfm?id=2151054.2151070
9. LiFi link price. https://www.lifitn.com
10. Aschenbruck, N., Ernst, R., Martini, P.: Indoor mobility modelling. In: 2010 IEEE Globecom Workshops, pp. 1264–1269 (2010)
11. CISCO visual networking index: forecast and trends, 2017–2022 (2018)
12. Koonen, A.M.J. et al.: In-house networks using multimode polymer optical fiber for broadband wireless services. Photon. Netw. Commun. **5**, 177–187 (2003)
13. Aladeloba, A.O., Woolfson, M.S., Phillips, A.J.: WDM FSO network with turbulence-accentuated interchannel crosstalk. J. Opt. Commun. Netw. **5**(6), 641–651 (2013)
14. Wu, X., Soltani, M.D., Zhou, L., Safari, M., Haas, H.: Hybrid LiFi and WiFi networks: a survey. IEEE Commun. Surveys Tutorials **23**(2), 1398–1420 (2021)
15. Ahmad, R., Soltani, M.D., Safari, M., Srivastava, A.: Reinforcement learning-based near-optimal load balancing for heterogeneous LiFi WiFi network. IEEE Syst. J. 16(2), 3084–3095 (2021)
16. Zeng, C., Licardie, S.: Systems and methods for performing layer one link aggregation over wireless links (2015). US Patent 9,125,084
17. Zhang, T., Sun, X., Wang, C.: On optimizing the divergence angle of an FSO based fronthaul link in drone assisted mobile networks. IEEE Internet Things J. **9**(9), 6914 - 6921 (2022)
18. Series G: Transmission systems and media, digital systems and network, Quality of service and performance
19. G.987.3: 10-gigabit-capable passive optical networks (XG-PON): transmission convergence (TC) layer specification (2020). https://www.itu.int/rec/T-REC-G.987.3
20. Wu, X., Safari, M., Haas, H.: Access point selection for hybrid Li-Fi and Wi-Fi networks. IEEE Trans. Commun. **65**(12), 5375–5385 (2017)
21. FSO link price. https://www.gigabit-wireless.com/tag/free-space-optics/"
22. IEEE draft standard for service interoperability in ethernet passive optical networks. IEEE P1904.1/D3.2, January 2013 pp. 1–826 (2013)
23. Arokkiam, J.A., Brown, K.N., Sreenan, C.J.: Optimised QoS-aware DBA mechanisms in XG-PON for upstream traffic in LTE backhaul. In: 2016 IEEE 4th International Conference on Future Internet of Things and Cloud Workshops (FiCloudW), pp. 361–368 (2016)
24. Series G: Transmission systems and media, digital systems and networks, Quality of service and performance, G.1010, ITU-T (2001)
25. Basnayaka, D.A., Haas, H.: Hybrid RF and VLC systems: Improving user data rate performance of VLC systems. In: 2015 IEEE 81st Vehicular Technology Conference (VTC Spring), pp. 1–5 (2015)
26. Beyranvand, H., Lim, W., Maier, M., Verikoukis, C., Salehi, J.A.: Backhaul-aware user association in FiWi enhanced LTE-A heterogeneous networks. IEEE Trans. Wirel. Commun. **14**(6), 2992–3003 (2015)

27. Bhatnagar, M.R., Ghassemlooy, Z.: Performance analysis of Gamma–Gamma fading FSO MIMO links with pointing errors. J. Lightwave Technol. **34**(9), 2158–2169 (2016)
28. Bianchi, G., Tinnirello, I., Scalia, L.: Understanding 802.11e contention-based prioritization mechanisms and their coexistence with legacy 802.11 stations. IEEE Netw. **19**(4), 28–34 (2005)
29. Castafion, G., Campuzanol, G., Tonguz, O.: High reliability and availability in radio over fiber networks. In: 2007 Mobile Networking for Vehicular Environments, pp. 25–30 (2007)
30. Chen, C., Basnayaka, D.A., Haas, H.: Downlink performance of optical attocell networks. J. Lightwave Technol. **34**(1), 137–156 (2016)
31. Chen, C., Soltani, M.D., Safari, M., Purwita, A.A., Wu, X., Haas, H.: An omnidirectional user equipment configuration to support mobility in lifi networks. In: 2019 IEEE International Conference on Communications Workshops (ICC Workshops), pp. 1–6 (2019)
32. Chow, C.W., Yeh, C.H., Wang, C.H., Wu, C.L., Chi, S., Lin, C.: Studies of OFDM signal for broadband optical access networks. IEEE Journal on Selected Areas in Communications **28**(6), 800–807 (2010)
33. Chowdhury, M.Z., Hasan, M.K., Shahjalal, M., Hossan, M.T., Jang, Y.M.: Optical wireless hybrid networks: trends, opportunities, challenges, and research directions. IEEE Commun. Surveys Tutorials **22**(2), 930–966 (2020)
34. Coura, D., Silva, J., Segatto, M.: A bandwidth scalable OFDM passive optical network for future access network. Photon Netw Commun 18, 409–416 (2009)
35. Crovella, M.E., Bestavros, A.: Self-similarity in World Wide Web traffic: evidence and possible causes. IEEE/ACM Trans. Netw. **5**(6), 835–846 (1997)
36. Cvijetic, N., Qian, D., Hu, J.: 100 gb/s optical access based on optical orthogonal frequency-division multiplexing. IEEE Commun. Mag. **48**(7), 70–77 (2010)
37. Davey, R., Kani, J., Bourgart, F., McCammon, K.: Options for future optical access networks. IEEE Commun. Mag. **44**(10), 50–56 (2006)
38. Dehghani Soltani, M., Wu, X., Safari, M., Haas, H.: Access point selection in li-fi cellular networks with arbitrary receiver orientation. In: 2016 IEEE 27th Annual International Symposium on Personal, Indoor, and Mobile Radio Communications (PIMRC), pp. 1–6 (2016)
39. Dehghani Soltani, M., Wu, X., Safari, M., Haas, H.: Bidirectional user throughput maximization based on feedback reduction in LiFi networks. IEEE Trans. Commun. **66**(7), 3172–3186 (2018)
40. Douik, A., Dahrouj, H., Al-Naffouri, T.Y., Alouini, M.S.: Hybrid radio/free-space optical design for next generation backhaul systems. IEEE Trans. Commun. **64**(6), 2563–2577 (2016)
41. Gao, Y., Sun, X., Dai, L.: IEEE 802.11e EDCA networks: modeling, differentiation and optimization. IEEE Trans. Wirel. Commun. **13**(7), 3863–3879 (2014)
42. Ghazisaidi, N., Maier, M.: Fiber-wireless (FiWi) access networks: challenges and opportunities. IEEE Network **25**(1), 36–42 (2011)
43. Gliese, U., Norskov, S., Nielsen, T.: Chromatic dispersion in fiber-optic microwave and millimeter-wave links. IEEE Trans. Microwave Theory Tech. **44**(10), 1716–1724 (1996)
44. Gu, Z., Lu, H., Zhu, Z.: On throughput optimization and bound analysis in cache-enabled fiber-wireless networks. IEEE Trans. Veh. Tech. **69**(8), 9068–9082 (2020)
45. Guo, H., Wang, Y., Liu, J., Kato, N.: Super-broadband optical access networks (oans) in 6G: vision, architecture, and key technologies. IEEE Wirel. Commun. **29**(6), 152–159
46. Gupta, A., Sharma, N., Garg, P., Alouini, M.S.: Cascaded FSO-VLC communication system. IEEE Wirel. Commun. Lett. **6**(6), 810–813 (2017)
47. Haas, H., Chen, C., O'Brien, D.: A guide to wireless networking by light. Progr. Quantum Electron. **55**, 88–111 (2017). https://www.sciencedirect.com/science/article/pii/S0079672717300198
48. Hammouda, M., Akln, S., Vegni, A.M., Haas, H., Peissig, J.: Hybrid RF/LC systems under QoS constraints. In: 2018 25th International Conference on Telecommunications (ICT), pp. 312–318 (2018)
49. Islim, M.S., Ferreira, R.X., He, X., Xie, E., Videv, S., Viola, S., Watson, S., Bamiedakis, N., Penty, R.V., White, I.H., Kelly, A.E., Gu, E., Haas, H., Dawson, M.D.: Towards 10 Gb/s orthogonal frequency division multiplexing-based visible light communication using a GaN violet micro-LED. Photon. Res. **5**(2), A35–A43 (2017). https://opg.optica.org/prj/abstract.cfm?URI=prj-5-2-A35
50. Jaffer, S.S., Hussain, A., Qureshi, M.A., Mirza, J., Qureshi, K.K.: A low cost PON-FSO based fronthaul solution for 5G CRAN architecture. Opt. Fiber Technol. **63**, 102500 (2021)
51. Jain, R.K., Chiu, D.M.W., Hawe, W.R., et al.: A quantitative Measure of Fairness and Discrimination. Eastern Research Laboratory, Digital Equipment Corporation, Hudson (1984)
52. Jeon, S.W., Kim, Y., Park, C.S.: Long-reach transmission experiment of a wavelength division multiplexed-passive optical networks transmitter based on reflective semiconductor optical amplifiers. Opt. Eng. **51**(1), 015008-1–015008-5 (2012). https://ui.adsabs.harvard.edu/abs/2012OptEn..51a5008J/abstract
53. Jivkova, S., Kavehrad, M.: Shadowing and blockage in indoor optical wireless communications. In: GLOBECOM '03. IEEE Global Telecommunications Conference (IEEE Cat. No.03CH37489), vol. 6, pp. 3269–3273 (2003)
54. Kanonakis, K., Tomkos, I., Krimmel, H.G., Schaich, F., Lange, C., Weis, E., Leuthold, J., Winter, M., Romero, S., Kourtessis, P., Milosavljevic, M., Cano, I.N., Prat, O.: An OFDMA-based optical access network architecture exhibiting ultra-high capacity and wireline-wireless convergence. IEEE Commun. Mag. **50**(8), 71–78 (2012)
55. Kantarci, B., Mouftah, H.T.: Availability and cost-constrained long-reach passive optical network planning. IEEE Trans. Reliab. **61**(1), 113–124 (2012)
56. Karoti, N.M., Paramita, S., Ahmad, R., Bohara, V.A., Srivastava, A.: Improving the performance of heterogeneous LiFi-WiFi network using a novel link aggregation framework. In: 2022 IEEE Wireless Communications and Networking Conference (WCNC), pp. 2322–2327. IEEE, Piscataway (2022)
57. Kashef, M., Ismail, M., Abdallah, M., Qaraqe, K.A., Serpedin, E.: Energy efficient resource allocation for mixed rf/vlc heterogeneous wireless networks. IEEE J. Sel. Areas Commun. **34**(4), 883–893 (2016)
58. Kaur, R., Gupta, A., Srivastava, A., Chatterjee, B.C., Mitra, A., Ramamurthy, B., Bohara, V.A.: Resource allocation and QoS guarantees for real world IP traffic in integrated XG-PON and IEEE802.11e EDCA networks. IEEE Access **8**, 124883–124893 (2020)
59. Kaur, R., Srivastava, A., Chatterjee, B.C., Mitra, A., Ramamurthy, B.: Performance analysis of fairness oriented dynamic bandwidth algorithm in integrated Fiber-Wireless architecture based on XG-PON and Wi-Fi. In: 2018 20th International Conference on Transparent Optical Networks (ICTON), pp. 1–4 (2018)
60. Komine, T., Nakagawa, M.: Fundamental analysis for visible-light communication system using led lights. IEEE Trans. Consum. Electron. **50**(1), 100–107 (2004)
61. Kramer, G.: Ethernet Passive Optical Networks. McGrawHill, New York (2005)
62. Lee, C.H., Sorin, W.V., Kim, B.Y.: Fiber to the home using a PON infrastructure. J. Lightwave Technol. **24**(12), 4568–4583 (2006). https://opg.optica.org/jlt/abstract.cfm?URI=jlt-24-12-4568

63. Li, Z., Nirmalathas, A., Bakaul, M., Cheng, L., Wen, Y.J., Lu, C.: Application of distributed raman amplifier for the performance improvement of WDM millimeter-wave fiber-radio network. In: 2005 IEEE LEOS Annual Meeting Conference Proceedings, pp. 579–580 (2005)
64. Lim, C., Nirmalathas, A., Bakaul, M., Gamage, P., Lee, K.L., Yang, Y., Novak, D., Waterhouse, R.: Fiber-wireless networks and subsystem technologies. J. Lightwave Technol. **28**(4), 390–405 (2010)
65. Loh, T.H., Cheadle, D., Wu, X., Haas, H., Liu, J., He, P., Bi, Y.: Link performance evaluation of 5G mm-wave and LiFi systems for the transmission of holographic 3D display data. In: 2019 13th European Conference on Antennas and Propagation (EuCAP), pp. 1–4 (2019)
66. Misic, J., Rashwand, S., Misic, V.B.: Analysis of impact of TXOP allocation on ieee 802.11e EDCA under variable network load. IEEE Trans. Parallel Distrib. Syst. **23**(5), 785–799 (2012)
67. Nirmalathas, A., Gamage, P.A., Lim, C., Novak, D., Waterhouse, R.: Digitized radio-over-fiber technologies for converged optical wireless access network. J. Lightwave Technol. **28**(16), 2366–2375 (2010)
68. Niu, H., Ngo, C.: Diversity and multiplexing switching in 802.11n MIMO systems. In: 2006 Fortieth Asilomar Conference on Signals, Systems and Computers, pp. 1644–1648 (2006)
69. Ogawa, H., Polifko, D., Banba, S.: Millimeter-wave fiber optics systems for personal radio communication. IEEE Trans. Microwave Theory Tech. **40**(12), 2285–2293 (1992)
70. Pan, G., Lei, H., Ding, Z., Ni, Q.: 3-d hybrid vlc-rf indoor iot systems with light energy harvesting. IEEE Trans. Green Commun. Netw. **3**(3), 853–865 (2019)
71. Perahia, E., Stacey, R.: Next generation wireless lans: throughput, robustness, and reliability in 802.11n (2008). https://api.semanticscholar.org/CorpusID:60885047
72. Priyadarshani, R., Bhatnagar, M.R., Ghassemlooy, Z., Zvanovec, S.: Effect of correlation on BER performance of the FSO-MISO system with repetition coding over Gamma–Gamma turbulence. IEEE Photon. J. **9**(5), 1–15 (2017)
73. Rahaim, M.B., Vegni, A.M., Little, T.D.C.: A hybrid radio frequency and broadcast visible light communication system. In: 2011 IEEE GLOBECOM Workshops (GC Wkshps), pp. 792–796 (2011)
74. Ravneet Kaur, A.S.: QoS-aware integrated fiber-wireless standard compliant architecture based on XGPON and EDCA (2018)
75. Rimal, B.P., Van, D.P., Maier, M.: Mobile-edge computing versus centralized cloud computing over a converged FiWi access network. IEEE Trans. Netw. Service Manag. **14**(3), 498–513 (2017)
76. Sakarindr, P., Ansari, N.: Security services in group communications over wireless infrastructure, mobile ad hoc, and wireless sensor networks. IEEE Wirel. Commun. **14**(5), 8–20 (2007)
77. Sala, D., Gummalla, A.: Pon functional requirements: services and performance. IEEE 802.3ah Ethernet in the First Mile meeting, Portland, Oregon (2001)
78. Sarigiannidis, A.G., Iloridou, M., Nicopolitidis, P., Papadimitriou, G., Pavlidou, F., Sarigiannidis, P.G., Louta, M.D., Vitsas, V.: Architectures and bandwidth allocation schemes for hybrid wireless-optical networks. IEEE Commun. Surveys Tutorials **17**(1), 427–468 (2015)
79. Sarkar, S., Dixit, S., Mukherjee, B.: Hybrid wireless-optical broadband-access network (woban): a review of relevant challenges. J. Lightwave Technol. **25**(11), 3329–3340 (2007)
80. Sauer, M., Kobyakov, A., George, J.: Radio over fiber for picocellular network architectures. J. Lightwave Technol. **25**(11), 3301–3320 (2007)
81. Schmuck, H.: Comparison of optical millimetre-wave system concepts with regard to chromatic dispersion. Electron. Lett. **31**, 1848–1849(1) (1995). https://digital-library.theiet.org/content/journals/10.1049/el_19951281
82. Sharma, S., Madhukumar, A., Swaminathan, R.: Effect of pointing errors on the performance of hybrid FSO/RF networks. IEEE Access **7**, 131418–131434 (2019)
83. Shaw, W.T., Wong, S.W., Cheng, N., Balasubramanian, K., Zhu, X., Maier, M., Kazovsky, L.G.: Hybrid architecture and integrated routing in a scalable optical–wireless access network. J. Lightwave Technol. **25**(11), 3443–3451 (2007)
84. Singh, P., Gupta, A., Bohara, V.A., Srivastava, A.: Cost effective hybrid FSO-wireless architecture for broadband access network. In: 2022 International Conference on Optical Network Design and Modeling (ONDM), pp. 1–6 (2022)
85. Singh, P., Gupta, A., Bohara, V.A., Srivastava, A.: QoS-aware reliable architecture for broadband fiber-wireless access networks. IEEE Syst. J. **16**(4), 5753–5764 (2022)
86. Solmaz, G., Turgut, D.: A survey of human mobility models. IEEE Access **7**, 125711–125731 (2019)
87. Soltani, M.D., Kazemi, H., Safari, M., Haas, H.: Handover modeling for indoor Li-Fi cellular networks: the effects of receiver mobility and rotation. In: 2017 IEEE Wireless Communications and Networking Conference (WCNC), pp. 1–6 (2017)
88. Song, X., Cheng, J.: Subcarrier intensity modulated MIMO optical communications in atmospheric turbulence. J. Opt. Commun. Netw. **5**(9), 1001–1009 (2013)
89. Stefan, I., Burchardt, H., Haas, H.: Area spectral efficiency performance comparison between VLC and RF femtocell networks. In: 2013 IEEE International Conference on Communications (ICC), pp. 3825–3829 (2013)
90. Stefan, I., Haas, H.: Hybrid visible light and radio frequency communication systems. In: 2014 IEEE 80th Vehicular Technology Conference (VTC2014-Fall), pp. 1–5 (2014)
91. Sugumaran, S., Lakshmi, D.N., Choudhary, S.: An overview of FTTH for optical network. In: Advances in Smart Communication and Imaging Systems, pp. 41–51 (2021)
92. Tatarko, M., Ovseník, L., Turán, J.: Properties of hybrid FSO/RF link with 60 GHz RF backup link. In: 2013 36th International Convention on Information and Communication Technology, Electronics and Microelectronics (MIPRO), pp. 495–497 (2013)
93. Wang, K., Machuca, C.M., Wosinska, L., Urban, P.J., Gavler, A., Brunnström, K., Chen, J.: Techno-economic analysis of active optical network migration toward next-generation optical access. J. Opt. Commun. Netw. **9**(4), 327–341 (2017)
94. Wang, Y., Wu, X., Haas, H.: Analysis of area data rate with shadowing effects in Li-Fi and RF hybrid network. In: 2016 IEEE International Conference on Communications (ICC), pp. 1–5 (2016)
95. Way, W.: Subcarrier multiplexed lightwave system design considerations for subscriber loop applications. J. Lightwave Technol. **7**(11), 1806–1818 (1989)

Index

V. A. Bohara, Swaminathan R, *Optical and Wireless Access Networks*, Textbooks in Telecommunication Engineering, https://doi.org/10.1007/978-3-031-89696-5

MIX
Papier aus verantwortungsvollen Quellen
Paper from responsible sources
FSC® C105338

If you have any concerns about our products,
you can contact us on
ProductSafety@springernature.com

In case Publisher is established outside the EU,
the EU authorized representative is:
Springer Nature Customer Service Center GmbH
Europaplatz 3, 69115 Heidelberg, Germany

Printed by Libri Plureos GmbH
in Hamburg, Germany